IET TRANSPORTATION SERIES 20

Shared Mobility and Automated Vehicles

Other related titles:

Shared Mobility and Automated Vehicles

Responding to socio-technical changes and pandemics

Edited by
Ata M. Khan and Susan A. Shaheen

The Institution of Engineering and Technology

Published by The Institution of Engineering and Technology, London, United Kingdom

The Institution of Engineering and Technology is registered as a Charity in England & Wales (no. 211014) and Scotland (no. SC038698).

The Institution of Engineering and Technology
Michael Faraday House
Six Hills Way, Stevenage
Herts SG1 2AY, United Kingdom

www.theiet.org

British Library Cataloguing in Publication Data
A catalogue record for this product is available from the British Library

ISBN 978-1-78561-862-8 (hardback)
ISBN 978-1-78561-863-5 (PDF)

Typeset in India by MPS Limited
Printed in the UK by CPI Group (UK) Ltd, Croydon

Contents

About the editors

Ata M. Khan is a professor emeritus in Civil and Environmental Engineering at Carleton University, Ottawa, Canada. His research interests focus on intelligent systems, automation in driving, policy and planning, and safe, efficient, and sustainable transportation. This research is being applied to urban and multimodal transportation. His honors and awards include the Lifetime Achievement Award, Canadian Institute of Transportation Engineers CITE and Carleton University Academic/Research Award, Fellowship of the Institute of Transportation Engineers (ITE), and Fellowship of the Canadian Society for Civil Engineering. He has carried out research/consulting projects for the United Nations; federal, provincial, and municipal governments in Canada; and private sector companies/societies in the USA, Japan, and Canada. Ata M. Khan has authored or co-authored one textbook, 11 Chapters in books, and 175 papers in refereed journal and conference proceedings.

Susan A. Shaheen is a professor in Civil and Environmental Engineering at renowned UC Berkeley, USA, where she co-directs the Transportation Sustainability Research Center (TSRC) of the Institute of Transportation Studies. She has authored 79 journal articles, over 150 reports and proceedings articles, over 30 book chapters, and co-edited three books. She received the 2017 Roy W. Crum award from the Transportation Research Board (TRB) for her distinguished achievements in transportation research. She is the current Chair of TRB's Executive Committee and served as Vice Chair in 2020. In May 2016, she was named one of the top 10 academic thought leaders in transportation by the Eno Foundation.

Preface

Over the last decade, shared mobility services—an alternative to private vehicle use—have experienced notable growth in a number of sectors including transportation network companies (TNCs, also known as ridehailing and ridesourcing), carsharing, microtransit, and shared micromobility (scooters and bikes). Smartphone and app-based technologies have served as catalysts. While current technology and access methods are actively used by many shared mobility customers, these services are often not well integrated with other modalities. Mobility on Demand (MOD) and Mobility as a Service (MaaS) platforms have emerged in recent years to provide an integrated platform for seamless routing, booking, and payment across modes. Many in the public sector are motivated to integrate shared mobility services with public transit and other modes (e.g., biking and walking) to foster a multimodal transportation network that can compete with private vehicle ownership and use, with the goal of providing more accessible, socially equitable, and sustainable transportation options.

Both the public and private sectors are interested in fostering more advanced forms of shared mobility, ranging from TNCs to carsharing and shared micromobility to microtransit and shuttles. These evolving services are anticipated to be available on-demand—accessible through flexible pick-up and delivery locations—as well as more affordable, environmentally friendly, and socially equitable in the future. Among other features, shared mobility services are expected to complement and fill gaps in the public transit network by extending its catchment area in lower density areas (e.g., suburbs, rural areas, and smaller cities). In the future, shared mobility services could also integrate more pooling options, as well as electric, connected, and automated vehicle technologies. In recent years, public and private sector stakeholders have focused on the role of micromobility, both shared and privately owned, to support a more integrated transportation network.

In the future, the shared mobility ecosystem will undergo a substantial transformation. A notable development in transportation is electrification. Cities have already adopted electrification policies or will do so in the near future. Many automakers have declared their commitment to an "all-electric" future for new vehicles, and their participation in full-electric, mobility services is increasing. The concept of vehicle right sizing and new form factors, alongside micromobility (e-scooters and e-bikes), represents another important trend toward vehicle specialization and adaptation, better reflecting the land use and built environment context. Another major change in urban mobility will be the use of connected and automated vehicles (or "autonomy") supported by smart city and intelligent transportation system initiatives—all of which will benefit from the availability of next-generation communication technologies.

Automation in driving technologies continues to progress and combined with on-demand dispatching applications shared automated vehicles (SAVs) could eventually become a mainstream reality. It is important to note, however, that lower levels of vehicle automation are already providing notable benefits (e.g., automated acceleration and steering to ease driving and reduce collisions). While the pathway and timeline to full autonomy are highly uncertain for all vehicle types, low-speed SAVs are increasingly being tested and deployed across the globe.

While pooling of shared-ride services could help to reduce vehicle miles/ kilometers traveled, congestion, and emissions, it is important to recognize that shifting individuals toward sharing rides with strangers is a significant challenge, particularly due to increased travel times, personal preference, and concerns about virus transmission. Pricing and incentives, along with curb management, could provide key strategies for shifting behavior but must ensure socially equitable outcomes. Today, many planners and decision makers are focused on better understanding how to manage the impacts of sharing, e-commerce, and automation, including how to pay for access to roads and the curb. What types of access will be prioritized, and how will it be priced? Will automated vehicles and SAVs reshape cities or will cities shape how vehicles fit into the urban fabric? Could algorithms ultimately dictate the access policies of cities eager to raise capital, and how will personal privacy be protected? The story of digitization and transportation finance will play an important role in managing the evolution of vehicle services and access. Again, social equity should be central in these developments to ensure fairness, particularly among underserved populations.

Labor concerns are also critical in the evolution toward the full automation of vehicles and shared mobility services, ranging from taxis to public transit. While this may present a notable challenge moving forward, great potential exists for transitioning the workforce toward vehicle monitoring and enhanced customer service (e.g., concierge, door-to-door services for children and older adults, etc.).

Prior to the COVID-19 pandemic, electrification, automation, and mobility service developments were fast paced. The slowing economy, reluctance to share vehicles due to public health considerations, and lockdowns have altered this landscape, adding to uncertainty about consumer demand, business models, and technological development. The future of mobility systems will require coordination and strategic actions on the part of many stakeholders (governments, policy-makers and planners, private sector businesses, consumers, insurance providers, developers, public health experts, and more). Cities have already started to focus on strategic and tactical initiatives to best accommodate electric automated vehicles, including SAVs (e.g., Toronto, Canada). To support next-generation mobility initiatives, more research is needed to guide these approaches to maximize their social and environmental benefits, including transportation equity and sustainable business models. For example, urban governments will need access to data to effectively integrate SAVs into the future transportation system, alongside active transportation. Governments will need to consider multiple transition phases during which electrification and automation are introduced and deployment scenarios (ranging from private AV domination to full sharing models and a mixture of fleets and private ownership).

This book is intended to serve as an up-to-date resource for students, researchers, and professionals interested in the future of mobility, particularly SAVs, across the globe. It covers the socio-technical factors of shared mobility services and automated vehicles. It also focuses on the critical transition period in which automated and nonautomated vehicles must share access to rights-of-way (roads, parking, and the curb) with each other, pedestrians, and micromobility devices.

The authors have contributed 18 chapters and an additional chapter that presents literature for further reading. The subjects covered include:

- The seismic shift in transportation (trends, issues, services, effects, and innovations);
- Policies and regulatory factors (social sustainability, accessibility, and social equity);
- Concept level system design, high-level architecture;
- Planning (rights-of-way management, planning principles, and prioritizing social good);
- Multimodal relationships (SAVs and high capacity public transit);
- Design of systems with nonautomated and automated vehicles (electrification and future proofing infrastructure);
- Demand models (SAVs, private automated mobility, first/last mile, and public transit);
- Balancing demand and supply (vehicle availability and fast charger availability);
- Operations and management for nonautomated and automated vehicles;
- Implementation context and long-term outcome on urban development;
- Economic factors and business models;
- Impacts (environmental sustainability and land use); and
- Future directions (strategic directions and tactical actions).

The chapter on "future directions" describes eight strategic directions and tactical actions for shaping shared mobility systems of the future. As a part of future directions, measures to adapt to the longer-term effects of the global pandemic are explored.

We expect that both the breadth and focused coverage of topics in this book will help to educate students as well as meet the informational needs of a wide range of transportation professionals and individuals in the public, private, non-profit, and academic sectors. The subjects included in the book span engineering, public policy and management, urban studies, economics and finance, sociology, and environmental impact disciplines. Students enrolled in undergraduate and graduate courses in these disciplines, as well as in multidisciplinary programs in sustainable development and technology-society-environment, will benefit from the balanced coverage of these subject areas. Researchers interested in addressing future mobility challenges will find answers to many questions. This book is also aimed at professionals and academics interested in gaining knowledge on various facets of future shared mobility systems, including SAVs.

Susan A. Shaheen, Department of Civil and Environmental Engineering, University of California, Berkeley, USA. Ata M. Khan, Department of Civil and Environmental Engineering, Carleton University, Ottawa, Canada.

Author contributions

The authors confirm contribution to this chapter. All authors reviewed and approved the manuscript.

Declaration of conflicting interests

The authors declared no potential conflict of interest with respect to publication of this chapter.

The views expressed are those of the authors.

Chapter 1

Introduction

Susan A. Shaheen[1] and Ata M. Khan[2]

Socio-economic and technology forecasts suggest that shared mobility will play a growing role in society due to increased consumer demand and the need for more equitable and environmentally sustainable transportation options [1–3]. Changes in lifestyle and societal trends are fostering on-demand shared modes as an alternative to private mobility [2], although demand for transportation network company (TNC) services and pooled rides has been dampened due to COVID-19 and virus transmission concerns. Sustainability goals, along with reduced battery costs, are steering transportation toward an electric-drive future [4]. Furthermore, automated vehicles (AVs) designed for on-demand shared mobility systems are predicted to become more mainstream over the coming decade(s). These services and technologies, along with supportive policies, could offer a more attractive alternative to private vehicle ownership in a range of land-use and built-environment contexts [1]. However, their attractiveness is tied to public policy (e.g., road pricing and curb management), along with a willingness to share rides with strangers.

Affordable shared mobility services have the potential to strengthen the position of high-capacity public transit operations by providing an integrated, curb-to-curb transportation system [5,6]. Over time, shared automated vehicles (SAVs), such as on-demand microtransit, could also help to fill mobility gaps in lower density areas that are unable to efficiently provide fixed-route public transit [7]. While the vision of SAVs offers great potential, there are notable technological hurdles to deploying fully automated services in diverse environments, along with a number of socio-economic issues (e.g., willingness to share, curb management, pricing, social equity, and labor considerations). It is important to note that autonomy can provide benefits, even if not fully driverless, such as automated acceleration and steering to ease driving and reduce collisions. This book aims to advance knowledge of shared mobility systems, particularly those that provide on-demand services that employ automation and electric vehicle (EV) technologies. It also aims to serve as a resource to help shape future shared mobility systems that maximize societal and environmental benefits, including social equity.

[1]Department of Civil and Environmental Engineering, University of California, Berkeley, CA, USA
[2]Department of Civil and Environmental Engineering, Carleton University, Ottawa, Canada

This book presents current-day knowledge of shared mobility systems and also covers future directions. It builds upon current understanding by focusing on the evolution of non-AV services toward SAVs. While the future is inherently unclear, it can benefit from scenario analyses and should be revisited as it evolves to ensure that outcomes maximize the public good. Given the growing trend toward vehicle electrification, we also focus on charging infrastructure and shared EV requirements. Further, the requirements for a post pandemic world are explored in discussing the evolution of the transportation system.

A range of interrelated topics are covered in this volume with the goal of serving as a reference for transportation students, researchers, professionals, and policymakers across the globe. This book aims to provide planning, design, and implementation knowledge of shared vehicle systems with an eye toward future mobility services. Following the introductory chapter, the book examines: (1) advances made in mobility on demand and shared mobility systems, (2) policy and regulatory frameworks, (3) planning and design within a multimodal travel environment, (4) technology, (5) demand and supply models, (6) operations and management, (7) economic factors and business models, (8) urban form implementation contexts, and (9) environmental impacts.

1.1   Scope of the book

This book is based on eight interrelated topics. The first and primary focus of this volume is evolving shared mobility systems, which can offer travelers safe, secure, efficient, and equitable access. Policy and regulatory frameworks that enable shared mobility services and define their operating structures are the emphasis of the second topic. The third theme is social and racial equity, which addresses equitable access for all travelers. The fourth explores environmental and financial sustainability, including business models and key economic factors. The fifth focuses on transportation planning and land use. The sixth theme examines system design, operations, and management. The seventh explores implementation contexts, urban development, and environmental and other impacts. In light of COVID-19, we also address the longer-term effects of global pandemics on SAV services (e.g., hygiene and sanitation) in the eighth topic area.

We briefly introduce each theme below. Chapters 2–17 provide in-depth coverage of each topic. In Chapter 18, we explore future directions, conclusions, and logical inferences based on the evidence presented.

1.2   Advanced, safe, secure, and efficient mobility

The present-day ecosystem of shared mobility services is dynamic, reflecting the ongoing evolution of technological and service advancements, including vehicle right sizing, multimodal mobility, and the potential for pooling (as well as reluctance to share). This ecosystem includes: (1) carsharing; (2) microtransit and shuttle services; (3) ridesharing (carpooling and vanpooling); (4) app-based for-hire ride services (referred to as transportation network companies (TNCs), ridesourcing, or ridehailing); (5) taxis and pedicabs; (6) shared micromobility (scooter sharing and bikesharing); and (7) courier network services (deliveries). Shared micromobility includes traditional bikes, e-bikes, and e-scooters. It is comprised of a wide range of service models that meet diverse traveler needs, such as station-based and dockless bikesharing and standing electric scooter sharing. Conventional public transit (buses and rail) and shared urban air mobility services are part of this ecosystem. While shared mobility focuses on passenger mobility, it also includes last-mile delivery services such as app-based deliveries (commonly referred to as courier network services), robotic delivery, drones, and other last-mile delivery innovations.

The application of electronic and wireless technologies including apps, which are typically associated with intelligent transportation systems (ITS), have improved the capabilities and efficiency of shared mobility services, particularly over the last decade. Many on-demand mobility services are made possible by the specialized application of these technologies. For instance, shared micromobility data, collected via a software tool called the Mobility Data Specification (MDS), are made possible by information technology [8].

The electrification of transportation modes has already started and is diversifying. Battery-powered e-bikes and e-scooters are now serving the shared micromobility market across the globe. Large fleets of electric cars intended for on-demand carsharing services have been introduced in North America and Europe [9]. According to recent announcements, steps have been taken for moving into an all-electric future, including large-scale EV charging investments [4].

Beyond mobile apps, ITS, and electrification, future technological advancement in shared mobility includes vehicle automation, along multiple levels—ranging from partial to full autonomy—in a diverse range of operating environments. Advanced algorithms are envisioned to optimize on-demand operations and management, and cost-efficiency improvements are predicted from automation and reduced labor costs.

With respect to autonomy, a brief introduction to AVs is recommended. The reader is advised to refer to publications by SAE International and the U.S. Department of Transportation for further detail [10,11]. According to SAE International, vehicle automation is classified according to five levels. In Level 1, only one primary control function (e.g., self-parking or adaptive cruise control) is automated. Level 2 automation provides full control of specific functions, such as accelerating, braking, and steering. With Level 2, the driver is expected to monitor driving and be prepared to immediately resume control at any time. In Level 3, the driver can engage in nondriving tasks for a limited time, and the automation function handles situations requiring an immediate response. The driver is expected to be prepared to intervene almost immediately when prompted. With Level 4 automation, a human operator does not need to control the vehicle when operating under specific conditions. Vehicles operating at Level 5 are designed to drive in all environments without human control.

In this book, AVs in shared mobility are assumed to operate at Levels 4 and 5, and travelers will not be expected to play a role in safe vehicle driving. However, in

SAV system design, a human monitor can intervene remotely, if required. Level 5 is likely more than a decade away, particularly in mixed-use traffic environments. It is important to note that partial autonomy can be beneficial in improving safe operations among shared mobility providers. In addition to safety and efficiency requirements, SAVs are expected to have security safeguards as described in the final chapter of this book. Given the broad scope of this volume, the reader is advised to search for the material of interest in Chapters 2–18.

1.3   Governance: policies and regulations

Many public policies have been favorable to shared mobility services due to their documented and/or potential social and environmental benefits. TNCs are offering shared mobility services around the world, although not always without some issues. A well-known concern was the labor problem in London (U.K.), but that was resolved. In other parts of the globe, labor issues are continuing to evolve (e.g., California, New York City).

At present, the vast majority of these services are offered by the private sector. Shared mobility operators are expected to apply and follow the guidelines of city, state, and transportation-related authorities. Following a check on applicable regulations, insurance provisions, and the availability of suitable infrastructure, a permit may be granted. Some jurisdictions may impose restrictive conditions for safety or protecting other stakeholder interests (e.g., social equity and labor). In some cases, a license may be withdrawn (along with fines when applicable), if the private sector operator does not comply with the required policies.

This topic and policy-related facets of a number of other subjects including "social equity and justice" are featured in Chapters 3–6 and Chapters 17 and 18 of the book. The broad scope of policies includes data ownership and sharing issues; privacy concerns; and impacts on labor, vehicle miles/kilometers, congestion, and emissions.

1.4   Social equity and justice

Public policy aims to enhance social equity and justice in transportation. At a basic level, social equity refers to fairness with respect to the distribution of benefits and costs. Transportation has been the subject of equity analysis for many decades, with research focusing on a range of issues including the local impacts of large transportation infrastructure projects (e.g., highway construction in low-income urban neighborhoods); equity assessment of how transportation is subsidized and taxed (e.g., the use of local sales taxes versus tolls to fund transportation projects); and how underlying land use and transportation planning decisions affect accessibility for various demographic groups (e.g., lower-density auto-dependent development impacts on senior, disabled, and low-income mobility) [12]. The notion of justice goes beyond fairness and accessibility (e.g., ease of accessing a destination in terms of travel time and costs) by focusing on the intentional inclusion of impacted communities/individuals and restoration.

This theme focuses on enhancing accessibility and equity of services for individuals who are unbanked or underbanked; suffer discrimination based on race, gender, or cultural factors; and have a physical or mental disability. A potential strength of shared mobility services is to address gaps in transportation needs that cannot be cost effectively filled with conventional public transit services, including rural or suburban locations and late-night mobility. Chapters 3–6 and Chapters 17 and 18 focus on this topic.

1.5 Sustainability: environmental and financial factors

In addition to ensuring socially equitable outcomes, shared mobility services should strive to achieve environmental and financial sustainability. The electrification of shared modes can help to reduce emissions and improve the operational efficiency of shared vehicles in the traffic stream, which can include eco-routing practices. In recent years, there has been a growing focus on reducing traffic congestion that may be caused by TNC ride services. Chapters 2, 17, and 18 are key sources of information on environmental sustainability.

Financial sustainability is also necessary for providing shared mobility services. Economic factors and business models are addressed in Chapter 16. Another aspect of financial sustainability is the potential to charge a fee for the use of public infrastructure (e.g., curb management). Chapter 3 discusses usage-based pricing policies, and Chapter 18 provides the rationale for pricing infrastructure use by private sector companies.

1.6 Planning: transportation system and land use

This broad theme serves as an overall framework for integrating a number of subjects covered throughout Chapters 3–18. This includes transportation and land use planning, multimodal options, integration of shared mobility into transportation system modeling, and guiding favorable land-use development subjects. Demand models treat both non-AVs and SAVs. Models are described for balancing demand and supply factors. Chapters 4–18 cover various facets of the planning theme.

1.7 Design, operations, and management

New understanding of system design, operations, and management is needed to support shared EV services. Included in this topic is the task of adapting infrastructure for future services. Naturally, system requirements become more challenging when electric AVs are integrated into the market. Concept-level design and high-level architecture are instrumental in advancing system design for shared mobility services, including non-AVs and AVs. More information is needed to better support system operations and management functions. Since shared mobility services are typically deployed in built-up areas, infrastructure modifications are critical. Greater understanding of how to optimize pick-up and drop-off locations, vehicle staging areas, and battery charging infrastructure is needed. Chapters 3–18 provide insights into these topics.

1.8   Implementation context, urban development, and other impacts

To support policy and planning decisions for shared mobility, studies of the implementation and impacts are needed. While urban development impacts are uncertain due to limited real-world experience, this book aims to better understand the implementation context and to foster favorable environmental and urban development outcomes. Chapters 14, 15, 17, and 18 provide key reference material on this theme.

1.9   Adapting to the long-term effects of the global pandemic

The global pandemic resulted in more work-from-home/telework due to concerns about virus transmission and the need for social distancing. These short-term effects have adversely impacted the use of public transportation and commercial space across cities. While the short-term effects of COVID-19 are more clear, the longer-term impacts are highly uncertain and should be revisited as they evolve [13–15]. We address the need for hygiene and sanitation practices related to virus transmission in Chapter 3, which highlights policy and regulatory considerations. In Chapter 18, we focus on future directions and study measures to support SAV evolution and adaption in the context of longer-term pandemic recovery.

1.10   Contents of the book

This volume explores shared mobility systems, policy and regulatory factors, social equity and justice, planning and design methods, financial and environmental sustainability, implementation contexts, and strategies to maximize benefits. The final chapter explores future directions, including measures for adapting to longer-term pandemic recovery.

The book is organized into 18 chapters and an additional chapter includes literature for further reading.

- Chapter 1, *Introduction*, describes the scope of this book in the form of eight focus areas, presents chapter-by-chapter contents, and identifies the intended audience.
- Chapter 2, *Navigating seismic shifts in transportation*, covers trends, issues, and provides a starting point for understanding shared mobility services and

their impacts. Converging innovations and their likely effects are described and potential opportunities are identified for smart cities, public transportation, and other stakeholders. It also acknowledges that shifting individuals toward pooled rides will likely remain a challenge.

- Chapter 3, *Policy and regulatory environment: shared automated vehicles,* covers the policy and regulatory frameworks commonly used in jurisdictions with shared mobility services. This chapter features key considerations applicable to permitting services, scope, and conditions. It also features data sharing, privacy, and data ownership issues; congestion, vehicle miles/ kilometers, and emission impacts; and labor and social equity policy considerations.

- Chapter 4, *Concept level design: high-level architectures,* describes how the existing shared mobility services (non-AVs) operate and how to transform them for SAV services. This focuses on the concept level, supported by high-level architectures for shared mobility services. Requirements for non-AVs and SAVs also are examined.

- Chapter 5, *Shared mobility: managing rights-of-way, developer incentives, and planning principles,* is focused on planning shared mobility systems and services. This chapter provides an overview of the ways in which shared mobility can impact planning and policymaking. Topics covered include curb management and public rights-of-way, vehicle right sizing and adaptability, the relationship between shared mobility and developer zoning regulations, shared mobility and the planning process, the role of shared mobility and the built environment, applications in the suburban context, stakeholder and public involvement in the planning process, and planning considerations for cities and public agencies in an automated future.

- Chapter 6, *Shared mobility services: prioritizing social good,* describes how to prioritize the public good in shared mobility services. Socio-demographics of shared mobility users are reviewed. Social and racial equity challenges that impact access and the use of shared modes are described, including policies for mitigating many of these challenges. The STEPS (Spatial—Temporal— Economic—Physiological—Social) framework to transportation equity is explored. STEPS can aid service providers and public agencies in identifying, understanding, and overcoming equity challenges in transportation access. Social equity and access considerations in planning a multiphase transition/ evolution toward highly automated vehicles are discussed.

- Chapter 7, *Multi-modal relationships: shared and automated vehicles and high-capacity public transit,* explains how shared mobility modes relate to one another and their interaction. The promise of a more integrated transportation system is highlighted, merging access to services offered across modes to help address capacity, directness, and distance by combining the advantages of SAVs and public transit. The potential of SAVs to enhance public transit systems by improving efficiency and filling gaps (and not competing with them) is discussed. This chapter explores whether and how the transportation system can maintain the benefits of public transit (congestion relief, social equity, and environmental sustainability), while also improving the quality of public transport services.

- Chapter 8, *Design of systems with nonautomated electric vehicles,* describes system designs for nonautomated EVs in providing safe and efficient shared mobility services. Design requirements are defined, and measures to meet them are discussed. Traveler service frequency requirements and supplier efficiency objectives should guide design within urban contexts. Application of ITS and advanced communications will continue to play a key role in system design. The potential of shared mobility services to advance the objectives of Mobility as a Service (MaaS) is also addressed.

- Chapter 9, *Design of systems with automated and electric vehicles,* covers AV system designs. This includes an advanced form of on-demand shared mobility in which driverless vehicles can travel to a requested pick-up location and deliver a traveler to the desired location. SAVs can park and charge at a number of stations in an urban area. When requested, vehicles can be in a free-floating mode, at a station, or staging area. These multistation and free-floating systems do not require customers to travel to a vehicle. A TNC-type of service could be implemented with automated EVs. EV services will require access to fast charging infrastructure supported by cities.

- Chapter 10, *Demand for shared mobility to complement public transportation: human driven and automated vehicles,* presents demand models for shared mobility services, which complement public transportation. Models include human driver and automated vehicle cases. The demand models address the first- and last-mile service gap, especially in low-density areas. The models support understanding of the on-demand ride service impacts on travel demand and social welfare, complementing public transit under different scenarios. Both crowdsourced human driven vehicles (e.g. Uber and Lyft) and centrally operated SAVs are modeled, and the influence of a fare discount on demand and modal shift is also investigated. A case study of the Oakville road network in Ontario, Canada is examined employing real-world data.

- Chapter 11, *Demand for shared mobility to replace private mobility using connected and automated vehicles,* describes another modeling study on the demand for shared mobility using connected and automated vehicles. This chapter examines how the introduction of shared connected and automated vehicles (SCAVs) could affect travel demand, social welfare, and traffic congestion. The model adapts an agent-based daily process and develops a central dispatching system, implemented on an in-house traffic microsimulator. This model considers a dual-sided market in which demand and SCAV fleet size change endogenously. To estimate SCAV fleet size, changing traffic conditions are considered, and the model assumes two available transport modes: private AVs and SCAVs. The resulting model is applied to downtown Toronto, Canada using real-world data.

- Chapter 12, *Matching demand and supply under uncertainty,* addresses the need to harmonize shared mobility demand and supply with electric AVs in an urban transportation environment. This chapter recognizes the need to accommodate a variable party size with a SAV service, while also providing sufficient EV charging stations and vehicle parking. On the demand side, SAV operators must develop algorithms that assign vehicles with the required number of seats and the shortest available wait times. On the supply side, algorithms must also be employed to direct SAVs to optimal EV charging locations, while accounting for traffic congestion. This chapter underscores the challenge of meeting SAV system supply and demand needs under uncertain operating conditions.

- Chapter 13, *Operations and management*, describes the requirements for operating and managing EV fleets in nonautomated and automated shared mobility services. Advanced methods and required technologies are described. Key operational and management changes in shared mobility operations due to vehicle autonomy are explored. These include system components, reservations, access, dispatching, ride matching, routing, fleet monitoring, and payment. The authors explain that most changes in SAV fleet operations are transformative, including the transition from drivers to algorithmic decision making, availability of more on-board sensors, and superior computing and communication capabilities. The cost effectiveness of operations and management changes is also discussed.

- Chapter 14, *Impacts on the public realm: understanding the context*.
- Chapter 15, *Impacts on the public realm: the potential for a positive outcome*, complement one another. Chapter 14 provides an overview of the AV-built environment context, and Chapter 15 describes the potential for a positive outcome in transitioning toward AV deployment. These chapters discuss the typological context and high-level spatial impacts of AVs, along with emerging mobility scenarios. Key determinants affecting SAV fleet size for a given population include spatial criteria, which impact street design, parking, and land use. Observations are provided for smaller cities that are leading the way in TNC integration and pooled ride services in more structured public transit environments.

- Chapter 16, *Economic factors*, covers concepts, methods, and applications for providing answers to economic feasibility questions. In this chapter, building blocks for business models are quantified, including cost and revenue estimates for fleets consisting of vehicles of various sizes (or right sizing for adaptability). Observations are drawn for the economic feasibility of SAV systems by employing an after-tax rate-of-return analysis for various business models.

- Chapter 17, *The impacts of shared and automated mobility*, includes a comprehensive review of shared and automated mobility impacts. The chapter reviews findings from shared mobility studies including ridesharing (carpooling and vanpooling); carsharing; shared micromobility (bikesharing and scooter sharing); and TNCs. It concludes with a discussion of SAVs and their potential impacts while acknowledging the challenges associated with vehicle pooling and encouraging people to share rides.

- *Chapter 18, Future directions: maximizing the social and environmental benefits of shared and automated mobility services,* describes future strategic approaches and tactical actions for shared mobility and SAV services. Innovative mobility services and technologies, socio-demographics, and mobility trends are expected to influence the future development of urban transportation and land-use environments. Converging mobility innovations— including shared mobility, first and last-mile connections, digitization and fare payment integration, the commodification of transportation services, connectivity and autonomy, and vehicle electrification—will impact future mobility. Nevertheless, their evolution is highly uncertain, which we will need to revisit over time. In this chapter, eight focus areas are explored, including: technology, governance, social equity, sustainability, system planning, design, implementation, and public health. These cover service innovations, the need to update policy and regulatory practices, the importance of social and racial equity, the need for environmental and financial sustainability, the role of transportation and land use planning, adapting infrastructure for future use by SAVs, transforming operations and management tasks, implementation contexts and impacts, and adapting to the longer-term impacts of the global pandemic. The contents of this chapter recognize that AVs will gain acceptance in real-world applications in a phased evolution over time, which requires access to shared mobility usage data. These data are needed to support the transition from non-AVs to highly automated services. It is important to note that pooled shared mobility services (non-AV) and SAVs must also overcome the reluctance of some travelers to share rides. The final focus area of this chapter addresses the longer-term response and recovery to the global pandemic.
- Further readings are included at the end of the book and brief annotations are also provided.

1.11   Audience for the book

This book provides a broad lens to the topic of shared mobility and shared automated vehicles or SAVs. It encompasses both shared mobility with non-AVs and SAVs in a range of environments, as well as EV-based systems. We expect that the breadth of cross-cutting subjects covered in this volume will contribute to the diverse educational objectives of students, transportation professionals, and public agencies.

The topics span engineering, public policy and management, urban studies, economics, sociology, and environmental impact disciplines. In most topic areas, there are multiple chapters that provide different perspectives on each subject. When applicable, models and algorithms are described and illustrated.

Students enrolled in undergraduate and graduate courses in a wide variety of disciplines, as well as multidisciplinary programs focused on "sustainable transportation," and "technology, society, and the environment" could benefit from this book.

University researchers and consultants interested in addressing future mobility challenges will find a discussion of many unanswered questions focused on the future of mobility. This book may also be of interest to transportation professionals and public agencies interested in various facets of shared mobility systems served by nonautomated and automated EVs.

Author contributions

The authors confirm their contribution to this chapter. All authors reviewed and approved the manuscript.

Declaration of conflicting interests

The authors declared no potential conflict of interest with respect to the publication of this chapter.

References

[1]   Boston Consulting Group (The) (BCG). *The reimagined car: Shared, autonomous and electric*. Boston, MA; December 2017.

[2]   International Transport Forum (ITF). *Shared mobility, innovation for liveable cities*. Corporate Partnership Board Report, OECD. Paris; 2016.

[3]   Zmud, J., Ecola, L., Phleps, P., and Feige, I. *The future of mobility scenarios for the United States in 2030*. (RR-246-ifmo). 2013. Rand Corporation Institute for Mobility Research. www.rand.org.

[4]   General Motors. "GM technology paves the way for an all electric future." *Technology*; 14 June 2020. (https://www.gm.com)

[5]   Federal Highway Administration FHWA). *Integrating shared mobility into multimodal transportation planning: Improving regional performance to meet public goals*. FHWA-HEP-18-033, U.S. DOT, Washington, D.C.; 2018.

[6]   Feigon, S. and Murphy, C. *Broadening understanding of the interplay among public transit, shared mobility, and personal automobiles*. Transit Cooperative Research Program; Transportation Research Board; National Academies of Sciences, Engineering, and Medicine; 2018.

[7]   Shaheen, S., Cohen, A., and Ismail Zohdy, I. *Shared mobility: current practices and guiding principles*. FHWA-HOP-16-022; April 2016.

[8]   Descant, S. "Los Angeles DOT in federal lawsuit over scooter data." *Government Technology News*. The Future of Community Design; 22 June 2020.

[9]   Volkswagen. "WeShare launched in Berlin as full-electric service." *Volkswagen News. Wolfsburg/Berlin*; 27 June 2019. Available from https://www.volkswagen-newsroom.com.

[10]   SAE International. *J3016_201806: Taxonomy and definitions for terms related to driving automation systems for on-road motor vehicles*. Warrendale: SAE International; 15 June 2018.

[11]   U.S. Department of Transportation. *Preparing for the future of transportation, automated vehicles 3.0.* Washington, D.C.; October 2018.

[12]   Shaheen, S., Bell, C., Cohen, A., and Yelchuru, B. *Travel behavior: Shared mobility and transportation equity.* U.S. Department of Transportation. Report # PL-18-007. 2017. https://www.fhwa.dot.gov/policy/otps/shared_use_mobility_equity_final.pdf

[13]   AVIN Ontario. *The auto sector and the COVID-19 pandemic, recovery support and opportunities.* AVIN Specialized Reports. Toronto; June 2020.

[14]   Descant, S. "The pandemic will mean big, lasting changes for urban mobility." *Government Technology, Future Structure, Transportation*; 08 May 2020. https://www.govtech.com)

[15]   McKinsey & Company. *The impact of COVID-19 on future of mobility solutions.* Global Management Consulting Company; May 2020. https://www.mckinsey.com

The views expressed are those of the authors.

Chapter 2

Navigating seismic shifts in transportation

Susan A. Shaheen[1] and Adam P. Cohen[2]

Over the past century, the world's population has grown exponentially from 2 billion in 1927 to 7 billion in 2012, with the United Nations (UN) expecting the figure to increase to 9.7 billion by 2050 and 11.2 billion by 2100 [1]. The UN further estimates that the majority of this population growth will occur in cities, and two-thirds of the world population will be based in urban areas by 2050 [2]. This growth is leading to the development of approximately 30 global megaregions that link various urban centers into "megalopolises" that typically exceed 10 million people with an economic output of at least 250 billion USD [3,4]. These demographic trends will continue to place considerable strains not only on urban infrastructure but also on the environment. Global growth, economic, environmental, and technological forces are changing the way people travel and consume goods and services. Technology and consumer preferences are transforming rapidly as the transportation network becomes more connected, data driven, demand responsive, shared, and automated.

2.1 Setting the stage: trends disrupting mobility

In recent years, socio-economic forces—coupled with advancements in technology, social networking, location-based services, wireless services, and cloud technologies—are contributing to converging trends in mobility innovation and disruption. Key trends disrupting the mobility marketplace are as follows.

Technological trends:

- Growth of cloud computing, location-based/satellite navigation services, and mobile technologies;
- Expansion of data collection, sharing, aggregation, and re-dissemination through crowd-sourced, private, and public sector sources facilitated through application programming interfaces (APIs) and other third-party tools; and

[1]Department of Civil and Environmental Engineering, and Transportation Sustainability Research Center, University of California, Berkeley, CA, USA
[2]Transportation Sustainability Research Center, University of California, Berkeley, CA, USA

- Ongoing development and deployment of advanced algorithms, machine learning, and artificial intelligence (AI) that enable on-demand and flexible route service offerings, electrification, and automation.

Socio-demographic trends:

- Heightened environmental awareness about emissions and carbon footprints;
- Growth of megaregions as economic centers and transportation corridors;
- Changes in land use and shifts toward urbanization and reduced interest in car ownership; and
- Demographic changes, such as rising life expectancies, an aging population, and retiring in place (particularly in industrialized nations).

Mobility trends:

- Increasing demand and urban congestion, reduced transportation funding, and the critical need to maximize existing infrastructure capacity;
- Growing popularity of shared mobility (including delivery services);
- Increasing consumer interest in on-demand transportation options; and
- Commodification of passenger travel, goods, and services driven by the growth of e-commerce and app-based mobility and delivery services.

These trends coupled with global growth and the need to reduce greenhouse gas (GHG) emissions are contributing to future seismic shifts in transportation.

2.2   Converging innovations

Mobility needs, consumption, and travel behavior are changing [5–9]. These changes are contributing to five converging mobility innovations. Each innovation is noted below in Figure. 2.1.

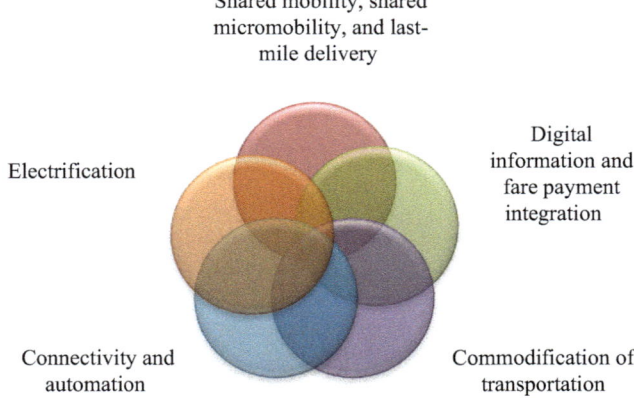

Figure 2.1   Five converging mobility innovations

Shared mobility, shared micromobility, and last-mile delivery: Shared mobility is an innovative transportation strategy enabling users to gain short-term access to transportation modes on an "as-needed" basis. The ecosystem of shared services continues to grow and includes an array of services such as carsharing, microtransit, app-based for-hire ride services (sometimes referred to as transportation network companies (TNCs), ridehailing, or ridesourcing); moped-style scooter sharing; shuttles; taxis; advanced air mobility/on-demand air mobility; and public transportation. Shared mobility also encompasses shared micromobility, the shared use of a bicycle, scooter, or other low-speed mode. Shared micromobility includes various service models and transportation modes that meet the diverse needs of travelers, such as station-based and dockless bikesharing, and standing electric scooter sharing. Increasingly, the future will include new and improved form factors in this space (e.g., three wheelers). Shared mobility is also inclusive of last-mile delivery services including app-based deliveries (commonly referred to as courier network services), robotic delivery, drones, and other last-mile delivery innovations.

Electrification: Electric drive vehicles (EVs) and electric devices (e.g., scooters, e-bikes, etc.) that use one or more electric or traction motors for propulsion can reduce GHGs and other emissions, mitigating many of the transportation-related impacts associated with increased urbanization in cities. Lower pollution and maintenance requirements are contributing to increased investment, improved performance (increased range and reduced charge times), and the growing popularity of EV technology.

Automation: Automated vehicles that are capable of sensing the environment and moving with little or no input have the potential to improve safety and increase vehicle occupancy (with policy levers). Vehicle automation also has the potential to create new and exciting opportunities for public transportation such as cost savings, automated pick-up, drop-off, and charging, and more economical and convenient demand-responsive services. At present, lower levels of vehicle automation are common, which is leading to higher levels of vehicle automation that require less human control.

Digital information and fare payment integration (digitization): With a growing number of mobility innovations, there is a demand for data-enabled technologies that aggregate modes, facilitate multimodal trip planning, and integrate payment. A growing number of digital information and fare payment services are increasingly offering seamless information and payment connectivity among different transportation modes. These services can help bridge information gaps, make multimodal travel and public transit more convenient, and enhance decision making with dynamic and real-time information throughout an entire journey.

Commodification of transportation: Increasingly, consumers are assigning economic values to modes and engaging in multimodal decision-making processes based on a variety of factors including cost, travel time, wait time, number of connections, convenience, and other attributes. Rather than making decisions between modes, mobility consumers can make decisions among modes, in essence "modal chaining" to optimize routes, travel time, and cost. Additionally, digital information and fare integration coupled with the commodification of transportation services is contributing to new on-demand access models such as mobility on demand (MOD) and mobility as a service (MaaS). Later in this chapter, the authors explore both the concepts and current differences between them.

Each innovation is discussed in greater detail in the sections that follow. It is important to note that while there is a growing body of knowledge about many of the innovative services and technologies described in this chapter, they are evolving rapidly. This is likely to be a constant for some time into the foreseeable future.

2.3   Shared mobility, shared micromobility, and last mile delivery

Shared mobility—the shared use of a vehicle, bicycle, or other travel mode—is an innovative transportation strategy that enables users to have short-term access to a transportation mode on an as-needed basis. Shared mobility includes various service models and transportation modes that meet the diverse needs of travelers. Shared mobility can include roundtrip services (vehicle, bicycle, or other mode is returned to its origin); one-way station-based services (vehicle, bicycle, or other mode is returned to a different designated station location); and one-way free-floating services (vehicle, bicycle, or other mode can be returned anywhere within a geographic area) [10,11]. Common and emerging shared mobility services include: bikesharing, carsharing, courier network services (CNS), microtransit, ridesharing (carpooling/vanpooling), scooter sharing, TNCs, and advanced air mobility/on-demand air mobility. Shared mobility also encompasses a variety of active and low-speed motorized modes, such as station-based and dockless bikesharing and scooter sharing, sometimes collectively referred to as shared micromobility. Table 2.1 provides definitions of common and emerging shared mobility and last mile delivery services.

In recent years, passenger and courier shared mobility services have grown rapidly due to technological advancements; evolving consumer patterns (both mobility and retail consumption); and changing perspectives toward transportation, car ownership, and urban lifestyles. Shared mobility can help bridge gaps in the transportation network, such as first- and last-mile connections to public transportation, late-night transportation, and service for low-density communities. As such, shared mobility can have a transformative effect in bridging gaps in the public transportation network by extending geographic coverage and the service times of public transit services. Additionally, a larger pool of travelers and modal options enables a "network effect," where modal options in closer proximity to one another add collective value.

In addition to motorized passenger mobility services, both personally owned and shared micromobility services are gaining popularity. Micromobility includes various active transportation and low-speed motorized modes such as station-based and dockless bikesharing and scooter sharing, traditional bicycles, and electric

Table 2.1   Definitions of common and emerging shared mobility and last mile delivery services. Adapted from [10–12]

Service	Definition
Bikesharing (also known as shared micromobility)	Provides users with on-demand access to bicycles at a variety of pick-up and drop-off locations for one-way (point-to-point) or roundtrip travel. Bikesharing users access bicycles using one of three bikesharing models: (1) station-based bikesharing (users access bicycles via unattended stations); (2) dockless (users may access (unlock) a bicycle and park it at any location within a predefined geographic region); and (3) hybrid bikesharing systems (users may check out and return bicycles either through a station or nonstation location). Bikesharing fleets are commonly deployed in a network within a metropolitan region, city, neighborhood, employment center, and/or university campus
Carsharing	Individuals gain the benefits of private-vehicle use without the costs and responsibilities of ownership. Individuals typically access vehicles by joining an organization that maintains a fleet of cars and light trucks deployed in lots located within neighborhoods and at public transit stations, employment centers, and colleges and universities. Typically, the carsharing operator provides gasoline, parking, and maintenance. Generally, participants pay a fee each time they use a vehicle
Courier network services (CNS)	Provides for-hire delivery services for monetary compensation via an online application or platform (such as a website or smartphone app) to connect couriers using their personal vehicles; bicycles; or scooters with freight (e.g., packages and food)
Drones	A short-range unmanned aerial vehicle (or UAV) that can transport small packages, food, or other goods
Microtransit	Privately or publicly operated, technology-enabled transit services that typically use multi passenger/pooled shuttles or vans to provide on-demand or fixed-schedule services with either dynamic or fixed routing. This also can be referred to as "pooling"
Ridesharing (also known as carpooling and vanpooling)	Facilitates formal or informal shared rides between drivers and passengers with similar origin–destination pairings. This also can be referred to as "pooling"
Robotic delivery	Offer short-range unmanned ground-based delivery of packages, food, or other goods using a small conveyance robot
Scooter sharing (also known as shared micromobility)	Users gain the benefits of a private scooter without the costs and responsibilities of ownership. Individuals typically access scooters by joining an organization that maintains a fleet at various locations. The scooter operator usually provides gasoline, parking, and maintenance. Generally, participants pay a fee each

(Continues)

Table 2.1   (Continued)

Service	Definition
	time they use a scooter. Scooters can be accessed via unattended stations or accessed (unlocked) and returned (parked) to any location within a predefined geographic region. Scooter sharing includes two types of services: Standing electric scooter sharing using shared scooters with a standing design with a handlebar, deck, and wheels that is propelled by an electric motor. The most common scooters today are made of aluminum, titanium, and steel Moped-style scooter sharing using shared scooters with a seated-design, either electric or gas powered, that generally have a less stringent licensing requirement than motorcycles designed to travel on public roads
Taxis	Provide prearranged and on-demand vehicle services for compensation through a negotiated price, zone pricing, or a taximeter. Trips can be made by advance reservations (booked through a phone, website, or smartphone application), street hail (by raising a hand or standing at a taxi stand or specified loading zone), or e-Hail (dispatching a taxi driver using a smartphone application)
Transportation network companies (TNCs) (also known as ridehailing and ridesourcing)	Provides prearranged and on-demand transportation services for compensation, which connect drivers of personal vehicles with passengers. Smartphone mobile applications facilitate booking, ratings (for both drivers and passengers), and electronic payment. TNCs also includes "ridesplitting," in which customers can choose to split a ride and fare in a TNC vehicle (where available). This also can be referred to as "pooling"
Advanced air mobility (AAM)/ on-demand air mobility	A broad concept focusing on emerging aviation markets and use cases for on-demand aviation in urban, suburban, and rural communities. AAM includes local use cases of about a 50-mile radius in rural or urban areas and intraregional use cases of up to a few hundred miles that occur within or between urban and rural areas. Urban and rural use cases are sometimes also referred to as urban air mobility (UAM) and rural air mobility, respectively

bikes, skateboards, segways, hoverboards, and unicycles. Between 2010 and 2018, 207 million shared micromobility trips have been completed in the United States, with 84 million trips completed in 2018 alone (National Association of City Transportation Officials (NACTO), 2019) [13]. NACTO [13] estimates that

36.5 million trips were taken on station-based bikesharing, representing an increase of 9% from 2017 while 9 million trips were taken on dockless bikesharing. NACTO also estimates that 38.5 million trips were taken on shared standing electric scooters [13]. As of May 2018, the United States had 261 bikesharing operators (station-based and dockless) with more than 48,000 bicycles (Russell Meddin, unpublished data). Roland Berger estimates that the global market size of bikesharing will increase from €6B Euros in 2019 to €7–8B Euros by 2021 [14]. A study by Shaheen *et al.* [15] estimates that there were an estimated 24,492 standing electric scooters being shared across 10 US cities (Austin, Columbus, Dallas, Denver, Detroit, Kansas City, Portland, San Antonio, San Francisco, and South Bend) (Shaheen, Cohen, Dowd, & Davis, 2019). Micromobility has the potential to offer communities an array of potential individual and community benefits such as increased mobility, greater environmental awareness, and increased use of active transportation and nonvehicular modes. The provision of rights-of-way and curb space into planning, street design, and policy is key to reducing modal conflicts (particularly with motorized modes) and ensuring safe, convenient, and multimodal access for all transportation users. With careful planning and public policy, it also has the potential to enhance accessibility and quality of life in cities [16].

Finally, last-mile goods delivery is equally important as changing consumer patterns disrupt traveler behavior. Advancements in courier services (technologies and service models) are transforming consumer behavior and disrupting both supply and trip chains, for example, linking a series of destinations in one single-origin based trip [17]. A person's decision to change their consumption preferences from driving to the store on the way home from work to having goods delivered to them could contribute to notable changes in traditional travel behavior [12]. Increasingly, there is a growing recognition that goods delivery can serve as a substitute for personal trips to access goods and services as consumers take advantage of "just-in-time delivery."

2.4   Electrification and automation

More recently, the convergence of electrification and automation with shared mobility is predicted to have a transformative effect on goods access and mobility. Shared modes that use electricity as their sole source of propulsion powered by clean energy can further reduce air pollutants and GHG emissions, mitigating many of the transportation-related impacts associated with increased urbanization in cities. Lower pollution and maintenance requirements are contributing to increased investment, improved performance, and the growing popularity of electric drive vehicles. Increased range, wireless charging innovations, and hydrogen infra-structure are already being deployed by multiple automakers, which will increase the demand and economies of scale for electric drive vehicles, making them more affordable for mass market and fleet deployment.

The growing expansion of electric drive technologies is leading up to one of the most groundbreaking developments in automotive history—the automated vehicle. SAE International, a global mobility standards organization, has established five levels of vehicle automation. Level 1 describes vehicles that automate only one primary control function (e.g., self-parking or adaptive cruise control). Level 2 describes a vehicle with automated systems with full control of specific vehicle functions such as accelerating, braking, and steering. With Level 2, the driver must still monitor driving and be prepared to immediately resume control at any time. Level 3 allows the driver to engage in nondriving tasks for a limited time. With Level 3, the vehicle will handle situations requiring an immediate response; however, the driver must still be prepared to intervene within a limited amount of time when prompted. With Level 4, a human operator does not need to control the vehicle as long as the vehicle is operating in the specific conditions it was intended to function. Level 5 describes vehicles capable of driving in all environments without human control.

Over the next 30+ years, Level 5 fully automated vehicles will be increasingly deployed in the marketplace, offering the ability to improve roadway safety by removing the most common cause of vehicle collisions—human error. Additionally, these vehicles could improve the use of existing infrastructure by re-routing around traffic to underused routes (eco-routing); reducing vehicle spacing through platooning and potentially lane width; and shifting goods movement to off-peak times (e.g., overnight).

Additionally, by automating driving tasks, shared, automated, and electric mobility services could become much more cost effective, efficient, and convenient than human-driven, privately owned vehicles [18]. According to one study, shared automated vehicles (SAVs) could reduce the per mile cost of transportation by 68% [19]. Another study found that shared automated EVs would be more cost effective than private vehicle ownership when deployed in a fleet system with usage between 40,000 and 70,000 miles annually, common for US taxis [20]. Operationally, the cost savings coupled with the convenience of e-Hail, pickup, drop-off, self-park, and self-charging could be a driving force toward reducing private vehicle ownership (particularly if coupled with curb management, street design, and pricing policies). Mobility consumers may opt to use these services as eValets for point-to-point mobility, as first-and-last mile connections to public transportation, or both. In either case, if e-Hailing becomes more economical and convenient to travel, many households could reduce the number of vehicles they own or postpone a purchase entirely.

SAVs have the potential to reduce vehicle ownership and provide innovative opportunities to lower cost and offer flexible public transportation systems. Shaheen *et al.* [12] identify a number of possible SAV use cases including:

- **Closed campus**—SAVs could provide short-distance, point-to-point travel in closed campus environments that can be easily mapped by software. These locations include theme parks, resorts, malls, business parks, college campuses, airport terminals, construction sites, downtown centers, real estate developments, gated communities, industrial centers, and others.
- **First mile/last mile connectivity**—Traditionally, public transit has been limited by fixed routes and fixed schedules. Due to these limitations, travelers may find it difficult to complete the first- or last-mile of their journey using

public transit. SAVs may be able to help bridge first- and last-mile gaps in the public transportation network.

- **Low-density service**—SAVs have the potential to provide lower cost and more frequent or responsive transit strategies in rural, exurban, and low-density suburban areas where low ridership and high labor costs often contribute to inefficient or cost prohibitive fixed route service
- **Off-peak/late night service**—Similarly, SAVs may be able to complement public transit by providing service during off-peak times, especially late at night when service is difficult and costly to provide; and
- **Paratransit**—Paratransit services could be provided by SAVs to meet the needs of people with disabilities; nevertheless, human assistance may still be required.

In addition to leveraging opportunities for passenger mobility, drones, robotic delivery, and automated vehicles offer opportunities for unmanned on-demand delivery options. Automated deliveries (vehicles and drones) could support the e-commerce trend of reducing brick-and-mortar storefront size, while simultaneously increasing the need for warehouses and urban logistics hubs to facilitate last mile goods storage, dispatch, and on-demand delivery [17].

2.5 Digital information, fare integration, and the commodification of transportation (digitization)

Together, these converging innovations are transforming how people access and consume mobility, goods, and services. In cities around the world, innovative and emerging mobility services are offering residents, businesses, travelers, and other users more options to access mobility, goods, and services. On both sides of the Atlantic, two parallel approaches to multimodal access to public and private transportation services are emerging [21]. In North America, consumers are assigning economic values to transportation services and making mobility decisions (including the decision not to travel and instead have a good or service delivered) based on cost, travel and wait time, number of connections, convenience, and other attributes [12]. On the other side of the Atlantic in Europe, apps allow travelers to sign up for mobility services in one package are gaining popularity [22]. Together these two approaches are about providing travelers with more seamless travel options (i.e., routing, booking, and payment) for all trip segments to improve the user experience and enable more informed transportation choices.

2.5.1 Mobility on demand (MOD)

In recent years, mobility on demand (MOD) is gaining popularity among mobility consumers. MOD is an innovative concept based on the principle that transportation is a commodity where modes have economic values that are distinguishable in terms of cost, journey time, wait time, number of connections, convenience, and other attributes. MOD enables consumers to access mobility, goods, and services on demand by dispatching or using shared mobility, delivery services, and public transportation strategies through an integrated and connected multimodal network [12,23].

MOD promotes choice in personal mobility, leverages emerging and existing technologies and big data capabilities, encourages multimodal connectivity and system interoperability, and promotes new business models that enhance traveler experience. MOD has three major guiding principles: (1) traveler centric and consumer driven, (2) data connected and platform independent, and (3) multimodal and mode agnostic. Technology enables an interoperable and multimodal transportation MOD ecosystem. MOD, as envisioned by the US Department of Transportation, culminates in the management of supply and demand across mobility services through an integrated transportation systems management and operations approach that is coordinated among the public and private sectors and the traveling public.

MOD also encompasses decision support systems to (1) aggregate real-time, historic, and predicted system condition information; (2) analyze alternative response strategies to address current or predicted problems; (3) assess the tradeoffs associated with strategies that support a number of operational objectives that vary dynamically; and (4) produce recommended strategies for implementation by system operators to guide and influence consumer choice [12]. Figure 2.2 visually depicts MODs role in integrating multimodal transportation operations and management to optimize the supply and demand of the transportation network.

2.5.2   Mobility as a Service (MaaS)

In Europe, another evolving concept known as mobility as a service (MaaS) is gaining popularity. Fundamentally, MaaS restructures the mobility distribution chain by integrating the products and services of mobility providers and supplying them to users as a single service. Typically, a digital platform creates and manages trips that users can pay for via a single account. A distinguishing feature of MaaS is giving users the option to purchase MaaS products, such as monthly subscription plans that best fit a user's or household's needs. These subscriptions can include a certain amount of each transportation service (e.g., public transportation, bike-sharing, carsharing, taxis, etc.) and are similar to other service bundles, such as mobile phone plans, where the user pays one price for the combination of multiple service elements (e.g., talk, text, data, roaming, long distance, etc.). Brokering travel with suppliers, repackaging, and reselling it as a bundled package is a distinguishing characteristic of MaaS [24–26].

In Gothenburg, Sweden, the first MaaS deployment known as UbiGo operated as a pilot between November 2013 to April 2014. UbiGo repackaged existing transportation services (e.g., public transit, taxi, bikesharing, and carsharing) into a one-stop, monthly, paid subscription service for the entire household (including children) [22]. This pilot included 195 people (173 adult participants and 22 children under the age of 18 years old). UbiGo subscriptions started at approximately €135 or 185 USD per month at the time of the trial, although the average subscription was approximately €200 or 280 USD per month. The pilot program

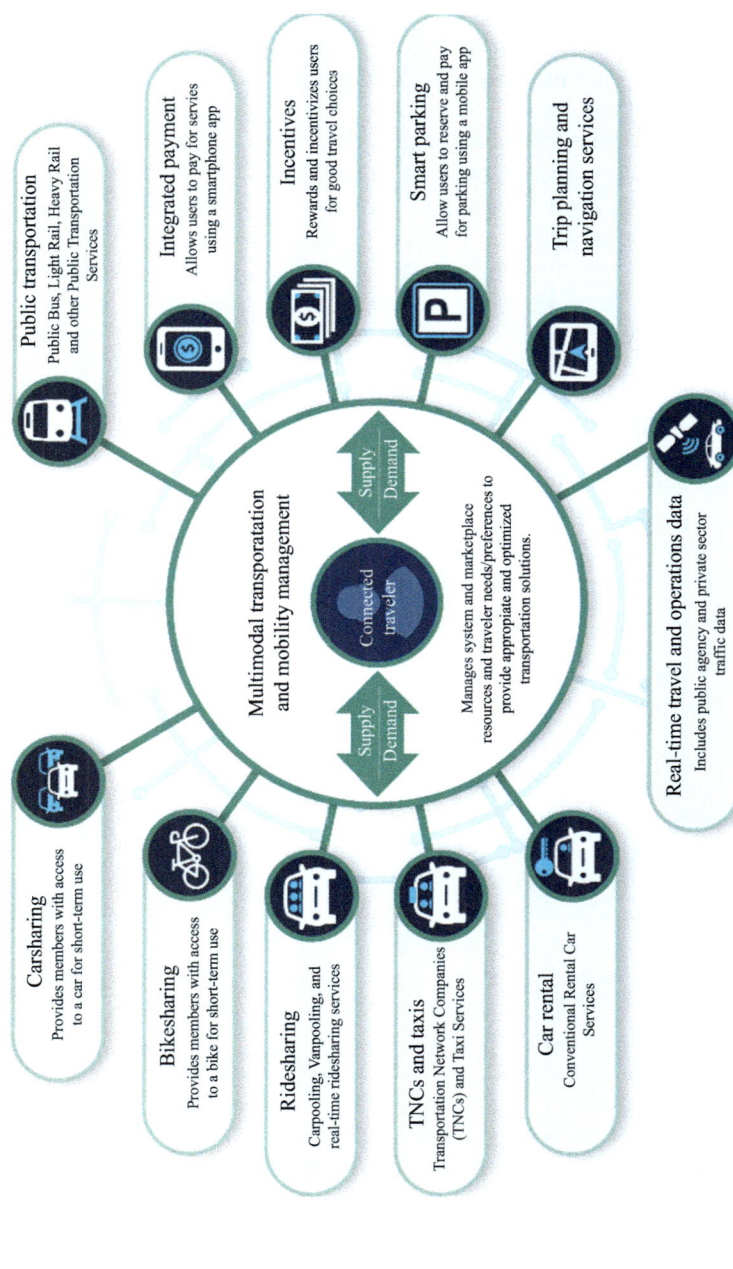

Figure 2.2 *Mobility on demand and multimodal transportation and mobility management. Source: U.S. Department of Transportation*

contributed to a reduction in household vehicle ownership and the increased use of bikesharing, carsharing, public transportation, and taxis. Based on a self-reported survey of other modes available, 50% of participants walked, 16% used a private bicycle, and 9% used a private vehicle at least three to five times per week [22]. More recently, UbiGo relaunched with a pilot in Stockholm in March 2018; however, the findings of this pilot have not been released yet.

2.5.3   Differences between MOD and MaaS

MOD differs from another concept, MaaS, in that MOD focuses on the commodification of passenger mobility and goods delivery and transportation systems management, whereas MaaS primarily focuses on passenger mobility aggregation and bundling services. Figure 2.3 illustrates the similarities and differences of MOD and MaaS. Specifically, MaaS is about integrating existing and innovative mobility services into one single digital platform where customers purchase mobility service packages tailored to their individual needs (ranging from per trip fares to bundled subscription mobility services). In contrast, MOD leverages passenger mobility and goods delivery services to enhance access while simultaneously leveraging MOD to achieve transportation system operational improvements through transportation network managers balancing supply and demand to match changing conditions across the transportation system.

    More research is needed to better understand the potential opportunities, best practices, and lessons learned for deploying MaaS and MOD services.

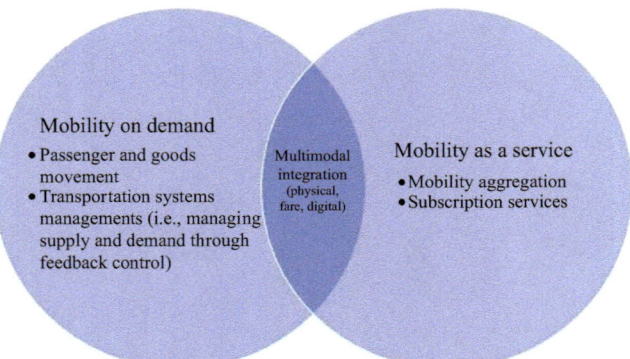

Figure 2.3   Similarities and differences of MOD and MaaS. Adapted from [16]

2.6   Conclusion

In the coming decades, converging innovations and technologies are likely to play a transformative role in transportation. In particular, the commodification of transportation coupled with vehicle automation will likely result in fundamental changes to cities by altering the built environment, costs, commute patterns, and modal choice. Naturally, vehicle automation will not inherently solve today's transportation challenges. To solve these challenges, the convergence of these mobility innovations requires thoughtful planning and public policies that balance societal goals with commercial interests. To harness and maximize the social and environmental benefits of these innovations, we need to prepare for this transition. Figure 2.1 shows the nature of longstanding modal relationships that have existed in transportation for years, which presents several potential opportunities for smart cities, public transportation, and other stakeholders:

- **Leveraging mobility innovations to overcome social equity challenges, not exacerbate them:** The convergence of these mobility innovations has the opportunity to enhance access and mobility for underserved communities, but it could also exacerbate existing barriers and increase inequality. For example, SAVs may be able to address spatial and temporal inequality in areas with limited alternatives to private vehicle ownership by providing additional mobility options for an entire trip or first- and last-mile connections to public transportation. The strategic placement of SAVs in communities underserved by public transportation could reduce inequities by providing innovative mobility options that have greater coverage and service availability than existing options. However, not all users may have access to a smartphone or debit/credit cards that are commonly required for payment as part of app-based and on-demand mobility services. In the future, it will be critical that policymakers ensure equitable access to all transportation services by addressing the needs of particular neighborhoods and populations, including users with special needs, as well as providing access options for the digitally impoverished and unbanked and underbanked communities [27].
- **New opportunities for first- and last-mile connections to expand the catchment area of public transportation:** In the future, SAVs could provide first- and last- mile connections to public transportation, such as bus rapid transit and rail transit, to expand the catchment area of high-quality fixed-route services. Coupled with on-demand access and dynamic routing, SAVs present the opportunity to make this a viable reality.
- **Changes in parking demand and opportunities for infill development:** Reduced vehicle ownership due to SAVs could result in changes in parking needs, particularly in urban centers. The reduction or the complete elimination of the need for parking minimums and their related costs increase the viability of development, creating the potential for more affordable residential units. In addition, the repurposing of urban parking has the potential to create new opportunities for infill development and increased densities. Infill development could also create higher densities to support additional public transit ridership and allow for the conversion of bus transit to rail transit in urban cores.
- **Potential to harness environmental benefits through pooling:** While SAV impacts remain uncertain, many practitioners and researchers predict higher efficiency, affordability, and lower GHGs. However, the number of personally

owned AVs may determine to some extent SAV demand. More importantly, SAV impacts will also depend on sharing levels (concurrent or sequential) and the future modal split among public transit, SAVs, and pooled rides. It is possible that SAV fleets could become widely used without very many pooled rides. Thus, single-occupant vehicles will continue to dominate the majority of vehicle trips (e.g., users could access a shared fleet without pooling). It is also feasible that pooled rides could become more common, if automation makes route deviation more efficient, cost effective, and convenient, along with pricing policies that encourage pooling (e.g., cordon and curbside pricing). It is important to note that it has been challenging to encourage individuals to share rides due to a variety of factors including personal safety, convenience, increased travel times, and personal preference. In addition, there was a heightened concern about vehicle sharing due to the global pandemic and virus transmission. While the environmental and travel behavior impacts of SAVs are still unknown, policy and planning tools will be critical to encouraging more pooling and guiding pooled SAV adoption [27].

- **Encouraging pooling and sustainable outcomes through pricing, incentives, and access policies:** Underpriced and overcrowded roadways create a "tragedy of the commons" where individual users acting independently and rationally, according to their own self-interest, behave contrary to the common good of maximizing road efficiency. In the future, single-occupant SAVs could continue to dominate the majority of vehicle trips, if users access SAVs with limited pooling. To minimize the risks associated with this scenario, policymakers should consider pricing and curb/road access policies that adjust access and price based on occupancies and vehicle right sizing and propulsion (e.g., EVs). Public agencies may be able to improve roadway performance by providing incentives for pooling and varying pricing and access by the time-of-day, roadway demand, and congestion [27].
- **Incorporation of shared mobility, micromobility, and SAVs into street design and management:** With the growth of active transportation and on-demand and flexible transportation options (e.g., shared mobility, micromobility, last mile delivery, etc.), public agencies should consider policies to guide shared mobility, micromobility, and SAV development through street design and the allocation of public rights-of-ways (e.g., parking, curb space management, dedicated lanes for micromobility and SAVs, and loading zones). Curb management, street design, and the allocation of public rights-of-way for shared mobility and micromobility today can support the growth of multimodal transportation and the development of intermodal mobility hubs today, which can be transitioned for SAVs in the future [27].
- **Need for multimodal curb space management:** Today, the curbside infrastructure in cities is being disrupted, particularly as innovative mobility options and the demand for these services expand in the urban environment, alongside COVID-19 trends (e.g., slow streets, parklets, outdoor dining, and pickup delivery). The once overlooked and underappreciated curb is now in high demand and impacting public transit and vulnerable road user populations

(e.g., cyclists and pedestrians). Planners are faced with growing competition for curb space and the need to focus on street design and curbside planning. Along with street design, cities should consider implementing a range of policies and plans for managing the curb including (1) fees for access and use; (2) prioritizing access for public transportation, cyclists, pedestrians, older adults, youth, and disabled populations; (3) geofencing to limit access to the curb; and (4) pricing to reflect key priorities (e.g., high-occupancy vehicles, walking and cycling, clean vehicles, parklets, slow streets, etc.).

- **Opportunities for public transportation innovation:** Automated transit vehicles have the opportunity to reduce the operating costs of public transportation. These savings could be passed onto riders in the form of increased service (more routes and reduced service headways) as well as lower and more equitable fares. Increased service and lower fares could make public transit more competitive than other modes and result in increased ridership [27].

While the impacts of these converging mobility innovations on travel behavior and communities are far from certain, technology has the potential to enhance the traveler experience and public transport effectiveness, allowing existing services to become more dynamic, demand-responsive, and automated. Nevertheless, technology (digital and fare payment integration, automation, and electrification) alone will not drive behavioral acceptance of vehicle sharing and pooling. Policy and planning are needed to encourage the adoption of ridesharing and shared-ride services, including pricing, incentives, and infrastructure enhancements. Finally, the impacts of the global pandemic on social equity, public transit, shared mobility, telework/work-from-home, active transportation, and e-commerce will likely have notable impacts on the future of mobility for years to come.

Author contributions

The authors confirm contribution to this chapter. All authors reviewed and approved the manuscript.

Declaration of conflicting interests

The authors declared no potential conflict of interest with respect to publication of this chapter.

References

[1]   United Nations. *World population projected to reach 9.7 billion by 2050 with most growth in developing regions, especially Africa*. Retrieved from United Nations: https://www.un.org/en/development/desa/population/events/pdf/other/10/World_Population_Projections_Press_Release.pdf; 29 July 2015.

[2]   United Nations. *World urbanization prospects.* New York City: United Nations; 2014.

[3]   Florida, R. *The real powerhouses that drive the world's economy.* Retrieved from CityLab: https://www.citylab.com/life/2019/02/global-megaregions-economic-powerhouse-megalopolis/583729/; 28 February 2019.

[4]   Lyons, R., and Gottman, J. *78-a-geographer who saw a northeast megalopolis.* Retrieved from The New York Times: https://www.nytimes.com/1994/03/02/obituaries/jean-gottman-78-a-geographer-who-saw-a-northeast-megalopolis.html; 2 March 1994.

[5]   Shaheen, S., and Cohen. A. *Mobility on demand: Three key components.* Retrieved from Move Forward: https://www.move-forward.com/mobility-on-demand-three-key-components/; 27 February 2018.

[6]   Galinsky, E. *Flexibility: Central to an effective workplace.* Retrieved from SHRM: https://www.shrm.org/hr-today/trends-and-forecasting/special-reports-and-expert-views/pages/ellen-galinsky.aspx; 1 September 2016.

[7]   Kolko, J. *Seattle climbs but Austin sprawls: The myth of the return to cities.* Retrieved from The New York Times: https://www.nytimes.com/2017/05/22/upshot/seattle-climbs-but-austin-sprawls-the-myth-of-the-return-to-cities.html; 22 May 2017.

[8]   Reagan, C., and Picker, L. *It's more than Amazon: Why retail is in distress now.* Retrieved from CNBC: https://www.cnbc.com/2017/05/05/its-more-than-amazon-why-retail-is-in-distress-now.html; 05 May 2017.

[9]   Koettl, J. *Boundless life expectancy: The future of aging populations.* Washington, D.C.: The Brookings Institution. Retrieved from The Brookings Institution; 2016

[10]   Cohen, A., and Shaheen, S. *Planning for shared mobility. Chicago.* American Planning Association; 2016.

[11]   Shaheen, S., Cohen, A., and Zohdy, I. *Shared mobility: Current practices and guiding principles.* Washington, D.C.: U.S. Department of Transportation; 2016.

[12]   Shaheen, S., Cohen, A., Yelchuru, B., and Sarkhili, S. *Mobility on demand operational concept report.* Washington, D.C.: U.S. Department of Transportation; 2017.

[13]   National Association of City Transportation Officials (NACTO). *84 million trips taken on shared bikes and scooters across the U.S. in 2018.* Retrieved from National Association of City Transportation Officials (NACTO): https://nacto.org/2019/04/17/84-million-trips-on-shared-bikes-and-scooters/; 17 April 2019.

[14]   Schönberg, A., Dyskin, A., and Ewer, K. *Bike sharing 5.0: Market insights and outlook.* Berlin: Roland Berger; 2018.

[15]   Shaheen, S., Cohen, A., Dowd, M., and Davis, R. *A framework for integrating transportation into smart cities: State of the practice in 20 U.S. cities.* San Jose, CA: Mineta Transportation Institute; 2019.

[16]   Shaheen, S., and Cohen, A. *Shared micromobility policy Toolkit: Docked and dockless bike and scooter sharing.* Washington, D.C.: The Smart Cities Lab and the International Council on Clean Transportation; 2019.

[17] Shaheen, S., and Cohen, A. *Seven goods delivery innovations: Transforming transportation & consumer behavior.* Retrieved from Move Forward: https://www.move-forward.com/seven-goods-delivery-innovations-transforming-transportation-consumer-behavior/; 12 December 2018.

[18] Greenblatt, J., and Saxena, S. "Autonomous taxis could greatly reduce greenhouse-gas emissions of US light-duty vehicles." *Nature*, 860–863; 2015.

[19] Corwin, S., Vitale, J., Kelly, E., and Cathles, E. *The future of mobility: How transportation technology and social trends are creating a new business ecosystem.* Retrieved from Deloitte Insights: https://www2.deloitte.com/insights/us/en/focus/future-of-mobility/transportation-technology.html; 24 September 2015.

[20] Chao, J. *Autonomous taxis would deliver significant environmental and economic benefits.* Retrieved from Berkeley Lab: https://newscenter.lbl.gov/2015/07/06/autonomous-taxis-would-deliver-significant-environmental-and-economic-benefits/; 06 July 2015.

[21] Cohen, B. *Mobility as a service (MaaS) and mobility on demand (MOD) via blockchain.* Retrieved from Medium: https://medium.com/iomob/mobility-as-a-service-maas-and-mobility-on-demand-mod-via-blockchain-64e36a2f6676; 27 April 2018.

[22] Sochor, J., Arby, H., Karlsson, M., and Sarasini, S. "A topological approach to Mobility as a Service: A proposed tool for understanding requirements and effects, and for aiding the integration of societal goals." *Research in Transportation Business and Management.* 2016; 3–14.

[23] Shaheen, S., and Cohen, A. *Prioritizing people, public transport, and pooling: Transitioning to shared automated vehicles.* Retrieved from Move Forward: https://www.move-forward.com/prioritizing-people-public-transport-and-pooling-transitioning-to-shared-automated-vehicles/; 22 May 2018.

[24] Matyas, M., and Kamargianni, M. "The potential of mobility as a service bundles as a mobility management tool." *Transportation.* 2018; 1–18.

[25] Durand, A., Harms, L., Hoogendoorn-Lanser, S., and Zijlstra, T. *Mobility-as-a-Service and changes in travel preferences and travel behaviour: A literature review.* KiM Netherlands Institute for Transport Policy Analysis; 2018.

[26] Hietanen, S. "Mobility as a Service: The new transport paradigm." *ITS & Transport Management Supplement. Eurotransport.* 2014; 2–4.

[27] Shaheen, S., and Cohen, A. "Is it time for a public transit renaissance?" *Journal of Public Transportation.* 2018; 67–81.

The views expressed are those of the authors.

Chapter 3

Policy and regulatory environment: shared automated vehicles

*Susan A. Shaheen[1], Adam P. Cohen[2], and
Adam D. Stocker[2]*

This chapter examines the policy implications of shared automated vehicles (SAVs). It discusses the current state of automated vehicle (AV) and SAV policy in the United States (U.S.) and assesses possible future technology deployment and adoption scenarios. The discussion explores a variety of topics related to SAVs and recommends policy approaches regarding these topics. The authors examine shared mobility policies and developments to compare and contrast possible SAV approaches. The policy topics covered in this chapter include: (1) passenger safety; (2) data sharing; (3) mitigating for externalities, such as congestion, vehicle miles/ kilometers traveled (VMT/VKT), and emissions; (4) labor implications; (5) SAV ownership and business model considerations; (6) public transit and SAVs; and (7) policy options for federal, state, and local governments. The chapter also includes a four-phase transition framework for policymakers to consider the shift to privately owned AVs and SAVs. It is important to note that shifting travel behavior toward pooling can be challenging due to a number of factors, such as personal safety, convenience, increased travel times, and personal preference. Further, there has been heightened concern about vehicle sharing due to the global pandemic and virus transmission. While the future of SAVs and policy approaches are uncertain, policies should be considered to effectively manage SAVs moving forward.

3.1 Introduction

The convergence of shared mobility, increasing levels of vehicle automation, and mobile technologies have the potential to revolutionize the way people travel. This convergence of trends and technologies could result in both privately owned automated vehicles (AVs) and shared automated vehicle (SAV) fleets (business to

[1]Department of Civil and Environmental Engineering, and Transportation Sustainability Research Center, University of California, Berkeley, CA, USA
[2]Transportation Sustainability Research Center, University of California, Berkeley, CA, USA

consumer, government to consumer, and peer-to-peer) that are accessed on an as-needed basis by travelers. Some analysts predict that the demand for SAV services could increase rapidly due to low expected per-mile/kilometer costs, increased comfort, and ease of travel due to the ability to multitask and divert attention away from driving tasks [1,2]. However, increased SAV demand could lead to unintended consequences including increases in traffic congestion, decreases in public transit use, and labor concerns. Thoughtful policies will be needed to mitigate the potential negative impacts of SAVs and to encourage positive societal outcomes. While some research has focused on AV policy recommendations more generally, not as much work has focused on SAV options, specifically.

This chapter considers possible societal, travel behavior, and environmental effects of SAVs, both positive and negative. It is organized into eight sections. Section 3.1 provides a summary of SAE (previously known as the Society of Automotive Engineers) International's levels of vehicle automation, a review of a four-phase framework in the transition to driverless vehicles, and an examination of U.S. AV policies. Section 3.2 explores passenger safety in an SAV context. In Section 3.3, we discuss the role of data and data sharing. Section 3.4 reviews the potential impacts of vehicle automation on congestion, VMT/VKT, emissions, and labor. Sections 3.5 and 3.6 explore potential SAV business models and opportunities for public–private partnerships in public transportation. In Section 3.7, we consider policy options for the public sector. The chapter concludes with final thoughts on how to support SAVs as they become more mainstream in cities and regions.

3.1.1   Levels of automation

Although AVs are often discussed in the popular media as monolithic, there are distinct levels of vehicle automation with varying functional capabilities that reflect operations in specific geographical areas or road conditions but possibly not others. SAE has defined five levels of automation from Level 0 to Level 5 [3]. They define the level of control needed from the human operator or provided by the vehicle. SAE's levels of automation are as follows:

- **Level 0:** Vehicles are not automated, and drivers perform all of the tasks.
- **Level 1:** Vehicles automate only one primary control function (e.g., self-parking or adaptive cruise control).
- **Level 2:** Vehicles have automated systems with full control of specific vehicle functions, such as accelerating, braking, and steering, but drivers must still monitor driving and be prepared to immediately resume control at any time.
- **Level 3:** Vehicles allow drivers to engage in nondriving tasks for a limited time. Vehicles will handle situations requiring an immediate response; however, drivers must still be prepared to intervene within a limited amount of time when prompted to do so.
- **Level 4:** A human operator does not need to control the vehicle as long as it is operating in the specific conditions in which it was intended to function.
- **Level 5:** Vehicles are capable of driving in all environments without human control.

At present, most SAV efforts are targeting Level 4 automation, where a human operator does not need to control the vehicle as long as it is operating within specified conditions (defined by SAE as the Operational Design Domain (ODD)). Throughout most of this chapter, we are envisioning SAVs that are capable of Level 4 or 5 automation. This is because Level 4 and 5 AVs can operate without a human driver present inside the vehicle, which is critical for the business viability of SAV services and introduces new policy implications that differ from those where a human operator is present in the vehicle.

3.1.2 Four-phase transition framework

Although analyst predictions vary, AVs and fleet-based SAVs will not become commonplace overnight. It is likely that there will be a lengthy transition period with increasing AV penetration levels on U.S. roadways. The different "phases" of this transition have important implications when developing public policy around AVs and SAVs. We present a four-phase transition framework to illustrate distinct possible stages of an AV transition (both privately owned and shared fleet SAVs). The four phases of increasing AV and SAV penetration based on the ODD are defined as follows: (1) Present day, (2) Specific-ODD automation, (3) Citywide-ODD Level 4 automation, and (4) Proliferation of Level 4+. The ODD represents the specific conditions an automated driving system is designed to function in, including limitations such as geography, traffic, speed, and roadway typology. Table 3.1 below describes the implications and possible services that may emerge during each of these four phases considering both privately owned AVs and shared SAV fleets. Note that while we refer mostly to passenger AVs in Table 3.1 and throughout this chapter, many experts predict that freight and goods movement (especially on highways) could be one of the first widespread AV applications [4]. We do not provide a year prediction for when each phase may occur due to the uncertain nature of AV technology development and deployment.

Phase one outlines present day vehicle automation technology and AV and SAV applications. At present, there are new vehicle models that have Level 2 automation (e.g., highway adaptive cruise control and lane keeping), and there are a growing number of SAV pilots using both low-speed AVs and conventional vehicle AVs in a testing phase or serving select groups of passengers. As of December 2019, there were 17 active SAV pilots in the U.S. [5]. Phase two describes possible conditions as AV technology slowly improves toward being able to operate across more environments and automated features become more common in new vehicle models. During this phase, there may be an increase in SAV pilots and use cases, with potentially more pilots emerging on select city streets and some beginning to remove physical operators for remote ones. For personally owned vehicles, Level 2 and 3 features may become more commonplace in new vehicle models, and these vehicles may become slightly more common on U.S. highways. Phase three represents a critical point at which AVs and SAVs could see rapid growth and adoption. When Level 4 automation reaches a point where in-vehicle operators are able to be safely removed from the vehicle itself (and replaced with remote

Table 3.1 Four-phase AV and SAV transition framework

Operational characteristics and safety requirements	Phase 1: Present day	Phase 2: Specific-ODD automation	Phase 3: Citywide-ODD Level 4	Phase 4: Proliferation of Level 4+
Operational design domain (ODD)	Highway and defined areas	Highway, defined areas, some city streets	Highway, major cities, and metro areas	Highway, many cities, and metro areas
In-vehicle supervision required	Yes	In most situations	No	No

Ownership model	Phase 1: Present day	Phase 2: Specific-ODD Automation	Phase 3: Citywide-ODD Level 4	Phase 4: Proliferation of Level 4+
Privately owned AVs	Small penetration of Level 2 automation on select new vehicle models (e.g., Tesla Autopilot, and Nissan Pro-PILOT)	Level 2 and 3 features continue to roll out slowly in new vehicle models. May see greater penetration of personally owned AVs used for highway driving	Some private ownership of Level 4 vehicles. AV penetration will depend on upfront and operating costs of automated technology. However, AV retrofit kits could increase private AV ownership rates	Decreasing cost of AV technology may make ownership possible for a greater portion of the population. Private AV ownership may be segmented by land-use context due to cost differences (e.g., AV ownership in suburban/rural areas, SAV services in dense cities)
Shared fleet SAVs	Small but growing number of low-speed SAV pilots and testing (e.g., EasyMile, May Mobility) and SAV pilots and testing with conventional vehicles (e.g., Waymo, GM/Cruise)	Additional SAV pilots emerge, most likely at low speeds and serving specific use cases and geographical areas (first- and last-mile to public transit services, office parks, downtown circulators, etc.)	Removal of human operator in-vehicle and introduction of SAV services in major cities and metro areas with high rates of travel demand. Potential for vehicle ownership reductions, modal shifts away from public transit, and personal vehicle driving	Shared fleet SAV services gain more ridership, expand to more cities, and into some suburban areas. Possibility for a greater reduction in privately owned vehicle rates but also competition with public transit ridership

operators in the case of an emergency), the economic competitiveness of SAV services could fuel an expansion in their adoption across most major U.S. cities and metropolitan areas. Passenger demand for these services could increase rapidly as well, similar to what has occurred with transportation network companies (TNCs, also known as ridehailing and ridesourcing) over the last decade [6]. Private ownership of Level 4 AVs may increase quickly as well, especially among frequent drivers, although the penetration rates will likely depend on the price premium. Phase four outlines the proliferation of Level 4 automation without in-vehicle supervision. The decreasing cost of AV technology as it matures may increase demand for private AV ownership, especially in more suburban and rural areas where shared fleets may not fare as well as they do in urban areas due to lower demand densities. However, SAV services could expand to more cities and suburban areas as the technology matures and could begin to become more competitive with personal vehicle ownership and public transit.

Phase three is one of the most critical for SAV/AV policies to address scaling effects and changes due to in-vehicle supervision. Safety regulations related to the removal of in-person operators will be needed, along with policies to mitigate the potential negative impacts of greater AV and SAV uptake (e.g., congestion, emissions, personal safety, etc.). As demonstrated through numerous attempts to institute road usage charging (RUC) in major U.S. cities, it has been politically challenging to charge the public for road use, if people are accustomed to using them at no cost [7]. In addition, there may be other issues with alternative usage-based taxation models, including public concern regarding location-based data collection. Since AVs could have a potentially disruptive impact on communities, it is important that policymakers carefully consider the role of policy in guiding sustainable and socially equitable outcomes, along with balancing public goals with commercial interests. We explore policy options in the context of the four transition phases later in this chapter. Next, we discuss the current state of AV policies in the U.S. at the federal, state, and local levels.

3.1.3 Current U.S. AV policies

At present, U.S. AV policies and legislation are at early stages, with most existing enacted policies at the state and local levels related to road safety, vehicle design requirements, liability and insurance, and operational areas. In this section, we examine current U.S. policy developments at the: (1) federal, (2) state, and (3) local levels.

3.2 Federal

While there are no binding federal AV-specific laws at present, there have been multiple guideline documents over the last few years that provide frameworks and suggestions for U.S. AV development, testing, and deployment. In September 2016, the National Highway Traffic Safety Administration (NHTSA) released the document titled the "Federal Automated Vehicles Policy" document, under the

Obama administration. The document establishes a 15-point framework for AV regulation in the U.S. and provides recommendations on safety, data sharing, privacy, cybersecurity, and ethical considerations, among others [8]. In September 2017, under the Trump administration, NHTSA released a second version of the document titled "Automated Driving Systems 2.0 (ADS 2.0)." This shortened version decreased the safety self-assessment from 15 to 12 key areas. ADS 2.0 clarifies that AV companies do not need to wait for Federal approval to test or deploy AVs. In October 2018, the U.S. Department of Transportation released a third version of the document titled "Automated Vehicles 3.0 (AV 3.0)," which builds upon but does not replace the ADS 2.0 guidelines. AV 3.0 expanded the scope to include all surface on-road transportation systems and incorporated feedback from a variety of stakeholders and public agencies [9]. However, all three of these guideline documents remain voluntary. While there are no enacted laws at the federal level, at present, many have been passed at the state level, which we describe below.

3.3   State

Wong and Shaheen [10] conducted a review of state-level AV stakeholder forums in the U.S. (e.g., committees, task forces) and strategic actions (e.g., programs, initiatives) initiated through legislation, executive orders, or state agencies (e.g., departments of transportation, departments of motor vehicles) between 2014 and 2019. Twelve states have developed stakeholder committees, nine have formed task forces, four have developed working groups, and nine have established a variety of initiatives, commissions, and councils. A handful of other states have commenced studies and implemented other initiatives, such as creating a connected and automated vehicle (or CAV) position or office, establishing AV vision plans, and engaging other stakeholders through technology teams and meetings. AV stakeholder forums and strategic actions address a diverse set of AV focus areas, but the most common concerns are safety and testing as well as implementation and regulatory issues, with less attention dedicated to AV implications on the environment, public health, social equity, land use, public transit, goods movement, and emergency response.

Between 2017 and June 2020, the National Conference of State Legislatures reported hundreds of legislative developments across the U.S. During this period, 45 state legislatures drafted 456 bills covering various aspects of AV safety, planning, and implementation. Ninety-four of these bills were enacted/adopted, 216 bills are pending, and 145 have failed to be ratified. One bill was vetoed and five received line-item vetoes, voiding specific provisions within the bill. Of the bills that were ratified:

- Twenty-two states have enacted 35 laws establishing terms and definitions related to AVs;
- Twenty-two states have enacted 30 laws related to commercial operations;
- Five states have established six laws pertaining to infrastructure (both digital and physical) to support vehicle automation;

- Eight states have established 12 laws addressing insurance and liability issues;
- Three states have passed five laws focused on licensing and registration;
- Fourteen states have adopted 20 laws related to AV operations on public roads; and
- Eleven states have enacted 24 laws addressing vehicle testing.

Additional legislation covers other topics, such as data privacy and demonstration programs. Some of the laws passed include multiple topics related to AVs [11]. According to the Governor Highway Safety Association (GHSA), 12 states are now allowing testing or deployment of an AV without a human operator in the vehicle [12].

State AV laws primarily regulate liability and insurance, licensing, registration, traffic rules, and infrastructure. States have undertaken a number of approaches to overseeing AV testing and deployment on public roads. California has implemented a permitting and registration process that requires companies to register with the state and track performance and AV disengagements [13]. Disengagements occur when an AV must switch from an automated to manual mode due to a system failure or safety emergency. Arizona's updated 2018 executive order requires operators to work with the state Department of Public Safety to submit a law enforcement interaction protocol. The order also requires additional criteria to test or operate AVs on public roads without a driver [14].

Only a few states have laws that focus on SAVs. For example, Nevada's Assembly Bill 69 authorizes an excise tax on SAV services at 3% of the total fare charged on each ride. Tennessee's Senate Bill 1561 establishes a one cent-per-mile tax on AV passenger vehicles and a 2.6 cent-per-mile tax for automated trucks. Tennessee's revenue is shared between the general fund, the transportation department, and local government entities according to a legislature-mandated formula [15]. However, it is not clear if either of these bills is being enforced, at present.

3.4   Local

Some local municipalities with AV testing or pilots in their jurisdictions have developed regulations around AV operations. One of the more comprehensive AV policy programs is administered by the City of Boston. In October 2016, Boston's mayor issued an executive order that established a multiphase AV testing program in the city. The city requires a memorandum of understanding (MOU) with any AV company wishing to test vehicles in Boston, which regulates the time, place, and manner of testing. AV companies nuTonomy and Optimus Ride both participate in the city's program. Initially, AV testing was restricted to the South Boston Waterfront, but it has since been expanded citywide for nuTonomy [16].

In the future, as AVs and SAVs develop, the public sector must focus regulation beyond liability and insurance, licensing, registration, traffic rules, infrastructure, and operating areas and consider laws governing safety, data sharing, congestion and emissions, roadway and curb usage, taxation, social equity outcomes, labor implications, and other topics of concern. Federal, state, and local regulators will have to work together to carefully define their respective purviews related to AVs and SAVs to ensure that these emerging technologies and services benefit the entire population and do not lead to inequitable outcomes.

In the sections that follow, we explore SAV policy options pertaining to a number of different topics including passenger safety, data sharing, mitigating for externalities, public transit and SAVs, and options at different levels of governance. Decision makers should consider these areas in crafting and enacting legislation pertaining to SAVs moving forward.

3.5   Passenger safety

While most AV safety policies focus on vehicle operations on public roads, far fewer have examined noncrash safety risks for SAV passengers. These risks could include victim crime (e.g., crime between passengers); property crime (e.g., defacing the vehicle); information sensitivity concerns (e.g., breach of privacy, identity theft, and location privacy); and others. Naturally, perceived safety risks could pose a barrier to SAV uptake, since passengers may opt for private (single-occupant) SAV rides or travel in a personally owned AV due to personal safety concerns. This could diminish some of the potential SAV positive impacts, including reduced congestion and emissions [17], and it is a topic that warrants further examination.

Since SAVs reflect a confined environment with strangers without a guardian (driver), some personal safety risks could arise. This section examines the possibility of SAV passenger crimes, how to reduce them, and how to assure passengers they will be safe onboard [18]. Examining potential SAV safety issues and policy strategies is critical to ensuring SAV growth and use in the future.

Since TNCs have a similar service model to future SAVs, we discuss the safety aspects of TNCs. There have been numerous cases of crimes between TNC passengers and/or drivers [19,20]. Routine activity theory suggests that certain environmental factors can lead to offenders targeting victims. These factors include the level of exposure, the presence of a capable guardian, the proximity to potential offenders, the attractiveness of potential targets, and the definitional properties of specific crimes [21–23].

Because shared rides typically represent an enclosed environment with little exposure, routine activity theory suggests there is some risk of offense [23–25]. With current TNC services, drivers provide oversight and act as a deterrent to crime that may otherwise occur between passengers pooling inside the same vehicle [23]. The platform's data collection also can be seen as a deterrent to internal crime since it obtains information for all parties. Since SAVs will not have a driver to act as informal guardians, there may be increased crime risks between passengers. In contrast, there have been cases of TNC drivers committing offenses, so SAVs would eliminate the risk of driver-on-passenger crime risks [26]. Data reporting and screening are critical to ensuring safety in shared ride services. Uber was banned in London from 2017 to 2018 in part for their lack of driver screening and reporting serious crimes;

however, this ban was lifted if the company now reported crimes directly to the police [27–29]. Misreporting or incomplete information on crimes can degrade the ability to deter offenders since there could be a low likelihood of repercussions. Safety data reporting, while protecting passenger privacy, will be critical to SAV success.

There are a number of steps that regulators could take to increase SAV safety. These include: (1) mimicking the presence of a guardian in the vehicle, (2) defining personal space, and (3) creating a precedent of prosecuting SAV crimes. One deterrent to crime could be increasing SAV visibility through vehicle design (e.g., large transparent windows, cabin lighting, etc.) and internal monitoring [17,30]. Setting "territorial props" in the form of arm rests or physical dividers could distinguish a passenger's personal space, which is important for passengers to feel physically and emotionally safe [17,31]. Setting SAV-specific design requirements could help to ensure perceptions of SAV safety among the traveling public.

A video monitoring system could act, or at least be perceived, as an informal guardian and alert authorities if an emergency occurs [17,30,31]. Video footage, combined with passenger data, could serve as primary evidence of an SAV crime— making it challenging for offenders to evade punishment. A button on an app or physically in a SAV could alert emergency authorities and the platform's monitoring staff; this could help to deter crime between passengers [17,30]. Some TNCs are already implementing rudimentary forms of this type of monitoring system. For example, Uber recently added an in-app emergency button that calls 911 responders [32,33]. Perceived control over environmental conditions can mitigate the stress of objectionable stimuli [17,31].

The global pandemic is also a reminder that sanitation of SAVs will be critical to ensuring public safety. While specific hygiene measures are beyond the scope of this chapter, methods that reduce contact and airborne spread of infectious diseases will be key. In July 2020, the World Health Organization (WHO) issued a statement relating to the spread of COVID-19 that "Short-range aerosol transmission, particularly in specific indoor locations, such as crowded and inadequately ventilated spaces over a prolonged period of time with infected persons cannot be ruled out." While numerous strategies exist to reduce contact transmission (i.e., cleaning vehicle surfaces), less emphasis has been placed on reducing aerosol transmission in shared vehicles. As SAVs are developed and deployed, consideration about vehicle sanitation and ventilation of enclosed spaces will be essential to mitigate the spread of infectious disease.

Ultimately, requiring some form of safety, security, and sanitation systems in all SAVs could enhance personal safety and security. Deterrents may vary among operators and/or vehicles. Finally, reporting incidents in a quick and efficient manner and sharing safety data with public sector regulators will be important for creating trust for pooled SAV rides.

3.6 Data sharing

It is important that travel data from SAVs are shared with the governmental entities and potentially with the public (e.g., academics and third parties). Without key information on where, when, and who SAVs are serving, transportation planners will be unable to develop evidence-based policy to effectively manage demand and systemwide impacts resulting from this emerging mode.

At present, many U.S. cities do not have access to TNC trip data, which limits their understanding of TNC travel impacts and ability to craft regulations that address the social and environmental impacts of these services [34]. Private-sector TNCs view these data as valuable assets and are hesitant to provide data to outside parties due to competitive and privacy concerns. While privacy concerns around personally identifiable information (PII) and proprietary interests are legitimate, travel data from TNCs and SAVs should strike a balance between protecting user privacy and providing useful information to transportation planners and regulators. In this section, we review some present-day examples of shared mobility data sharing, the advantages and disadvantages of these approaches, and options for SAVs moving forward.

There are a handful of examples of data sharing between TNCs and public agencies in a few major U.S. cities. The New York City Taxi and Limousine Commission (TLC) receives and discloses origins and destinations of TNC trips to the granularity of a "TLC taxi zone" [35]. In 2019, the city of Chicago received access to release TNC vehicle, driver, and trip data on a quarterly basis. Trip origins and destinations are aggregated by census tract; start and end times are rounded to the nearest 15 min [36]. In May 2018, the Los Angeles Department of Transportation released a standard data format focused on collecting and sharing trip data, fleet status, and parking verification in real time between cities and shared mobility operators, called the mobility data specification (MDS) [37]. Although the specification is still evolving, real-time data could potentially help cities manage curb and lane restrictions dynamically and by applying preset rules. As of July 2020, the American Civil Liberties Union (ACLU) and Uber were in litigation with the City of Los Angeles arguing that MDS violates California's Electronic Communications Privacy Act by making location data sharing a requirement for an operating permit [38]. The ACLU argues that MDS violates constitutional civil liberties because the data could be used by law enforcement to identify user behavior without a warrant.

Real-time data sharing has the potential to help cities manage and regulate SAVs. A standardized data format could be helpful for communities and reduce the burden on operators if data are shared in the same format across multiple jurisdictions. However, strategies are needed to protect user privacy. For example, real-time public application programming interfaces (APIs) for developers could simply show the location of available (with no passengers) SAVs whereas trip data shared with cities for transportation planning purposes could be de-identified and aggregated to an appropriate geographic scale (census block, etc.).

In response, the public sector is slowly adopting privacy legislation. In California, the state's Consumer Privacy Act (CCPA) took effect on January 01, 2020. Under CCPA, consumers can request that a business does not sell their data, request the deletion of their data, and sue companies for violations. CCPA protects a variety of sensitive data, including but not limited to location data, demographic information, veteran status, biometric information, and browsing history [39].

Regularly occurring and appropriately defined data sharing practices should be adopted if the full benefits of SAVs are to be realized. Finally, there may be opportunities for certain forms of data sharing between SAV operators in order to produce societally beneficial outcomes. One example of this could be to require SAV operators to share information on the pool of passengers requesting shared rides in real time. Matching passengers who request shared rides becomes more challenging if there are a greater number of operators in a given market. In this sense, high levels of competition between many SAV operators could lead to lower matching (and therefore pooling) rates in the absence of a policy intervention. Therefore, policies that encourage or require SAV operators to share passenger shared-ride requests in real time could allow for better matching at higher pooling rates at the regional level. If such a sharing mechanism were implemented, it could greatly improve the environmental and economic performance of SAVs and could lead to more societally beneficial outcomes.

3.7   Mitigating externalities

In crafting public policies around SAV services, it is important to consider how to mitigate the potential negative impacts and encourage societally beneficial outcomes. In this section, we discuss possible approaches for mitigating traffic congestion, VMT/VKT, and GHG emissions due to SAV services. We also examine labor and social equity implications and provide related policy options.

3.7.1   Mitigating for congestion, VMT/VKT, and emissions

The potential for induced demand (i.e., increased demand due to reduced travel time, costs, or both) and modal shift effects due to the uptake of SAVs could lead to increases in congestion, VMT/VKT, and emissions. There is some evidence that these impacts may already be occurring with present-day nonautomated TNC services. For example, a study of Uber and Lyft's impacts on traffic congestion in San Francisco showed that weekday vehicle hours of delay increased by 62% compared to 22% in a counterfactual scenario without TNCs between 2010 and 2016 [40].

A number of studies that assess the impact of TNC services on modal shift have found that passengers are either: (1) substituting a trip they formerly made with another transportation mode (public transit, driving, walking, biking, etc.) or (2) making a new trip they otherwise would not have made without the availability of TNC services (induced demand). There are conflicting conclusions regarding the extent to which TNCs compete with public transit. While some studies conclude that TNCs are largely not substituting public transit trips [41–43], several others suggest that a significant portion of travelers substitute TNCs for public transit, biking, and walking [44–48]. Past surveys show that the degree to which TNCs substitute for other travel modes varies by city and the built environment. Denser cities like New York City, Boston, and San Francisco exhibited some of the highest proportions of passengers who would have used public transit for their last TNC trip had TNCs been unavailable.

Study findings on TNC impacts on public transportation ridership vary, possibly due to local differences in public transit service, urban density, and the built environment. Hall *et al.* [49] examined the impact of Uber's entry on public transit ridership between 2014 and 2015 across the 196 U.S. Metropolitan Statistical Areas (or MSAs) where Uber was operating. This study found that Uber is a complement for the average public transit agency, increasing ridership by 5% after two years. In contrast, Graehler *et al.* [50] found that the entry and presence of TNCs cumulatively decreased heavy rail ridership by 1.29% per year and bus ridership by 1.70% per year in a study examining data from 2002 to 2018. Gerhke *et al.* [46] found that passengers with lower incomes and those who possess a weekly or monthly public transit pass were more likely to have substituted TNC services for public transit. In addition, relatively low TNC service costs, low TNC trip times, poor weather, and unavailability of public transit were also predictive of public transit substitution.

A few studies have assessed TNC impacts on VMT and tripmaking decisions. The most comprehensive studies have employed trip-level TNC activity data in San Francisco [51,52] and New York City [53,54] to analyze mileage, trip metrics, and impacts. Schaller [53] conducted an analysis with publicly available taxi and for-hire vehicle trip and mileage data in New York City. This study found, after accounting for mileage declines in yellow cabs and personal vehicles, TNCs and other on-demand ride services (including Uber, Lyft, Via, Gett, and Juno) contributed 600 million additional miles (or 969 million kilometers) of vehicle travel to the city's roads between 2013 and 2016. These additional miles equated to an estimated 3.5% increase in citywide VMT and a 7% increase in VMT in Manhattan, western Queens, and western Brooklyn in 2016. Another study conducted by Schaller [54] found that usage rates among taxis and TNC vehicles declined in New York City between 2013 and 2017, while the number of unoccupied taxi and TNC vehicles increased by 81% over this time period. Schaller [54] also found that total taxi and TNC weekday mileage in the central business district (or CBD) increased by 36% from 2013 to 2017. The San Francisco County Transportation (SFCTA) has conducted two studies of TNC impacts on the City of San Francisco. SFCTA collected TNC trip data from one month in late-2016 and found that TNC trips made up 15% of average weekday vehicle trips within San Francisco and 9% of average weekday person trips within the city. In terms of mileage, this study found that TNCs represented 20% of the average weekday intra-San Francisco VMT (trips that originate and end within city limits only) and 6.5% of total VMT (including regional trips starting or ending within city limits) on an average weekday. SFCTA also found that around 20% of all TNC miles were deadheading (or zero-occupancy travel) miles. However, neither of these three studies assess the impact of pooled TNC services.

AVs and SAVs could lead to increased use or induced demand due to a variety of factors. Harper *et al.* [55] found that the introduction of AVs would increase the total annual light-duty VMT by 14%. Sixty-five percent of this VMT increase would be from current adult nondrivers, while 16% is from older drivers without a medical condition, and 19% is from adult drivers with travel-restrictive medical conditions. Harb *et al.* [56] found an 83% overall VMT increase (in part due to older adults and children being able to use a vehicle without driving) and a 21% increase in VMT associated with deadheading.

These effects may intensify with automation due to decreased costs and increased travel ease with AVs or SAVs, potentially leading to more trips taken and miles/kilometers driven, as AVs/SAVs could continue to drive throughout a city or back home rather than parking. As such, policy strategies aimed at curbing congestion, VMT/VKT, and emissions are needed prior to the rollout of SAVs. For example, the California Clean Miles Standard and Incentive Program (Senate Bill 1014) is a GHG reduction policy targeted at TNCs, including SAVs. The bill directs the California Air Resources Board to establish regulations including establishing a 2018 GHG emission baseline on a per-passenger-mile basis by January 01, 2020 and setting annual GHG reduction targets for TNCs by January 01, 2021. The bill also directs the California Public Utilities Commission to implement the TNC regulations and track compliance. The bill also requires TNCs to develop their own GHG emission reduction plans by January 01, 2022 and every two years thereafter to meet the GHG reduction requirements with implementation beginning in 2023.

Some U.S. cities are already exploring usage-based pricing mechanisms that are expected to mitigate traffic congestion/VMT. For example, New York City recently added a $2.75 per-trip fee for TNC trips that begin, end, or pass through most of Manhattan [57], and the state legislature approved a plan that will charge all drivers entering downtown Manhattan, although the exact details are yet to be determined [58]. The TNC fee includes a discounted per-passenger fee of $0.75 for pooled rides, providing incentives for TNC passengers to elect shared ride options.

While these mitigation measures and those that are being considered by other U.S. cities are positive steps forward, other strategies may need to be considered to ensure beneficial outcomes when SAVs become more commonplace. A number of different pricing strategies should be considered to mitigate increases in congestion and VMT/VKT including:

- **Trip-based pricing:** This entails fees that are applied to each trip taken using a particular transportation service, which could be applied to SAV services (especially single-occupant trips) to encourage the use of more sustainable modes like active transportation or public transit.
- **Mileage-based pricing:** This includes policies where fees are paid on a per-distance basis with the typical aim of reducing congestion and VMT/VKT and to provide additional infrastructure funding. Since transportation services, including SAVs, comprise some percentage of deadheading miles, mileage-based strategies could incentivize operators to reduce these empty miles. Reducing deadheading miles is critical to mitigating the potential negative impacts of AVs and SAVs on congestion and VMT/VKT, as one study demonstrated that AVs could more than double vehicle travel to, from, and within dense, urban cores since cruising could become less costly than parking [59].
- **Occupancy-based pricing:** Fees could be charged based on the number of passengers in a vehicle, which could in turn encourage more pooling and

increase vehicle occupancies, possibly reducing congestion and VMT/VKT. Such pricing strategies and user incentives (e.g., discounted rides) could help to shift behavior toward pooling [60]. The discounted per-passenger fee for pooled TNC rides in New York City is one example of this. Similar policies could be enacted for future SAV services.

- **Spatiotemporal pricing:** This entails pricing charged when entering or circulating in specified areas or during certain times of the day. Areas can include CBDs, particular lanes, or designated curb spaces. These strategies could keep AVs from entering certain busy downtown or other areas during peak times and could help mitigate congestion. However, these pricing strategies could have a variety of unintended social equity implications, such as impacting workers with particular schedules and displaced workers who are forced to travel across pricing zones due to a lack of affordable housing in close proximity to one's workplace [61].

In addition to strategies that attempt to reduce congestion and VMT/VKT, approaches that encourage alternative fuel technologies are needed to reduce potential SAV emission increases. While present-day electric vehicle (EV) TNC rates are low, largely due to the use of a driver's personal vehicle and low EV ownership rates nationally, this could change as SAV services come online. If SAVs are operated by a service provider as opposed to individual contractors with personally owned vehicles, there may be an added incentive to use more fuel-efficient vehicles. For example, one study found that a fleet of SAVs driven between 40,000 and 70,000 miles (or 64,374 and 112,654 kilometers) per year (typical for U.S. taxis) would be more cost effective with a fuel cell or battery electric vehicles compared to gasoline-powered vehicles due to much lower maintenance costs primarily [1]. This study also found that a fleet of shared automated electric vehicles with right-sized vehicles by trip, in combination with a 2030 low-carbon electricity grid, could reduce per-mile GHG emissions between 63% and 82% compared to a privately owned hybrid vehicle fleet [1].

These considerations and others should be examined by stakeholders and policymakers when assessing strategies for mitigating congestion, VMT/VKT, and emissions due to future SAV services.

3.7.2   Social equity

Social equity can be difficult to analyze because several types of challenges could impede a user's access to AVs/SAVs [62]. For example, shared mobility typically requires a smartphone, mobile Internet access, and/or a credit or debit card. As a result, shared mobility and SAV services in the future could exclude access to digitally impoverished, low-income, and unbanked and underbanked users. Additionally, service availability may be limited or unavailable for low-density and rural areas, older adults, and people with disabilities. However, AVs and SAVs could create opportunities to enhance access and social equity by providing increased mobility options (e.g., lower fares and a greater variety of routes); increased travel speed and reliability; first- and last-mile connections; and expanded coverage to historically underserved users or communities. Legislation and regulation could play an important role in transportation equity by preventing and mitigating technological and access barriers. Thoughtful policymaking is needed to help ensure that SAVs serve the public good, while also supporting private industry.

Shaheen *et al.* [63] developed the Spatial, Temporal, Economic, Physiological, and Social (STEPS) equity framework to better understand the potential barriers that travelers using shared mobility may face. This framework can also be applied to understand potential social equity issues associated with AVs and SAVs. Table 3.2 defines each part of the STEPS framework, as well as opportunities and challenges a SAV network may face when addressing social equity concerns.

For more on social equity, refer to Chapter 6, *Shared mobility services: prioritizing social good.*

3.7.3   Labor implications

As of 2018, approximately four million Americans were employed as professional drivers, including delivery and heavy tractor-trailer drivers, bus drivers, taxi drivers, chauffeurs, and others [64,65]. In addition, as of 2019, there were an estimated 900,000 U.S. Uber drivers, although many likely work on a part-time basis [66]. Moreover, about 11.7 million American workers use vehicles as part of their jobs, including first responders and those involved in construction trades, repair and installation, and personal home care aides. These "on-the-job" drivers may benefit from greater productivity and better working conditions provided by AVs, whereas professional drivers (e.g., taxis and TNCs) may be more negatively impacted [67].

AVs and SAV services will likely lead to the removal of certain professional driving jobs. One study simulated various adoption and technology diffusion scenarios and found that AVs could directly eliminate 1.3–2.3 million jobs in the U.S. over the next 30 years [68]. The authors predict that employment disruptions will not become substantive until after 2030 and will be gradual, with about 100,000 U.S. jobs disrupted per year during the peak impact years. While some workers may be negatively affected by AVs, there could be new jobs prompted by AVs due to overall growth in transportation use, new labor inputs to support the AV sector, and increased purchases of other goods and services by consumers who spend less on transportation [68].

A variety of policy options and efforts should be considered to mitigate disruptive job losses as a result of AV and SAV adoption [62]. These include efforts to design new jobs that take advantage of skills that workers in occupations disrupted by AVs already encompass. For example, former professional drivers might have skills related to traffic monitoring, truck maintenance, or remote operations. Some companies, like Phantom Auto, are developing teleoperation technologies so that remote drivers can take control of AVs, as appropriate [69]. Drivers may also be needed to assist with door-to-door services for individuals needing extra assistance, such as older adults and passengers with disabilities. Remote monitoring and operations may play an important role for SAV services and could be a source of

Table 3.2   SAVs and the STEPS framework. Adapted from [5]

	Definition	SAV opportunities	SAV challenges
Spatial	Spatial factors that compromise daily travel needs (e.g., excessively long distances between destinations, lack of public transit within walking distance)	• Public transit agency and first- and last-mile partnerships • Cost-effective SAV service for low-density areas • Land use changes	• Higher operating costs in lower-density exurban and rural settings • Limited curb space for increasing mobility on demand (MOD) and SAV services
Temporal	Travel time barriers that inhibit a user from completing time-sensitive trips, such as arriving at work (e.g., public transit reliability issues, limited operating hours, traffic congestion)	• Dynamic, on-demand transportation service • SAV availability in late night and early morning	• Wait time and travel time volatility on congested roadways • Unpredictable wait times due to supply fluctuations
Economic	Direct costs (e.g., fares, tolls, and vehicle ownership costs) and indirect costs (e.g., smartphone, Internet access, and credit card access) that create an economic hardship or preclude users from completing basic travel	• SAV subsidies for low-income users • Multiple payment options (e.g., cash, card, and mobile) • Mobility hubs or SAV vehicles with public Wi-Fi access	• Disruption to existing revenue streams, for example, parking, traffic violations • High cost for long-distance/peak-demand trips • Maintaining affordability while providing livable wages
Physiological	Physical and cognitive limitations that make using standard transportation modes difficult or impossible (e.g., older adults and people with disabilities)	• Older adult or child-focused, Americans with Disabilities Act (or ADA) compliant, and SAV services • Voice-activated app features	• Maintaining legacy technology access • Ensuring adequate training, if chaperones or chauffeurs are needed for the SAV ride
Social	Social, cultural, safety, and language barriers that inhibit a user's comfort with using transportation (e.g., neighborhood crime, poorly targeted marketing, lack of multi-language information)	• App interfaces that minimize or eliminate socio-demographic profiling • Targeted outreach to low-income and underserved individuals • Network information in the user's native language	• Attracting underserved groups to SAV services • Providing security at unmanned SAV stations, hubs, or pickup locations

new employment in the AV industry. Given increasing levels of automation across a range of industries, broader sector policies such as wage insurance (providing compensation; if wages are lowered due to AVs); universal basic income (recurring pay to all members of society to meet basic needs); portable benefits (i.e., benefits connected to individuals vs. employers); or others may be needed moving forward. Ultimately, thoughtful mitigation strategies must be developed to ensure more equitable impacts for those who will be most negatively affected by the proliferation of AVs and SAV services.

3.8   SAV ownership and business model considerations

While many experts believe that SAVs could displace personal vehicle ownership among residents in denser urban areas due to lower per-mile/kilometer travel costs [70], there is less certainty about the ownership and business models that might be employed in future SAV operations. For example, a SAV fleet could be owned by: (1) a private business (business to consumer or B2C) or public-sector entity, (2) individuals (a peer-to-peer or P2P network), or (3) a hybrid model that incorporates a third-party operator and private individuals [71]. While there are similar regulatory concerns in SAV operations (i.e., safety, data sharing, and negative externalities), each SAV model presents unique policy challenges.

If SAVs are owned by a business (i.e., B2C), specific regulatory policies may be needed to prevent aggressive competition between businesses that could lead to negative externalities, such as increased congestion or emissions. Alternatively, if SAVs are primarily owned by one business or a single public sector entity (or monopoly), additional policies may be needed to ensure quality standards. A current example of a TNC monopoly is the Chinese company Didi. The service has been faulted with illegal operations, public safety concerns, weak emergency management procedures, and insufficient protection of PII [72]. If SAVs are owned by individuals and accessed in a P2P network, insurance and liability laws are needed to protect AV owners from damages occurring during commercial use. A related example is California Assembly Bill 1871, which allows hosts to share private vehicles through a P2P carsharing platform (e.g., Getaround, Turo) without invalidating an owner's auto insurance policy [73]. Other states, such as Oregon and Washington, have adopted similar legislation. Generally, this legislation prohibits insurance carriers from penalizing individuals who share/rent their personal vehicles by dropping or spiking their insurance premiums [74]. In addition, legislation will be needed to assess SAV liability in the case of an AV crash, including the vehicle owner, manufacturer, or software developer [75]. Liability might depend on the type of P2P business model, such as a third-party operator that provides access to individually owned AVs for short-term use on-demand via an app or a decentralized SAV model where AV hosts and guests arrange and facilitate payment through a public open-source ledger, such as those that employ blockchain technology [74]. Finally, if SAVs operating on the same network are jointly owned by a third-party entity and individuals (a hybrid model), all of the concerns mentioned previously will need to be addressed, including ensuring a quality standard across the fleet.

Existing policies that govern shared mobility business models (e.g., B2C and P2P) could be similar in SAV operations in the future. In addition, increasing levels of automation in shared vehicles could further blur the lines between public transit and private mobility services. In Section 3.9, we explore the potential benefits and challenges of public and private SAV operations and related policy considerations.

3.9   Public transit and SAVs

There is growing concern among transportation planners and decision makers that AV and SAV uptake could displace and pull ridership from traditional public transit systems, like buses and rail. The lower potential labor costs and more easily adjustable supply of SAVs compared to present-day public transit could compete with short-distance public bus and rail trips. In contrast, SAVs could also provide first- and last-mile connections to supplement public transit, if funding flexibility and appropriate incentives are established, especially for longer-distance trips. Public transit agencies could partner with SAV operators to develop a complementary framework for a seamless transportation system. A complementary public-private partnership (P3) among public transit agencies, local governments, and private SAV providers could take on many forms depending on the: (1) desired structure of public control/involvement, (2) services provided by private SAVs, (3) land-use context, and (4) level of SAV industry competition. A P3 could provide the public with consistent, high-quality, and affordable mobility if coordination between public and private sector entities is achieved.

There are tradeoffs between private- versus public-operated SAV services. A privately operated SAV could provide demand-based, flexible transportation since competing private operators would be incentivized to innovate their operations to cut costs or raise the service quality. A publicly operated SAV may have higher staffing costs but might not have the incentive to make operations as efficient. Brown [76] provided insights on Lyft's ability to provide mobility across every area of Los Angeles regardless of income and racial differences. Compared to the inflexible, limited services of buses and sometimes discriminatory nature of taxis and TNCs, SAVs may provide an opportunity for higher quality transportation services in a greater range of geographic areas, regardless of neighborhood demographic composition.

In the future, SAVs could be employed as an arm of public mobility, enabling public transit agencies to provide service in different land-use contexts and at different times of the day. This flexibility would offer a variety of possibilities to complement or substitute existing public transit services. For instance, SAVs could provide first- and last-mile linkages, fill in service gaps, replace underused short-distance routes, and guarantee services to persons with disabilities through supportive public policies. Long-distance, fixed-route public transit services with dedicated rights-of-way would still provide travel time and cost savings to travelers. In cases where private SAV operators would not choose to operate, some form of P3 or public subsidy may be needed.

Several possible types of partnerships between public entities and private SAV providers could be offered, such as:

- **First- and last-mile connections to public transportation:** The public partner subsidizes SAV trips to or from public transit stops or stations. The SAV service provider geofences the designated transit stops or stations to ensure the discount only applies to eligible trips.
- **Low-density service:** The public partner subsidizes SAV trips anywhere within a designated zone that targets low-density areas that cannot efficiently support fixed-route bus service.
- **Off-peak service:** The public partner subsidizes SAV trips during off-peak times that cannot efficiently support fixed-route transit service.
- **Paratransit:** A public agency leverages SAVs (or the underlying technology, such as ride dispatch algorithms) to improve, supplement, or replace the public transportation agency's existing paratransit service.

For example, GoDublin, an on-demand ride service was deployed in Dublin, California in January 2017. This program provided discounted pooled rides (e.g., UberPool and Lyft Shared rides) to replace low-performing bus lines, provide first- and last-mile connections to rail commuters, and serve persons with disabilities. As part of the pilot program, the Livermore Amador Valley Transit Authority (or LAVTA) provided half of the fares (up to $5 US) for Uber, Lyft, or local taxi rides as long as they are within city boundaries [77]. To enable this pilot, a comprehensive review of the bus lines was conducted. In May 2020, the GoDublin program was expanded to include the nearby communities of Pleasanton and Livermore and rebranded as Go Tri-Valley, with the same subsidy.

This is one example of how a P3 can help to fill gaps in suburban and smaller urban environments. Other land-use contexts (urban and rural) may need different approaches. For example, a dense urban environment may not require a P3, if the public transit agency can provide cost-effective services to residents living in higher-density areas. In more suburban and rural environments, it could be challenging to provide quality public transit or SAV services to the entire area. Since fixed-route public transit services can provide efficient scale between common destinations, SAVs could provide first- and last-mile services to fixed-route stations and fill gaps between less frequent origins and destinations. In more rural areas, SAVs have the potential to provide higher quality mobility services given reduce labor costs and the ability to more flexibly navigate sparse areas [78]. In a study of the rural–urban divide of TNC use in Pennsylvania, researchers estimated that limited TNC use in rural areas resulted from the inconsistent incomes of TNC drivers who mostly worked part-time hours [79]. Since SAVs could greatly reduce labor costs and address inconsistent hours of operation, their implementation may be based more on hardware availability in rural areas and public subsidies. In the future, it is difficult to say what type of industry model might emerge for SAV services in a diverse range of land-use contexts—perfect competition, a cartel, or a monopoly, etc.—but public policy will be needed to safeguard the public good.

3.10   Policy options for federal, state, and local governments

While AV regulations in the U.S. currently exist at the state level primarily, additional policies at all governance levels (federal, state, and local) are needed to ensure AVs and SAVs lead to socially and environmentally beneficial outcomes. Since no binding policies or laws have yet been enacted at that federal level directly related to AVs, more clarity is needed on vehicle design certification and safety. At the state level, policies should go beyond defining operational characteristics and pilot project approval to include licensing, alternative funding mechanisms, and safety reporting. Furthermore, local governments will have to become more intimately involved in partnerships, data sharing, and the mitigation of negative externalities. While we explore a range of policy strategies for SAVs at the federal, state, and local levels in this section of the chapter, there are tradeoffs among these approaches that each governing body should carefully consider. In addition, many actions will require close coordination among different levels of government for desired outcomes to be achieved. We explore several unique SAV policy considerations at different governance levels in Sections 3.10.1–3.10.3.

3.10.1   Federal

To date, NHTSA has released three versions of their AV policy guidelines [9]. While these documents are not binding, they establish voluntary guidance that has been helpful in creating understanding between industry and the federal government. Moving forward, the federal government should consider providing policies beyond voluntary guidance in the following areas:

- Safety and auditing,
- Vehicle design and technology certification,
- Liability clarity,
- Privacy and cybersecurity,
- Interoperability standards, and
- Career development programs.

Ensuring AV safety is a principal area of federal AV policy. NHTSA should consider developing a performance-based safety certification process, which is not overly prescriptive of the technological approach yet still upholds appropriate safety standards. In cases where safety-related auditing is required, the National Transportation Safety Board (NTSB) or another federal entity could provide expert investigations. The federal government should also consider supplying and upholding vehicle design and technology standards to ensure an acceptable quality standard among AVs on U.S. roadways. This could require modifications to the Federal Motor Vehicle Safety Standards (FMVSS) or new regulations altogether. Clarity will be needed to identify who is liable in the case of a crash, as well as passenger privacy and cybersecurity breach guidance. Passenger privacy is especially important in the context of SAVs, where a third-party company or individuals may have access to precise location data about passengers.

Data on AV pilots and testing will likely play a key role in increasing consumer acceptance. In June 2020, NHTSA unveiled the Automated Vehicle Transparency and Engagement for Safe Testing (AV TEST) initiative. The program provides a voluntary online public platform for automakers and AV operators to share automated driving system on-road testing activities in an effort to raise public awareness and community acceptance of vehicle automation [80]. Additionally, interoperability standards should be considered by federal agencies in the context of vehicle to vehicle (V2V) connectivity and roadside infrastructure or vehicle to everything (V2X), where appropriate, to achieve better traffic management, safety, and environmental outcomes than would be possible without this level of connectivity. Finally, public sector entities at the federal and other governance levels should consider developing and implementing career development programs that retrain displaced workers due to vehicle automation. While most federal options relate to AVs and SAVs, some topics for state and local governments relate more directly to SAVs, which we discuss in greater detail below.

3.10.2   State

While a number of states have passed AV legislation related to liability and insurance, licensing, registration, traffic rules, and infrastructure, additional policies are needed as the industry matures and AVs become more prevalent on state roadways. States should consider providing further clarification and regulations in a number of areas including:

* Roadway design and infrastructure,
* Liability and insurance,
* Licensing,
* Reporting,
* Traffic rules,
* Funding mechanisms, and
* Partnerships.

For instance, states could provide design standards related to AVs, including connected vehicle technology standards and curbside drop-off zone design. States should work closely with appropriate federal regulatory bodies to provide clarity around whether AV passengers, operators, manufacturers, or technology developers are liable in the case of a crash. Requirements for insurance providers also should be established by state regulatory entities. Licensing should continue to be handled by state authorities since companies or individuals holding AV licenses would be responsible for actions on state roadways. Reporting and permitting requirements for AV testing and deployment should be developed to ensure public transparency around AV and SAV operations. California, for example, has developed reporting requirements for companies operating on state roads. California requires companies to disclose vehicle collisions within ten business days of an incident and an annual report summarizing AV disengagements that require human intervention during testing [13]. State governments should also ensure that traffic rules are followed and coordinated with federal regulations, as appropriate. Several states are considering the introduction of AV and SAV pricing policies, such as road charge alternatives. Mechanisms, such as VMT/VKT-, congestion-, or occupancy-based pricing, could become more feasible if a growing portion of travel occurs using SAV services in contrast to personally owned vehicles. It is possible that usage-based fees could become more politically feasible if AVs and SAV services spur a public conversation regarding alternative methods of payment for public roadways. Finally, states should consider partnerships with other public entities (e.g., other states, cities, metropolitan planning organizations [MPOs], etc.) and the private sector to ensure that AV technology continues to develop safely and is implemented to achieve positive societal and environmental outcomes.

3.10.3   Local

Cities and other local agencies will need to ensure that AVs and SAV services meet local goals and integrate smoothly into the transportation landscape. While some AV testing and SAV pilot projects operate under specific local policies, at present, local governments will need to develop additional policies in the following areas:

- Curb space and other rights-of-way allocation,
- Pilot programs,
- Data sharing, and
- Mitigating for negative externalities.

While AV roadway design standards should be coordinated at the state level to ensure uniformity, the use of curb space and other rights-of-way should be determined by local governments to ensure these policies reflect local priorities. At present, a variety of AV and SAV pilot programs and testing are implemented in coordination with local governments, and this should continue in the future.

Data sharing regulations should occur at the regional level, which could be coordinated and collected by a MPO to ensure that transportation planners can understand and plan for AVs and SAV services. Robust data-sharing agreements that protect individual passenger privacy yet allow planners and other city officials to understand travel patterns would be helpful in ensuring AV/SAV benefits can be realized by the private sector and public agencies. Finally, cities and regional agencies should safeguard that AVs and SAVs do not contribute to unwanted negative externalities (e.g., increasing congestion and emissions). Data sharing will be critical for effectively mitigating negative externalities since it is challenging to develop informed policies without understanding where and when vehicles and services are operating. In addition, careful consideration should be given to what governance level is responsible for crafting policies to address the potential negative externalities of AVs and SAVs. For instance, most TNC legislation is set at the state level, which pre-empts many local municipalities from creating policies for their particular region and managing negative TNC externalities [81]. Consequently, state and local regulators will need to coordinate closely in delegating externality-focused policy responsibilities to ensure that the correct balance is reached between statewide coordination and local considerations.

3.11 Conclusion

The future of ground transportation is facing a potential seismic shift with SAV services. While AVs and SAVs will not become commonplace overnight, there will be key phases of technological development as AVs evolve to handle more challenging road types and conditions. While the exact timelines are highly uncertain, AVs and SAVs will begin to emerge and could become more commonplace over the next decades, eventually dominating ground transportation.

Although AV policies currently exist in the U.S., most are enacted at the state level. Additional policies will be needed to manage the complexities that arise as AVs become more common and SAV networks begin serving more passengers. While the policies needed to govern SAVs are unclear at this early stage, some best practices can be garnered from shared mobility. Although best practices and understanding in this area are still evolving, some lessons are beginning to emerge that can help to inform future SAV policy. While TNCs have been reluctant to share ridership data with public agencies and the general public due to proprietary and privacy concerns, empirical evidence is mounting that TNCs increase congestion and emissions during peak travel periods. If similar effects occur with SAVs, it will be critical for the public sector to have access to travel data to help inform their planning and decision making. Not surprisingly, some policy considerations will be unique to SAVs and AVs. For instance, it will be critical to ensure passenger safety in the absence of a human driver in a SAV service and to mitigate the widespread labor impacts of AVs and SAVs. In the future, policy will need to be carefully employed to ensure the public good, particularly as AVs and SAVs mainstream and further impact travel behavior and a wide range of socio-economic factors.

Author contributions

The authors confirm contribution to this chapter. All authors reviewed and approved the manuscript.

Declaration of conflicting interests

The authors declared no potential conflict of interest with respect to publication of this chapter.

References

[1] Greenblatt, J., and Saxena, S. "Autonomous taxis could greatly reduce greenhouse-gas emissions of US light-duty vehicles." *Nature*, 2015; 860–863.

[2] International Transport Forum. "Shared mobility: Innovation for livable cities." in *Organization for Economic Cooperation and Development (OECD)*, Paris, France; 2016.

[3] SAE International, *SAE J3016 Levels of Driving Automation*, Warrendale: SAE International; 2019.

[4] Davies, A. *Self-driving trucks are now delivering refrigerators*; 13 November 2017. [Online]. Available: https://www.wired.com/story/embark-self-driving-truck-deliveries/.

[5] Shaheen, S., Cohen, A., Broader, J., Farrar, E., Hoban, S., and Auer, A. "Mobility on demand and automated driving systems: A framework for public-sector assessment." Washington, D.C.: National Cooperative Highway Research Program (NCHRP); 2020.

[6] Iqbal, M., *Uber revenue and usage statistics*; 10 May 2019. [Online]. Available: https://www.businessofapps.com/data/uber-statistics/.

[7] Bayen, A., Shaheen, S., and Forscher, E. *An equitable and integrated approach to paying for roads in a time of change*. Berkeley, CA: University of California, Berkeley; 2019.

[8] U.S. Department of Transportation. Federal Automated Vehicles Policy, Washington, D.C.: National Highway Traffic Safety Administration; 2016.

[9] National Highway Traffic Safety Administration. U.S. Department of Transportation Releases *"Preparing for the future of transportation: Automated vehicles 3.0"*; 04 October 2018. [Online]. Available: https://www.nhtsa.gov/press-releases/us-department-transportation-releases-preparing-future-transportation-automated.

[10] Wong, S., and Shaheen, S. *Synthesis of state-level planning and strategic actions on automated vehicles: Lessons and policy guidance for California*. California Resilient and Innovative Mobility Initiative; 2020.

[11] National Conference of State Legislatures. *Autonomous vehicles state bill tracking database*; 2020. [Online]. Available: https://www.ncsl.org/research/transportation/autonomous-vehicles-legislative-database.aspx.

[12] Governor Highway Safety Association. "Governor highway safety association"; 2020. [Online]. Available: https://www.ghsa.org/taxonomy/term/521.

[13] California Department of Motor Vehicles. "Driverless testing of autonomous vehicles"; 19 December 2019. [Online]. Available: https://www.dmv.ca.gov/portal/dmv/detail/vr/autonomous/auto.

[14] Ducey, D. "Executive Order 2018-04"; 1 March 2018. [Online]. Available: https://azgovernor.gov/sites/default/files/related-docs/eo2018-04_1.pdf.

[15] Edelstein, S. "Some U.S. states consider taxing self-driving cars to raise cash"; 21 August 2017. [Online]. Available: https://www.thedrive.com/tech/13653/some-u-s-states-consider-taxing-self-driving-cars-to-raise-cash.

[16] City of Boston. "Autonomous vehicles: Boston's approach"; 28 December 2019. [Online]. Available: https://www.boston.gov/transportation/autonomous-vehicles-bostons-approach.

[17] Sanguinetti, A., Kurani, K., and Ferguson, B. "Is it OKA to get in a car with a stranger? Risks and benefits of ride-pooling in shared automated vehicles." Davis, CA: University of California, Davis; 2019.

[18] Lavieri, P., and Baht, C. "Modeling individuals' willingness to share trips with strangers in an autonomous vehicle future." *Transportation Research Part A: Policy and Practice*; 2019. pp. 242–261

[19] National Limousine Association. "Ride responsibly"; 02 January 2020. [Online]. Available: http://www.rideresponsibly.org/national-incident-report/.

[20] Taxicab, Limousine and Paratransit Association (TLPA). "Reported incidents involving Uber and Lyft"; 02 January 2020. [Online]. Available: http://www.whosdrivingyou.org/rideshare-incidents#deaths.

[21] Cohen, L., and Felson, M. "Social change and crime rate trends: A routine activity approach." *American Sociological Review*. 1979; 588–608

[22] Cohen, L., Kluegel, J. and Land, K. "Social inequity and predatory criminal victimization: An exposition and test of a formal theory." *American Sociological Review*. 1981; 504–524.

[23] Park, J. Kim, P. M. S., and Lee, B. *Offender or guardian? An Empirical analysis of ride-sharing and sexual assault.* Philadelphia, PA: Fox School of Business; 2017.

[24] Kirabo, J., and Owens, E.G. *One for the road: Public transportation, alcohol consumption, and intoxicated driving.* Cambridge, MA: National Bureau of Economic Research; 2010.

[25] Mulholland, S., and Dills, A. *Ride-sharing, fatal crashes, and crime. Southern Economic Journal.* 2018.

[26] O'Brien, S., Black, N., Devine, C., and Griffin, D. "CNN investigation: 103 Uber drivers accused of sexual assault or abuse"; 30 April 2018. [Online]. Available: https://money.cnn.com/2018/04/30/technology/uber-driver-sexual-assault/index.html.

[27] BBC News. "Uber London loses licence to operate"; 22 September 2017. [Online]. Available: https://www.bbc.com/news/uk-england-41358640.

[28] BBC News. "Uber to report crimes direct to police to boost safety"; 16 February 2018. [Online]. Available: https://www.bbc.com/news/business-43082729.

[29] Aleem, Z. "Why London is banning Uber from its streets"; 24 September 2017. [Online]. Available: https://www.vox.com/world/2017/9/24/16350064/why-london-banned-uber-explained.

[30] Chaudhry, B., Yasar, A.U.H., El-Amine, S., and Shakshuki, E. "Passenger safety in ride-sharing services. *Procedia Computer Science*. 2018; 1044–1050.

[31] Merat, N., and Madigan, R. "Human factors, user requirements, and user acceptance of ride-sharing in automated vehicles." *International Transport Forum*, Paris; 2017.

[32] Dent, S. "Uber unveils much-needed passenger safety features"; 12 April 2018. [Online]. Available: https://www.engadget.com/2018/04/12/uber-app-safety-features-driver-checks/.

[33] Moritz, K. "Where are the ridesharing companies for women? 28 February 2019. [Online]. Available: https://www.rewire.org/living/ridesharing-companies-women/.

[34] Berg, N. "Inside the transportation data tug of war"; 06 March 2019. [Online]. Available: https://theoverheadwire.com/2019/03/inside-the-transportation-data-tug-of-war/.

[35] New York City Taxi and Limousine Commission. "New York City Taxi and Limousine Commission"; 26 December 2019. [Online]. Available: https://www1.nyc.gov/site/tlc/about/tlc-trip-record-data.page.

[36] Freund, S. "Chicago first city to publish data on ride-hailing trips, drivers, and vehicles"; 15 April 2019. [Online]. Available: https://chicago.curbed.com/2019/4/15/18311340/uber-lyft-chicago-data-fares-drivers.

[37] City of Los Angeles. "Los Angeles Department of Transportation"; 31 October 2018. [Online]. Available: https://ladot.io/wp-content/uploads/2018/12/What-is-MDS-Cities.pdf.

[38] American Civil Liberties Union Southern California. "ACLU Southern California"; 08 June 2020. [Online]. Available: https://www.forbes.com/sites/bernardmarr/2020/06/29/introducing-the-mindboggling-flying-taxis-of-the-future/#2e536b6118dd.

[39] Hautala, L. "CCPA: Everything you need to know about California's new privacy law"; 25 December 2019. [Online]. Available: https://www.cnet.com/news/ccpa-everything-you-want-to-know-about-californias-new-privacy-law/.

[40] Erhardt, G., Roy, S., Cooper, D., Sana, B., Chen, M., and Castiglione, J. "Do transportation network companies decrease or increase congestion?" *Science Advances*. 2019; 1–11.

[41] Feigon, S., and Murphy, C. "Shared mobility and the transformation of public transit"; 2016. [Online]. Available: https://doi.org/10.17226/23578.

[42] Hampshire, R., Simek, C., Fabusuyi, T., Di, S., and Chen, X. "Measuring the impact of an unanticipated disruption of Uber/Lyft in Austin, TX": 2017. [Online]. Available: https://papers.ssrn.com/sol3/papers.cfm?abstract_id=2977969&download=yes.

[43] Feigon, S., and Murphy, C. *Broadening understanding of the interplay between public transit, shared mobility, and personal automobiles.* National Academies Press; 2018.

[44] Rayle, L., Dai, D., Chan, N., Cervero, R., and Shaheen, S. "Just a better taxi? A survey-based comparison of taxis, transit, and ridesourcing services in San Francisco." *Transport Policy*. 2016; 45: 168–178

[45] Clewlow, R., and Mishra, G. "Disruptive transportation: The adoption, utilization, and impacts of ride-hailing in the United States." Davis, CA: Institute of Transportation Studies; 2017.

[46] Gerhke, S., Felix, A., and Reardon, T. "Fare choices survey of ride-hailing passengers." Metropolitan Area Planning Council; 2018. pp. 1–19.

[47] New York City (NYC) Department of Transportation. "New York City mobility report"; 08 August 2019. [Online]. Available: http://www.nyc.gov/html/dot/downloads/pdf/mobilityreport-2018-print.pdf.

[48] Henao, A., "Impacts of ridesourcing - Lyft and Uber - On transportation including VMT, mode replacement, parking, and travel behavior." Denver, CO: University of Colorado at Denver, 2017.

[49] Hall, J., Palsson, C., and Price, J. "Is Uber a substitute or complement for public transit?"; *Journal of Urban Economics*; 2018, pp. 36–50

[50] Graehler, M., Mucci, A., and Erhardt, G.D. "Understanding the recent transit ridership decline in major US cities: Service cuts or emerging modes?" *Transportation Research Board 98th Annual Meeting*, Washington, D.C., 2019.

[51] SFCTA. "TNCs today: A profile of San Francisco transportation network activity"; 2017. [Online]. Available: https://www.sfcta.org/sites/default/files/2019-02/TNCs_Today_112917_0.pdf.

[52] SFCTA. "The TNC regulatory landscape: An overview of current TNC regulation in California and across the country"; 2017. [Online]. Available: https://www.sfcta.org/sites/default/files/2019- 03/TNC_regulatory_020218.pdf.

[53] Schaller, B. "Empty seats, full seats: Fixing Manhattan's traffic problem"; 2107. [Online]. Available: http://www.schallerconsult.com/rideservices/emptyseats.pdf .

[54] Schaller, B. "Unsustainable? The growth of App based ride services and traffic, travel and the future of New York City"; 2017. [Online]. Available: http://www.schallerconsult.com/rideservices/unsustainable.pdf .

[55] Harper, C. Hendrickson, C., Mangones, S., and Samaras, C. "Estimating potential increases in travel with autonomous vehicles for non-driving, elderly and people with travel-restrictive medical conditions." *Transportation Research Part C: Emerging Technologies*. 2016; 72: 1–9.

[56] Harb, M., Xiao, Y., Circella, G., Mokhtarian, P., and Walker, J. "Projecting travelers into a world of self-driving vehicles: estimating travel behavior implications via a naturalistic experiment." *Transportation* 45(6); 2018.

[57] Hu, W. "Your taxi or Uber ride in Manhattan will soon cost more." *The New York Times*; 31 January 2019.

[58] Durkin, E., and Aratani, L. "New York becomes first city in US to approve congestion pricing"; 1 April 2019. [Online]. Available: https://www.theguardian.com/us-news/2019/apr/01/new-york-congestion-pricing-manhattan.

[59] Millard-Ball, A. "The autonomous vehicle parking problem." *Transport Policy*. 2019;99–108.

[60] Lazarus, J., Bauer, G., Greenblatt, J., and Shaheen, S. *Bridging the income and digital divide with shared automated electric vehicles*, Berkeley, CA: Transportation Sustainability Research Center; 2021.

[61] Shaheen, S., Stocker, A., and Meza, R. *Social equity impacts of congestion management strategies*. Berkeley, CA: Transportation Sustainability Research Center; 2019.

[62] Ricci, A. "Socioeconomic impacts of automated and connected vehicles." *Transportation Research Board Conference Proceedings 5*, Washington, DC; 2018.

[63] Shaheen, S., Bell, C. Cohen, A., and Yelchuru, B. "Travel behavior: Shared mobility and transportation equity." Federal Highway Administration; 2017.

[64] U.S. Bureau of Labor Statistics. "U.S. Bureau of Labor Statistics"; May 2018. [Online]. Available: https://www.bls.gov/news.release/ocwage. t01.htm.

[65] Kalra, N. *What autonomous vehicles could mean for American workers.* Santa Monica, CA: Rand; 2017.

[66] Sianato, M. "'I made $3.75 an hour': Lyft and Uber drivers push to unionize for better pay"; 22 March 2019. [Online]. Available: https://www.theguardian. com/us-news/2019/mar/22/uber-lyft-ipo-drivers-unionize-low-pay-expenses#:~:text=According%20to%20Uber%2C%20there%20are,the%20expenses %20associated%20with%20driving.

[67] Beede, D., Powers, R., and Ingram, C. "The employment impact of autonomous vehicles." *Social Science Research Network*; 2017.

[68] Groshen, E., Helper, S., MacDuffie, J., and Carson, C. "Preparing U.S. workers and employers for an autonomous vehicle future." *Upjohn Institute*, Kalamazoo; 2019.

[69] Korosec, K. "Why self-driving car companies are eyeing this remote driving startup"; 12 June 2018. [Online]. Available: https://fortune.com/2018/06/12/ phantom-auto-self-driving-cars/.

[70] Litman, T., *Autonomous vehicle implementation predictions: implications for transport planning.* Victoria, Canada: Victoria Transport Policy Institute; 2019.

[71] Stocker, A., and Shaheen, S. "Shared automated vehicles: Review of business models." *International Transport Forum*, Paris; 2017.

[72] Tong, Q., and Ziyi, T. "Official frustration with Didi Gros, underscoring monopoly concerns"; 28 September 2018. [Online]. Available: https://www. caixinglobal.com/2018-09-28/official-frustration-with-didi-grows-underscoring-monopoly-concerns-101331166.html.

[73] Gorenflo, N. "California's P2P car-sharing bill signed into law"; 30 September 2010. [Online]. Available: https://www.shareable.net/ californias-p2p-car-sharing-bill-signed-into-law/.

[74] Shaheen, S., Martin, E., and Bansal, A. *Peer-to-peer (P2P) carsharing: Understanding early markets, social dynamics, and behavioral impacts.* Transportation Sustainability Research Center, Berkeley, CA; 2018.

[75] Insurance Information Institute. "Background on self-driving cars and insurance," 30 July 2018. [Online]. Available: https://www.iii.org/article/ background-on-self-driving-cars-and-insurance.

[76] Brown, A. "Ridehail revolution." Ridehail Travel and Equity in Los Angeles. Los Angeles, CA: University of California, Los Angeles; 2018.

[77] Wegner, C. "Mobility forward update and presentation; 05 July 2017. [Online]. Available: https://www.wheelsbus.com/wp-content/uploads/2017/ 06/WAAC_070517_File-for-Website.pdf.

[78] Hough, J., and Taleqani, A.R. "Future of rural transit." *Journal of Public Transportation*. 2018; 31–42.

[79] Joseph, R. "Ride-sharing: The rural urban divide." *Twenty-Fourth Americas Conference on Information Systems*, New Orleans; 2018.

[80] Shepardson, D., and Cooney, P. "U.S. will unveil data-sharing platform for autonomous vehicle testing"; 01 June 2020. [Online]. Available: https://venturebeat.com/2020/06/15/u-s-will-unveil-data-sharing-platform-for-autonomous-vehicle-testing/.

[81] Graves, Z., Rizer, A., and Kane, J. *Beyond legal operation: The next ridesharing policy challenges*. R Street, Washington, DC; 2018.

The views expressed are those of the authors.

Chapter 4

Concept level designs: high-level architecture

Matthew Barth[1], Peng Hao[2] and Shams Tanvir[3]

Shared mobility systems are expected to respond to socio-technical changes and pandemics. Given the future transportation market and other uncertainties, the resilience of designs in addressing multiple goals is a valued attribute. Previous chapters 2 and 3 of this book provide details of a possible seismic shift in transportation and the associated challenges posed by the policy and regulatory environment, including concerns about sharing rides with strangers and social/racial equity. There is of course the additional requirement to meet health requirements in the post-pandemic operating environment. This chapter covers the subject of concept-level design and associated high-level architecture for shared mobility systems of the future.

Most shared mobility systems are complex in terms of components, including their interactions and interfaces with the socio-economic and physical environment. Well-studied concept level designs of these systems are necessary for viewing the entire system, identifying the main components that would play a role in the service demand and supply interaction, and ensuring that interfaces with the overall urban system and the environment do not cause adverse effects. The high-level architectures defined at the concept design level lead to details on system planning, design, operations, maintenance, and management. Chapters 5–17 cover key elements of these subjects, and the final chapter (Chapter 18) is focused on future directions, including uncertainty and the need to prioritize social and environmental benefits in an evolving marketplace through dynamic policy and planning processes. Links to detailed coverage of these subjects are noted throughout this chapter.

4.1 Introduction

Throughout history, the movement of both people and goods has always been a critical component to human society. To enable this movement, societies have

[1]Department of Electrical and Computer Engineering, and Center for Environmental Research and Technology, University of California at Riverside, Riverside, CA, USA
[2]Center for Environmental Research and Technology, University of California at Riverside, Riverside, CA, USA
[3]Department of Civil and Environmental Engineering, California Polytechnic State University, San Luis Obispo, CA, USA

developed a wide variety of transportation systems, allowing their economies to grow and improving their overall quality of life. With the development of different transportation systems, travel demand has steadily increased as new opportunities continue to emerge within society. Along with these increased opportunities and economic growth, it is important to note that a number of negative consequences have also emerged in the form of transportation-related air pollution, greenhouse gas emissions leading to climate change, and other related environmental impacts.

The earliest transportation systems evolved as a direct extension of a traveler's needs. If a traveler wanted to get from points A to B faster than walking, they could use different types of *vehicles*, which have also evolved over time. It was a natural evolution that a traveler's needs could be quickly satisfied by possessing their own vehicles (e.g., bicycles and cars) and using those vehicles when desired. This "individual vehicle ownership" model has flourished across the world. However, for many, individual vehicle ownership was not economically feasible, so society also developed public transportation systems to satisfy societal travel demand. Albeit not as efficient and convenient as individual vehicle ownership, these public transportation systems, in their many forms, have and continue to play an important role in society.

It is well known that over the last several decades, there have been significant advances in communications, computing, and the sharing of information. This information technology revolution has certainly improved our different transportation systems as a whole, but it has also allowed for the development of new transportation system models that go beyond the personal vehicle ownership model and traditional public transportation models. Many forms of *shared mobility* have emerged, taking advantage of a variety of these technological advances. As the name implies, the root of shared mobility systems is the shared use of vehicles among multiple demand entities to complete a trip [1]. Specific technologies that have enabled innovative shared mobility models include the Internet, wireless communications, location determination systems (e.g. Global Navigation Satellite Systems such as GPS), electronic payment systems, and the emergence of smartphones that have pulled together all of these technology elements into the hands of the traveler [2].

Today, many shared mobility system models exist, and innovative models are evolving every year. In general, the overall goal is to pair travel demand with supply, that is, a variety of mobility systems, execute the operation, and complete the economic transactions. The goal of these systems is to efficiently move people (and goods), reduce the overall costs of travel and operation, and minimize any negative environmental impacts. Chapters 5, 6, and 7 provide further information on goals, planning, and transportation system options.

At the heart of any shared mobility system are five components that interact together: the traveler; some type of vehicle; the infrastructure used by the vehicle (e.g., roadway and parking); vehicle operators (e.g., drivers); and the overall transportation system managers and operators. This chapter addresses the high-level architectures of how these five components (or some subset of these components) interact in accomplishing the goals described above. Establishing these architectures is useful for a number of reasons:

- They help describe how a particular shared mobility system works;

- They identify what are the key interactions between components of the system, along with the data transactions that are necessary for the shared mobility system to work properly, aiding in the initial setup as well as the ongoing maintenance;
- They are useful for establishing key performance metrics (e.g., trip efficiency and VMT reductions) that should be measured for every shared mobility system; and
- They assist system operators to determine where opportunities may exist for automation and other future technology advances.

4.2   Shared mobility components and a general architecture

There are many forms of shared mobility, the common theme is that vehicles are shared among multiple users, either sequentially or concurrently, to satisfy travel demand [3]. There are many incumbent public transportation systems that exist today that are "shared" by travelers, such as traditional public bus and rail systems, car rental companies, taxis, and shuttles. In addition, a number of innovative shared mobility systems have emerged, those that have flourished due to recent advances in information technology. This is illustrated best with the shared mobility ecosystem diagram from [4], as in Figure 4.1.

In this diagram, well-established incumbent services are shown on the left, while the emerging shared services are shown on the right. There are some services that overlap both categories. Within these different service models, there are different variations that exist in terms of how they are actually implemented.

In addition to this shared mobility ecosystem diagram, various shared mobility "typologies" have been put forth, illustrating how shared mobility systems are defined and how they relate to one another. Examples include [5] describing a typology for early forms of carsharing, and [6] which describes typologies for Mobility-As-A-Service (MAAS) and Mobility-on-Demand (MoD), and [7] which addresses shared automated fleet systems. As emerging shared mobility systems are put forth, usually, the overall design and architecture are described in different levels of details typically on websites, primarily from the user's perspective (e.g., see [8]). There have also been numerous academic studies that have addressed different types of shared mobility, describing their designs, system operations, and key metrics of performance.

For the convenience of the reader, the following links to details of the topics covered in this chapter are noted next. Chapter 8 of this book covers the design of systems with nonautomated and electric vehicles, and Chapter 9 describes the design of systems with automated and electric vehicles. Chapters 10 and 11 describe demand models, and Chapter 12 reports an approach to matching demand and supply under uncertainty. Chapter 13 describes many of the specific interactions and activities that occur in the operations and management of a shared mobility system.

There are other facets of the broader system planning and design subject that are useful to know. These are impacts on the public realm (covered in Chapters 14

SHARED MOBILITY SERVICE MODELS

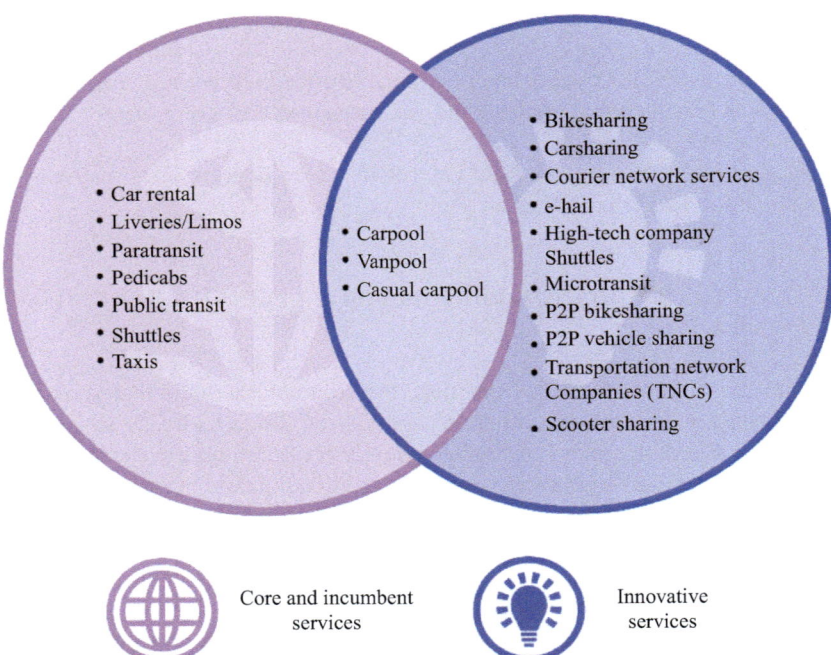

Figure 4.1   *Shared mobility ecosystem. Adapted from [4]*

and 15), factors that affect economic feasibility (described in Chapter 16), and urban development and environmental impacts (covered in Chapter 17). Chapter 18 discusses future directions to maximize the social and environmental benefits of shared and automated mobility services, particularly in light of uncertainty and evolving market forces.

In general, any shared mobility system typically has five interactive components, outlined below. In this description, we primarily focus on how we can move people, although there are various shared mobility models that exist specifically for moving goods.

Traveler—the traveler is the core element of what we want to move; it may consist of an individual or a group of people (e.g., a family traveling together), and the travel needs may be simple (i.e., point-to-point) or more complex (multiple trips). Today's travelers rely on real-time information, typically enabled by the Internet and communication devices (smartphone, information kiosk, home computer, etc.).

Vehicle—there are many types of vehicles or form factors that can be used as part of shared mobility including cars, buses, vans, bicycles, and various forms of micromobility (e.g., scooters, Segways, hoverboards, skateboards, unicycles, etc.).

Some shared mobility systems may be set up to only use one type of vehicle, while others may consist of a variety of vehicle types. Also note that vehicles are becoming increasingly connected and automated, enhancing the use and operations of various shared mobility models. When considering environmental metrics, it is also important to note the different sources of power that the vehicle uses, such as an internal combustion engine requiring liquid fuel or electric drive requiring electricity from batteries or fuel cells.

Infrastructure—the vehicles and users of shared mobility also interact with various forms of infrastructure including roadways, roadway management systems (e.g., traffic signals), parking, and potentially *vehicle stations*, where the shared vehicles are picked up or dropped off. These vehicle stations may have electric vehicle charging if the vehicles are electrically powered.

Vehicle operators—in many shared mobility systems, the vehicles are usually operated (i.e., driven) by the traveler themselves; however, in many transit and shared-ride services, there are other vehicle operators that come into play. These vehicle operators also play an important role in different shared mobility models and are a primary opportunity for automation.

System operators—a shared mobility system typically has system operators that manage and administer the system itself. The system operators may play an active role in the day-to-day operation, or they may simply set up the system at the beginning and then mildly monitor the overall use with time. In complex shared mobility systems, there are numerous tasks that system operators have to carry out including: monitoring the status of the vehicle fleet, responding to traveler demands, assigning vehicles or drivers, performing financial accounting, maintaining the vehicle fleet, forecasting demand and supply, attracting new customers, detecting anomalies and troubleshooting, performing data management, and evaluating overall performance metrics (see Chapter 13 for further details on overall system operation). A large number of research studies focus on these various tasks implemented by the shared mobility system operators.

All of these five components will have different degrees of interaction when implementing some kind of shared mobility system. Figure 4.2 shows these interactions in a general way. Some shared mobility systems will only use and depend on a subset of these interactions. As we analyze the different types of shared mobility systems, these interactions will be emphasized in different ways.

4.3   Shared mobility systems and their architectures

4.3.1   Incumbent shared mobility systems

For completeness, we quickly illustrate the shared mobility interaction architecture diagram for some of the standard incumbent systems, such as public transit (e.g., buses and light rail), taxis, and typical car rental services.

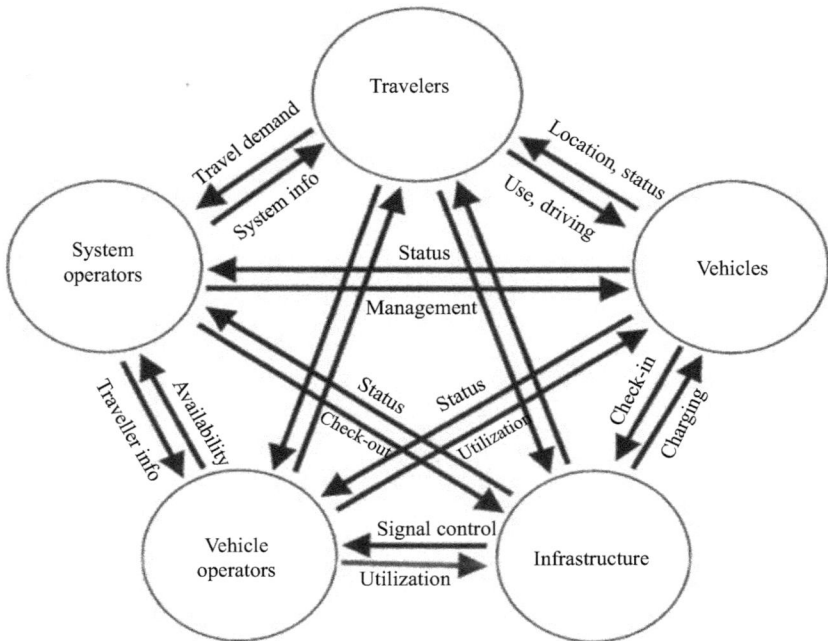

Figure 4.2   Generalized shared mobility interaction architecture

Public transit—for public transit such as bus and light-rail systems, there is an interaction between all of the different components defined in Section 4.2. The interaction diagram for these systems is shown in Figure 4.3. In comparison to the innovative shared mobility systems described in Section 4.3.2, the interactions between the different components are less dynamic. Once bus and light rail systems are set up (establishment of routes and timing, vehicle fleet, driver schedules, etc.), real-time information flowing between the different entities is not really necessary nor critical. This makes sense since these incumbent systems evolved prior to the information technology revolution. Nevertheless, modern bus and rail systems are now taking advantage of providing real-time information to passengers (e.g., NextBus technology, see [9]). Further, vehicle operators can be in touch with system managers about schedule adherence and making minor adjustments (e.g., see [10]).

Taxis—taxi services have been around for quite some time, even prior to the introduction of the automobile. In the simplest taxi systems (see Figure 4.4), a number of drivers in vehicles would roam the cities, picking up and dropping off passengers at various locations. Passengers would simply hail an available taxi at the curb, get in, ride to the desired location, pay the driver, and then exit the vehicle. The drivers would then move on to another passenger trip. With added communications, passengers can also call a centralized taxi service, the service

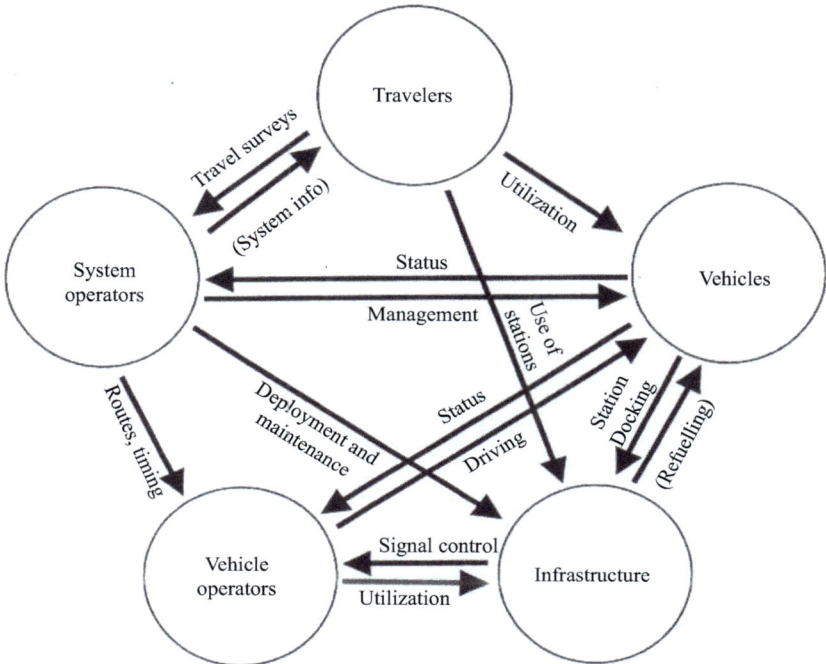

Figure 4.3   General public transit interaction architecture diagram

could then communicate with available taxis, and dispatch the taxis to the passenger locations. In many cases, the system management may be the drivers themselves, communicating via cellular phones. There can also exist "taxi stands," where taxis line up at an organized location that has a high level of travel activity, such as an airport or train station. Passengers would simply walk up to the taxi stand, make a trip request, and wait to be matched with a driver.

Similar to a car-based taxi service, there are also ***pedicab*** services in many places around the world. This for-hire ride service operates very similarly to a simple taxi services, except a cyclist (driver) transports traveler(s) on a tricycle. This system may service one passenger at a time, or potentially multiple passengers. These pedicab systems typically do not have centralized operations, in essence each cyclist is an independent operator.

Conventional car rental—Incumbent car rental systems started before the information technology revolution and simply involved a car rental company that provided vehicles to rent to travelers for different periods of time. In this simplest case, the vehicles would be "checked-out" from the car rental location and checked back in at the end of the rental period (see Figure 4.5). In this case, the system operators would likely be located at the car rental facilities. With added communication and

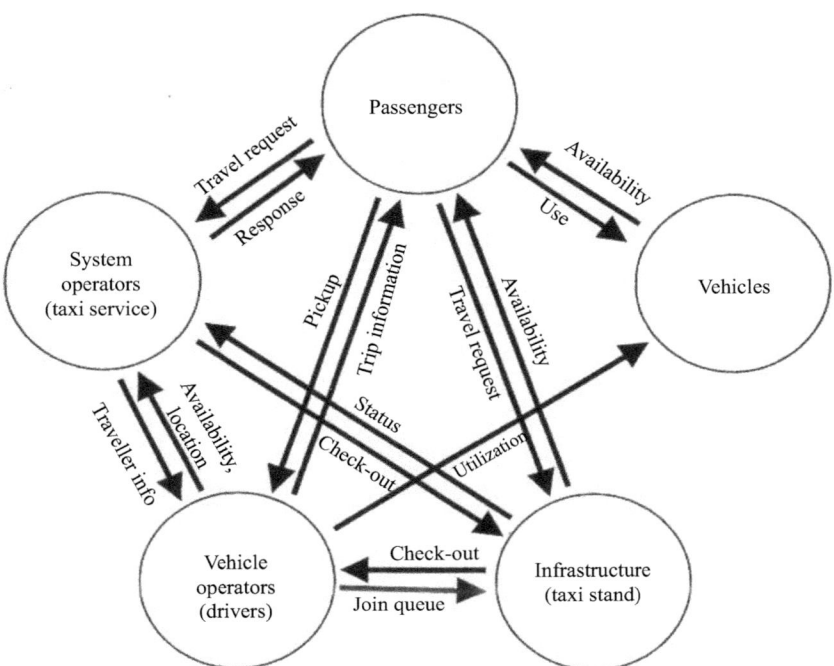

Figure 4.4   General taxi interaction architecture diagram

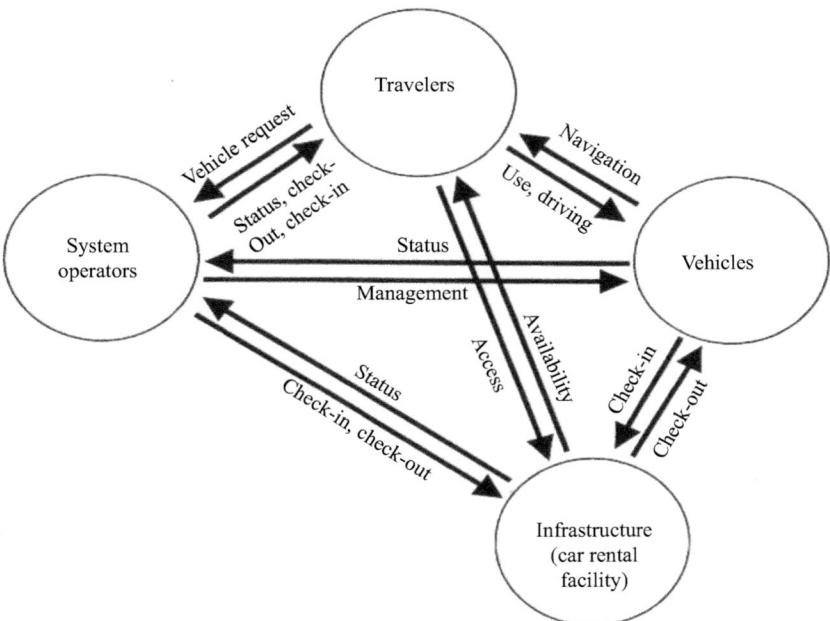

Figure 4.5   Conventional car rental interaction diagram

coordination between multiple locations, travelers can make requests (i.e., reservations) via the Internet or phone, and the system operators can be separate from the car rental facility (e.g., in a "back office" data handling location). During a check-out process, travelers usually get their choice of vehicle types, and the check-in process can happen at the same location, or at another affiliated location within the car rental agency. One of the key features of the car rental system is that the travelers themselves are the drivers, and take on the added activities (and freedoms) of trip scheduling, navigation, fueling, etc.

4.3.2   Innovative shared mobility systems

The U.S. Department of Transportation [11]), the American Planning Association [12]), and SAE International [13]) have now more or less converged on the definitions of the most common shared mobility system models. These models are briefly described here, together with a specific snapshot of their shared mobility interaction architecture diagram.

Bikesharing—this shared mobility model provides travelers with on-demand access to bicycles at a variety of pick-up and drop-off locations for either round-trip travel or one-way (point-to-point) travel. These systems are commonly deployed in a network within a metropolitan region, city, and/or university campus [4,13,14]. The bicycles are typically owned and maintained by the bikesharing organization.

Carsharing—similar to bikesharing, carsharing provides users with on-demand access to cars at a variety of pick-up and drop-off locations for either roundtrip travel or one-way (point-to-point) travel. Carsharing systems have been deployed around the world and are tracked by a number of websites (e.g., [8]). Similar to bikesharing, the cars themselves are typically owned and maintained by the carsharing organization.

Both bikesharing and carsharing systems can take on different forms and complexities. Some are station-based, where the vehicle (bikes or cars) are checked in and out of specific stations, in a multistation system. Other systems do not require stations or hubs, instead, the vehicles are picked up or dropped off at any possible parking location. These stationless or dockless systems lend themselves to one-way services (vehicles can be dropped off anywhere, not requiring returning to the same location); these one-way services are convenient for the user, but they add a high degree of complexity for the system operators, since vehicles occasionally need to be relocated to maximize the overall system efficiency [15,16]. There are many research papers that address these relocation issues; several of these are outlined in Chapter 13 of this book.

The overall interaction architecture for bikesharing and carsharing is shown in Figure 4.6(a) and (b). Figure 4.6(a) illustrates the typical interactions that occur for a station-based system, while Figure 4.6(b) illustrates the interactions for a stationless system. Similar to car rental shared vehicle models, the travelers themselves are the drivers, eliminating that aspect in the architecture diagram.

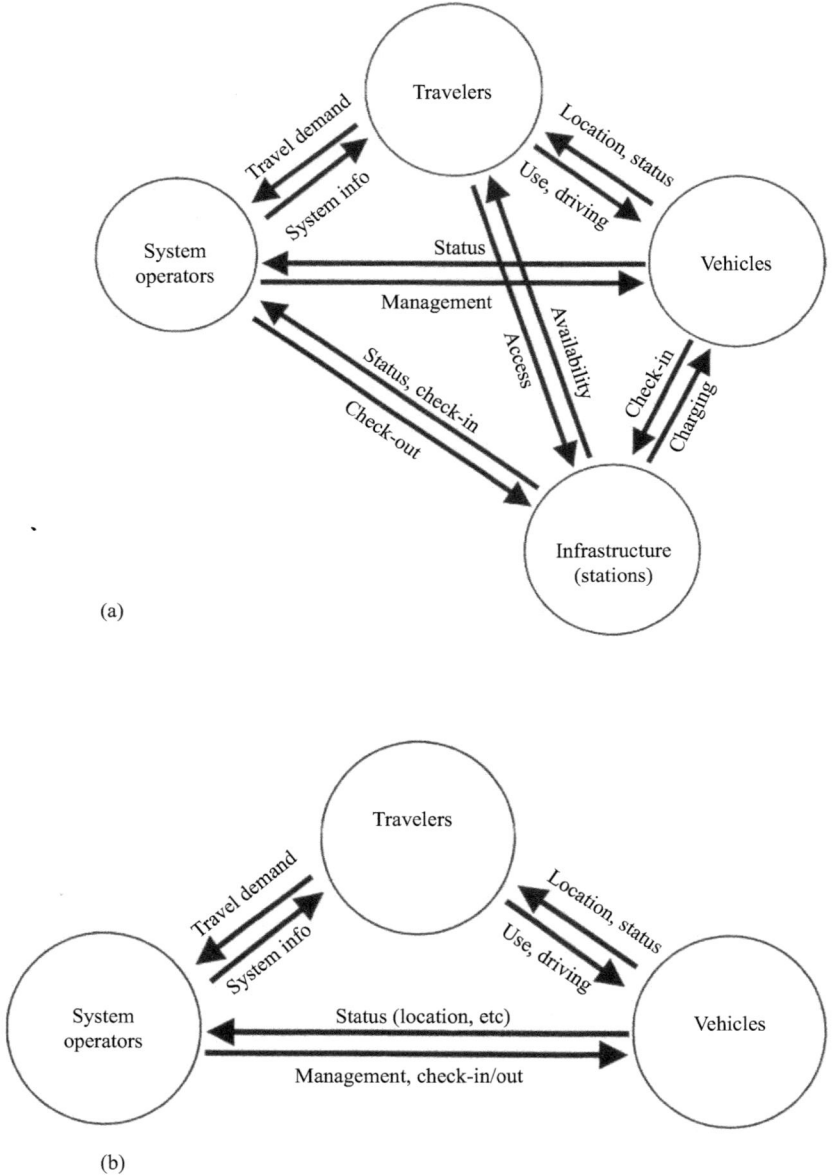

Figure 4.6   *(a) Bikesharing/carsharing interaction architecture diagram: station-based system. Note that micromobility sharing usually has a similar architecture to bikesharing and (b) bikesharing/carsharing interaction architecture diagram: stationless system. Note that micromobility sharing usually has a similar architecture to bikesharing*

The system operators typically rely on a cloud-based data system that manages the trip requests, the vehicle data, and various system parameters (see Section 4.4 for more details on data management).

Shared micromobility (e.g., scooter sharing)—similar to bikesharing, sharing of micromobility modes (e.g., scooters) is a growing market that provides travelers with an additional on-demand short-term mobility service [17]. In fact, bikesharing can be considered as the original form of shared micromobility, which now includes a number of other modes of travel such as scooters, skateboards, and hoverboards. All of these shared micromobility systems have an architecture that is shown in Figure 4.6(a) and (b) where the different modes of transport are typically picked up and dropped off in different zones.

Ridesharing—often referred to as carpooling or vanpooling, this transport service uses multipassenger/pooled shuttles or vans to provide the sharing of rides between similar origins and destinations [4,13,14]. Many companies are developing these types of vanpools to reduce overall vehicle miles traveled (VMT) associated with their company activities. The interaction diagram is illustrated in Figure 4.7(a). In this shared mobility system, the role of the system operator is quite low once the vehicles have been arranged and deployed, other than performing vehicle maintenance. The burden of the operations lies with the vanpool participants, who may rotate driving responsibilities. The vanpool participants can dynamically opt-in or out depending on their travel needs, using web-based scheduling tools.

Microtransit—this transport service can be privately or publicly operated, and similar to ridesharing, uses multipassenger/pooled shuttles or vans to provide on-demand or fixed-schedule services with either dynamic or fixed routes [4,13,14]. The interaction diagram is illustrated in Figure 4.7(b). Similar to ridesharing, the role of the system operator is low once the vehicles have been arranged and deployed in a fixed route system; however, in the case of a dynamic microtransit system, the system operator plays a much more active role.

Personal vehicle sharing—this shared mobility system is sometimes referred to as a "time-share" system where privately owned vehicle(s) are shared among users. The vehicles may be owned by one or more people in the organized group. The system "operators" in this case have a low-key role, simply providing the organizational resources to make this sharing possible, including taking care of vehicle safety, insurance, and providing web-based scheduling tools. The travelers would simply use the shared vehicle(s) based on the schedule. The interaction diagram for this case is shown in Figure 4.8.

Transportation network company (TNC), ridehailing, and ridesourcing—this type of shared mobility system has grown significantly in the past decade. These are well-organized services that provide travelers with rides that are either prearranged or on-demand, using a web browser or smartphone application. The rides are supplied by a number of independent drivers that use their personal, rented, or leased vehicles, who also make use of a smartphone application that is connected to

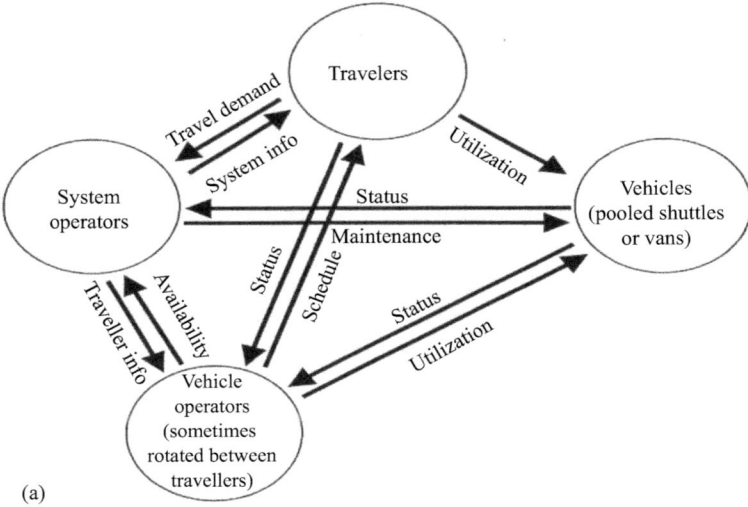

Figure 4.7(a)   Interaction diagram for ridesharing

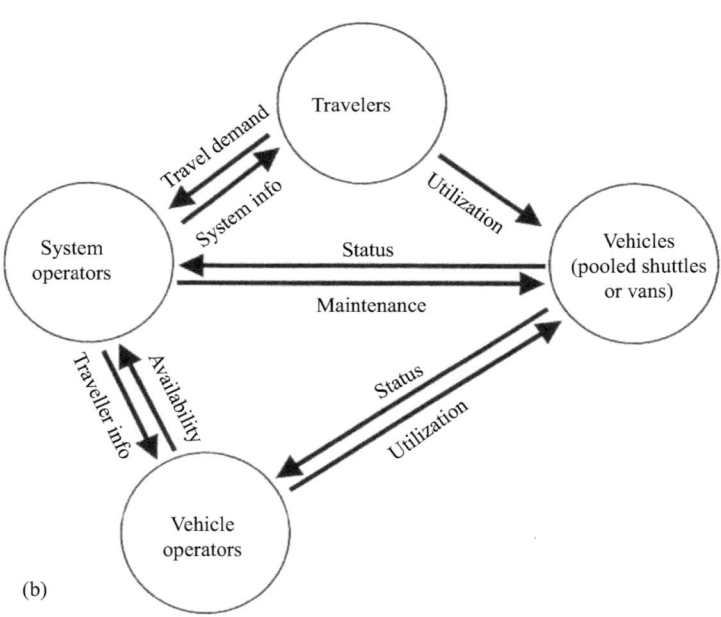

Figure 4.7(b)   Interaction diagram for microtransit

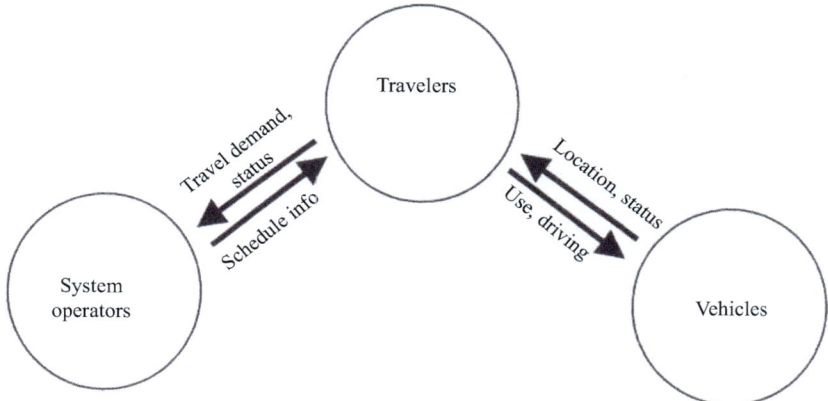

Figure 4.8   Personal vehicle sharing interaction diagram

a cloud-based management system. The smartphone applications together with the back-office management system take care of the traveler requests, bookings, the driver and vehicle selection, the vehicle routing, electronic payments, and rating systems for both drivers and passengers.

Many TNCs also offer "pooling" services for travelers who are willing to share their rides with other passengers who are heading in the same direction. The overall interaction architecture is shown in Figure 4.9. For the majority of trips, the passenger pick-ups and drop-offs can occur at any location, thereby greatly reducing the need for any organized physical infrastructure. However, due to the popularity of these services, cities, airports, and other locations have started designating specific locations where TNC drivers can pick-up and drop-off.

This type of shared mobility service relies heavily on digital information and communications. There are many algorithms that operate on large databases of user data, travel requests, and driver data, performing tasks such as setting pricing strategies, recruiting drivers, matching rides, as well as predicting travel demand both temporally and spatially. Overall system operations manage all of the costs and payments; because payments to drivers is a significant fraction of the costs of operations, many transportation network companies are looking to provide automated vehicles in their future fleets.

4.4   Data architecture and management

As described earlier, many of these shared mobility systems evolved due to the information technology advances that took place during the last several decades. In order to carry out the various interactions between the components described earlier, a robust data architecture is necessary that includes traveler applications, communication channels, database aggregation and management, and various active processes (a.k.a. agents) that act on various forms of data. As different

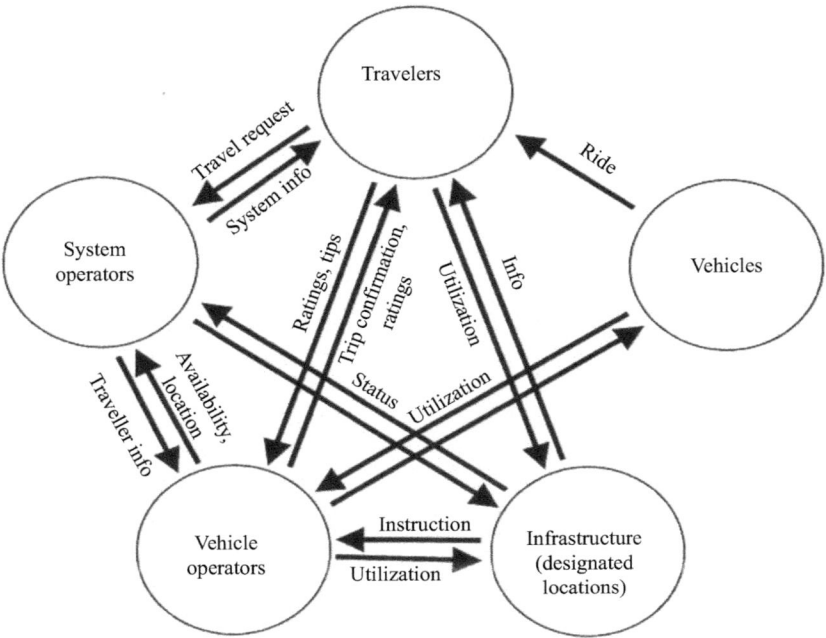

Figure 4.9   Typical transportation network company interaction diagram

shared mobility systems are described in the literature, a variety of data architectures are presented (e.g., see [18]).

4.4.1   Generalized data architecture

A generalized data architecture for shared mobility systems is shown in
Figure 4.10. This data flow architecture begins with travel requests made by travelers, typically made from a smartphone, web browser, or system kiosk. System
operators typically have smartphone applications or web services to provide an
interface for the traveler to the system. Once a trip is requested, supply-side processing occurs that evaluates the travel request information (e.g., current and
desired locations, number of occupants, vehicle type request, etc.) and carries out
some type of resource allocation algorithm. Shared mobility systems that rely on
drivers then provide information to the drivers, such as the request, locations, user
ratings, etc. If the shared mobility system does not use external drivers, then the
travelers themselves are directed to a vehicle's location, providing a vehicle check-
out process. It is important to realize that data not only flows between travelers and
drivers but also between the system and the vehicles themselves.

As these transactions are carried out, a dynamic database of travel demand,
users, vehicles, and drivers is managed by the system operations. A variety of

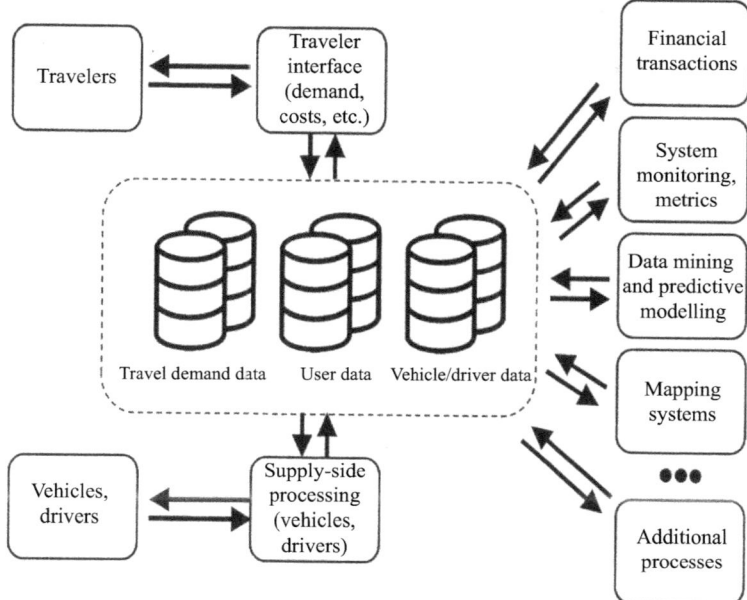

Figure 4.10   Generalized data flow diagram

processes typically occurs on this dynamic database, including financial transactions, the calculation of real-time system metrics, data mining and predictive modeling, and mapping functions (e.g., such as localization, routing, and geofencing).

Note that many dockless (free-floating) forms of shared mobility systems are increasingly using geofencing, which can be used for pricing and limiting locations of drop-offs. Many cities see this as being critical to better manage shared mobility in their communities.

Many of the shared mobility systems use cellular communications between users, drivers, vehicles, and the management system itself. Other communication systems can be used depending on the nature of the data exchange; for example, local Wi-Fi may be used at vehicle stations, and other low-latency applications may rely on dedicated short range communications (DSRC).

4.4.2   Data management

Data flowing between travelers, vehicles, drivers, and the system operations is a critical component of any shared mobility operation. These data are not only important for carrying out operations successfully, but they are also important for informing transportation decision making by local and regional transportation policy makers.

Over the years, there have been efforts to standardize many data elements associated with shared mobility operations [19–21]. By using standard formats, evaluation metrics across many shared mobility systems can be established. There are a number of efforts underway to provide open access to public transportation data, such as the OpenMobilityData platform, which employs the Mobility Data Specification [22]. Note that MDS is a data and API standard that allows a city to gather, analyze, and compare real-time and historical data from shared mobility service providers. An API allows the creation of applications that access data from another service or application. The specification also serves as a measurement tool that helps enable the enforcement of local regulations. In addition, MDS allows service providers and public agencies to communicate with each other about their services because it consists of two APIs: (1) a service provider API and (2) a public agency API. MDS includes data such as (1) mobility trips (and routes); (2) location and status of equipment (e.g., available, in-use, and out-of-service); and (3) service provider coverage areas. As of Fall 2019, cities using MDS included Los Angeles, CA; Santa Monica, CA; Austin, TX; and Ulm, Germany.

Another major concern with data is focused on privacy. It is paramount that data on shared mobility users are protected. This not only includes sensitive personal information on users (e.g., names, addresses, etc.) but also trip data where the trips may show home and work information and daily routine activities. Shared mobility system operators must anonymize their data if the data are going to be shared publicly for metric evaluation.

4.5   Summary

Understanding the interaction between travelers, vehicles, drivers (as needed), infrastructure, and system operators is critical for understanding how different shared mobility systems work. For system operators, a detailed architecture is not only necessary for setting up the system in the first place but also very important for debugging any problems as well as innovating new forms of operation. These architectures are also helpful when evaluating overall system efficiency and key metrics of performance. Transportation planners in many cities and regions need to understand these performance metrics so that they can best serve the transportation needs of their communities.

Further, these system architectures are useful in that they help system operators (and transportation planners) innovate for the future, particularly when it comes to vehicle electrification and vehicle automation. As we shift away from internal combustion vehicles toward electrifying our vehicles, the architecture and operations of shared mobility will likely change slightly to deal with refueling (e.g., EV charging and hydrogen for fuel cell vehicles) and potentially vehicle range [23]. Vehicle automation will be another game-changer in shared mobility (see, e.g., [24]), as it evolves. Not only can vehicle automation change shared mobility systems that rely on drivers, but they can also help with system operations themselves, if for example, a vehicle fleet needs to be occasionally "repositioned" to best satisfy demand. Although vehicle automation can offer many advantages, it is important to note that we are still many years away from having robust vehicle automation that can be used at scale in shared mobility systems.

Author contributions

The authors confirm contribution to this chapter. All authors reviewed and approved the manuscript.

Declaration of conflicting interests

The authors declared no potential conflict of interest with respect to publication of this chapter.

References

[1]   Shaheen, S., Chan, N., Bansal, A., and Cohen, A. *Shared mobility: Definitions, industry development, and early understanding.* 2015. Available from: http://innovativemobility.org/wp-content/uploads/2015/11/Shared Mobility_WhitePaper_FINAL.pdf.

[2]   National Academies of Sciences, Engineering, and Medicine (NAS). *Between public and private mobility: Examining the rise of technology-enabled transportation services.* Washington, DC: The National Academies Press. see https://doi.org/10.17226/21875.

[3]   Shaheen, S., Cohen, A., and Randolph, M. *Shared mobility policy playbook.* 2019. doi:10.7922/G2QC01RW, https://escholarship.org/uc/item/9678b4xs.

[4]   Shaheen, S. A., Cohen, A. P., and Zohdy, I. *Shared mobility: Current practices and guiding principles.* 2016. pp. 1–105. Retrieved from Washington, DC: U.S. Department of Transportation, Federal Highway Administration website: https://ops.fhwa.dot.gov/publications/fhwa-hop16022/fhwahop16022.pdf

[5]   Barth, M., and Shaheen, S. A. "Shared-use vehicle systems: Framework for classifying carsharing, station cars, and combined approaches." *Transportation Research Record: Journal of the Transportation Research Board.* 2002; 1791(1): 105–112.

[6]   Shaheen, S., and Cohen, A. "Mobility on demand (MOD) and mobility as a service (MaaS): Early understanding of shared mobility impacts and public transit partnerships." *Demand for emerging transportation systems.* Elsevier; 2020. pp. 37–59.

[7]   Hyland, M. F., and Mahmassani, H. S. Taxonomy of shared autonomous vehicle fleet management problems to inform future transportation mobility. *Transportation Research Record.* 2017; 2653(1) 26–34.

[8]   Shared Use Mobility Center (SUMC). Available from https://share-dusemobilitycenter.org/ [Accessed January 2021].

[9]   NextBus. *NextBus technology.* Available from https://www.nextbus.com/ [Accessed January 2021].

[10]   Transit Cooperative Research Program (TCRP). *Shared mobility and the transformation of public transit.* TCRP Report 188. Transportation Research Board, National Academies, 2016.

[11]   Federal Highway Administration (FHWA). *Shared mobility: Current practices and guiding principles.* FHWA Report # FHWA-HOP-16-022, 2016. Available from https://ops.fhwa.dot.gov/publications/fhwahop16022/fhwahop16022.pdf [Accessed January 2021].

[12]   American Planning Association (APTA). *Planning for shared mobility.* PAS Report 583, 2016. ISBN 978-1-61190-186-3, see Available from https://www.planning.org/publications/report/9107556/ [Accessed January 2021].

[13]   Society of Automotive Engineers (SAE). *Taxonomy and definitions for terms related to shared mobility and enabling technologies.* SAE Standard J3163_201809. 2018. Available from https://www.sae.org/standards/content/j3163_201809/ [Accessed January 2021].

[14]   Cohen, A., and Shaheen, S. Planning for shared mobility. 2016. Available from https://doi.org/10.7922/G2NV9GDD.

[15]   Barth, M., Todd, M., and Xue, L. "User-based vehicle relocation techniques for multiple-station shared-use vehicle systems." *Proceedings of the 83rd Annual Meeting of the Transportation Research Board, 2004.* Washington, DC.

[16]   Bruglieri, M., Colorni, A., and Luè, A. The relocation problem for the one-way electric vehicle sharing. *Networks.* 2014; 64(4): 292–305.

[17]   Shaheen, S., and Cohen, A. *Shared micromobility policy toolkit: Docked and dockless bike and scooter sharing.* 2019. DOI: 10.7922/G2TH8JW7. Available from Available from https://escholarship.org/uc/item/00k897b5.

[18]   Barth, M., Todd, M., and Shaheen, S. "Intelligent transportation technology elements and operational methodologies for shared-use vehicle systems." *Transportation Research Record: Journal of the Transportation Research Board.* 2003; 1841(1): 99–108.

[19]   Shaheen, S., Cohen, A., Zohdy, I., and Kock, B. *Smartphone applications to influence travel choices: Practices and policies.* FHWA. 2016a; 1–90.

[20]   NABSA. *Documentation for the General Bikeshare Feed Specification, a standardized data feed for bike share system availability.* 2019. Available from https://github.com/NABSA/gbfs [Accessed January 2021].

[21]   MDS. Mobility Data Specification. Available from https://www.openmobilityfoundation.org/about-mds/ [Accessed January 2021].

[22]   Plautz, J. *TransitScreen, MobilityData form open transit data platform.* 2019. Rapid Shift website: Available from http://www.rapidshift.net/transitscreen-mobilitydata-form-open-transit-data-platform/. https://ops.fhwa.dot.gov/publications/fhwahop16023/fhwahop16023.pdf [Accessed January 2021].

[23]   Chen, T. D., Kockelman, K. M., and Hanna, J.P. "Operations of a shared, autonomous, electric vehicle fleet: Implications of vehicle & charging infrastructure decisions." *Transportation Research Part A: Policy and Practice*. 2016; 94: 243–254.

[24]   Costlow, T. "Big data, big challenges." in *Autonomous vehicle engineering*. Society of Automotive Engineers. March 2018.

The views expressed are those of the authors.

Chapter 5

Shared mobility: managing rights-of-way, developer incentives, and planning principles

Adam P. Cohen[1] and Susan A. Shaheen[2]

The connection between shared mobility—the shared use of a vehicle, bicycle, scooter, or other mode—and transportation planning is not new [1]. Local governments and public transit agencies are increasingly confronting opportunities and challenges associated with shared modes such as (1) competition versus complementarity with public transportation, (2) how to manage curb space and public rights-of-way for a growing number of modal alternatives, and (3) how to plan multimodal transportation systems in a fast changing and evolving mobility ecosystem, particularly in light of the global pandemic. Shared mobility influences and is influenced by numerous facets of urban planning including:

- **Transportation and circulation**: Shared mobility can affect travel patterns such as modal choice, vehicle occupancy, and vehicle miles/kilometers traveled (VMT/VKT) (discussed in Chapter 17).
- **Curb space access**: Shared mobility can impact how the curb is used through a variety of modes and uses that can vary by block and times of day with an array of operational impacts on mobility, active transportation, and public transit operations.
- **Zoning and parking:** Shared mobility can affect land use-related planning factors such as zoning, land-use densities, parking demand, and parking minimums.
- **Shared mobility and the planning process:** Incorporating shared mobility into planning processes can be important for long-term infrastructure and capital planning and to help public agencies achieve a range of climate, environmental, and congestion mitigation goals.
- **Stakeholder and community engagement:** Both stakeholder and community engagement can be key in a number of aspects of planning for shared mobility such as converting general-purpose parking for shared mobility or the development of a planning document (e.g., a General Plan).

[1]Transportation Sustainability Research Center, University of California, Berkeley, CA, USA
[2]Department of Civil and Environmental Engineering, and Transportation Sustainability Research Center, University of California, Berkeley, CA, USA

In this chapter, we provide an overview of the ways in which shared mobility can impact planning and policymaking. First, in Section 5.1, we review shared mobility policies related to the curb and public rights-of-way management. In Section 5.2, the relationship between shared mobility and developer zoning regulations is discussed. In Section 5.3, we explore shared mobility and the planning process. In Section 5.4, we review the role of shared mobility and the built environment, including examples of how shared mobility can be employed in the suburban context. In Section 5.5, we discuss the importance of stakeholder and public involvement in the shared mobility planning process. Finally, in Section 5.6, we conclude the chapter with planning considerations for cities and public agencies in an automated future.

5.1 Shared mobility and the public rights-of-way

Shared mobility has the potential to offer numerous individual and community benefits such as increasing mobility, fostering environmental awareness, reducing greenhouse gas (GHG) emissions (discussed in Chapter 17), enhancing multimodal connectivity, and encouraging active transportation modes. Shared mobility can also have numerous impacts, some positive and negative, on public rights-of-way. Rights-of-way is a term used to describe the legal passage of people (and their means of transportation) along public and sometimes private property (the latter typically through licenses and easements) [1]. Rights-of-way often encompass most ground transportation facilities including streets, bicycle lanes, and sidewalks. A number of local governments and public agencies have developed a combination of formal and informal policies to dedicate public rights-of-way for shared mobility uses, such as curb space and parking. Common examples include (1) carsharing parking; (2) station-based bikesharing kiosks; (3) dedicated locations for the pickup and drop-off of a variety of dockless modes (e.g., scooters and bicycles); (4) loading zones for microtransit and transportation network companies (TNCs, also known as ridehailing and ridesourcing); and (5) policies for last-mile delivery (e.g., courier network services) and delivery robots/personal delivery devices [1]. Many of these policies cover a number of overarching issues such as:

- Defining a shared mode;
- Deciding if and how curb space—such as loading zones (for taxis and employer shuttles), parking, first responder access (e.g., for access to fire hydrants), accessibility (e.g., accessible curb ramps and drop-off points), and other rights-of-way—should be allocated;
- Deciding if there should be a different policy between for-profit and nonprofit mobility operators;
- Managing demand among multiple mobility operators for public rights-of-way;
- Determining the monetary value of the rights-of-way; and
- Addressing administrative issues such as permits, snow removal, curb and street cleaning, parking enforcement, and signage.

The most common policies involve the allocation and management of rights-of-way for shared mobility. We examine the role of local governments and public agencies in allocating rights-of-way in more detail in Section 5.1.1.

5.1.1   *Allocating rights-of-way and curb space*

The dedication of curb space and public rights-of-way for shared mobility is typically implemented through a combination of formal and informal processes. Once a local government or public agency has decided to dedicate public rights-of-way for a shared mode, many find themselves having to address eight specific issues:

1. Developing a process for allocating public space to shared mobility operators and balancing a variety of requests for curb space (e.g., shared mobility, outdoor recreation, dining, etc.);
2. Deciding if there should be limits on the amount of public space allocated (e.g., a specified amount of curb feet, number of parking spaces, and square footage);
3. Determining if fees or permits should be assessed for private use of the rights-of-way (and if so, how these costs will be determined and assessed);
4. Deciding if special signage and markings should be permitted to identify areas, such as special parking spaces and loading zones and who will be responsible for their installation and maintenance;
5. Identifying enforcement mechanisms to prohibit unauthorized activities (e.g., ticketing, booting, and impounding);
6. Ensuring public involvement and compliance with environmental justice principles (discussed in Chapter 6);
7. Preventing the block of Americans with Disability Act (ADA) access with respect to curb space management; and
8. Documenting the social, environmental, and transportation impacts (discussed in Chapter 17).

Each of these considerations is described in greater detail in Table 5.1.

5.2   Shared mobility and the development process

In addition to allocating rights-of-way and curb space within the public domain, public agencies also can encourage shared mobility in private developments through an array of policies aimed at easing zoning regulations and parking minimums to promote the inclusion of shared mobility in new developments [1]. Commonly referred to as incentive zoning, these policies typically include provisions that either directly reduce a required number of parking spaces in a new development or provide developers density bonuses in exchange for incorporating shared mobility into a new development. These incentives can be administered in a number of ways such as:

- **Zoning:** Institutionalizing parking reductions and/or density bonuses for the inclusion of shared mobility into local land-use laws and ordinances universally across an entire land use or zone type;

Table 5.1   Considerations when allocating public rights-of-way and dedicating curb space for shared mobility

Policy consideration	Description/example (if available)
Processes for allocating space	This can include a combination of formal and informal processes and may vary by mode. Formal processes can include using policies that are written, codified by local ordinances and zoning provisions, or negotiated through a formal request for proposal (RFP) process. Informal processes can include variances, special permits, and case-by-case approvals from either administrative staff or an elected council. **Example:** The Washington Metropolitan Area Transit Authority (WMATA) uses a RFP process to allocate carsharing parking
Limits on space allocated	Limited space, numerous multimodal needs, and competition among operators may cause public agencies to limit the amount of rights-of-way space dedicated to shared modes, specific operators, or both. Methods for allocating space can include limits on the number of parking spaces, linear feet or curb space, or vehicle permits, for example. Agencies with multiple shared modes vying for on-street curb space (e.g., carsharing, TNCs, micro-transit, bikesharing, and scooter sharing) may want to consider a comprehensive approach that allocates on-street space based on the ridership productivity and/or environmental impacts of each mode. In other cases, a city may cap the number of vehicles or equipment permitted rather than limit the amount of curb space or rights-of-way dedicated to a shared mode. **Example:** In August 2018, New York City capped the number TNC vehicles for a 12-month period [2]
Fees for use	Public agencies provide rights-of-way to shared mobility operators free of charge or at a reduced cost or charge market rates. **Example:** A few common methods for charging fees for use of the rights-of-way include (1) residential parking permit costs; (2) foregone meter revenue; (3) costs of providing parking (e.g., operations, administrative costs, overhead, and maintenance); and (4) the market rate for private or public off-street parking in a given parking district or municipal jurisdiction [1]. In San Francisco, the Municipal Transportation Agency (SFMTA) has developed a program to enable employer shuttle services to pay for use of loading zones, if certain guidelines are followed such as yielding to public buses and pulling to the front of the loading zone to make room for other vehicles [3]

(Continues)

Table 5.1   (Continued)

Policy consideration	Description/example (if available)
Painting, signage, and maintenance	Allowing shared mobility operators to paint markings on the ground for where vehicles and shared micromobility (e.g., bikesharing and scooter sharing) should operate or to place special signage to highlight their services can expand the use of alternative modes and minimize operational conflicts. Some public agencies may regulate the painting or signage to conform to local requirements (e.g., size, color, and material) and negotiate maintenance requirements (e.g., installation, graffiti removal, etc.). **Example:** In South Bend, Indiana, the city has established parking zones for dockless bicycles after the city received complaints from residents that bicycles were being "carelessly" parked throughout the jurisdiction. In response, the city has designated recommended parking zones labeled by epoxy paint on the curb. Although not required at present, users are being asked to park in the recommended areas [4]
Enforcement mechanisms	Enforcement is important to ensure that shared fleets are being parked correctly and to ensure other modes are not illegally parking in locations designated for shared fleets. **Example:** A few common methods for enforcement can include tickets, towing/impounding, and boots. For example, in Fall 2018, the City of Homewood, Alabama began impounding Bird scooters that were illegally placed on sidewalks and for doing business in the city without a business license. Fines and court costs are estimated at $371 per scooter. In nearby Birmingham, Alabama, the city has issued a cease and desist letter to scooter sharing operators that are operating a business on the public rights-of-way without permission, punishable by a fine of up to $500 and/or 180 days in jail for each violation and each day a violation continues [5]
Public involvement	Working with local neighborhoods and community groups can be a requirement of public agencies for service providers seeking to allocate rights-of-way and curb space. **Example:** In 2013, New York City and Citi Bike initiated over 400 public meetings related to the development of the bikesharing program and provided a website for public input, which received more than 10,000 suggestions about station locations and 55,000 responses in support of station proposals [1]
Ensuring social equity access for underserved communities	Ensuring equitable access for underserved communities, people with disabilities, and unbanked

(Continues)

Table 5.1   (Continued)

Policy consideration	Description/example (if available)
	and underbanked users is critical. There have been a variety of efforts to guarantee services for these communities as well as ensuring that shared mobility does not block or impede access to or use of the transportation network. **Examples:** The San Francisco Municipal Transportation Agency (SFMTA) requires at least 20% of station-based bikesharing kiosks to be located in low-income communities to promote equitable service distribution [6]. In Santa Monica, the city regularly tickets improperly parked e-scooters that block ADA ramps and curb access [7]
Impact studies	Public agencies may require shared mobility operators to conduct impact studies documenting the transportation, social, and environmental effects of a system when considering the allocation of public rights-of-way. These studies can take place at the time of the initial application during regular intervals after the rights-of-way have been granted or both [1]

- **Overlay zoning:** Providing parking reductions and/or density bonuses for the inclusion of shared mobility into specific areas known as "overlay zones." Overlay zones provide an additional layer of zoning standards applied over part of a zoning district or multiple zoning districts;
- **Conditional use permits:** Permitting land uses under specifically approved conditions. Conditional use permits may be appropriate where zoning may otherwise prohibit shared mobility from legal operations. For example, a zoning ordinance may prohibit commercial activities in a residential zone, but a conditional-use permit could provide specific exceptions for shared mobility services, such carsharing, to operate in residential neighborhoods;
- **Variances:** Special permission granted to a land owner on a case-by-case basis; and
- **Discretionary review:** A process that permits local officials and planning commissions (or other designated bodies) to review specific development proposals and attach detailed conditions or deny approval.

5.2.1   Parking reductions and substitution

Due to the potential of shared mobility to support transportation demand management goals, a number of local jurisdictions have allowed parking reductions and parking substitutions for the inclusion of shared mobility into development projects. Parking reductions involve reducing the number of required parking spaces in a new development, and parking substitution entails substituting general-use parking for shared modes, such as carsharing parking, bikesharing kiosks, and other shared modes.

Parking reduction is a policy that can be employed in urban areas with high housing or parking construction costs. This can help reduce housing costs (by reducing per-unit construction expenses) and can encourage neighborhood revitalization by making it easier for projects to be financially viable through higher capitalization rates on real estate projects. Unlike parking reduction, which is a policy that focuses on new construction, parking substitution can be applied to both new and existing developments. Substituting private vehicle parking for shared modes can contribute to an overall network effect for mobility on demand (MOD) by enhancing access to a network of alternative modes across an urban area.

In some cities with comprehensive public transit services, this is taken further. Parking requirements are eliminated for some uses, and parking maximums are designated for others. For example, within the commercial core or "C-3" zone of San Francisco, there are zero parking requirements for commercial uses and maximums of 0.25 spaces/unit for residential. This can be increased to 0.5 spaces/unit with a conditional use permit. In addition, for the sale of residential units, any related parking must be "unbundled," such that the sale of the unit is separate from the sale of the parking space, thus exposing one of the formerly "hidden" costs of private vehicle ownership. When employed in urban areas with robust public transit access, these strategies can further support a modal shift away from private vehicle use.

5.2.2   Density bonuses

In addition to parking policies, local governments can also allow density bonuses for the inclusion of shared mobility into a new development project. Policies that allow increased density can include: (1) greater floor-to-site area ratios, (2) more dwelling units permitted per acre, and (3) greater height allowances [1]. Policies that provide density bonuses for developers leverage a similar principle of aiming to make development more lucrative for developers and real estate investors. Rather than reducing per-unit or overall project costs, these policies increase the overall cash flow of development projects. Allowing density bonuses for the inclusion of shared mobility can be an attractive strategy for cities wanting to increase overall urban density, residential density, or both. This policy can be particularly well suited for encouraging brownfield redevelopment because these parcels are often more expensive to repurpose due to the costs commonly associated with environmental remediation (thereby necessitating additional project incentives to encourage redevelopment) [1].

5.2.3   Incentive zoning in practice

In practice, incentive zoning is typically incorporated into local municipal codes. In Seattle, for example, the city's municipal code allows for a reduction of up to 5% of a development project's required total parking spaces with the inclusion of a city-recognized carsharing program. Seattle's ordinance reduces the number of required spaces by one space for every parking space leased by a carsharing program. For developments requiring 20 or more parking spaces that also provide carsharing parking, the number of required spaces may be reduced by the lesser of three required parking spaces for each carsharing space or 15% of the total number of required spaces (Seattle Municipal Code § 23.54.020). To qualify for the latter provision, there must be an agreement between the property owner and the carsharing operator filed and approved by the city and recorded with the deed.

In Indianapolis, the city adopted a revised zoning and subdivision ordinance in 2016 that offers developers up to a 35% cumulative parking reduction for the inclusion of transportation demand management provisions, such as:

- **Shared vehicle, carpool, or vanpool spaces:** The minimum number of required off-street parking spaces may be reduced by four for each shared vehicle, carpool, or vanpool space provided. Each shared space counts toward the minimum number of required parking spaces.
- **Electric-vehicle charging stations:** The minimum off-street parking requirement may be reduced by two parking spaces for each electric vehicle charging station provided. Each charging station counts toward the minimum number of required parking spaces. Although this provision does not specifically apply to shared mobility, it may be applicable for shared fleets, such as electric carsharing charging stations.
- **Bicycle parking:** For every five bicycle parking spaces provided in excess of the required bicycle parking spaces (or where no bicycle parking is required), the minimum number of required off-street parking spaces may be reduced by one or up to a maximum of five. Although this provision does not explicitly mention shared mobility, it may be applicable for bikesharing parking, such as a designated pick-up/drop-off zone or kiosk.
- **Proximity to public transportation:** The minimum number of off-street parking spaces required for any development may be reduced by 30%, if the developer builds within a quarter mile of a sheltered public transit stop or public transit corridor. The minimum number of off-street parking spaces required may be reduced by 10%, if the development is between a quarter mile and a half mile of a stop or public transit corridor.

Finally, Vancouver, Washington offers reduced transportation impact fees and residential density bonuses for the inclusion of shared mobility and alternative transportation in the city's public transit overlay district. Impact fee reductions are granted on a percentage basis for each alternative transportation measure employed (e.g., 1% reduction for the inclusion of carpool/vanpool parking and 1% reduction for connecting a project to an existing or future regional bike trail).

5.3   Shared mobility and the planning process

Incorporating shared mobility into the planning process can help public agencies enhance access and mobility, and establish goals and policies to guide future growth and infrastructure development. By incorporating shared mobility into planning processes, public agencies can document the role of shared mobility and its impacts on travel behavior and evolve transportation forecasts and models [1]. Additionally, public agencies can leverage understanding of the positive social and environmental impacts of shared mobility to (1) increase infrastructure efficiency, (2) mitigate congestion and air pollution, and (3) incorporate shared mobility into future planning and policy-decision activities. At the local level, incorporating shared mobility into planning processes can include a range of plans, such as comprehensive, community, and specific plans.

- **Comprehensive planning** (also known as General or Master Plans) establishes a set of long-term goals and policies that communities use to guide development decisions. These plans typically establish a high-level policy vision and often focus on key elements including transportation, land use, housing, conservation and climate, open space, noise, and public safety. Within the context of transportation or innovative mobility sections, comprehensive plans can catalog the growing number of shared mobility options within a community.
- **Community planning** (also known as a Subarea Plan) focuses on a smaller area, such as a neighborhood. The purpose of community plans is to concentrate on more specific issues and allow public agencies to identify the locations, service availability, and coverage gaps of shared mobility services within particular neighborhoods.
- **Specific planning** (also known as Functional Plans) is the most detailed and used to implement particular planning provisions. Specific plans generally include special development standards that apply to limited geographical areas. In some cases, specific plans can be used in lieu of zoning ordinances. Specific plans are a good way for public agencies to illustrate how shared mobility can be deployed at specific sites and support urban design that connects people and places through a cohesive mobility vision.

Figure 5.1 includes examples of how different jurisdictions are incorporating shared mobility into these three levels of planning.

As exemplified in these examples, shared mobility can help public agencies achieve an array of short- and long-term goals. In a comprehensive plan, shared mobility can support smart growth strategies that (1) encourage densification and infill development, (2) provide transportation choices that bridge first-and-last mile gaps, and (3) enhance overall mobility. At a micro level, incorporating shared mobility into specific, functional, and site plans can encourage tactical mobility improvements, such as incorporating shared micromobility into a transit-oriented development site plan. At a more macro level, incorporating shared mobility into the planning process can help communities reimagine automobile-centered "edge cities" or suburbs by providing innovative and financially sustainable mobility options to complement traditional public transportation.

5.4 Shared mobility and the built environment

While shared mobility is most often associated with dense urban areas, there are many possible applications for shared mobility to serve suburbs and edge cities. In

Figure 5.1   Examples of how cities are incorporating shared mobility into planning processes

---

Seattle's comprehensive plan

In Seattle, Washington, the city updated its comprehensive plan establishing key policy goals pertaining to shared mobility. The first policy goal is designating space in the public rights-of-way to accommodate multiple travel modes, including public transit, freight movement, pedestrians, bicycles, general-purpose traffic, and shared transportation options. The second policy involves: allocating rights-of-way, sharing space among modes and uses, prioritizing space for shared and shorter duration uses, and implementing parking and demand management strategies to reduce conflicts and maximize the efficiency of the existing rights-of-way. Finally, the plan encourages the development of programs and facilities, such as bikesharing to encourage that short trips are made by walking or cycling.

Rohnert Park's community plan

Rohnert Park, California incorporated shared mobility into its Priority Development Area (PDA) plan goals, policies, and implementation actions. The plan encourages carsharing and bikesharing programs within the PDA through partnership with carsharing and bikesharing service providers. It commits to studying the feasibility of implementing carsharing and bikesharing services at the city's SMART rail station and city center.

Santa Monica's downtown community plan

In Santa Monica, California, the city has developed a downtown community plan that includes shared mobility policy goals of "[promoting] the development of uses and facilities that enable and encourage mobility by alternative modes to the automobile; these include businesses for sale, service, rental, and sharing, of bicycles, as well as rideshare, flex vehicle leasing and rental services." The plan also identifies the following action item: "Create curb space for new mobility modes as part of a coordinated approach such as bike corrals, ride sharing and ride hailing, car share, and shuttles."

Seattle's transportation strategic plan

Seattle's transportation strategic plan (TSP) encourages support for the city's carsharing operators and encourages continued municipal support through on-street and off-street parking, incentives for new development to offer carsharing parking in buildings, and promoting education and outreach of carsharing.

North America, the first shared mobility initiatives (carsharing and bikesharing) launched in 1994. Initial deployments of shared mobility emphasized walkable, high-density, mixed-use urban locations. Over the past decade, shared mobility has expanded to an increasing array of built environments (Figure 5.2), such as

- **City center:** A development framework with the highest concentration of jobs comprised of central business districts and surrounding neighborhoods;
- **Suburban:** A built environment characterized by high levels of low-density residential uses with fewer jobs than residences;
- **Edge city:** An urbanization pattern presenting some features of city center employment mixed with suburban form. Edge cities tend to have large concentrations of office and retail space often paired with multifamily residences;
- **Exurban:** Low-density residential development within the commute shed of a larger and denser urbanized area; and

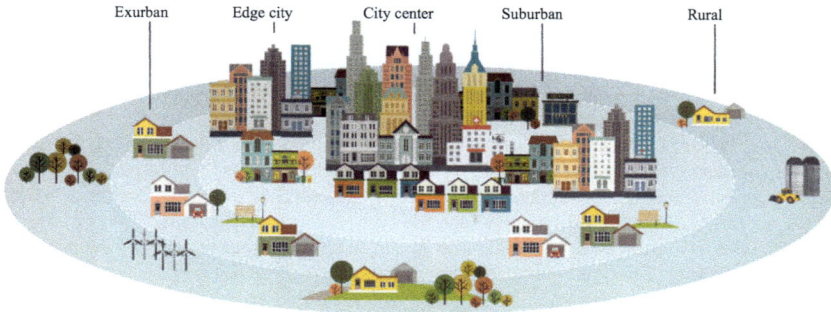

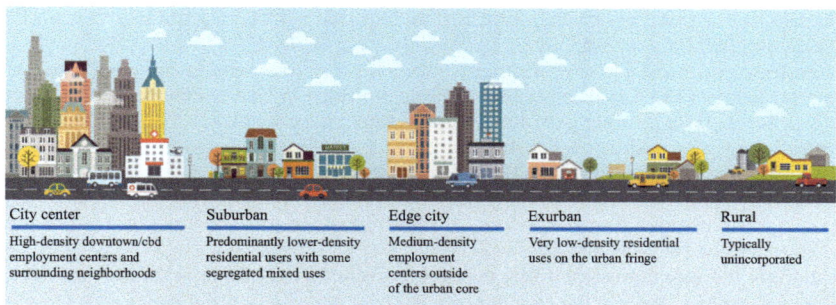

Figure 5.2   Five common development typologies for shared mobility applications. Adapted from [8].

- **Rural:** The lowest density development pattern characterized by low-density light industrial, agricultural, and other resource-based employment.

Understanding the role of the built environment can help public agencies plan for shared mobility and leverage incentives, such as incentive zoning and the allocation of public rights-of-way, to encourage the expansion of shared mobility into less dense urban environments. Figure 5.3 includes some examples of how shared mobility can be applied in the suburban context.

5.5   Stakeholder and public involvement

Public and stakeholder engagement can be key to ensuring that shared mobility meets the needs of its intended users and use cases. Additionally, from a planning perspective, public and stakeholder involvement not only helps to ensure a fair and equitable planning process, but this can also reduce opposition, provide public agencies and mobility operators with valuable information on community and stakeholder concerns, reduce conflict among stakeholders, and help jurisdictions comply with public-agency environmental justice requirements [1]. Thus, public and stakeholder engagement can provide both the opportunity to participate in

Figure 5.3   Examples of how shared mobility may be employed in the suburban context

---

First- and last-mile connections to public transportation
Shared micromobility (e.g., bikesharing and scooter sharing) may also be able to facilitate connections between public transportation hubs (typically a rail station) and adjacent employment or residential centers (e.g., a transit-oriented or other development within one to three miles of the mobility hub). One study found that station-based bikesharing in the Twin Cities was increasing rail ridership in the outer suburban regions of the metro where trips were longer. This study suggests that in locations where public transportation is sparse, bikesharing may complement public transit and facilitate access to-and-from existing rail lines [9]. The EasyConnect pilot deployed electric bicycles, nonmotorized bicycles, Segway HTs, eLockers, smart parking, and an online web portal for reservations, payment, and access to a shared mobility hub located at the Pleasant Hill Bay Area Rapid Transit (BART) station in the East San Francisco Bay Area [10]. This field test provided early understanding and lessons learned for how shared micromobility could be employed at suburban rail mobility hubs. The EasyConnect pilot results indicated higher vehicle demand by day users (e.g., lunch, business meetings, and errands) rather than commuters. The availability of low-speed modes for mid-day activities, however, may have contributed to greater public transit use and carpooling—as individuals no longer needed to rely upon a private vehicle for mid-day tripmaking. Although this study did not account for modal shifts, the results suggest net benefits for commuters and day users at the employment site due to reduced vehicle travel and increased physical activity [10].
In addition to shared micromobility, public transit-based carsharing vehicles (or station cars) can be used to travel between suburban public transportation hubs and job centers (e.g., office parks), retail centers (e.g., shopping malls and power centers), and private residences. During commute hours, residents in bedroom communities typically drive shared vehicles to a mobility hub (typically rail transit), park, and commute by public transit. Inbound (typically reverse) commuters pick up carsharing vehicles and drive them to suburban employment centers. In the evening, the cycle is reversed. At the end of each weekday, home-based commuters take the carsharing vehicles home, having dedicated access to the vehicles on evenings and weekends.
Two pilots, CarLink I and CarLink II, demonstrated the first- and last-mile concept at two separate locations in the San Francisco Bay Area. CarLink I, which was based at the Dublin/Pleasanton BART station in the East Bay, provided employees at Lawrence Livermore National Laboratory access to 12 compressed natural gas Honda Civics to drive between the BART station and their work location and use during the day. Other members could pick up one of the shared cars at the BART station at the end of the day and use it during the evenings and weekends. CarLink I resulted in a net commute reduction of approximately 20 vehicle miles per commuter per day [11,12]. CarLink II introduced 18 ultra-low emission vehicles and included 100 members serving commuters using the California Avenue Caltrain station in Palo Alto and employees working for companies at or near the Stanford Research Park. CarLink II resulted in an average daily reduction of 23 vehicle miles, reduced single occupant vehicle modal share by 23%, and reduced vehicle ownership of nearly 6% [13]. In the future, shared automated vehicles (SAVs) and automated shuttles may be able to serve this use case (i.e., first-mile and last-mile connectivity and vehicle access in suburban locations for day trips, evenings, and weekends).
Planned unit developments
Shared mobility can be used in planned unit developments (PUDs). These developments are typically built by one or more developers comprised of a mixture of land uses and densities. Sometimes PUDs are designed to be self-sufficient, striking a balance between residential, office, and retail; however, more often than not they are biased toward one use

Figure 5.3   (Continued)

with the other uses serving as amenities. Although mixed-use, in general, these developments are auto-centric with most uses more than a five- to 10-min walk away, and street designs that often discourage walking and/or limit efficient public transit services (e.g., cul-de-sacs, circles, etc.). If biased toward a specific use, they will still generate commuter traffic, either outbound for residential or inbound for office.

Carsharing, for example, can provide enhanced mobility and connectivity within the geographic confines of the PUD. One example of this is the application of electric vehicle (EV) carsharing in the senior adult community of Rossmoor in Walnut Creek, California. Rossmoor is a gated community of approximately 9,500 senior residents, spanning 72.3 km^2 acres with 6,700 residential units in three cooperatives, 12 condominiums, and one single-family-detached home development consisting of 63 individual homes. The cost per unit ranges from $100,000 for cooperatives to $500,000 for condominiums, and it can be over $1 million for single-family homes. The community offers amenities that cater to the "active adult" including hobby shops, 200 clubs, golf courses, tennis courts, hiking trails, open space, and a fitness complex. At least one resident in each household is required to be 55 years of age or older to live at Rossmoor. As of 2017, 34% of Rossmoor's population was between the age of 75 and 84 years of age, while 25% is 85 years of age and older. Rossmoor provides its own transportation services paid for by a monthly fee that ranges between $550 and $800, depending on housing type. A total of 443 Rossmoor residents participated in a survey that ran from December 2009 through May 2011 to assess potential interest in an EV carsharing program. The survey results indicate that 30% of all respondents were interested in participating in an EV carsharing program, although this has not yet been implemented at this location [14].

Closed campus mobility

Shared mobility can also provide mobility in closed-campus applications, such as suburban business parks and college campuses [13,15]. For example, a 2015 study of Zipcar in the college/university market found that a lower percentage of suburban university carsharing users sell a private vehicle (compared to urban campuses) due to land-use constraints and fewer mobility options. As a result of suburban carsharing, 36% of suburban university carsharing users reported using public transit less, and 7% use it more often. Seventeen percent of suburban university carsharing users indicated that they would not have taken their last trip had carsharing been unavailable. This compares to 15% who would have borrowed a car or gotten a ride from a family member or a close acquaintance to complete their last trip or 21% who would have used public transportation. Seventy percent of respondents said that carsharing improved their quality of life, e.g., more varied experiences, greater accessibility and flexibility, and the ability and enhanced convenience to go out of town, run errands, and making retail trips [15]. Other forms of shared mobility, such as microtransit, shuttles and bikesharing, can provide mobility to campus communities, covering distances that may not be easily walkable.

Microtransit and shuttles can be used to connect passengers to public transit, employment centers, and retail or entertainment destinations. These modes also can provide mobility options within office parks (and other closed campus environments). These types of services can be particularly well suited to relatively dense but geographically separated areas (e.g., connecting public transit (rail) stops, office parks, retail destinations, and multifamily residential). A study of employer shuttles in the San Francisco Bay Area found that employer shuttles reinforce a jobs-housing imbalance by enabling individuals to live farther from work; however, more research is needed on the impacts of shuttles in the suburban context.

(Continues)

Figure 5.3   (Continued)

Public transit replacement

TNCs may be well suited for areas where there is insufficient demand for fixed-route public transit services. A number of public agencies have partnered with TNCs, such as Lyft and Uber, to facilitate first- and last-mile connections or replace underperforming or inefficient public transit services. For example, in August 2016, the Livermore Amador Valley Transit Authority in the San Francisco East Bay Area launched a 1-year pilot program to subsidize Lyft, Uber, and taxi fares within two pilot areas in Dublin, California [16]. The public transit agency eliminated one underperforming bus route that was attracting an average of five riders per hour at a subsidy of $15 per rider. Similar pilots have been tested elsewhere around the United States, including a pilot in Centennial, Colorado to provide free Lyft rides to and from the community's light rail station [17]. This pilot was credited with reducing first- and last-mile costs per trip from an average of $21 to an average of $4.70 [18]. Similarly, the City of Monrovia, California offers subsidized TNC rides to fixed-route public transportation and the city's Old Town; both areas have high levels of peak congestion and parking shortages [19].

---

decisions about shared mobility planning and implementation that affect them. A few examples of how stakeholder and public involvement can be incorporated include:

- Providing business owners and residents of a particular parcel or block the ability to provide public comment on proposed changes to on-street rights-of-way;
- Soliciting public and stakeholder input in the design and implementation of developer incentives for the inclusion of shared mobility (e.g., reduced parking minimums in exchange for the inclusion of shared mobility in transportation demand management initiatives); and
- Encouraging civic participation on policy issues related to shared mobility.

Typically, the nature and format of the stakeholder and public engagement process varies by agency and local government according to policy and custom. For example, Washington, DC typically solicits feedback on shared mobility through neighborhood councils. Other cities and public agencies may use public hearings, town hall meetings, and staff or administrative review processes. As a general rule, a collaborative process is often a best practice because it can reduce conflict (and litigation) among stakeholders, while simultaneously advancing shared goals [1].

5.6   Conclusion

Shared mobility is an innovative transportation strategy that is continually evolving and reshaping mobility. Shared mobility has the potential to help planners and policymakers achieve a variety of policy goals, such as encouraging active and shared transportation modes. As such, shared mobility has the potential to support multimodality, improve first- and last-mile access, increase transportation options in lower-density built environments, and enhance mobility for populations with specific needs. While the longer-range impacts of the global pandemic are uncertain, rights-of-way policies and incentive zoning that supports the inclusion of shared and personal micromobility could be key in helping reduce VMT/VKT and GHG emissions, particularly if it becomes difficult to rebuild the mode share of public transportation and other pooled services. As automation becomes more mainstream, shared mobility will continue to have a transformative impact on transportation access and options.

How public agencies manage curb space and the rights-of-way will continue to be at the forefront of urban mobility policy. Policymakers will have to develop policies that fairly manage the demand for access to rights-of-way (e.g., outdoor dining and other commercial curb use, loading zones for private shuttles, taxis, paratransit, microtransit, TNCs, delivery services, and public transportation; parking for carsharing and SAVs and infrastructure for shared micromobility, including bikesharing and scooter sharing).

The convergence of mobility services, shared modes, electrification, and automation will not only transform how people travel but also how streets are designed and the ways in which land uses are planned, zoned, and developed. While the impacts of automation and other emerging technologies on auto ownership, parking, and travel behavior remain to be seen, planning for shared and automated mobility can minimize disruption, encourage shared use, and guide sustainable transportation outcomes. As electrification and automation become more mainstream, policymakers will need to rethink traditional notions of access, mobility, and automobility and reimagine the public rights-of-way for SAVs, automated delivery, and electric micromobility. In an automated future, cities may be able to repurpose on-street parking for other uses (such as wider sidewalks, bicycle lanes, and loading zones for SAVs) and replace some parking with infill development, such as housing. Today's planning and policies can lay the longer-term groundwork for reconsidering parking minimums, replacing overbuilt parking, building complete streets, and adding infill development. Planning for this transformation today will be critical to preparing for this socio-technological change and to maximize public benefits.

Author contributions

The authors confirm contribution to this chapter. All authors reviewed and approved the manuscript.

Declaration of conflicting interests

The authors declared no potential conflict of interest with respect to publication of this chapter.

References

[1]   Cohen, A., and Shaheen, S. *Planning for shared mobility*. Chicago, IL: American Planning Association; 2016. pp. 1–110.

[2]   Shapiro, A. *New York city just voted to cap Uber and Lyft vehicles, and that could make rides more expensive*. 2018. Available from https://www.cnbc.com/2018/08/08/new-york-city-votes-to-cap-uber-and-lyft-vehicles.html [Accessed 23 November 2018].

[3]   San Francisco Municipal Transportation Agency (SFMTA). *Commuter shuttle program*. 2016. Available from https://www.sfmta.com/sites/default/files/projects/2016/Commuter%20Shuttle%20Program%20Mid%20Term%20Status%20Report.pdf [Accessed 23 November 2018].

[4]   Parrott, J. *LimeBike, South Bend ask users to park bikes in certain zones*. 2018. Available from https://www.southbendtribune.com/news/local/lime-bike-south-bend-ask-users-to-park-bikes-in-certain/article_6940234c-4e83-58db-9e27-3d044fdddc37.html [Accessed 30 May 2021].

[5]   Edgemon, E. *Homewood impounds scooters: Birmingham asks Bird to leave*. 2018. Available from https://www.al.com/news/birmingham/index.ssf/2018/09/homewood_impounds_scooters_bir.html [Accessed 23 November 2018].

[6]   San Francisco Municipal Transportation Agency (SFMTA). Bikeshare. 2018. Available from https://www.sfmta.com/getting-around/bike/bike-share [Accessed 23 November 2018].

[7]   Sisson, P. *As scooters multiply, Santa Monica plans regulations to address community concerns*. 2018. Available from https://la.curbed.com/2018/6/7/17438168/santa-monica-electric-scooters-bird-regulations-lime [Accessed 23 November 2018].

[8]   Shaheen, S., Cohen, A., Yelchuru, B., and Sarkhili, S. *Mobility on demand operational concept report*. Washington, D.C.: U.S. Department of Transportation; 2017. pp. 1–164.

[9]   Martin, E., and Shaheen, S. "Evaluating public transit modal shift dynamics in response to bikesharing: A tale of two U.S. cities." *Journal of Transport Geography*. 2014; 41: 315–324.

[10] City of Monrovia. *GoMonrovia*. 2019. Available from https://www.cityofmonrovia.org/your-government/public-works/transportation/gomonrovia [Accessed 7 October 2019].

[11] Shaheen, S., and Wright, J. "The CarLink II pilot program: Testing a commuter-based carsharing model." *Proceedings of IEEE Intelligent Transportation Systems (Cat. No.01TH8585)*. 2001; pp. 1067–1072.

[12] Shaheen, S. *Dynamics in behavioral adaptation to a transportation innovation: A case study of CarLink–A smart carsharing system*. Berkeley, CA: University of California; 1999. pp. 1–472.

[13] Shaheen, S., and Novick, L. "Framework for testing innovative transportation solutions: Case study of CarLink, a commuter carsharing program."

Transportation Research Record: Journal of the Transportation Research Board. 2005;1927(1): 149–157.

[14] Shaheen, S., Cano, L., and Camel, M. "Exploring electric vehicle carsharing as a mobility option for older adults: A case study of a senior adult community in the San Francisco Bay Area." *International Journal of Sustainable Transportation.* 2013;10(5): 406–417.

[15] Stocker, A., Lazarus, J., Becker, S., and Shaheen, S. *North American College/University market carsharing impacts: Results from Zipcar's College travel study.* Berkeley, CA: University of California; 2016. pp. 1–44.

[16] Cuff, D. *Dublin: Uber, Lyft to partner in public transit.* 2016. Available from https://www.eastbaytimes.com/2016/08/18/dublin-uber-lyft-to-partner-in-public-transit/ [Accessed 23 November 2018].

[17] Aguilar, J. *Centennial teams up with Lyft for free rides to light rail station.* 2016. Available from https://www.denverpost.com/2016/08/15/lyft-centennial-team-up-for-free-rides-light-rail-station/ [Accessed 23 November 2018].

[18] Centennial Innovation Team and Fehr & Peers. *Go Centennial final report.* Centennial: City of Centennial; 2017. pp. 1–110.

[19] Rodier, C., Shaheen, S., Blake, T., Lidicker, J., and Martin, E. *EasyConnect II: Integrating transportation, information, and energy technologies at the Pleasant Hill BART Transit Oriented Development.* Berkeley, CA: University of California; 2010. pp. 1–123.

The views expressed are those of the authors.

Chapter 6

Shared mobility services: prioritizing social good

Susan A. Shaheen[1] and Adam P. Cohen[2]

Shared mobility—the shared use of a vehicle, bicycle, or other mode—has the potential to help overcome a variety of transportation equity challenges. Shared mobility offers opportunities to bridge social equity gaps, including increasing mobility for users who are unable to access private vehicles and reducing transportation expenditures [1]. However, there are also a number of social equity challenges that may inhibit use by a variety of travelers.

In 2018, we identified six common equity challenges [2]:

1. *Affordability*: "It's too expensive."
2. *Predictability*: "Will dynamic or surge pricing make it too expensive?"
3. *Availability*: "The services aren't available in my neighborhood."
4. *Payability*: "I don't have an acceptable payment method."
5. *Accessibility*: "The service isn't accessible for my medical condition."
6. *Techno-ability*: "I don't have a smartphone or a data plan."

While shared mobility is mainstreaming, the demographics of shared mobility users often differ from the general population. Users tend to be younger, have higher levels of educational attainment, higher incomes, and are less diverse than the general population. Additionally, access to the Internet, smartphones, and banking services—a prerequisite for many shared mobility modes—tend to be lower among older adults, low-income individuals, rural households, and minority communities [1]. This chapter is organized into five sections. In Section 6.1, we review the demographics of shared mobility users (who is and who is not using shared mobility). In Section 6.2, we discuss common social equity challenges impacting access and use of shared modes, including some policies for mitigating many of these challenges. In Section 6.3, we present the STEPS (spatial—temporal—economic—physiological—social) framework to transportation equity [1]. This framework can aid service providers and public agencies in identifying, understanding, and overcoming social equity challenges impacting transportation access. In Section 6.4, we discuss social equity and access considerations for a

[1]Department of Civil and Environmental Engineering, and Transportation Sustainability Research Center, University of California, Berkeley, CA, USA
[2]Transportation Sustainability Research Center, University of California, Berkeley, CA, USA

shared automated vehicle (SAV) future. Finally, in Section 6.5, we conclude with thoughts on the need for thoughtful planning for a multiphase transition to highly automated vehicles.

6.1   Demographics of shared mobility users

Generally, shared mobility users tend to be younger, have higher levels of educational attainment, higher incomes, and are less diverse than the general population. Older adults, low-income households, and rural and minority communities have historically been less likely to use shared mobility. Additionally, access to the Internet, smartphones, and banking services—a prerequisite for many shared modes—tends to be lower among many of these groups, which may be a contributing factor to lower usage among these communities [1]. Finally, shared mobility may not be universally accessible to people with physical and cognitive disabilities.

Table 6.1 summarizes the demographics from six of one of the authors' shared mobility impact studies: three North American studies of carsharing (roundtrip, one-way, and peer-to-peer models); a North American study of station-based bikesharing; a study of microtransit in Kansas City; and a study of transportation network companies (TNCs, also known as ridehailing and ridesourcing) in the San Francisco Bay Area. For example, a minimum of two-thirds of all respondents identified as Caucasian in four of the studies, which collected demographics on race and ethnicity. All of the studies had more than half of the respondents indicating that they had at least a 4-year college degree. Many of the studies had a high percentage of respondents who were under 35 years of age (with a majority under the age of 55 across all studies). Similarly, while the studies did have some lower-income users with annual household incomes under $35,000, a greater percentage of households had upper incomes exceeding $100,000 annually.

Other studies of shared mobility tend to echo these demographic patterns. A study of City Carshare by Cervero and Tsai [3] found that the program tended to attract younger, Caucasian, and upper-income users. Forty-three percent of City CarShare's members were between the ages of 25 and 34, compared to just 28% of the city's population (among those 20 years and older). Caucasians comprised 81% of the study population compared to approximately 50% for the entire city. Members' median annual income was $57,000, with over 90% working in white collar jobs (both above the city's average) [3].

In another example, a study of Capital Bikeshare found that 78% of its casual (short-term) users were Caucasian compared to 4% Hispanic, 5% African American, and 8% Asian. The study also found that 80% of its annual members were Caucasian with smaller percentages of annual members representing minority communities. The study also found that more than 80% of its annual members and casual users had a 4-year college degree or higher, and more than three-fourths of its annual members and casual users were between 18 and 44 years old [4].

The primary exception to these general user trends is studies documenting the demographics of carpooling. Teal found that carpooling participants were more

Table 6.1 Some shared mobility user demographics

Mode	Race/ethnicity	Income	Educational attainment	Age
Roundtrip carsharing (N. America) [8]	80%–87% Caucasian 1%–10% Hispanic/Latino 1%–5% African American	21% earned >$100K 23% earned <$40K (US)	81% had a 4-year degree or higher	35% ages under 30 31% Ages 30–40
One-way carsharing (N. America) [9]	80%–87% Caucasian 1%–10% Hispanic/Latino 1%–5% African American	35%–56% earned >$100K 7%–17% earned <$35K (US)	72%–96% had a 4-year degree or higher (across five cities)	48%–64% ages under 35 32%–41% ages 35–54
Peer-to-peer (P2P) carsharing (N. America) [10]	67% Caucasian 3% Hispanic/Latino 3% African American	30% earned >$100K 21% earned <$35K (US)	86% had a 4-year degree or higher	73% ages under 35 23% ages 35–54
Microtransit (Kansas City) [11]	89% Caucasian 6% Asian 6% African American	50% earned >$100K 6% earned <$35K	100% had a 4-year degree or higher	55% ages under 35 39% ages 35–54
Station-based (docked) bikesharing (N. America Multi-City Studies) [12]	74%–92% Caucasian 1%–5% Hispanic/Latino 1%–2% African American	29%–39% earned >$100K 9%–26% earned <$35K (US)	55%–89% had a 4-year degree or higher (two studies)	37%–54% ages under 35 36%–51% ages 35–54
TNCs (SF Bay Area) [13]		38% earned >$100K 9% earned <$30K	81% had a 4-year degree or higher	73% ages under 35 25% ages 35–54

Note: Percentages may not add up to 100%.

likely to have lower incomes and be the second worker in a household [5]. Some studies have shown that carpooling can provide job access to households with lower incomes and households with more workers than vehicles [6]. More recent data from the National Household Travel Survey and the American Community Survey show that ridesharing (i.e., carpooling and vanpooling) users tend to have lower incomes, and Hispanics and African Americans carpool more than other racial and ethnic groups. These surveys and other studies indicate that ridesharing may serve an important role in enhancing mobility in low-income, immigrant, and nonwhite communities where travelers are more likely to be unable to afford personal automobiles and obtain a driver's license [6,7]. It is important to note that more recent shared mobility sociodemographic data (e.g., shared micromobility, TNCs) appear to reflect a shift toward a more diverse user population in many locations. More information on the travel behavior and environmental impacts of these studies can be found in Chapter 17.

6.2   Common transportation equity challenges

Broadly speaking, shared mobility equity challenges typically include four areas: (1) access for persons with disabilities, (2) service for un- and under-banked households, (3) low-income affordability [2], and (4) potential algorithm bias. Each of these are explained in greater detail below.

6.2.1   *Access for persons with disabilities*

In many industrialized countries, there is a growing older adult population. An estimated one in five Americans currently lives with a disability. In the United States, by 2045, the number of Americans over the age of 65 will increase to 77% [14]. As populations across the world age, the number of persons with disabilities will continue to increase. Shared mobility has the opportunity to lower the cost and diversify the range of assisted modes to users with cognitive and physical challenges. However, rapid technology change can create unforeseen access challenges for disabled users, if specific needs are not considered [1]. The American Disability Act (ADA) is a U.S. civil rights law that prohibits discrimination based on disability. It affords similar protections against discrimination to Americans with disabilities as the U.S. Civil Rights Act of 1964, which made discrimination based on race, religion, sex, national origin, and other characteristics illegal. Additionally, ADA requires public entities (e.g., public transit providers) and businesses that provide public accommodations (e.g., transportation services) to provide reasonable accommodations (i.e., wheelchair access) or equivalent transportation services (i.e., paratransit) to people with disabilities.

Common challenges can include the lack of: (1) ADA accessible carsharing vehicles in some fleets, (2) ADA accessible TNC vehicles in some markets, and (3) ADA accessible micromobility equipment. Another concern is micromobility equipment blocking ADA curbs and wheelchair ramps.

Mandated by federal law, demand-responsive service provides support to passengers with limited mobility through dial-a-ride or paratransit services. However, these services often have limited geographic coverage, require advance scheduling, and are generally very expensive to operate on a per trip basis (compared to public transportation and other modes). Demand-responsive, shared mobility services for people with limited mobility may be one way of complying with U.S. federal requirements and potentially offering users enhanced services (e.g., reduced wait times) and at reduced cost (e.g., to the end user, public agency, or both). However, shared mobility also can raise a number of social equity concerns, particularly around the lack of demand-responsive services for passengers with limited mobility when shared modes do not offer accessible services or equivalent accessible alternatives [15].

Shaheen *et al.* [1] discuss a few strategies and examples of policies that can help overcome access challenges for persons with disabilities, such as:

- **ADA consumer protections in local, state, and federal regulations and legislation**: Consumer protections for people with disabilities at the local, state, and federal level are key, particularly with a growing number of transportation services. For example, Texas has implemented a statewide TNC law that prohibits drivers from discriminating on the basis of a passenger's or potential passenger's location or destination, race, color, national origin, religious belief or affiliation, sex, disability, or age. This law also prohibits TNC drivers from refusing to provide service to a potential passenger with a service animal. Furthermore, the law prohibits a company from imposing an additional charge for the transport of individuals with physical disabilities due to those disabilities [16].
- **Wheelchair accessible vehicles/equipment**: Providing wheelchair accessible vehicles and equipment for users is important to expanding ADA access. For example, Uber offers two programs: UberWAV and UberASSIST. UberWAV offers passengers with disabilities a dispatch service to wheelchair accessible vehicles. UberASSIST drivers receive special training by third-party organizations to help riders access and egress vehicles, as well as appropriate handling for wheelchairs, walkers, and scooters [17]. In Portland, Oregon, the city's bureau of transportation has launched an adaptive bicycling pilot project as part of its BIKETOWN Bike Share program [18].
- **Wheelchair accessible service fund**: This policy involves establishing a funding mechanism for wheelchair accessible service and equipment. In Seattle, Washington, taxis and TNC operators pay a $0.10 per ride surcharge for all rides originating in the city "to offset the higher operational costs of wheelchair accessible taxi ('WAT') services for owners and operators, including but not limited to the vehicle costs associated with purchasing and retrofitting an accessible vehicle, extra fuel and maintenance costs, and the time involved in providing wheelchair accessible trips" [19].
- **Application programming interface (API) integration with paratransit**: Paratransit services may operate more cost effectively by developing smart

dispatch systems that identify certain trips that may be better served with shared mobility. A number of public agencies are pursuing partnerships and initiatives for shared mobility service providers to participate in paratransit pilots and to build software extensions that enable shared mode dispatching. For example, Lyft developed a product that employs automatic "smart dispatch" to assign rides that are best served by paratransit and TNCs. This allows telephone dispatch operations, which negate the smartphone requirement (Emily Castor, unpublished data, 2016).

- **Policies to prohibit shared modes from blocking ADA access**: Shared modes can unintentionally block ADA accessibility when services obstruct loading zones, curbs, ramps, and other ADA infrastructure. For example, a number of cities have received complaints about dockless shared micromobility (bicycles and scooters) impeding sidewalks and wheelchair ramps [20]. Some cities and public agencies have developed policies with recommended or required parking policies for dockless bikes or scooters to prevent this. For example, Seattle has developed a policy for curbside management to guide where dockless bicycles should be parked in urban areas, which prohibits bicycles from being parked on corners, driveways, or curb ramps or in a way that blocks access to buildings, parking meters, benches, bus stops, or fire hydrants. Seattle also requires dockless bike-sharing companies to move improperly parked bicycles and correct parking violations within two hours of a problem being reported during normal business hours [21].

6.2.2   Service for un- and under-banked households and low-income affordability

Economic inclusion can represent a notable barrier for some travelers. Many shared mobility services require debit/credit cards for payment and, in some cases, collateral for vehicles or equipment. This can be a barrier for consumers who are underbanked or unbanked or households without a bank account or debit/credit card. To assess the inclusiveness of the banking system, the U.S. Federal Deposit Insurance Corporation (FDIC) conducts a biennial survey of household access to the bank system. In 2017, the FDIC found that 6.5% (approximately 8.4 million) of U.S. households were "unbanked" and did not have access to an account at a financial institution [22].

The survey also found that 18.7% (approximately 24.2 million) of U.S. households were "underbanked" meaning that a household had a checking or savings account but also obtained financial products and services outside of the banking system [22]. In addition to banking access, the survey also asked questions about access to credit card payment methods. Sixty-nine percent of households surveyed had a credit card from Visa, MasterCard, American Express, or Discover [22]. However, the use of one of these credit products was much lower among unbanked households compared to underbanked and fully banked households. The survey found that only 7.2% of unbanked households had a credit card, compared to 60% of underbanked households and 76.3% of fully banked households.

The study also found that a household's availability and use of mainstream credit products varied widely across demographic and socioeconomic groups [22]. Generally, households that are lower income, less educated, black or Hispanic, younger, older, or with a working-age disabled member(s) were less likely to use mainstream credit cards. The study also identified related racial and ethnic demographic differences that could have been contributing factors. For example, the study found that 36% of African American households and 31.5% of Hispanic households had no mainstream credit compared to 14.4% of Caucasian households. At all income levels, African American and Hispanic households were more likely not to have mainstream credit. The study concluded that racial and ethnic differences in bank account ownership and socioeconomic and demographic characteristics beyond income can account for some, but not all, of the racial and ethnic differences in the likelihood of limited or no mainstream credit.

According to the FDIC, two reasons why households may not have mainstream credit are (1) households are not interested in having credit or (2) they do not appear to be creditworthy. According to the study, approximately one in six households with no mainstream credit showed interest in having credit. The study also suggests that staying current on bill payments is one potential indicator of creditworthiness. Among households with no mainstream credit, who showed interest in it, 46.7% were "current" on their bills. This provides some insight to account for these findings [22].

Additionally, shared mobility typically employs a pay-as-you-go pricing model that can be expensive (and sometimes costlier in comparison to walking, cycling, and public transportation) according to a New York Times critique of shared mobility [23]. For example, in Helsinki, Finland, the Kutsuplus program, a municipal demand-responsive transit service in the urban core, was critiqued as having too few vehicles and too large a service area, creating higher user fees. The program had a service area of 100 square kilometers (approximately 38 square miles) with a fleet of 10 shuttles (expanding later to 15 vehicles). This resulted in higher program costs and fares with a base fare of $4.75 plus $0.60 per kilometer (about $0.97 per mile) [15,24,25].

Additionally, the use of surge or demand-responsive pricing by service providers also can create affordability challenges. Proponents of dynamic pricing argue that surge pricing can help to increase supply and manage demand. However, critics note that uncertain roundtrip costs (e.g., the risk of surge pricing on the return leg of a journey) can make these services less desirable or reliable, particularly among low- to moderate-income user groups [15].

It is important that un/underbanked and low-income communities have equitable service. This includes multiple payment methods; affordable mobility options (e.g., subsidies and discounted membership); equivalent travel modes; comparable hours and frequency of service; and similar wait times. Shaheen *et al.* [1] discuss a few strategies and examples of alternative payment options that can help overcome under- and unbanked household access to mobility options [15]. These include:

- **Cash payment**: This method allows users to pay membership and usage fees with cash, either at the point-of-sale or at third-party locations. For example, in Philadelphia, Indego bikesharing has partnered with PayNearMe, an electronic transaction network that allows bikesharing users to make online purchases using cash at nearby retail chains [26].
- **Use of prepaid cards**: This method allows users to purchase prepaid cards at a retail store using cash or another payment method. For example, Uber sells gift cards at retail locations as a strategy for unbanked customers [27].
- **Direct carrier billing** (to mobile services or another utility provider): A number of shared mobility providers have proposed linking fobs and memberships to mobile phones to allow low-income users to pay for memberships and usage fees on their phone bill to eliminate the need for a credit or debit card [28]. By billing usage to a person's phone bill, the bill can be paid at a utility retail storefront, potentially negating the need for both data access and a credit/debit card.
- **Payment with debit cards** (in place of credit cards): A number of shared mobility providers accept debit card payment (in lieu of a credit card); however, this payment method may require placing a security hold on the account requiring account funds.
- **Linking to family accounts**: This method allows relatives and minors to link to another family member's account (e.g., a child does not have a credit card, but a parent has a user account with a credit card on file).
- **Subsidies, vouchers, and other promotions for low-income households**: This includes a variety of strategies to overcome requirements of having a banking account or mobile device and methods to reduce or eliminate the shared mobility costs for low-income households (e.g., providing vouchers or complimentary rides for unbanked users). For example, the Pinellas Suncoast Transit Agency provides subsidies to paratransit and TNC operators when providing services to low-income users and persons with disabilities.
- **Partnerships with housing authorities and nonprofit organizations**: Partnerships with housing authorities and nonprofits is another mechanism to help deliver mobility services to unbanked and low-income households. For example, City CarShare (now defunct) previously provided subsidized memberships and user fees for low- to moderate-income users who applied through one of six project partners serving these demographics [15].
- **Partnerships with financial institutions**: Service providers also can partner with financial institutions to expand banking access for travelers. Capital Bikeshare's Bank on DC is a partnership between Capital Bikeshare and two financial institutions to provide low-cost "starter" bank accounts to prospective bikesharing users and raise consumer awareness about the benefits of bank products. New York City's Citi Bike, Ithaca CarShare, and Chicago's IGo (now defunct) have implemented similar partnerships to reach unbanked and underbanked users in their regions [29].

6.2.3   Digital poverty

While some shared modes can be accessed without a smartphone, shared mobility increasingly requires a smartphone and high-speed data packages to access services, multimodal aggregation, on-demand trip planning, booking, and payment. Lack of mobile Internet access can inhibit travelers from using shared modes. Since many shared mobility services are accessed by a smartphone, even the lack of familiarity with and access to mobile and web technology can preclude users from accessing the potential benefits of these services.

One notable social equity concern is the lower rate of smartphone and mobile data usage among older adults, lower-income households, and persons with disabilities; this is often referred to as the digital divide or digital poverty [1,15,30]. According to a recent study by Anderson *et al.* [30], 11% of Americans do not use the Internet. The percentage of Americans who do not use the Internet is notably higher among underserved populations including older adults, rural residents, low-income households, and individuals without a high school diploma. Of the U.S. adults who do not use the Internet, Anderson *et al.* [30] found that 35% had less than a high school diploma, 34% were 65 years of age or older, 22% lived in rural communities, and 19% had an annual household income of less than $30,000 [30].

Lack of Internet access can be a barrier to low-income and rural households who may not be able to afford or may lack mobile data coverage to access shared mobility modes. Alternatives such as digital kiosks, telephone services, and nontech access (such as street hail) can help overcome these challenges. Shaheen *et al.* [15] discuss a few strategies and examples of alternative payment options that can help overcome accessibility challenges associated with digital poverty (Table 6.2).

Table 6.2   Strategies and examples of alternative payment and access options. Adapted from Shaheen et al. [15].

Strategy	Description
Cash payment at point-of-sale and Payment through third parties	Policies that enable cash payment can help overcome the interrelated challenges of having to make an electronic payment using a credit or debit card and a mobile device
Partnerships with third-party web platforms	Increasingly, more partnerships with third-party providers are emerging that allow users to dispatch services without a smartphone or an account with the mobility provider. For example, Lyft and Uber both offer services allowing users to login and book a rider from any website (not requiring a smartphone app). Both services also offer APIs that enable third-party integration. For example, Uber has a partnership with

(Continues)

Table 6.2   (Continued)

Strategy	Description
	Amazon that allows users to add an Uber "Alexa" skill and dispatch Uber from Amazon's Echo device [31]
Telephone dispatch	A number of service providers offer telephone dispatch and concierge services to enable access without a smartphone. For example, GoGo Grandparent is a service that has been designed to allow people to use Lyft and Uber without a smartphone. Uber offers Uber Central, a platform that enables its business account holders to request, manage, and pay for rides for its employees and guests, for instance. ITNAmerica, a volunteer ride service, offers telephone ride coordination for all of its in-network affiliates
Digital kiosks	Digital Wi-Fi kiosks have the potential to connect un-phoned and digitally impoverished residents with app-based mobility options. For example, CIVIQ Smartscapes, a smart city communications manufacturer announced the launch of WayPoint at the 2016 American Public Transportation Association conference. WayPoint is a digital kiosk that provides the public with real-time transportation information, wayfinding, and service announcements. The WayPoint system has optional features such as Wi-Fi, charging ports, emergency 911 intercoms, and integration with shared mobility services. The system is currently being deployed in New York City. Another manufacturer, Sidewalk Labs, offers a product with similar features, which is currently being deployed in Columbus, Ohio as part of the U.S. Department of Transportation's Smart City Challenge grant

6.2.4   *Potential algorithm bias*

The growth of shared mobility and AVs is increasingly being enabled by artificial intelligence (AI) and algorithms that have the potential to interact with or worsen pre-existing social inequities (e.g., spatial, racial, gender, income, etc.). According to a number of studies, TNC and other shared mobility service operations have come under scrutiny regarding pricing algorithms and the choices of human drivers [32–34].

More research and policy may be needed to: (1) identify the potential inequities and biases that AI could be learning in the transportation context; (2) develop strategies to prevent, mitigate, and correct these inequities/biases; and (3) discover opportunities for AI to prevent and even correct historic inequities (i.e., "equitable AI" or "e-AI"). For example, AI could be used to prioritize traveler routing for underserved communities and neighborhoods with limited public transit service to enhance access to employment, healthcare, and other critical services. AI also could be used to provide more affordable and stable pricing for transportation services in low-income communities. While AI presents risks for exacerbating historic inequities, it also creates opportunities to support a more socially equitable transportation network.

6.3   STEPS to transportation equity

Shaheen *et al.* [15] published the STEPS to Transportation Equity framework, which stands for Spatial – Temporal – Economic – Physiological – Social, to help service providers and public agencies identify, understand, and overcome spatial, temporal, economic, physiological, and social barriers [1]. Each of these considerations are described below:

Spatial factors include challenges, such as lack of service availability in a particular neighborhood, excessively long distances between destinations, and the lack of public transit within walking distance [1,2]

Temporal barriers include time factors that can inhibit a user from completing time-sensitive trips (e.g., such as arriving at work) or making trips late at night when there are limited or no public transit options [1,2]

Economic challenges include direct costs, such as fares, tolls, and vehicle ownership costs, as well as indirect costs (e.g., smartphone, Internet, and credit card access) that create economic hardships or preclude users from completing basic travel [1,2]

Physiological considerations include physical and cognitive limitations that make using standard transportation modes difficult or impossible for certain individuals, such as children, older adults, and persons with disabilities [1,2]

(Continues)

(Continued)

 **Social** factors include social, cultural, safety, and language barriers that create challenges for travelers. Examples of social barriers can include neighborhood crime, poorly targeted marketing, and the lack of multilanguage information [1,2]

By applying this framework, the public and private sectors and key stakeholder groups can identify and remove barriers to ensure that driverless vehicles are accessible to everyone. Identifying and understanding various social equity challenges related to shared mobility and AVs is a key step to helping to ensure access and equity in service for all users.

6.4   Social equity and access considerations for an SAV future

To date, policy discussions surrounding AVs typically focus on safety, hackability, liability, and environmental and traffic impacts. More recently policymakers have begun to discuss the potential impacts of driverless vehicles on social equity, accessibility, and society. More planners and policymakers should consider developing comprehensive transportation equity policies to ensure SAVs are equally accessible and available to everyone (e.g., policies that ensure access for persons with disabilities, un- and under-banked households, underserved communities, households without access to smartphones or mobile data, etc.). Additionally, policymakers should consider policies that ensure driverless vehicles preserve and enhance access to jobs, healthcare, healthy food, and other critical services for all users [35]. Additionally, this should include policies that prevent discrimination and bias from artificial intelligence and other systems that impact or guide the operations of driverless vehicles [35]. In the future, decision makers will need to develop policies that respond to six core equity questions arising from vehicle automation, as follows.

6.4.1   What are the spatial impacts of AVs/SAVs and how will this impact access and mobility?

Suburbanization has been one of the great underlying trends impacting transportation during the twentieth century. While early suburbs were often built around railroad and streetcar lines, post-World War II suburbanization has become primarily an auto-driven phenomenon. In many cities, urban centers declined as development patterns focused on personal vehicle ownership; the marketing of suburbia as a residential location; and the building of highways that manifested in strip malls, suburban retail and employment centers, and low-density housing. In many regions, this development pattern led to an overreliance on private vehicles, which has resulted in high costs to households, public health, and the environment. For example, it is not uncommon for low-income households to spend upward of 30% of their income on transportation. For those without a private vehicle, limited job access, education, and health care can be a barrier to upward mobility.

Shared and automated mobility has the potential to enhance access and mobility for underserved communities, but it could also exacerbate existing barriers and increase inequality. SAVs may be able to address spatial inequality in areas with limited alternatives to private vehicle ownership by providing additional mobility options for an entire trip or first- and last-mile connections to public transit. The strategic placement of SAVs in communities underserved by public transit could reduce inequities by providing additional mobility options that have greater coverage and service availability than existing options.

In the longer term, automation may result in fundamental changes to our built environment. Reduced vehicle ownership due to SAVs could impact parking needs, particularly in urban centers. The repurposing of urban parking has the potential to create some opportunities for infill development and increased densities. While SAVs may compete with public transit, infill development could create higher densities to support more public transit ridership in urban core locations [33,34].

6.4.2 How will public transportation be impacted in an AV future?

Concerns that the introduction of SAVs could reduce demand for public transit and may encourage increased vehicle use are real. However, just as SAVs have the potential to reduce driving costs, automated transit vehicles have the opportunity to lower operational costs and pass these savings onto riders through lower fares. Reduced operational costs and lower fares could allow public transit agencies to increase the number of routes or service frequency, making public transit more competitive than other modes [36]. While the impacts of automation on public transit are uncertain, leveraging it to reduce overhead costs and improve public transport efficiency is an important consideration [37].

Additionally, vehicle automation could change historic relationships between public and private transportation services. In the future, public transit agencies may opt to provide more flexible demand-responsive service in smaller vehicles (right sizing), while others may elect to pursue such systems through public-private partnerships [37]. The emergence of SAVs could give rise to the development of hybrid quasi public–private transportation systems that could result in a range of partnerships that vary by the region [37].

6.4.3   How can driverless vehicles help overcome temporal barriers to access?

Late-night transportation options are critical to meeting the travel needs of late-night/early-morning commuters, particularly those without an automobile who need access to employment and other locations. In some cases, riders may have access to public transit services for the start of their shift, but service may be unavailable at the end of it. Late-night transportation services can serve an important equity role, particularly since those who benefit most from late-night service are households working second- and third-shift jobs, many of whom are low-wage earners and for whom these services are a mobility lifeline to employment [38]. As such, late-night transportation can represent a critical economic ladder of opportunity for low-income households.

Today, in many communities, access to public transit during late-night or early-morning hours is limited. Many public transit agencies stop running at or before midnight. While some agencies have implemented late-night services designed to meet the transportation needs of night-time commuters, these are usually much more limited than during the day and may have notably higher operating costs due to lower ridership and route productivity.

In an AV future, both public and private transportation services may be able to more cost effectively offer late-night transportation services, such as demand-responsive automated shuttles and more frequent fixed-route automated public transportation [38]. Additionally, SAV fleets also could be used to provide and/or replace late-night transit services where ridership is not sufficient enough to support public transit. A variety of partnerships and subsidies can be employed to make these options more affordable for late-night workers [38].

6.4.4   What should policymakers consider to better understand and respond to AV economic impacts on users and nonusers?

Vehicle automation will likely have notable impacts on people currently employed in the transportation field. Local and state governments should develop workforce training programs designed to prepare for and respond to a driverless future. This should include a broad program encompassing job training/re-training and job placement resources to minimize the potential adverse labor impacts of vehicle automation.

Additionally, for users, policymakers should ensure that SAVs are affordable and accessible for all users (e.g., low-income, un-/under-banked households). Public policy could play an important role in regulating fares, requiring multiple fare payment options, and providing subsidies for low-income and unbanked users to overcome economic hardships [2].

As AVs and SAVs come to market, more research is needed to understand the potential impacts this could have on job access (and other services, such as healthcare and healthy food) particularly for underserved populations. For example, it is unclear if the spatial and temporal distribution of AVs will ensure equal access to jobs, healthcare, and other critical services for all users. While AVs and SAVs have the potential to expand access to jobs and other services that may not exist today, new challenges could also emerge in a driverless vehicle future. For example, service providers may let unused vehicles roam in low-income neighborhoods waiting for fares (causing congestion, aesthetic concerns, or other adverse impacts) or may charge fleets overnight when demand may be lower thereby limiting late-night transportation services. Decision makers will have to be proactive and consider developing policies that support equitable access to AVs and ensure that services create economic opportunities for job access, healthcare, healthy food, and other critical services.

6.4.5   What policies are needed to ensure access for users with physiological and other special needs?

In a driverless vehicle future, transportation service providers will need to consider what types of services and technologies may be needed to provide a completely accessible journey from a traveler's origin and destination that accommodates an array of physical, cognitive, and other special needs. This could include ensuring that SAVs provide accessible apps (e.g., ensuring apps have features that can be used by people with visual, auditory, physical, cognitive, and other disabilities) and fleets (e.g., dispatching a wheelchair accessible vehicle and requesting a vehicle with an attendant that has specialized training to provide access, egress, and other assisted services). Additional policy considerations could include whether personal attendants or nurses for older adults and persons with disabilities should be required to pay a fare to accompany a passenger requiring assistance and how private transportation service providers can best provide reasonable accommodations. Planners and decision makers also need to ensure that persons with disabilities and special needs have equitable services (e.g., service availability, comparable wait and journey times, etc.). This could include requiring universal design for all SAV fleets to accommodate access, riding, and egress for people with disabilities. For example, policy makers could require that SAV service providers charge a fee earmarked to provide accessible transportation services (e.g., paratransit) and telephone-dispatch services. Innovative and emerging technologies may also be able to improve the quality of automated assistance through voice-activated mobility app features, robotic arms, artificial intelligence, and other technologies.

6.4.6   What will be the impacts of AVs/SAVs on social inclusion (e.g., social isolation, social interaction, public health, etc.)?

In an AV future, certain communities may confront barriers to access (e.g., low-income communities, minorities, and users with limited English proficiency). Whether vehicle automation will help overcome and contribute to social isolation is unclear. The public and private sectors can work together to encourage broad social inclusion in automated transportation systems by: (1) providing alternative methods of service access (e.g., digital kiosks, etc.); (2) developing innovative mobility programs to enhance access and mobility for a variety of market segments (e.g., children, prenatal mothers, veterans, etc.); and (3) encouraging mobility services that improve access to jobs, healthcare, and education for all members of society [39]. Incorporating service features that are designed to minimize and overcome social challenges (e.g., targeted outreach to underserved communities, apps that limit discriminatory outcomes, multilingual apps and marketing, etc.) is critical. Service providers may also consider shifting to performance-based community engagement metrics, which require ongoing assessment and refinement of outreach/service offerings until social equity and racial justice goals are achieved [1]. Finally, when considering social inclusion, it is important that the public and private sectors ensure that algorithms, artificial intelligence, and other automated systems support broad equitable access and social inclusion, and they do not intentionally or unintentionally reinforce or introduce discrimination or bias in the delivery of transportation services.

6.5   Conclusion

Access to transportation is integral to enhancing opportunities for employment, education, health care, and recreation. SAVs will not fundamentally solve today's transportation challenges. To solve them, AVs require thoughtful planning and public policies that balance societal goals with commercial interests. To harness and maximize the social and environmental benefits of highly automated vehicles, we need to prepare for a multiphase transition toward highly automated vehicles today. If driverless vehicles are thoughtfully implemented with access and social/racial equity in mind, automation could expand access to resources for users of all ages, genders, races, abilities, and incomes. However, public agencies will need to actively pursue policies to ensure that driverless vehicles do not reinforce existing disparities in access and mobility. Identifying and understanding the various transportation equity challenges related to driverless vehicles is the first step to help ensure equitable service, accessibility, and affordable transportation for all and to prevent discrimination and correct mobility injustices. In the future, it will be critical that policymakers ensure equitable SAV access for all neighborhoods and users, including access options for people with disabilities and digitally impoverished and underbanked communities. The public and private sectors, along with key stakeholders (e.g., nongovernmental organizations, community-based organizations, and foundations) can partner to help overcome these challenges by understanding these issues and implementing tailored strategies to overcome each challenge through intentional inclusion [33].

Author contributions

The authors confirm contribution to this chapter. All authors reviewed and approved the manuscript.

Declaration of conflicting interests

The authors declared no potential conflict of interest with respect to publication of this chapter.

References

[1]   Shaheen, S., Bell, C., Cohen, A., and Yelchuru, B. *Travel behavior: Shared mobility and transportation equity.* Washington, D.C.: U.S. Department of Transportation; 2017.

[2]   Shaheen, S., and Cohen, A. *Mobility on demand and transportation equity.* Retrieved from Move Forward: https://www.move-forward.com/mobility-on-demand-and-transportation-equity/. 15 March 2018.

[3]   Cervero, R. and Tsai, Y.H. *San Francisco City CarShare: Travel-demand trends and second-year impacts.* Berkeley, CA: University of California; 2003. Retrieved from https://cloudfront.escholarship.org/dist/prd/content/qt4f39b7b4/qt4f39b7b4.pdf?t=kro6ez&v=lg.

[4]   Borecki, N., Buck, D., Chung, P., *et al. Virginia Tech Capital Bikeshare Study.* 2012. Retrieved from https://nacto.org/wp-content/uploads/2012/02/CaBiReport13Dec-2.pdf

[5]   Teal, R. "Carpooling: Who, how, and why." *Transportation Research Part A.* 1987; 203–214.

[6]   Shaheen, S., Cohen, A., and Bayen, A. *The benefits of carpooling: The environmental and economic value of sharing a ride.* 2018. Retrieved from https://cloudfront.escholarship.org/dist/prd/content/qt7jx6z631/qt7jx6z631.pdf?t=ph0r4f&v=lg

[7]   Liu, C., and Painter, G. "Travel behavior among Latino immigrants: The role of ethnic concentration and ethnic employment." *Journal of Planning Education.* 2012; 62–80.

[8]   Martin, E., and Shaheen, S. *Greenhouse gas emission impacts of carsharing in North America.* San Jose, CA: Mineta Transportation Institute; 2010.

[9]   Martin, E., and Shaheen, S. *Impacts of Car2go on vehicle ownership, modal shift, vehicle miles traveled, and greenhouse gas emissions: An analysis of five North American Cities.* Berkeley, CA: University of California; 2016.

[10] Shaheen, S., Martin, E., and Bansal, A. *Peer-to-peer (P2P) carsharing: Understanding early markets, social dynamics, and behavioral impacts.* Berkeley, CA: University of California; 2018.

[11] Shaheen, S., Stocker, A., Lazarus, J., and Bhattacharyya, A. *RideKC: Bridj pilot evaluation: Impact, operational, and institutional analysis.* Berkeley, CA: University of California; 2016.

[12] Shaheen, S., Martin, E., Chan, N., Cohen, A., and Pogodzinski, M. *Public bikesharing in North America during a period of rapid expansion:*

Understanding business models, industry trends, and user impacts. San Jose, CA: Mineta Transportation Institute; 2014.

[13]   Rayle, L., Dai, D., Chan, N., Cervero, R., and Shaheen, S. "Just a better taxi? A survey-based comparison of taxis, transit, and ridesourcing services in San Francisco." *Transport Policy.* 2016; 168–178.

[14]   U.S. Department of Transportation. *Beyond traffic 2045.* Washington, D.C.: U.S. Department of Transportation; 2016.

[15]   Shaheen, S., Cohen, A., Yelchuru, B., and Sarkhili, S. *Mobility on demand operational concept report.* Washington, D.C.: U.S. Department of Transportation; 2017.

[16]   State of Texas. *Sec. 2402.112. Nondiscrimination; Accessibility.* 29 May 2017. Retrieved from Transportation Network Companies Law: https://www.tdlr.texas.gov/tnc/tnclaw.htm#2402112

[17]   Uber. *Accessibility at Uber.* Retrieved from https://accessibility.uber.com/. 30 November 2018.

[18]   City of Portland. *Adaptive BIKETOWN pilot.* Retrieved from City of Portland. 2018. https://www.portlandoregon.gov/transportation/73371.

[19]   City of Seattle. *Transportation network companies.* 2018. Retrieved from Business Regulations: http://www.seattle.gov/business-regulations/taxis-for-hires-and-tncs/transportation-network-companies/tnc-companies

[20]   Berg, S. *Lime's first day in Idaho sees scooters blocking Meridian sidewalks, wheelchair ramps.* 28 September 2018. Retrieved from Idaho Statesman: https://www.idahostatesman.com/news/local/community/west-ada/article219 213920.html

[21]   Shaheen, S., and Cohen, A. *How dockless bikesharing is transforming cities: Seven policy recommendations to minimize disruption.* 28 June 2018. Retrieved from Move Forward: https://www.move-forward.com/how-dock less-bikesharing-is-transforming-cities-seven-policy-recommendations-to-mini mize-disruption/

[22]   Federal Deposit Insurance Corporation. 2017 *FDIC national survey of unbanked and underbanked households.* 22 October 2018. Retrieved from https://www.fdic.gov/householdsurvey/. Also see Federal Deposit Insurance Corporation. *FDIC national survey of unbanked and underbanked house-holds.* October 2018. Retrieved from https://www.fdic.gov/householdsurvey/ 2017/2017report.pdf

[23]   Manjoo, F. *The Uber model, it turns out, doesn't translate.* 23 March 2016. Retrieved from The New York Times: https://www.nytimes.com/2016/03/ 24/technology/the-uber-model-it-turns-out-doesnt-translate.html

[24]   Barry, K. *New Helsinki bus line lets you choose your own route.* Retrieved from Wired: https://www.wired.com/2013/10/on-demand-public-transit/. 11 October 2013.

[25]   Shared-Use Mobility Center. *What killed Kutsuplus? 3 takeaways for cities pursuing mobility-on-demand.* 03 May 2016. Retrieved from Shared-Use Mobility Center: https://sharedusemobilitycenter.org/news/killed-kutsuplus-3-takeaways-cities-pursing-mobility-demand/

[26] PayNearMe. *Philadelphia bike share program becomes first in U.S. to launch with cash payment option.* 23 April 2015. Retrieved from Market Wired: http://www.marketwired.com/press-release/philadelphia-bike-share-program-becomes-first-us-launch-with-cash-payment-option-20127 00.htm

[27] Uber. *Give the gift of Uber.* 2018. Retrieved from Uber: https://www.uber.com/gift-cards/

[28] Schmitt, A. *Why isn't bike-share reaching more low-income people?* 03 October 2012. Retrieved from StreetsBlog USA: https://usa.streetsblog.org/2012/10/03/why-isnt-bike-share-reaching-more-low-income-people/

[29] Kodransky, M., and Lewenstein, G. *Connecting low-income people to opportunity with shared mobility.* New York, NY: Institute for Transportation & Development Policy and Living Cities; 2014.

[30] Anderson, M., Perrin, A., and Jiang, J. *11% of Americans don't use the internet: Who are they?* 5 March 2018. Retrieved from Pew Research Center: http://www.pewresearch.org/fact-tank/2018/03/05/some-americans-dont-use-the-internet-who-are-they/

[31] Edwards, A. *Request Lyft or Uber without a smartphone: How to call a ride on a computer.* 05 March 2018. Retrieved from Ridester: https://www.ridester.com/request-uber-lyft-online/

[32] Ge, Y., Knittel, C.R., MacKenzie, D., and Zoepf, S. *Racial and gender discrimination in transportation network companies.* NBER Working Paper No. w22776. 31 October 2016. Retrieved from SSRN: https://ssrn.com/abstract=2861708

[33] Stark, J., and Diakopoulos, N. 'Uber seems to offer better service in areas with more white people that raises some tough questions'. *The Washington Post:* 10 March 2016. Retrieved from https://www.washingtonpost.com/news/wonk/wp/2016/03/10/uber-seems-to-offer-better-service-in-areas-with-more-white-people-that-raises-some-tough-questions/

[34] Pandy, A., and Caliskan, A. *Iterative effect-size bias in ridehailing: Measuring social bias in dynamic pricing of 100 million rides.* 22 June 2020. Retrieved from https://arxiv.org/abs/2006.04599

[35] Shaheen, S., and Cohen, A. *Convergence of sharing and automation: Need for proactive public policy and research understanding.* 20 December 2018. Retrieved from Move Forward: https://www.move-forward.com/convergence-of-sharing-and-automation-need-for-proactive-public-policy-and-research-understanding/

[36] Shaheen, S., and Cohen, A. *Prioritizing people, public transport, and pooling: Transitioning to shared automated vehicles.* 22 May 2018. Retrieved from Move Forward: https://www.move-forward.com/prioritizing-people-public-transport-and-pooling-transitioning-to-shared-automated-vehicles/

[37] Shaheen, S., and Cohen, A. "Is it time for a public transit renaissance? Navigating travel behavior, technology, and business model shifts in a brave new world." *Journal of Public Transportation.* 2018; 67–81.

[38]   Shaheen, S., and Cohen, A. *Late-night transportation: How two public agencies are filling service gaps through mobility on demand.* Retrieved from Move Forward. January 10, 2019

[39]   Shaheen, S., and Cohen, A. *Five strategies to enhance transportation equity in shared mobility.* 16 October 2018. Retrieved from Move Forward: https://www.move-forward.com/five-strategies-to-enhance-transportation-equity-in-shared-mobility/

The views expressed are those of the authors.

Chapter 7

Multimodal relationships: shared and automated vehicles and high-capacity public transit

Yonah Freemark[1], Neema Nassir[2], and Jinhua Zhao[3]

Today's urban transportation system does not function adequately for most of its users. Those who take transit are frequently beset by infrequent service, crowded trains, delayed buses, and inadequate coverage. Those who drive or are driven are stuck in traffic, contributing to climate change, and subject to the ever-present possibility of committing vehicular carnage. Those who walk or bike are confined to poorly maintained, often discontinuous, dangerous, and circuitous routes. Few receive the transportation services that adequately meet their needs, and some are excluded from the transportation system altogether due to their inability to pay, their age, or their physical abilities.

Part of the problem is that no mode, by itself, provides the capacity necessary to carry all people over all the necessary distances and with reasonable directness, which requires serving the full array of origins and destinations. Transportation network companies (TNCs), for example, offer direct door-to-door services but have limited capacity to carry large numbers of passengers; typical public transportation options, on the other hand, are high capacity but not direct. The introduction of automated vehicle (AV) technology, which we expect to be largely an extension of today's private automobile and TNC system, will not by itself alter this dichotomous relationship. In Figure 7.1, we compare several of the key advantages of AVs and public transit. A transportation system that is made up primarily of these two modes would offer options that are demand responsive and offer economies of scale, respectively—but not both at the same time.

In this chapter, we consider the potential for a more integrated transportation system that merges access to services offered across modes in order to effectively bridge the gap between the three demands of capacity, directness, and distance, and that combines the respective advantages of AVs and transit as enumerated in Figure 7.1. Our examination specifically considers whether the advent of AV technology—particularly shared

[1]Metropolitan Housing and Communities Policy Center, Urban Institute, Washington, DC, USA
[2]Department of Infrastructure Engineering, Faculty of Engineering and IT, University of Melbourne, Melbourne, Australia
[3]Department of Urban Studies and Planning, Massachusetts Institute of Technology, Cambridge, USA

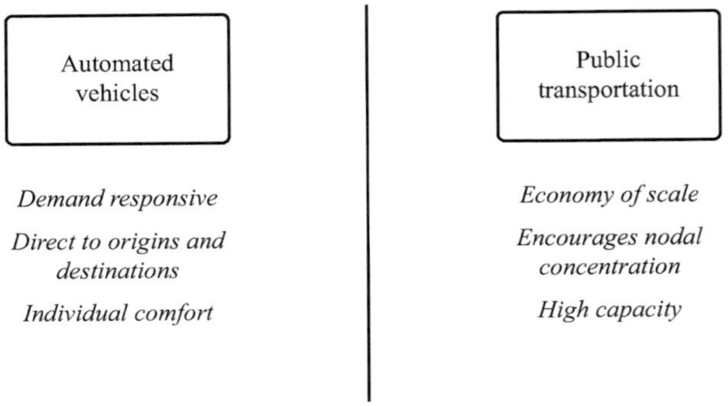

Automated vehicles	Public transportation
Demand responsive	*Economy of scale*
Direct to origins and destinations	*Encourages nodal concentration*
Individual comfort	*High capacity*

Figure 7.1   The key advantages of automated vehicles versus public transportation

automated vehicles (SAVs)—can be used as a lever to enhance transit systems, not only by improving their efficiency but also by filling the major gap that has limited transit's success in attracting riders, especially in the United States: directness. Thus the fundamental motivation for our text is an effort to understand whether and how the transportation system can maintain the capacity benefits of transit (and thus its congestion-relieving and sustainable characteristics) while improving services to more people.

Our contribution is premised on the assumption that AV technology will have a major influence on the workings of the transportation system, but that its influence can be shaped. Simply layered on top of today's options, it will have both positive and negative consequences. It promises to improve certain aspects of the ground transportation system, potentially expanding options in underserved neighborhoods, offering new connections for historically excluded populations, and reducing transportation-related externalities, such as casualties and pollution. An AV system can theoretically respond to changes in system demand more quickly and with fewer implementation obstacles than a human-driven one. On the other hand, AVs may reinforce the negative attributes of the current network, encouraging more vehicle use, congestion, and pollution. Evidence from TNCs thus far, as we shall show, suggests that these impacts are not to be dismissed—and that neither TNCs nor AVs will replace the high-capacity benefits of transit.

In this chapter, we emphasize that public transportation is, and will continue to be, an important element of the ground transportation system, but that its role too, can and should be altered in the context of AVs for an improved urban transportation future that serves more people more effectively. We begin by parsing out how transit interacts with shared vehicles operated by TNCs today, examining how their markets intersect and their effects on one another. We review recent scholarship to demonstrate that the manner in which TNCs have been rolled out has largely been in competition, rather than complementary, with transit. Second, we consider the potential for integration between these two modes in the context of autonomy, a technological change that will ultimately affect TNCs, transit systems, and privately owned cars (though we do not focus on the latter in this chapter).

Third, we point to ways in which AV technology offers the opportunity to expand transit's reach by allowing the provision of smaller vehicles in more places and encouraging intermodal links. By taking advantage of automated technology to allow real-time, fleet-based decisions about the network, we show how an integrated system can advance a transportation system that responds best to the constrained urban right-of-way. And finally, we consider several key questions related to a potential integrated system, such as what behavioral and regulatory norms would have to be upended to make such integration possible.

7.1 Current relationship between TNCs and public transit

In this section, we describe the existing relationships between public transit systems and TNC services, with a focus on North America. We investigate the modal characteristics of each, highlighting the range of services both provide and their strengths and weaknesses in addressing mobility needs. We then investigate how their passenger markets and operational patterns interact.

7.1.1 Modal characteristics

We begin by defining transit: the provision of fixed-route services using vehicles (usually buses or trains) that have the capacity to carry many people at once (see Reference [1] for more detail on alternative definitions of transit). Over 95% of all transit vehicle miles in the United States, as defined by the Federal Transit Administration [2], are operated in such a fashion, and about 99% of riders use transit in this way. Most transit services in developed countries are planned by public sector organizations, which subsidize them using funds generated from other sources. Many services are operated by public-sector agencies, though this was historically not the case, and is not always the case today. Transit typically serves two primary purposes, which sometimes intersect. First, it offers transportation in dense areas with high travel demand, for which frequent transit vehicles provide the highest-capacity and most cost-efficient option available to a wide range of riders (sometimes referred to as "mass transit"); and second, it offers a minimum level of service with infrequent, circuitous routes for people and neighborhoods with limited alternative options ("social transit") [3].

Nontransit shared vehicles, on the other hand, typically provide point-to-point services without a fixed route. Carpooling is a traditional way of accomplishing this; passengers headed to the same destination, such as school or work, share a trip. But its use in the United States has fallen from about 20% of commuting trips in 1980 to less than 10% today, a consequence of increased affordability of and access to private automobiles [4,5]. Even so, online carpooling services such as BlaBlaCar, operating on a peer-to-peer transaction model, have increased in popularity recently [6–8].

For the remainder of this section, we focus on the market of services provided by third parties at a cost through taxis or TNCs (such as Lyft and Uber), which typically occurs through mobile phone-based applications ("ride-hailing" or "ridesourcing"). Unlike transit, these services are offered in four-door automobiles, minivans, and sport-utility vehicles; they are not planned, subsidized, or provided by the public sector; and vehicles are generally owned by drivers, not as a fleet. TNC companies allow users to select from an array of services, some mimicking taxis in that only a single passenger or a group of passengers who know one another share a trip, or departing from the taxi precedent by allowing individuals unknown to one another to share trips.

This latter service, sometimes referred to as "pooled" rides, can be operated in at least two ways. The most common approach, offered through services such as Lyft Line and UberPool, is to connect multiple passengers whose origins and destinations may not be exactly shared, but which are "on the way" of one another. A traveler going from point 1 to point 3 may pass near or through point 2; in the process, a driver may be able to pick up a passenger going from point 2 to point 4, a trip which may pass near or through point 3. This system may cause some inconvenience depending on just how overlapping the two routes are with one another, and the willingness of passengers to share space. Another approach has been piloted by ride-hailing services such as Via and involves customers walking to a nearby pickup location before being picked up on an "optimized" route that reduces detours to pick up shared rides [9,10].

Figure 7.2 maps out the spectrum of transportation services, noting how conventional transit differs from paratransit (optimized-route services operated or funded by transit agencies, typically serving disabled riders), conventional TNC service, and pooled TNC service across 10 characteristics: route, area served, schedule, span of service, vehicle size, trip sharing, service planning, operations, affordability, and funding. This figure provides a view of the differences between these modes, but this should be seen as illustrative since the reality is that transportation services are location-specific and in flux. Key to Figure 7.2 is the differing norms by which passengers interact with each mode. Passengers using transit understand that they will be in a shared environment, receiving scheduled service on predetermined routes; people using taxis (or their privately owned vehicles) expect isolation and on-demand, door-to-door service. Given these differing norms, an integrated AV-transit system, in which people move between services in different-size vehicles would undoubtedly require the cultivation of a new norm among passengers.

Transit is designed to encourage road-space and vehicular efficiency (its specific characteristics depend on mode and location). Its fixed routes, schedules, service only at certain times of the day, low costs, and large vehicles are in theory meant to move the largest number of people while consuming the smallest possible footprint in the transportation network (not that transit always achieves this goal). Transit services attempt to achieve these objectives by typically being planned, subsidized, and operated by public-sector entities. TNCs, on the other hand, are operated to encourage operational flexibility. On-demand routes, schedules, high costs, and small vehicles respond to individual needs. This is typically done through private decision-making related to service planning and funding, and

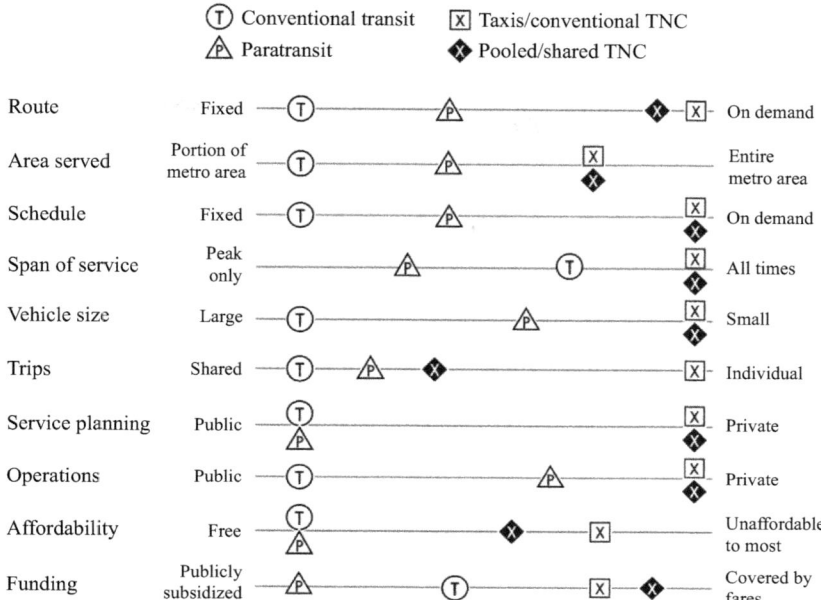

Figure 7.2   Spectrum of ground transportation services, across an array of characteristics

public involvement is limited to regulations over issues such as medallions or driver qualifications. In most metropolitan areas, transit and TNC operations reflect these opposing goals and are largely separated. Few people take multimodal trips incorporating TNC and transit links. The dual advantages of transit efficiency and operational flexibility, referenced in Figure 7.1, thus largely have yet to be combined. Figure 7.2, for example, illustrates that current options fail to offer on-demand, affordable, publicly planned services that meet the transportation needs of an entire metropolitan area.

Even so, the availability of shared ride-hailing services demonstrates that the dichotomous view of transportation services—taxi (individuals riding alone or with known others from their chosen origins to their chosen destinations) versus transit (groups of people who do not know one another riding between fixed stops)—is evolving into something that begins to fill the gaps in the spectrum presented above. This is perhaps made most evident by certain demand-responsive offerings used to provide transportation either for people unable to use the conventional transit system or for "last-mile" connections intended to extend its reach. Transit agencies such as the Massachusetts Bay Transportation Authority [11], for example, have contracted with TNCs to provide such services that are sometimes as affordable for passengers as conventional transit. In the case of Altamonte Springs, Florida [12], these offerings have replaced the transit system.

Moreover, transit is increasingly being contracted out; the share of services directly operated by public agencies in the United States declined from 81% in 2002 to 70% in 2018 [13]. Similarly, the rise of public subsidies for TNCs plus the decision of some cities to impose specific taxes and regulations on TNCs (such as New York's limitation on the number of ride-hailed vehicles. Wodinsky [14] suggests that these services are increasingly within the purview of public decision making. From a passenger perspective, the incorporation of transit and TNC information on the same application, an approach being piloted in several cities, means that many may experience them as part of the same transportation "ecosystem."

The deployment of AV technology will make TNC services even less clearly differentiable from the transit system, and vice versa, blurring the spectrum. For example, rather than a difference between transit and TNCs being whether their routes are fixed, an integrated system could incorporate both types of routes. This blurring of the modal spectrum provides an opportunity for the transportation system to become more integrated among the services available and thus achieve higher levels of capacity, directness, and distance.

7.1.2 Passenger markets interactions

We so far lack adequate information about how, exactly, AVs will impact the public transportation system; it remains to be seen whether they will be rolled out as TNCs (with separate fares and no public-sector planning), like the private automobile system (individual car ownership), or as part of the transit system (such as with larger vehicles). Nevertheless, the rise of TNCs over the past decade provides evidence for what we might expect from the perspective of interactions with both passenger markets, which we describe here, and transportation operations, which we describe in the section 7.1.3. TNCs have addressed some of the problems with transit networks, offering new, low-wait-time options for underserved areas at an arguably higher comfort level; in this way, they may be seen as providing a net benefit for the urban transport network. At the same time, we show that most scholarship indicates that TNCs have competed with transit, reducing ridership, and in the process increased congestion, particularly in dense neighborhoods. If operated similarly, AVs would likely produce similar results.

TNCs have become particularly attractive in North American cities for a specific audience. Surveys point to a ride-hailing user base that is disproportionately young, wealthy, and well-educated [15–19]. This is partly a reflection of the fact that such services are typically much pricier than equivalent transit options (more than four times as expensive on average—for pooled rides—in one Chicago study; see [20], and their use requires access to, and competency in, phone-based applications. TNC passenger markets, then, are different from those of most transit systems, whose clientele is on average poorer than the population overall. The gradual transfer of riders from transit to TNCs, which we describe below, has likely widened this difference.

TNC use varies based on location. In the United States overall, 70% of ridership is in just nine metropolitan areas, generally the country's densest [18]; and use there is concentrated in dense neighborhoods [16] with high levels of local activity in terms of employment or attractions [15]. Transit use, of course, is equally concentrated in dense communities. The environments where the two modes work most effectively, then, are the same, meaning that if there is a limited passenger market, they become competitors (increasing use of one means reduced use of the other). Higher TNC prices become acceptable to many customers because trip times, especially outside of downtown, can be shortened dramatically through ride-hailing—even when shared [20]. Competition between modes is stimulated by the minimal integration between the two. Intermodal trips combining transit and TNC segments are complicated by the lack of a common fare card, the logistical difficulty of planning a trip that includes both, and the reality that arriving at a transit station may involve a long wait for a bus or train.

Demonstrating the competitive nature of the TNC-transit relationship, Jin *et al.* [21] and Circella and Alemi [16] find that a large share of TNC users previously used transit; Schaller [18] demonstrates that 60% of customers would have otherwise ridden transit, walked, biked, or not taken the trip at all without a TNC option. In other words, TNCs act as substitutes for transit service. Graehler *et al.* [22] note that this competitive relationship grows over time, with transit (especially bus services) losing an increasing number of passengers after a TNC service is introduced.

If there is a competitive relationship between transit and TNCs *in general*, the two modes can be complementary (reinforcing the use of one another). Though Hall *et al.* [23] find that TNC service reduces transit ridership among cities with already high transit use, they find that it increases transit ridership *on average*, and specifically in cities where transit is currently least effective. Nelson and Sadowsky [24] identify a short-term increase in transit ridership after TNC entry. Whether TNCs are competitive or complementary depends on temporal and spatial effects. For example, ride-hailing is frequently used for non-home-work trips [19,21], situations where transit often does not offer convenient frequency. In a neighborhood comparison in Chengdu, China, Kong *et al.* [25] show that in suburban communities with poor transit coverage, riders used TNCs to fill last-mile gaps rather than replacing transit altogether.

In some cases, TNCs are neither complementary nor competitive with transit. They may be independent (their use having no effect on the other), such as in cases where there is no transit service, either because neighborhoods are too sprawling or because it is too late at night. TNC use may also replace personal automobiles [26], since they allow people not to have to look for, or pay for, parking [18]. These examples suggest avenues for thinking of TNCs—and, in the future, AVs—as a mechanism to broaden urban transportation options, rather than compete with transit.

Are TNCs the cause of the drop in ridership among U.S. transit agencies that has occurred over the past few years? In 2017, per-capita ridership declined to its lowest level since 1997, and buses have suffered most (though in some cities, such as Seattle, transit increased). While buses accounted for 64% of U.S. ridership in 2002, they had fallen below rail in use 15 years later [27]. Transit ridership declines produce a vicious cycle, because they reduce revenues, which in turn force transit agencies to either hike fares or worsen service, both of which further dissuade transit use. But TNCs are only one of several potential explanations for ridership decline. Boisjoly *et al.* [28] point to reductions in transit service offered, and share Manville *et al.*'s [29] argument that increased access to motor vehicles (particularly among low-income families) has disincentivized transit use.

Even so, the weight of the evidence suggests that transit and TNCs are competitors for at least a portion of the passenger market—that portion concentrated in the densest neighborhoods and the largest cities. In these cases, transit loses ridership where is it most effective. As we describe below, this relationship could worsen in the context of AVs if they are rolled out similarly to TNCs.

7.1.3   Operational impacts

If there is competition between TNCs and transit to attract passengers, there is also competition between the two modes over space. This problem is aggravated in dense urban areas that already have limited room and which consequently are congested. These are the places where transit is most useful, since it is simply more effective from a spatial perspective to travel collectively, and it is possible to move far more people to a single point on trains and buses than in cars, whether personally driven, driven as part of TNC networks, or operated as automobile-sized AVs [1].

TNCs impact transit operations in several ways. First, if they move people out of transit into smaller, less spatially efficient cars, they increase traffic, assuming demand remains constant. Second, if they increase traffic, they slow speeds and reduce road-network reliability, therefore making street-running transit trips take longer, inconveniencing passengers and increasing the cost for transit agencies to provide a certain level of service. And third, if they use curb lanes for pick up and drop off, they may stand in the way of buses attempting to make similar moves.

To consider the net impact of TNCs, we propose an index of road-space usage efficiency: passenger miles traveled (PMT) divided by passenger-car-equivalent vehicle miles traveled (VMT). A higher index is preferable. This index can be written as follows:

$$R_c = \sum P_c / \sum V_c$$

where R_c is the road-space usage efficiency for a specific corridor c; P_c is the total passenger miles traveled; and V_c is the passenger-car equivalent VMT (e.g., a bus might have a road-space occupancy of three times that of a car). If ten passengers ride one mile on one bus whose footprint is three times that of a car, their index would be $10/3 = 3.3$. If those passengers all moved to individual TNCs (unshared), their index would be $10/10 = 1$; if they moved to two-passenger shared TNCs, their index would be $10/5 = 2$. In order to be useful, the index must include nonrevenue miles. We will return to this index later in this chapter, but note that a shift from transit to TNCs not only reduces the index for the transit system (since it is still running, but emptier), but adds new low-index automobiles to the street.* We emphasize that this is just one metric among many needed to assess how various services have affected urban transportation; for example, it might be useful for cities to develop a measure of geographical coverage based on different modes.

But this tool for analyzing road-space usage efficiency is particularly important now because evidence suggests that ride-hailing has already been an important contributor to increased VMT—thus reduced efficiency [17]. A New York City study showed TNCs put 2.8 miles on the road for each mile of personal driving removed; even shared TNCs produced more VMT than previously [18]. Another study showed TNCs produced 83.5% more VMT than would have occurred otherwise [30], partly a product of the additional driving TNC drivers undertake to pick up passengers and while waiting for dispatch, when they often continue driving in order to avoid paying for parking. In San Francisco, Castiglione *et al.* [31] show TNCs contributed half of the increase in congestion that occurred between 2010 and 2016—thus increasing vehicle hours of delay and VMT, and reducing average street speeds (employment and population growth accounted for the remainder of the change). Street speeds declined from 24 mph on average to 20.9 mph.

Reductions in street speeds had detrimental impacts on bus service in San Francisco, with average local bus speeds declining by almost 8% between 2010 and 2016 [13]. Though scholarship is limited on TNC impacts on bus stops and lanes, anecdotal evidence suggests that this, too, is a cause of transit delay [32]. As a result, operators have to spend more money (in hours) to provide the same service (in miles). Passengers suffer from lengthier commutes. The incentive for many choice riders, then, is to stop using buses. They may shift to TNCs or private automobiles. Either shift worsens bus service.

7.2 Knowns and unknowns, needs and necessities

In the section 7.1, we show that under the current regulatory regime TNCs are competing with transit for a significant portion of the market and for limited street space. In doing so, TNCs increase traffic. This produces a vicious spiral of reduced surface transit effectiveness that pushes more riders from transit to TNCs, further adding to congestion and pollution. We assume that AVs will take on many of the characteristics of current TNCs, albeit at lower costs. Thus, if submitted to the same regulatory and behavioral norms as today, AVs may exacerbate TNCs' negative effects—though the degree to which these future services are pooled, or SAVs, will influence the magnitude of these impacts. In this section, we summarize

---

*It is also important to emphasize that what matters is *passenger* miles, not total occupancy. If we were applying this index to privately owned cars, a car with a parent driving a child to soccer practice would only count for one passenger, not two. Henao and Marshall [30] find TNC vehicular passenger occupancy of just 0.8 when accounting for deadheading. The average passengers per mile for New York City buses is 7.7 (Federal Transit Administration, Reference 13), producing an index of 2.6, thus more than three times as spatially efficient.

scholarship on the interaction between transit and AVs. We argue that though AVs will extend access to urban transportation for many, left unregulated, they will also diminish the effectiveness of the transit system.

7.2.1   Transit automation and AVs

The London Underground Victoria Line has featured automated train control since 1968. Fully driverless operations have been used for fixed-guideway rail systems since the early 1970s, and major cities from Dubai to Vancouver collectively carry millions of passengers every day on such lines, of which there are now more than 1,000 km in operation [33]. Transit automation is not a new phenomenon, though its relevance has grown.

In this article, however, we focus on *automated* vehicle (AV) technology. As with automated transit, AV takes the human out of the driver's seat, relying on computers to make decisions about how to navigate and when to accelerate and decelerate. The technology could be applied to vehicles of any size. Yet unlike automated transit, AV allows operation outside of fixed-guideway, grade-separated environments that are typically the domain of high-capacity metro systems. AVs, theoretically, will be able to drive on whatever transportation rights-of-way human-driven cars use today, such as highways or city streets. AVs thus represent a dramatic geographical and service expansion of automation. It also suggests new options for transit service, since it means that routes could be adjusted on the fly.

Automated buses are being trialed throughout the world, typically in the form of shuttles that carry about ten people each, operating in special areas such as largely pedestrianized business districts or touristic zones [34]. As this technology matures, it will be extended to the 40-to-60-foot buses that account for the majority of the American surface-transit rolling stock.

The extension of automated technology to buses comes in the context of the development of 2-to-8-passenger automobile-size AVs. If European cities have largely focused on automated shuttle trials, U.S. cities and states have been working with companies like General Motors and Waymo to test smaller-vehicle AVs, which could provide services similar to what is available today via TNCs—both ride-hailing pickups for individuals and people they know, and pooled rides combining journeys between strangers.

7.2.2   Economics of AV operations and future impacts on transit

To what degree will automated technology alter mobility costs? Public transportation—particularly bus transit—is labor-dependent. In cities like Austin, labor accounts for more than 45% of operating expenses [35]. Bus system productivity in the United States has expanded slowly, thereby increasing labor costs as a share of expenses over time [36]. The result is a growing difficulty on the part of agencies to maintain cost-effective transit systems, and a need for rising subsidies to support minimum levels of service.

There is limited scholarship on the relative cost-efficiency and mobility benefits of automated versus human-driven transit, though Cohen *et al.* [37] provide an overview of recent experience with automated trains. They find that automated lines allow for a staffing reduction of 30%–70%, with a smaller reduction in operating expenses (due to automated-system supervisors earning more than drivers they replace). Such systems are more reliable, allow higher train frequency, and carry more passengers per train due to the lack of cabs. This experience is suggestive of changes that could accompany automated transit generally. A move to electrified, automated buses could, for example, reduce overall capital and operating expenses by about a third, saving transit systems millions of dollars that could be redistributed elsewhere [35].

Many models used to evaluate automobile-sized AV systems, assume a single operator, for example, Wen *et al.* [38], an assumption that may bias results because it produces greater network efficiency estimates than may actually occur with competitive operators, as is the case with TNCs today. This same assumption, however, highlights one of the key differences between AVs and cheaper human-driven TNC services: with AVs, the entire network can be controlled as an ensemble, allowing much quicker responses to new demands through movement of vehicle location. Human-driven vehicles, on the other hand, require the individual judgement and driver compliance—and their personal interests may not be aligned with the interests of the system as a whole. Thus AVs could improve the level of service offered on the network. Yet an open-market AV system may suffer from considerable overlap between operators, which may mean more-than-optimal VMT and a larger fleet. These models are also limited in that the actual rollout of AVs remains speculative; modeling must advance to incorporate changes such as fleet size elasticity and vehicle repositioning [39].

One major question affecting the cost of AVs is the degree of sharing they engender. First, will be vehicles largely be privately owned, and thus left to sit idly most of the day (as is the case for most privately owned cars in the United States now), or will they mostly be operated by third-party service providers (like TNCs)? The latter option could lower passenger-mile costs below those of traditional automobiles because, even with higher capital costs, operational costs would be much lower than those of today's TNCs because the cost of paying drivers would be eliminated [40]. This reduction in costs would attract significantly more riders [41]. Second, if the vehicles are owned by third parties, will they largely be operated with individuals and known companions riding alone (like taxis), or will they serve multiple unknown people at the same time (like transit or pooled TNCs)? This latter approach, which we refer to as SAVs, could reduce costs even further by maximizing vehicular occupancy.

Models that consider the introduction of SAV systems suggest that it is possible to reduce the automobile fleet's size if linked to vehicle repositioning to serve appropriate trips [42,43]. This, in turn, could reduce the need for parking and allow for urban redevelopment in its place. At the same time, cheaper AV service would increase VMT, specifically if rides are not shared [44], and particularly in urban cores, where parking is now often costly [45]. Empty unshared cars will circulate around, either waiting for their owners to "pick them up" or in pursuit of long-away parking lots [46].

A mode shift toward SAVs at the primary travel mode may be unlikely. Of those surveyed by Cools *et al.* [47], 60% agreed they would prefer owning a private AV, rather than using TNC services exclusively, and only 47% agreed they would be willing to share AVs with strangers. Even so, those levels would be an improvement over the 91.5% of U.S. households who own at least one vehicle and the 76% of commuters who drive alone to work, according to 2017 U.S. Census data.

If regulated in the same manner as TNCs today and not integrated into the larger-vehicle transit system, AVs (whether shared or not) offering cheaper-than-TNC service can be expected to have several impacts on transit. First, an increase in VMT from AVs would slow the road network, therefore slowing surface transit. Second, the ability to avoid paid parking would encourage people to opt-out of traditional transit services, since their overall costs would decline. And third, as we detail below, expanded accessibility to travel for the young, the old, the disabled, and others who currently rely on transit, would, in turn, reduce transit ridership. In other words, an AV system that simply replicates the TNC status quo would magnify its deleterious effects on transit.

7.2.3   Equity, accessibility, and environmental sustainability

If unregulated automobile-sized AV service may challenge traditional transit operations, it is nonetheless worth emphasizing the potentially large benefits of AV rollout from the perspective of equity and accessibility. By lowering costs compared to TNCs, SAVs could increase mobility for many low-income people who do not currently have automobile access, easing links to jobs and services for people who live in a society largely built around automobiles and who are currently stuck using inadequate transit [48].

Increased mobility in much of the United States is unlikely to be provided anytime soon through traditional transit expansion—though automated technology may make doing so somewhat easier—because of the high cost of moving large vehicles around low-density communities. If two planning goals are efficiently transporting people while serving those who are least able to afford travel, SAVs may fill an important gap [49]. A similar argument can be made for AVs improving access for people who are left out of the automobile-dominated transportation system for other reasons, such as being too old, too young, or physically disabled.

Nevertheless, if automobile-sized AVs are deployed as a continuation of the current privately owned automobile system or current TNC business models, these equity benefits will be undermined by increasing traffic and continued inequality in access. Low-income individuals who continue riding the bus as AVs clog the streets (because AVs remain unaffordable) will suffer from reduced mobility. Eased long trips for some, moreover, may spur urban sprawl and decrease easy connections to jobs and services, further widening disparities in access.

Finally, the environmental impacts of automobile-sized AVs remain unclear. Though AV technology is frequently discussed in association with transportation-system electrification, most of the automobile-sized AVs currently being tested are powered by fossil fuels. Increased VMT from them could increase particulate emissions while worsening climate change. Here, too, eased access to far-off areas could be deleterious since low-density development is also associated with higher carbon emissions than neighborhoods designed around pedestrians and transit access.

We thus stand at a crossroads. AVs could benefit people who have been left behind by both the automobile and transit systems. Yet they could also worsen certain aspects of urban life, including congestion, pollution, and inequality. In Section 7.3, we articulate an approach to integrating transit and SAVs that attempts to reconcile the contrasting features of this new technology.

7.3   Opportunities for integration of SAVs with public transit

In this section, we articulate a vision for a future integrated AV and transit network that combines the positive elements of the two modes that we first identified in Figure 7.1. This vision, we believe, offers an opportunity to expand the directness of the transportation network—offering more useful access to more people—while maximizing its capacity. This, of course, is a speculative view, one whose manifestation would require policymakers to develop plans and regulations that orient new technology. But our hope is that, in offering a framework for how such integration can occur, we can best take advantage of AVs, rather than allow their introduction to worsen urban transportation systems.

Ainsalu *et al.* [34] identify two approaches that can be taken with regards to AV rollout: "pedestrian-friendly" versus "rider-friendly." The former focuses on building dense, walkable communities in which AVs complement a growing, productive transit system; this requires multimodal local planning. The latter prioritizes individual AV use, and the result would be more sprawl, congestion, and pollution. In some places, transit will thrive, being complemented by AVs; in others, it will struggle, suffering from the competition. Even so, we have identified almost universal consensus among leaders of U.S. city transportation and planning departments for integrating the two modes. In a representative survey of more than 120 cities with populations of 100,000 or more, we found 88.4% of officials agreed that they should redesign transit in the context of AVs [50].

Here, we return to the index of road-space usage efficiency we described in Section 7.3. In Figure 7.3, we show how optimal road-space efficiency can be achieved using different modes, depending on the number of passengers moved along a corridor. Given constrained street space, different vehicle types fulfill different demands. In this view, it is most efficient to move people by automobile (SAV) if one to four people are sharing a route; by shuttle, if 5 to 10 people are on the same path; and by bus otherwise. This theoretical model, of course, assumes

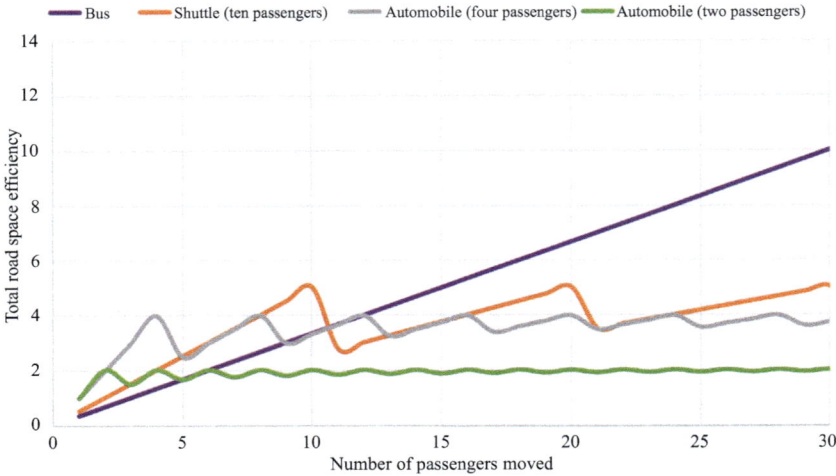

Figure 7.3   Road space efficiency by mode

that people are willing to share vehicles and that it is possible to allocate people to the "appropriate" vehicle when necessary. Despite this limitation, this model can help us identify which modes make the most sense in which circumstances.[†]

7.3.1   Synergies involved in the integration between SAV and transit

If we come to see automobile-sized AVs and transit as two transportation modes competing for riders and road space, our cities will suffer from increased traffic and transit systems will become increasingly difficult to operate effectively. Yet the advent of AVs offers an alternative: Expanding, rather than decreasing, transit's footprint through a multimodal travel system incorporating vehicles of many sizes [51,52]. This means shared, automated transportation as a spectrum of services adapted to the needs of travelers in different areas—from sedan-sized cars and minivans to shuttles and buses. To make this spectrum of services useful, integration details matter.

Yan *et al.* [53] and Shen *et al.* [54] show that SAVs could replace low-ridership bus routes and provide last-mile access to stations, in so doing reducing wait and travel times, while consuming fewer road resources and using public resources more judiciously. Wen *et al.* [55] emphasize how SAVs can serve low-density, suburban areas—neighborhoods where transit performs least effectively—

[†]It is worth emphasizing that this simplified model assumes three fixed vehicle sizes. With AVs, however, a wider variety of vehicular options may be available (especially if crash-safety standards are reduced due to fewer crashes); moreover, it is feasible to envision AVs traveling in platoons, which would reduce their relative footprint on the street network.

and find that doing so could reduce car use and increase transit ridership. An SAV-transit synergy means providing the level of service appropriate for the demand, in terms of vehicle size and frequency, altered based on geography and time. In doing so, the transportation system can meet the needs both of "mass transit"—moving the most people possible through limited space—and "social transit"—guaranteeing mobility and accessibility to all.

Even in an optimistic AV rollout scenario [49], many cities will continue to need high-capacity transit to serve the highest-demand areas. Major regions feature routes that carry more than 20,000 people per direction per hour—much more than is possible with automobile-sized SAVs [56]. The capacity limitations of such vehicles are two-fold: first, given limited room and little public support for highway expansion, existing roadways simply do not have the capacity to move an adequate number of vehicles. Second, pedestrian-accessible transit stations served by trains carrying up to 2,000 people each are more capable of responding to demands generated by dense employment zones than streets.[‡]

To supplement those high-demand routes, SAVs could offer first- and last-mile connectivity where origins and destinations are not within walkable distance of transit stations, or when temporal conditions merit it—for example, at night. Routes operated by larger automated shuttles and buses could be interlined. In the absence of an extensive rail network, fixed-route automated buses operated in high-demand corridors or line-hauls may also improve transit efficiency. With operational-cost savings, transit authorities could deliver more frequent services to a wider suburban geography, reduce passenger waiting and walking times, and improve ridership and fare revenue.

In lower-density areas, lower-capacity services would be most appropriate. Depending on the neighborhood, it might be reasonable to provide service using automated buses on fixed routes, automated shuttles, or automobile-sized SAVs. The latter two services could be offered on demand, with routes adjusted based on passenger requests, replacing low-productivity transit routes and add service in areas that currently have none. In the process, transit agencies will need to consider route redesigns, focusing on large-vehicle, fixed routes in areas with heavy demand and providing small-vehicle, and adjustable routes in areas with light demand. Attracting passengers to these services will require a significant change in norms among residents of such areas, now used to traveling alone. This may be the greatest challenge in the implementation of any integrated system.

Services may also be differentiated by speed and distance. For longer trips requiring faster speeds to maintain reasonable travel times, fixed-guideway, large-vehicle, and limited-stop services may be more appropriate than road-based SAVs. This is particularly true in cities with road capacity constraints, where high-speed

---

[‡]For example, if we envision an AV service provided by four-person vehicles the size of the Toyota Camry (4.9 m by 1.8 m), you would need 500 vehicles, taking up a total of 4,400 square meters (with no space between them) to provide the same capacity as a 2,000-person train. For comparison, a New York City subway train is about 446 square meters in surface area, so it is at minimum ten times as space efficient.

travel is impossible at peak hours, even on expressways. Identifying appropriate solutions for this sort of operational integration requires further analysis and will surely be location-specific.

7.3.2   Details of SAV-transit service integration

Integrating automobile-sized SAVs with automated shuttles and an automated transit system will require partnerships between modes to avoid overlapping services and congestion, and to promote multimodal trips [57]. We identify four strategies to encourage such partnerships.

First, information about demand, system use, and performance must be integrated between operators. Data are essential to optimize the provision of appropriate-scale vehicles along appropriate routes. Information integration allows the system to anticipate, and respond to, demand. The primary goal of information sharing is to guarantee high-quality options throughout a metropolitan area without overwhelming the system with unnecessary service. Today, TNCs and transit offer overlapping services in the same places and at the same times; the result is frequently empty seats and wasted service miles, for which we all pay the price, not only from service inefficiencies but also from the resulting additional congestion and pollution.

From a governance perspective, integrating access to information requires either a centralized control system—all shared transportation services operating under the same roof—or a series of cautious, well-executed contracts between providers. This means real-time, clearly defined data about where passengers are and where services are offered. And it means that operators must share information to one another and to users about services they plan to provide. A multimodal journey planner integrating transit options with access/egress alternatives can help riders make better travel choices by providing estimates for real-time travel and transfer times, cost, carbon footprint, and more.

Such information is closely guarded by TNC operators today, and the consequence is that planners considering whether to upgrade or downgrade transit routes operate on only partial knowledge. The impulse to hide data makes sense in a private sector where competition is the primary motivation. Yet transportation has an important role to play for society as a whole, and the negative externalities of such competition mean that AVs must be approached as a public service with freely shared information.

Second, pricing strategy and policies must be integrated. For riders to feel comfortable moving between multiple services, they need to be confident that their rides will reflect the service provided, no matter which type of vehicle they happen to be riding in. The question of setting fare policy is especially important from the perspective of ensuring equitable access. If automobile-sized SAVs, automated shuttles, and automated transit are all made available to low-income people or people in currently transit-inaccessible neighborhoods, for example, they will require subsidies; but those subsidies will only be reasonably allocated if all operators are on the same page about what fares to charge for a certain type of trip.

The pricing strategy of an integrated system will require a compromise between affordability to the full spectrum of the public, political acceptance, and ensuring adequate revenues to cover operations costs. Fare policies should encourage achieving the maximum road-usage index, as presented previously. Rider costs must be distributed so as to incentivize vehicle use that minimizes street-network impact, and that sometimes should require providing monetary incentives to transfer to higher-capacity vehicles on high-demand portions of their journeys. As such, a closed-circuit system where operators encourage customers to use just their services (e.g., transit riders stay on transit; Uber riders stay on Uber) is unacceptable.

A fully integrated system that shifts much of the population out of single-occupancy automobiles will produce beneficial knock-on effects, including lower pollution, less congestion, less sprawl, and more equitable access. These externalities will be difficult to incorporate into the network's revenue and subsidy structure, but these broader impacts must be integrated into its financial structure. This will require political initiative to make the link successfully.

Relatedly, all operators in the SAV and transit network must use a single set of ticketing technologies. This means that a passenger using a fare card, a cellphone-based ticket, or anything else accepted on one mode should be able to use that same technology to take a new trip or transfer onto any other mode.

Third, operations must be integrated to allow seamless access to the entire system. This means ensuring that various-sized vehicles are available as necessary at major transit nodes. It would be unreasonable to make changes to bus services that eliminate low-density routes unless users continue to have access (both physically and monetarily) to the new SAVs that replace them. It also means limiting the number of automobile-sized AVs in dense zones where their mass deployment would increase congestion rather than increase mobility.

Integrated operation could also include service coordination and timed transfers. By coordinating with train departure times, SAV itineraries could be optimized to reduce waiting at transfer points (first-mile), and vehicle rebalancing trips can coordinate with train arrivals and among operators to avoid service shortages or congestion due to over-rebalancing (last-mile). Central dispatching and passenger matching of vehicles among multiple SAV providers could further reduce vehicle idle times and VMT. For example, an Uber request could pool with a Lyft rider if the two operators were integrated, but this, again, would require altering the competitive mentality that now dominates among TNC operators.

Similarly, the business models of multiple operators require coordination. This may mean the development of a fully public system, with publicly supported and operated services at all scales, or it might mean a mix of public and private operations. Yet in all cases, all operators must follow the policy that supports ensuring adequate transportation service to the whole population. An integrated business model could incorporate shared incentives among different operators. For example, a passenger can be transferred from one SAV to another if this transfer would reduce VMT and both operators have enough incentive to coordinate and execute this transfer.

Fourth, and finally, regulations related to the system must be coordinated, perhaps requiring the development of a separate set of codes specific to this new network. Today, the public sector typically oversees transit planning and operations, determining which routes go where, how much labor is paid, and minimum levels of service. On the other hand, regulations over TNCs are far less explicit; other than in New York, no American city has mandated a minimum take-home pay for drivers. And operations are generally allowed wherever—with no consequences for operators if they provide insufficient service in certain neighborhoods.

In the context of an integrated system, service standards must be treated similarly across the full range of offerings. Similarly, riders' rights, such as their ability to resolve legal concerns or address poor service, must be put on an even playing field. Even with multiple operators, people using the system should be comfortable raising complaints and resolving problems in a uniform fashion.

Together, these strategies offer a framework for integrating transit networks with AVs, offering variously sized vehicles for different needs and opportunities. These strategies are rooted in a view of the transportation system as a single network. It is unreasonable and untenable, in our view, to continue separating modes into individual services experienced differently by riders and, in the process, furthering the problems of inequality, pollution, and congestion. At the same time, we are aware that such an integrated view is not universally supported, as it would limit potential profits and could reduce certain types of service innovation. Making decisions about specific aspects of these strategies requires a broad debate about what goals we prioritize for the transportation system, which we describe in Section 7.4.

7.4   Transit-oriented versus generic SAV deployment

In this chapter, we make the case for an integrated transportation system that unites traditional transit with future AVs of various sizes. This system would preserve many of the accessibility-expanding benefits of automobile-sized AVs while circumventing many of the negative products of a system that replicates the TNC experience. If AVs are deployed without being integrated with transit, we fear that cities will have unproductive and ineffective fixed-route bus and rail systems relegated to the least well off, congestion in dense neighborhoods, and higher levels of pollution. These are outcomes to be avoided.

The details of an integrated system, however, remain to be formulated. We have not touched on what are likely to be intense disagreements between levels of government about who should be regulating what, for example. Nor do we explore other important aspects of mobility, such as bikesharing or freight. We recognize that planning for AVs is at a preliminary stage. Nevertheless, there is considerable interest among cities to begin planning to take these vehicles into account [58]. We thus end this chapter by asking a series of questions that require further research and debate. We explore how a transit-oriented SAV system—the integrated network we believe cities should promote—differs from what we refer to as "generic"

SAV deployment, which would replicate current TNCs. We consider these differences across four levels: The social purpose expected from each, regulations related to and required for each, different agents involved and intrinsic behavioral differences among them, and the algorithmic specifications that pertain to each orientation of AV technology deployment. In order to advance an integrated system, certain norms will need to change: as a society, we will need to reconsider how we use the transportation system, and we will need to think about multiple options as a spectrum of services, not as a series of differentiated modes.

7.4.1 Who are we to serve?

Broadening access to the transportation system is an equity question as much as a mobility one. Generic SAV deployment might mean simply allowing TNCs to offer service wherever makes the most sense for their profit margins. SAVs would be accessible throughout metropolitan areas, but at significantly higher costs than transit, especially for longer trips, thereby excluding a portion of the population from their use. They would concentrate in dense neighborhoods, increasing congestion.

An alternative approach—transit-oriented SAV deployment—focuses on ensuring equitable access without worsening service on the existing system. To ensure affordability for everyone, one approach would be to use means-testing to determine user fares throughout the transportation system, providing subsidies for certain groups of people, such as low-income families, but doing so would create a daunting bureaucracy. Moreover, what level of subsidy is reasonable? Do we envision transportation costing the average person about as much of their salary as it costs to own and operate a personal car today (about $10,000 a year), or about as much as it costs to subscribe to a monthly unlimited transit pass (about $1,500 a year)? Do we need to also consider peoples' ability to convert monetary means into their real capabilities to access transport, jobs, and other services?

Let us also consider the question of geography. An integrated transit-SAV network, given its breadth, would serve many more people and trips than today's transit. It would connect people living and working in low-density, often unwalkable, suburban neighborhoods whose communities make them inhospitable to trains or buses. Should people in those areas receive equal service, at the same cost, as those in zones where providing collective services is more efficient, neighborhoods that are more walkable and friendlier to higher-capacity transit?

Finally, let us return to the issue of vehicle size. People living in lower-density neighborhoods would also be those receiving service with smaller vehicles, while people in dense neighborhoods ride in higher-capacity shuttles, buses, and trains for many of their trips. Is there a difference in comfort level between these options? Is it unfair to give some people access to a certain type of vehicle, while depriving it from others, just because of where they live?

7.4.2 How do we regulate?

The street transportation system today is largely divided into publicly planned, publicly operated transit systems; TNCs run by multinational corporations; and individuals driving privately owned automobiles. A generic SAV rollout would reinforce this modal disintegration. It could decrease the number of people driving personal automobiles due to lower costs of SAVs, but it would continue to promote competition among both individuals who own automobiles and private corporations providing services. The result would be overlapping services and a less efficient system.

An integrated transit-SAV network eliminates this division by transforming the relationship between public and private actors. First, it suggests treating automated, automobile-sized TNCs as the base level of a spectrum of shared transportation services and recommends integrating them in terms of pricing, planning, and operations. Second, it is founded on the principle that a full suite of mobility options will encourage a significant mode shift away from private automobiles.

This network requires a growing role for the public sector in planning, coordinating operations, and identifying sources of subsidies. Would this expansion of government's involvement be politically acceptable? Would it be acceptable for some countries, states, and cities to engage in developing a publicly supervised, integrated system, whereas others choose limited, if any, regulations over private AV operators?

7.4.3   How will riders behave?

The deployment of a generic SAV system would use price-discrimination mechanisms to encourage riders to choose shared vehicles, but also continue to allow them to choose to ride in AVs alone, as TNCs do today. It would provide door-to-door rides, doing little to diminish the automobile orientation of most neighborhoods. This would reinforce the experience that pervades in American cities; many people would continue to understand the transportation system as one that is experienced primarily by individuals moving alone or with people they know.

The integrated transit-SAV approach, on the other hand, prioritizes moving many people together and maximizing street capacity. As such, it would minimize levels of solo travel and try to move as many people as possible in shared vehicles of various sizes. It might encourage people to walk a few blocks to get onto an SAV, rather than receive door-to-door service. It would also incentivize multimodality through transfers between modes to maximize system efficiency.

The integrated strategy would attempt to alter the behavioral norms of transportation in the United States. But to what degree would the average person be willing to support travel on a sharing-heavy network, even if it were cheaper? Would concerns about interacting with people of different social groups and economic classes raise problems? Would people be willing to make walking a key part of their daily routines, or would they be hostile to this change? And would a transfer-based system make transportation inconvenient to many, even if it were faster?

7.4.4   What types of algorithms are needed?

The current TNC system responds to passenger demand in bursts. A major event, for example, may attract hordes of TNCs outside a stadium, waiting to pick people up. The result is ineffectiveness and congestion—the same outcome that would occur in the context of generic SAV rollout. These problems are magnified in the context of competition between operators.

An integrated transit-SAV network would engage an alternative approach, organizing passengers by final destination, perhaps, placing them first in larger vehicles and then having them transfer later in their journeys to smaller shuttles or automobile-sized vehicles. Such coordination may require advanced algorithms for addressing the vehicle routing problem with time windows (VRPTW) in real-time which is known to be computationally expensive [59]. There is a gap in heuristic solution algorithms that can solve VRPTW quickly enough for real-time vehicle dispatching in transit-SAV networks. Such formulations would also apply to first-mile services, when dispatching and routing decisions can be made with information on transit departure times to minimize transfer delays.

In order to account for and reduce the negative footprint of SAVs on the network, such as impacts on traffic, congestion-aware SAV routing approaches will be desired. System Optimal Dynamic Traffic Assignment algorithms can be utilized to optimize SAV paths in order to reduce path-choice externalities [60,61]. Congestion impacts can be considered in SAV dispatching decisions (vehicle-to-passenger assignment and ride-share choices) for a more effective minimization of SAV externalities [62]. SAV rebalancing decisions can also be optimized to reduce network impacts.

In creating these algorithms, transportation providers will have to consider what to prioritize: Access to all or congestion relief? Speedier service or higher efficiencies? Higher-paying customers or the public at large? Developing a clear sense of the public-policy goals that inform how these algorithms should work is a key element of any integrated transit-SAV system.

7.5 Conclusion: searching for new avenues for integration

In this chapter, we offer a vision for enhancing urban transportation. We show that TNCs have largely competed with, and undermined, transit systems, often to the detriment of city mobility. Left unchecked, AVs will replicate or worsen this experience. But by integrating transit systems with SAVs across a variety of vehicle sizes, cities will be able to respond more effectively to travel demand while improving access to the population. An integrated system will address many of transit's flaws—particularly its limited breadth due to its need to concentrate on the densest neighborhoods—while reducing the negative impacts of TNCs produced by their low capacity, such as increased traffic. But achieving an integrated system will not be easy; it will require real action by public officials and a commitment to developing a new approach to using the automobile. Challenging the norm that suggests that it is acceptable to occupy most of the public right-of-way with single-occupancy automobiles will require altering the behavioral assumption that people have the right to travel alone in virtually all circumstances. It will also require breaking the lock on access to data and decisions about service that many private transportation operators hold today.

Acknowledgments

The authors wish to thank Gabriel Sanchez Martinez, Zhejing Cao, Mary Rose Fissinger, and Yang Xin for reviewing drafts of the chapter.

Author contributions

The authors confirm contribution to this chapter. All authors reviewed and approved the manuscript.

Declaration of conflicting interests

The authors declared no potential conflict of interest with respect to publication of this chapter.

References

[1]   Watkins, K. "Does the future of mobility depend on public transportation?" *Journal of Public Transportation.* 2018; 21(1). 53–59.

[2]   Federal Transit Administration. 2017 National Transit Summary and Trends. Office of Budget and Policy. 2018. Available from https://www.transit.dot. gov/sites/fta.dot.gov/files/docs/ntd/130636/2017-national-transit-summaries-and-trends.pdf.

[3]   Delbosc, A. Currie, G. Nicholls, L. and Maller, C. "Social transit as mass transit in Australian suburban greenfield development." *Transportation Research Record: Journal of the Transportation Research Board.* 2016; 2543: 62–70.

[4]   Ferguson, E. "The rise and fall of the American carpool: 1970–1990." *Transportation.* 1997; 24(4): 349–376.

[5]   Polzin, S. "The decline of carpooling—Can App-based carpooling reverse the trend?." *Planetizen.* 2015, February 25. Available from https://www. planetizen.com/node/74522/decline-carpooling%E2%80%94can-app-based-carpooling-reverse-trend.

[6]   Farajallah, M., Hammond, R.G. and Pénard, T. "What drives pricing behavior in peer-to-peer markets? Evidence from the carsharing platform BlaBlaCar." *Information Economics and Policy.* 2019.

[7]   Shaheen, S., Cohen, A. and Zohdy, I. *Shared mobility: Current practices and guiding principles.* U.S. Department of Transportation. 2016.

[8]   Shaheen, S., Stocker, A. and Mundler, M. "Online and App-based carpooling in France: Analyzing users and practices—A study of BlaBlaCar." in

Gereon Meyer and Susan Shaheen (eds.). *Disrupting mobility: Impacts of sharing economy and innovative transportation on cities.* New York City: Springer; 2017. pp. 181–196.

[9]   Dahir, A.L. "Uber has launched its first bus service in Egypt." Quartz Africa. 05 December 2018. Available from https://qz.com/africa/1484896/uber-launches-first-bus-service-in-cairo-egypt/.

[10] Etherington, D. "Lyft shuttle is an experimental new Lyft line feature that works like a bus route." *TechCrunch.* 29 March 2017. Available from https://techcrunch.com/2017/03/29/lyft-shuttle-is-an-experimental-new-lyft-line-feature-that-works-like-a-bus-route/.

[11] Massachusetts Bay Transportation Authority. *On-demand paratransit pilot program.* 2019. Available from https://www.mbta.com/accessibility/the-ride/on-demand-pilot.

[12] Woodman, S. "Welcome to Uberville." *The Verge.* 01 September 2016. Available from https://www.theverge.com/2016/9/1/12735666/uber-alta-monte-springs-fl-public-transportation-taxi-system.

[13] Federal Transit Administration. *Monthly module adjusted data release November 2018.* 2019. Available from https://www.transit.dot.gov/ntd/data-product/monthly-module-adjusted-data-release.

[14] Wodinsky, S. "In major defeat for Uber and Lyft, New York city votes to limit ride-hailing cars." *The Verge.* 08 August 2018. Available from https://www.the-verge.com/2018/8/8/17661374/uber-lyft-nyc-cap-vote-city-council-new-york-taxi.

[15] Alemi, F. *What makes travelers use ridehailing? Exploring the latent constructs behind the adoption and frequency of use of ridehailing services, and their impacts on the use of other travel modes. Dissertation,* University of California Davis. 2018.

[16] Circella G., and Alemi, F. "Transport policy in the era of ridehailing and other disruptive transportation technologies." in Yoram Shiftan and Maria Kamargianni (eds.). *Advances in transport policy and planning: Volume I: Preparing for the new era of transport policies: Learning from experience,* Amsterdam: Elsevier. 2018.

[17] Clewlow, R.R. and Mishra, G.S. *Disruptive transportation: The adoption, utilization, and impacts of ride-hailing in the United States.* Research Report UCD-ITS-RR-17–07. University of California Davis Institute of Transportation Studies. 2017.

[18] Schaller, B. *The new automobility: Lyft, Uber and the future of American cities.* Schaller Consulting. 25 July 2018.

[19] Young, M. and Farber, S. "The who, why, and when of Uber and other ride-hailing trips: An examination of a large sample household travel survey." *Transportation Research Part A.* 2019; 119: 383–392.

[20] Schwieterman, J. and Smith, C.S. "Sharing the ride: A paired-trip analysis of UberPool and Chicago Transit Authority services in Chicago, Illinois." *Research in Transportation Economics.* 2018; 71: 9–16.

[21] Jin, S.T., Kong, H., and Sui, D.Z. "Uber, public transit, and urban transportation equity: A case study in New York City." *The Professional Geographer.* 2019.

[22] Graehler, M., Mucci, R.A., and Erhardt, G. "Understanding the recent transit ridership decline in major U.S. cities: Service cuts or emerging modes?' Conference Paper, Annual Meeting of the Transportation Research Board. 2019.

[23] Hall, J.D. Palsson, C., and Price, J. "Is Uber a substitute or complement for public transit?' *Journal of Urban Economics*. 2018; 108: 36–50.

[24] Nelson, E. and Sadowsky, N. "Estimating the impact of ride-hailing App company entry on public transportation use in major US urban areas." *B.E. Journal of Economic Analysis & Policy*. 2018; 19(1).

[25] Kong, H. Zhang, X. and Zhao, J. "How does ridesourcing substitute public transit?' Under review at *Journal of Transport Geography*. 2019.

[26] Hampshire, R.C., Simek, C., Fabusuyi, T., Di, X. and Chen, X. "Measuring the impact of an unanticipated suspension of ride-sourcing in Austin, Texas." Available at SSRN. 2017.

[27] Freemark, Y. "U.S. transit systems are shedding riders: Are they under threat?' *The Transport Politic*. 18 May 2018. Available from https://www. thetransportpolitic.com/2018/05/18/u-s-transit-systems-are-shedding-riders-are-they-under-threat/.

[28] Boisjoly, G., Grisé, E., Maguire, M., Veillette, M-P, Deboosere, R., Berrebi, E., and El-Geneidy, A. "Invest in the ride: a 14 year longitudinal analysis of the determinants of public transport ridership in 25 North American cities." *Transportation Research Part A*. 2018; 116: 434–445.

[29] Manville, M., Taylor, B.D., and Blumenberg, E. *Falling Transit Ridership: California and Southern California*. UCLA Institute of Transportation Studies. *Southern California Association of Governments*. 2018.

[30] Henao, A. and Marshall, W.E. "The impact of ride-hailing on vehicle miles traveled." *Transportation*. 2018.

[31] Castiglione, J., Cooper, D., Sana, B., et al. "TNCs and congestion." *Civil Engineering Reports*. 2018;1.

[32] Rudick, R. "Supervisor shocked to hear Uber and Lyft violate bike and transit lanes." *Streetsblog SF*. 26 September 2017. Available from https://sf. streetsblog.org/2017/09/26/supervisor-shocked-to-hear-uber-and-lyft-vio-late-bike-and-transit-lanes/.

[33] UITP. World Metro Figures: Statistics Brief. 2018. Available from https:// www.uitp.org/sites/default/files/cck-focus-papers-files/Statistics%20Brief%20-%20World%20metro%20figures%202018V4_WEB.pdf.

[34] Ainsalu, J. Arffman, V., Bellon, M., et al. "State of the art of automated buses." *Sustainability*. 2018; 10(9): 3118.

[35] Quarles, N., and Kockelman, K.M. "Costs and Benefits of Electrifying and Automating U.S. Bus Fleets." Transportation Research Board Annual Meeting. 2018

[36] Sarriera, J.M. Salvucci, F.P., and Zhao, J. "Worse than Baumol's disease: The implications of labor productivity, contracting out, and unionization on transit operating costs." *Transport Policy*. 2018; 61: 10–16.

[37] Cohen, J.M., Barron, A.S., Anderson, R.J., and Graham, D.J. "Impacts of unattended train operations on productivity and efficiency in metropolitan

railways." *Transportation Research Record: Journal of the Transportation Research Board.* 2015; 2534: 75–83.

[38] Wen, J., Nassir, N., and Zhao, J. "Value of demand information in autonomous mobility-on-demand systems." *Transportation Research Part A.* 2019; 121: 346–359.

[39] Hyland, M.F., and Mahmassani, H.S. "Taxonomy of shared autonomous vehicle fleet management problems to inform future transportation mobility." *Transportation Research Record: Journal of the Transportation Research Board.* 2017; 2653(1): 26–34.

[40] Polzin, S., *Implications to public transportation of emerging technologies.* National Center for Transit *Research.* 2016.

[41] Faisal, A., Yigitcanlar, T., Kamruzzaman, Md., and Currie, G. "Understanding autonomous vehicles: A systematic literature review on capability, impact, planning and policy." *The Journal of Transport and Land Use.* 2019; 12(1): 45–72.

[42] Ma, J., Li, X., Zhou, F., and Hao, W. "Designing optimal autonomous vehicle sharing and reservation systems: A linear programming approach." *Transportation Research Part C.* 2017; 84: 124–141.

[43] Zhu S., and Kornhauser, A.L. "The interplay between fleet size, level-of-service and empty vehicle repositioning strategies in large-scale, shared-ride autonomous taxi mobility-on-demand scenarios." Transportation Research Board Annual Meeting. 2017.

[44] Fagnant, D.J., and Kockelman, K.M. "Dynamic ride-sharing and fleet sizing for a system of shared autonomous vehicles in Austin, Texas." *Transportation.* 2018; 45(1): 143–158.

[45] World Economic Forum (WEF). *Reshaping urban mobility with autonomous vehicles: Lessons from the City of Boston.* In collaboration with the Boston Consulting Group. 2018.

[46] Millard-Ball, A. "The autonomous vehicle parking problem." *Transport Policy.* 2019; 75: 99–108.

[47] Cools, M., Rongy, C., and Limbourg, S. "Can autonomous vehicles reduce car mobility? Evidence from a stated adaptation experiment in Belgium." *Transportation Research Board Annual Meeting.* 2017.

[48] King, D.A., Smart, M.J. and Manville, M. "The poverty of the carless: Toward universal auto access." *Journal of Planning Education and Research.* 2019.

[49] Polzin, S. "Just around the corner: The future of U.S. public transportation." *Journal of Public Transportation.* 2018; 21(1): 43–52.

[50] Freemark, Y., Hudson, A.W., and Zhao, J. "Policies for autonomy: How American cities envision regulating automated vehicles." *Urban Science.* 2020; 4(4): 55.

[51] Currie, G. and Lies, D. "AVs, shared mobility, and urban transit futures." *Journal of Public Transportation.* 2018; 21(1): 19–30.

[52] UITP. *Autonomous vehicles: A potential game changer for urban mobility.* 2017.

[53] Yan, X., Levine, J., and Zhao, X. "Integrating ridesourcing services with public transit: An evaluation of traveler responses combining revealed and stated preference data." *Transportation Research Part C.* 2018.

[54] Shen, Y., Zhang, H., and Zhao, J. "Integrating shared autonomous vehicle in public transportation system: A supply-side simulation of the first-last mile service in Singapore." *Transportation Research Part A.* 2018; 113: 125–136.

[55] Wen, J., Chen, Y.X., Nassir, N., and Zhao, J. "Transit-oriented autonomous vehicle operation with integrated demand-supply interaction." *Transportation Research Part C.* 2018; 97: 216–234.

[56] Buehler, R. "Can public transportation compete with automated and connected cars?' *Journal of Public Transportation.* 2018; 21(1): 7–18.

[57] Tsay, S-P., Accuardi, Z., and Schaller, B. *Private mobility, public interest: How public agencies can work with emerging mobility providers.* Transit Center. 2016.

[58] Freemark, Y., Hudson, A.W., and Zhao, J. "Are cities prepared for autonomous vehicles? Planning for technological change by U.S. local governments." *Journal of the American Planning Association.* 2019; 85(2): 133–151.

[59] Miguel Andres Figliozzi. "An iterative route construction and improvement algorithm for the vehicle routing problem with soft time windows." *Transportation Research Part C.* 2010; 18(5): 668–679.

[60] Peeta, S., and Mahmassani, H.S. "System optimal and user equilibrium time-dependent traffic assignment in congested networks." *Annals of Operations Research.* 1995; 60: 81–113.

[61] Tajtehranifard, H., Bhaskar, A., Nassir, N., Haque, Md. M. and Chung E. "A path marginal cost approximation algorithm for system optimal quasi-dynamic traffic assignment." *Transportation Research Part C: Emerging Technologies.* 2018; 88: 91–106.

[62] Levin, M.W. "Congestion-aware system optimal route choice for shared autonomous vehicles." *Transportation Research Part C: Emerging Technologies.* 2017; 82: 229–247.

The views expressed are those of the authors.

Chapter 8

Design of systems with nonautomated electric vehicles

Ata M. Khan[1]

An advanced form of shared mobility service can be offered by accessing on-demand, with a cell phone App, nonautomated electric vehicles located at park and charge stations in an urban area. Another version of this service is the offer of the free-floating vehicle that can be accessed at a site other than a park and charge station. These multistation and free-floating vehicle sharing services require the customer to drive the vehicle. The ride-hailing service using electric vehicles driven by the transportation network company drivers can be enhanced with access to fast chargers and infrastructure support provided by urban governments for pick-up and drop-off tasks. These shared mobility services have the potential to meet the objectives of Mobility as a Service (MaaS).

This chapter covers the design of systems using nonautomated electric vehicles for providing shared mobility services safely and efficiently. Design requirements are defined and measures to meet these are advanced. The level of service requirement of the traveler and the efficiency objective of the supplier of service guide the designs within the urban travel context. Application of intelligent transportation system and advanced communication technologies play a role in system design.

8.1 Introduction

Shared mobility in the form of mobility on demand (MOD) and as a part of Mobility as a Service (MaaS) is attracting attention around the world [1–3]. Other notable developments are the emergence of transportation network companies (TNCs) and the intention of some original motor vehicle equipment manufacturers (OEM) to participate in the *MaaS* market [4,5]. As these developments are underway, city authorities are keen on determining how best to adapt land use, transportation network, and regulations for safe and efficient accommodation of shared mobility services [6,7].

[1]Department of Civil and Environmental Engineering, Carleton University, Ottawa, Canada

At present, the transportation service providers offer the following services with the use of nonautomated vehicles. In the car sharing service, a subscribing traveler can reserve and use a car for a specific trip. A number of companies provide this service, including Car2Go, Free2Move, Toro, and Zipcar [8]. The emergence of a "free-floating" car sharing system without rental stations was anticipated by the industry some years ago [9]. Volkswagen's WeShare system implemented in 2019 based on electric cars is a recent example that is in operation [9]. A customer uses an App to rent a car and can leave it anywhere within a well-defined rental area. The company's employees relocate it to a site of high demand such as a park and charge station. According to the future plan, customers will be offered incentives to charge cars at public charging locations [10].

Ride-hailing, another type of shared mobility service, gained popularity before the Covid-19 pandemic. The TNC provides this transportation on-demand (TOD) service using a smartphone APP to connect a traveler to a vehicle driver. A traveler requests a ride or a shared ride in a car driven by a driver who has a contract with a TNC. Examples include Uber, DiDi, Lyft, and Via [1,8]. A shuttle is yet another type of service, in which a traveler shares the use of a vehicle for a specific trip from an origin to a destination (e.g., a shuttle service between a rapid transit station and a medical campus or a shuttle service between a parking facility and an airport terminal).

Given the move to shared mobility using electric vehicles and the availability of advanced transportation and communication technologies, a number of developments are likely to take place in the postpandemic future. Of course, when safe and affordable automated vehicles become available, these developments will undergo further evolution.

8.2   Categories of services

Systems with nonautomated electric vehicles can provide two notable categories of service (Figure 8.1). In the first category, vehicles are driven by customers. Electric vehicles can be accessed by customers at a number of park/charge stations in the urban area. The stations feature parking stalls and some stalls have fast charging

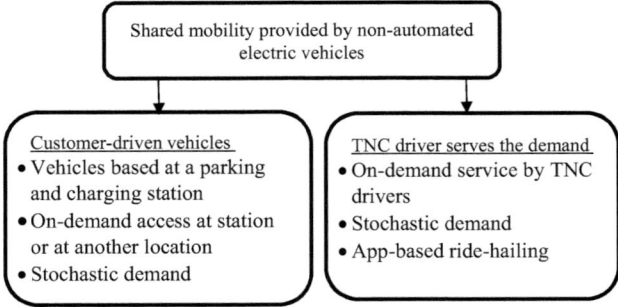

Figure 8.1   Shared mobility types

facilities. The vehicles are used for one-way or two-way trips. In the one-way trip service option, a customer need not bring a vehicle back to the origin location. A recent version of this category of service offers the option of free-floating cars that can be accessed at locations other than stations and need not be dropped off at a station either [5,10].

The vehicles and the station infrastructure (i.e., park/charge infrastructure) can be owned by a private sector entity or a public–private partnership. Private sector owner could be a TNC, the MaaS arm of an OEM, or any other entrepreneur. The TNC or any other entrepreneur can potentially purchase vehicles by benefiting from fleet discounts. When fully operational autonomous vehicles will become available, such a vehicle can pick up the customer at a specified location. The MaaS arm of an OEM can also invest in the network of park and charge stations in a manner similar to a TNC and will be in a position to assign a fleet of vehicles at these strategically located park and charge stations.

The second category of service is the electric vehicle version of the present ride-hailing service provided by TNC drivers. When not in service for a prolonged period within the 24 h cycle, the vehicle will be parked and charged at the driver's residence. During service periods, the driver can access fast-charging facilities in the urban area.

In order to fully participate in the mobility as a service, the TNCs are expected to include in their smartphone application public transit routes, times, and fares. This application of advanced technology will encourage complete trips from origin to destination. An outcome of this intelligent networking initiative will increase the use of their service and at the same time enhance public transit ridership [1,11,12]. Both ride-hailing and park and charge station approaches are expected to meet the requirements of first-mile, last-mile, and also the complete trip.

8.3   Design framework for shared electric vehicle system

8.3.1   Design variables

A system based on the park and charge multistation model and its adjunct free-floating car service require advances in planning and design methods. Also, intelligent transportation system and communications technologies play essential roles.

The shared electric vehicle system can be characterized by the following components:

1. Multistations
2. User access to a vehicle
3. Vehicle fleet
4. Park and charge stalls
5. Operations and control
6. Maintenance

Past experience indicates that the performance of a shared vehicle system is sensitive to variables shown in Figure 8.2 [13,14]. In the following parts of this

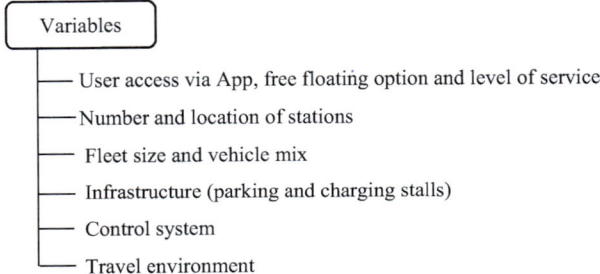

Figure 8.2   Variables

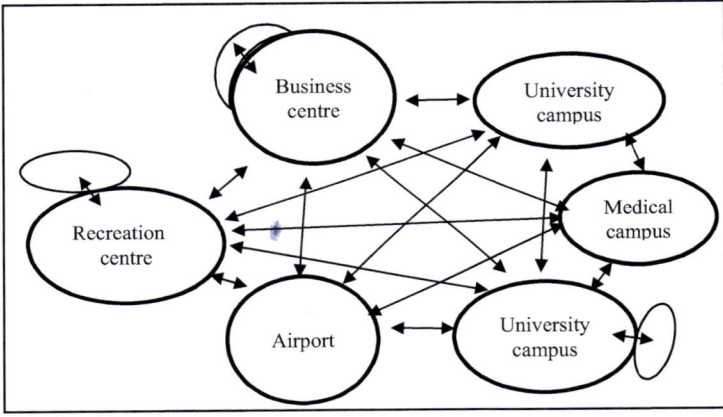

Figure 8.3   Multistation system. Adapted from [15]

chapter, the role of these variables in shaping shared vehicle system (SVS) design and operation is explained.

8.3.2   Multistation system

The multistation shared vehicle system is an organized way to provide access to vehicles in a well-defined service area [15] (see Figure 8.3). In this form of shared vehicle system, vehicles are used among multiple stations to go from one activity center to another. Such a system may be set up in any part of a metropolis, that is, university campus, resort and recreational areas, high access zones including medical campuses, shopping districts, employment centers, major railway and rapid transit stations, and airports.

In the latest form of the multistation system, customers use cars in a free-floating manner in terms of returning a vehicle at a location other than a station and the operations and maintenance personnel relocate the vehicle to a station that has vehicle deficiency and the battery can be charged at that site. The proponents of the free-floating car sharing system suggest that a user can hire and drop off cars directly at or very close to demand points [16].

With the use of a smartphone, a multistation system can offer on-demand mobility. Such a system features park and charge stalls and a fleet of vehicles that can be accessed at these stations. Since the vehicles are not automated, a sub-scribing customer has to drive a vehicle for a one-way or a two-way trip. That is, a vehicle can be accessed with the use of an App at one station and returned to the same or another station. As noted above, there is also the option of free-floating service which will permit the customer to drop-off the vehicle at a site other than a station. If a vehicle of the required specification is not available, a short waiting time may be necessary. A vehicle can be used by one customer or a party. If parties have a common origin and destination and they agree to share the ride, this will be possible.

For a specific niche market and service area, the number of stations can be found that have the potential to generate demand for service by subscribing members while keeping vehicle relocations and other service factors under control. The stations can be located at sites with high density of travelers.

8.3.3 User access

The operation of multistation SVS is designed for convenient and efficient use by the customer. The use of the system by the user involves three major steps. These are on-demand request with a cell phone, accessing a vehicle with credit card, and use of vehicle en route (Figure 8.4). Following acceptance of the on-demand request, the customer approaches the location where the vehicle is parked and access to the vehicle is gained with credit card. As a part of the access step, the user ID is confirmed and the state of battery charge (SOC) is checked. Also, an esti-mated range before new charging will be required is displayed. The user drives the nonautomated vehicle and the system monitors the SOC. At the end of the trip, the

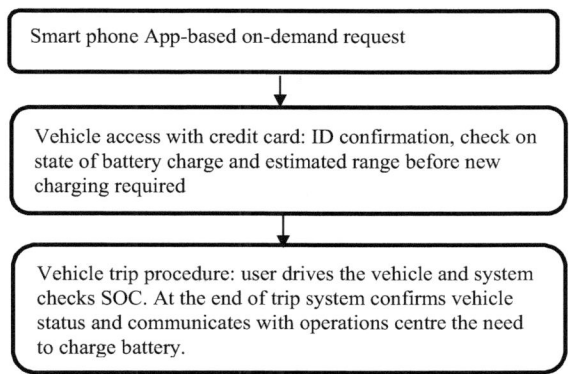

Figure 8.4   User and vehicle interaction

system confirms vehicle status and communicates with the operation center for the need to charge battery, if applicable.

For designing a modern SVS to serve the on-demand travel needs, details of system operation are required. In the earlier version of the multistation system, a number of issues were encountered, including slow kiosk-based registration, inaccurate estimates of vehicle imbalance among stations, and lack of availability of appropriate vehicle size. With intelligent technology applications supported by enhanced methods, these can be overcome.

The use of a smartphone App makes the kiosk-based registration obsolete. The App sends out the information on vehicle of specified size, the party size, and an estimate of time when the vehicle will be dropped off at the intended destination. The App will also provide the estimated wait time before a vehicle of specified size and required SOC will become available. If the customer accepts the access offer, the use of the credit card will enable quick start of the journey. While the customer is using the vehicle, the location and SOC are monitored by the operating system. At the end of the trip, the check-out process covers the automated billing to the customer account.

The control system updates the vehicle status and the need to charge battery, if applicable. From the user perspective, the only issue that could be encountered is the waiting time to access a vehicle of required size and necessary SOC for the trip. According to the literature, some companies are offering or intend to offer incentives to customers to charge batteries at public charging stations before returning the vehicle at the selected destination [10].

It is known that demand for travel is stochastic and predictions cannot be made with certainty. It is likely to be different at various stations and also at the same station at different times of the operation period. Therefore, the vehicles in the fleet may be disproportionately distributed within the multistation system with more vehicles at one station and fewer vehicles or no vehicle at another station. To correct this imbalance, there will be a need for relocation of excess vehicles from station(s) with surplus vehicles to station(s) with fewer or no vehicle. The effect of imbalance will be inefficient operation and the user demand will not be met at an acceptable level of service without corrective action.

8.3.4   *User expectations*

A knowledge of travel demand for specific purposes, its distribution among stations by time of day, and party or group size is obtainable from user need studies in the market to be served. Information on user sensitivity to average waiting time and queue length is useful for sizing the fleet and related components of the SVS. User socio-economic characteristics can guide the designer about user expectation of on-demand access and service quality. Users expect to access vehicles without much waiting time. On the supply side, the SVS owner aims to serve demand with minimum required resources including the fleet size. The balance of these objectives is studied in the form of vehicles/demand ratio, average waiting time, and queue length, among other factors. The number of customers (i.e., trip requests) ahead of a given customer quantifies the queue length.

8.3.5 Vehicle fleet and relocation

The fleet attributes are the number of vehicles of required sizes at each station at the start of daily operation. The vehicle fleet can be populated with electric vehicles of different sizes, such as one seater solo, two seater mini, and four seater sedan. A better match between vehicle size demanded by the customer and its availability contributes to improved demand and the economic viability of the system. Therefore, populating the vehicle fleet with appropriate vehicle mix constitutes a better design of the SVS.

Estimation of the fleet size for serving a niche market requires the use of a method that can balance the level of service offered to the user and operator's objective to avoid excessive costs on vehicles, relocating vehicle, and park and charge facilities. Therefore, a demand-supply balance of the system is useful in order to have just enough vehicles and infrastructure to achieve acceptable system efficiency and economic viability.

Since vehicles are not automated, relocation becomes necessary for correcting the vehicle imbalance among stations. A number of approaches have been suggested to solve the vehicle imbalance problem. Users can be offered an incentive to book more distant vehicles and drop off vehicles at a location other than their chosen destination. Also, paid relocation as well as demand pooling (i.e., to join a ride or give a lift to another user) were studied. However, according to the literature, these approaches have not been very effective [17].

Therefore, the SVS operator is left with little choice but to initiate timely relocation strategies. For this purpose, intelligent transportation and communication systems can assist in maintaining on a real-time basis, a dynamic database to track vehicles in the system and their battery state-of-charge (SOC). As a part of the relocation strategy, predictive models can help in estimating the need for relocation from locations of excess vehicles to other parts of the multistation system where there are likely to be needed. Since vehicles are not automated, relocation becomes necessary and will be carried out according to the current practice of transporting vehicles with a specially designed truck or a towing vehicle.

8.3.6 Parking stalls and other infrastructure

The infrastructure attributes are the number of parking stalls at each station, including stalls equipped with fast chargers. The parking stall-to-vehicle ratio is used in models as an indicator of infrastructure availability.

For accommodating vehicle imbalance throughout the multistation system, an appropriate number of parking stalls and vehicle charging facilities are needed at each station in order to operate the system efficiently. The location of stations with parking/charging facilities for electric vehicles calls for strategic planning.

Since the SVS described in this chapter is not intended to replace all or even a substantial portion of private mobility in the service area, and that only a well-identified market niche is to be served by the vehicle fleet, the number of parking stalls will be a function of vehicle fleet size and vehicle imbalance during the service period. In some shared vehicle situations, the system is intended only for employees of an organization that has offices across the city. However, providing service to subscribing members is a more likely future model. An open market scenario can be considered in case a public transit agency becomes a partner in a public-private business model.

8.3.7 Fast charger stalls

Since the station-based or free-floating electric car shared mobility services will be operating on a 24 h or another service duration basis, DC fast charging (DCFC) will be necessary to enable frequent recharging. If many transportation network company (TNC) drivers will not have access to a charger at home or at a publicly available fast-charging outlet, they may be offered a fast charging time slot within the SVS system at an agreed price. However, the first priority for fast charging will be assigned to the park-charge station-based vehicles. Another scenario is that if these vehicles are owned by the fleet owner, this fleet owner is likely to invest in a centralized charging infrastructure.

A number of research studies have resulted in detailed information on charging infrastructure [18–21]. The current thinking is that the commonly used 50 kW charger is not suitable for meeting future needs. On the other hand, very high power chargers such as 350 kW may not be economically feasible for many years in the future due to the limited battery capacity of vehicles that will require charging time. Also, such high power demand throughout the urban area may adversely affect the local electricity grid. However, a number of fast charger studies are favoring 150 kW or slightly higher power chargers for installation at multiple sites in urban areas.

A simulation study carried out by the Natural Resources Canada and Hydro Ottawa (Canada) found that under an optimistic scenario of electric vehicle presence in the fleet of vehicles in 2037, the total load of DCFC stations amounts to 2.4% of the annual peak load of the city. The DCFC clusters will cause an increase of 0.9% in the annual electricity consumption. The researchers concluded that the impact of clusters of DCFC stations on the bulk power supply in Ottawa will be fairly limited [18].

The above referenced simulation study concluded that since the impact of installing DCFC clusters will increase gradually, the local electricity utility can prepare for it. During the installation phase of clusters, actions can be taken to install grid connection equipment such as transformer wiring. Also, upgrades to the grid can be considered in order to cope with the worst-case scenario when all chargers will be in use throughout the city.

8.3.8 System operation, control, and maintenance

An operations and control subsystem of the SVS handles the tasks of monitoring all vehicles in the fleet, addresses incidents in the traffic network, resolves technology issues, enables customer access to the vehicle, checks battery state of charge, and decides on the distribution of vehicles among stations. If a free-floating car is left at a site other than a station, the station personnel will relocate it to a station in need of a vehicle. The control personnel can carry out these tasks with the assistance of intelligent transportation and advanced communication technologies. The tasks of vehicle allocation to customers, check-out and check-in processes, and updating the inventory of vehicles at specific sites are assisted with Apps. The control personnel employ algorithms to efficiently determine the redistribution or relocation needs of available vehicles among stations or to serve the special requirements of the free-floating part of the system. The maintenance subsystem handles a variety of tasks such as battery charging, unusual repairs, and vehicle and infrastructure cleaning.

8.4   Methods to support system design

8.4.1   Treatment of stochastic factors

Shared vehicle system operation will encounter uncertainties regarding customer demand for service, origin-destination pattern of trips served, and travel time required to complete a trip. Literature review shows that past researchers modeled demand as a modified Markov process and the customer or trip inter-arrival time was defined by a negative exponential distribution [14,22]. Among other approaches used is the activity-based micro-simulation to estimate the car sharing demand [23].

The routing and therefore the location of vehicles in real-time can be studied with stochastic dynamic simulation. In real-world operation, the advanced communication systems in association with geographic information system (GIS) can assist in locating vehicles in the network.

The random nature of travel time in the network can be studied with the use of the Bayesian method. The travel time estimates obtained from the City's traffic control center can be used in the Bayesian method as the initial estimates. States of travel time can be defined and their associated prior probabilities can be based on frequency analysis. In order to incorporate changes in traffic conditions, additional information may be acquired. This option in the Bayesian method enables a change from prior probabilities to posterior probabilities with the use of conditional probabilities. The conditional probabilities are a mechanism to apply a reliability factor to the estimated travel times between stations. Chapter 12 of this book covers the Bayesian method for treating uncertainty in travel time estimates.

8.4.2   Simulation and optimization techniques

For the analysis and evaluation of system design, simulation and optimization methods are available. The discrete event simulation technique is suitable for the study of shared vehicle system design and operation [14,16,22]. Commercial software is available for implementing the methodology. Applicable system measures of performance are used to assess alternative designs and operations. An iterative approach is employed to find the best design and efficient operation strategy for the shared vehicle system. Associated methods include Bayesian analysis for treating the stochastic nature of inter-station travel times in the network.

MATSIM, an open-source agent-based traffic simulation platform, can be used for SVS design. It can work with the dynamic traffic assignment method and can handle a large number of scenarios and detailed modeling of transportation network [17]. For additional simulation tools for SVS design, see Chapter 9 of this book.

A system optimization method suitable for use in SVS studies is the direct search method. A MATLAB® toolbox is available for the implementation of this method to SVS design and operations [24]. Other approaches that have been reported are Genetic Algorithm and mathematical programming methods. These are available from Mathworks [25].

8.5   Employee-dedicated shared vehicle system case study

8.5.1   General description

An SVS was modeled using discrete-event simulation and an associated Bayesian method. For realism, serving 100% of requests for a vehicle was not a design requirement. But, in order to serve a high percentage of requests, and at the same time to keep the value of the design variables within reasonable limits, the criterion adopted was to serve a minimum of 95% of the total requests while using the best values of measures of performance (MOP). The input control parameters such as fleet to served demand ratio, and parking stalls to fleet ratio were varied to find an efficient system design.

The simulation data were based on an employee-dedicated shared electric vehicle system in Ottawa (for Canadian Forces). In an earlier study, a total of 44 stations were investigated. For the case study described here, up to nine stations were studied for sensitivity analysis of the number of stations, but the number was narrowed down to five relatively high-demand stations that are located within a well-defined service area (Table 8.1). The party size of trip makers was included in the study, with an upper limit of five persons [13,22].

Free flow inter-station travel times sourced from the City of Ottawa traffic management center were used as the starting step. The modified travel times used in simulations were obtained by using the Bayesian method which treats the effect of stochastic travel conditions [22].

Table 8.1   Station and address in Ottawa (Canada)

Station number[a]	Address	Efficient number of parking and charging stalls[b]
1	100 Metcalfe	18
2	200 Elgin	18
3	140 O'Connor	18
4	National Defence HQ—101 Colonel By Drive	18
5	CFB Ottawa—Rockcliffe Park	18

[a]Five stations were selected out of 44 candidates.
[b]Efficient number of parking and charging stalls found from the simulation study.

Simulation model validation tests were performed. Selected measures of performance used include mean, mode, std. dev., co-variance (COV), minimum and maximum values of outputs, and 95% confidence interval. The variables tested were demand (trips) served, number of relocations, queue length (expressed as trips), and wait time. All tests were satisfactory. For example, the statistic COV that indicates the variation in sample data for a particular MOP was within the range of "very low" to "acceptable" [22].

8.5.2 *Location and size*

For designing an SVS, one objective is to define an efficient number of stations and their location within the market to be served. The variables that affect the decision on the number of stations are the number of relocations, wait time to access the vehicle, fleet size as a percentage of demand, and parking-to-vehicle ratio. For the Central Ottawa case study, the demand was increased linearly as the number of stations rose. Figure 8.5 shows results on the number of relocations that correspond to fleet expressed as a percent of the demand of 10.5% and parking to vehicle ratio of 2:1. Simulation results showed that five to seven stations were needed for efficient operations. For this range of stations, on average, about 30 to less than 60 relocations take place, while keeping the customer average wait time within acceptable level at about 11 min. In general, as expected, results suggest that as the number of stations increases the need to relocate vehicles rises at an increasing rate.

A comparison was made of the trips-to-stations ratio with a research study reported for a suburban location in the City of Lyon (France) [17]. The Lyon study used a multiagent and multimodal transport simulation model. In the case of Ottawa, the ratio of the trips/stations in percentage terms was about 84%. For one

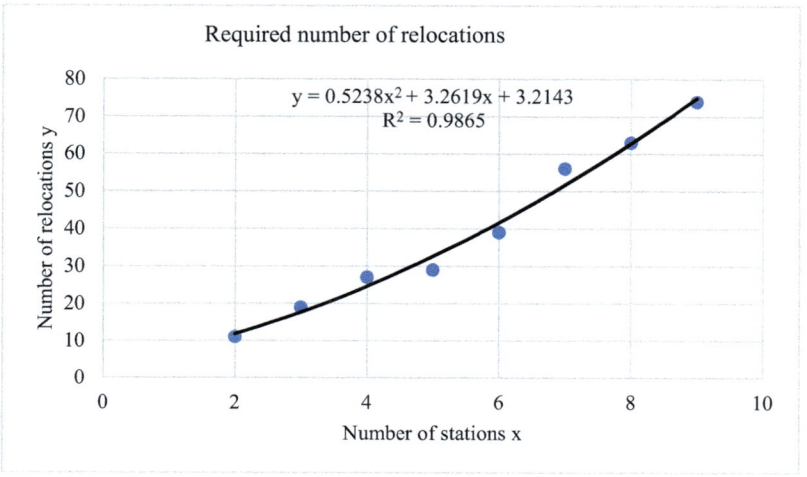

Figure 8.5 *Number of relocations as a function of number of stations. Note: Fleet as a percentage of demand 10.5% and parking to vehicle ratio 2:1*

scenario of the Lyon case study, the ratio was reported as 80% and for a second scenario, it was 91%.

The size of a station was defined as the number of stalls for parking and charging (usually called parking stalls) as a function of parking/vehicles ratio. For serving electric vehicles, the installation of chargers at a selected number of stalls was assumed. The parking/vehicle ratio was varied from 1.0 to 2.2 for the five-station system. As the ratio increased, the number of relocations and queue length dropped until ratio 2.1. But, at a ratio of 2.2, the number of relocations and queue length increased.

Another comparison was made with the City of Lyon case study, which found the fleet size to be 40% of total parking spaces. According to the authors of the Lyon research study, this percentage can be regarded as a rule of thumb, supported by the actual experience for Autolib in Paris [17]. In the Ottawa case study, the fleet amounted to 50% and later refined to 47.6% of parking stalls [22].

8.5.3   Demand and fleet

The system capacity is a function of the total number of available vehicles or fleet size. Using simulation or an optimization method, an effective fleet size can be found using the criteria of demand (trips) served, queue size (i.e., trips waiting to be served), and average waiting time. For the economic viability of the shared vehicle system, fleet size is an important factor.

Figure 8.6 shows the percent demand served as a function of fleet size (expressed as a percent of demand). This is consistent with real-world expectations. The dependent variable is percent demand served and the explanatory variable is fleet size as a percent of demand. The total system capacity in terms of the number of trips served increases as fleet size increases, but at a marginally declining rate. As an additional vehicle is added to the fleet at the margin, the corresponding contribution to serve trips is less than the previous addition to the fleet. That is,

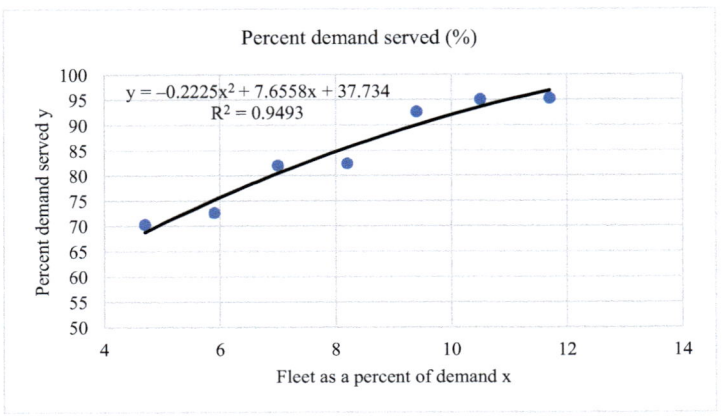

Figure 8.6   Demand served as a function of fleet size

total system capacity in terms of number of trips increases at a diminishing marginal rate. The regression equation that relates percent demand served as a function of fleet (expressed as a percent of demand) has a high coefficient of multiple determination (i.e., square of multiple correlation coefficient R^2).

8.5.4   Service factors

The sensitivity of service factors to variation in fleet size is illustrated in Figures 8.7 and 8.8. The effect of fleet size (as a percent of demand) on wait time and the number of relocations is relatively higher at lower values of fleet size as compared to higher fleet size values. For the case study area, at 10.5% fleet size,

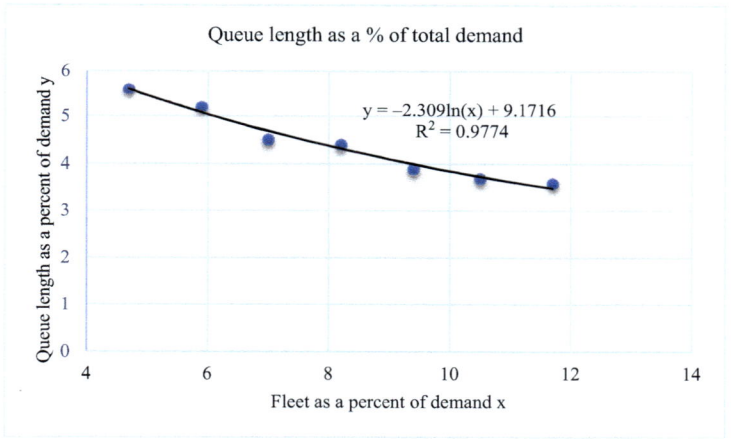

Figure 8.7   Effect of fleet size on queue length

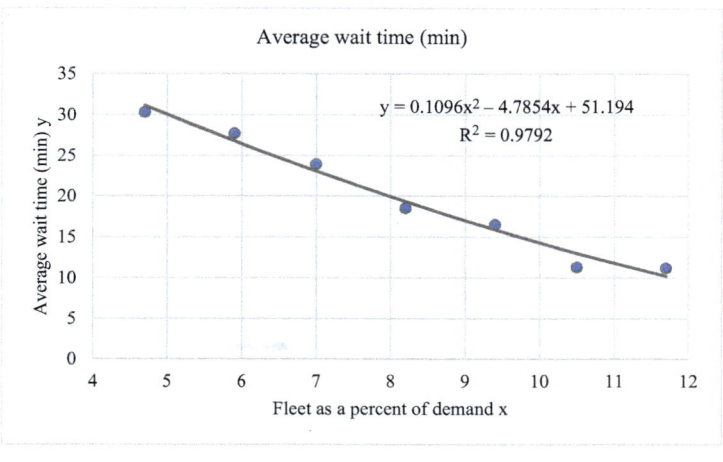

Figure 8.8   Effect of fleet size on average wait time

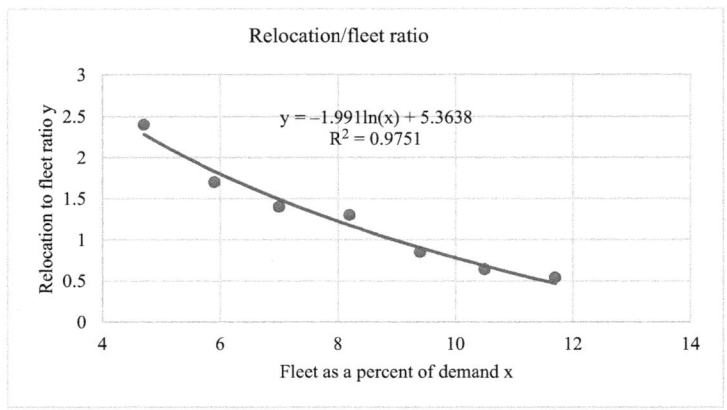

Figure 8.9   Effect of fleet size on relocations

the results appear good. At this level of fleet size (i.e., 45 vehicles) to serve a 427 employee trip demand for the 9 h duration of analysis, the length of the queue is 3.7% of total demand and the wait time is 11 min.

As expected, as the fleet size rises, the level of service improves. As shown in Figures 8.7 and 8.8, the queue length and average wait time decline with increasing fleet size. These effects are more pronounced at lower values of fleet expressed as a percent of demand, but the curves show a lower slope at higher fleet size.

Simulation of relocations within a five-station service area shows that as the fleet size (i.e., expressed as a percent of demand) increases, the relocation to fleet ratio declines (Figure 8.9). But, the drop in this ratio is relatively high at lower values of fleet size as compared to higher values of fleet size.

8.5.5   Serving higher demand levels

To examine the effect of higher travel demand on major SVS performance indicators, namely user average wait time, number of relocations, and total trips served, further simulations were carried out. The following observations are drawn from results presented in Figures 8.10–8.12. It should be noted that the fleet size is adjusted according to demand level.

- Since the fleet is expressed as a percent of demand, under high demand condition, the fleet size is higher as compared to lower base demand condition.
- Under high demand condition, the SVS experiences a minor drop in trips served at a lower fleet-to-demand ratio as compared to a higher fleet-to-demand ratio. However, under lower demand condition, due to constrained fleet size, the drop in served trips is significant. At high fleet-to-demand ratio, due to higher number of vehicles, the two scenarios show identical service in terms of trips served (Figure 8.10).

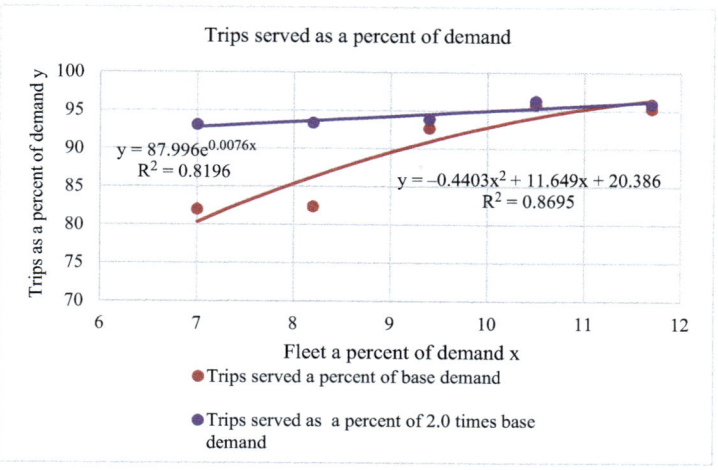

Figure 8.10   Effect of fleet size on demand served at two demand levels. Note: Fleet size is adjusted according to demand level

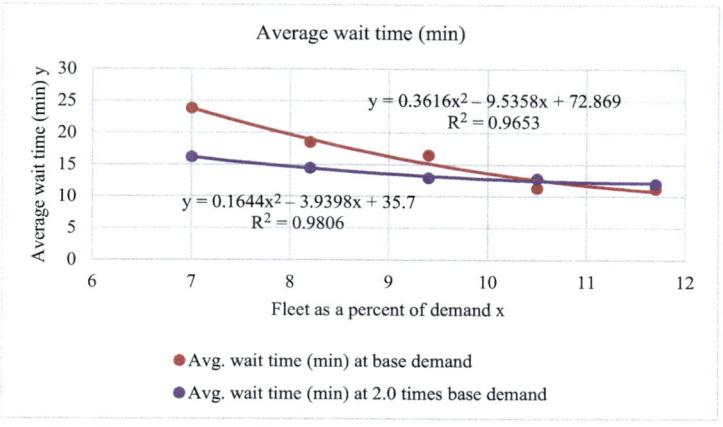

Figure 8.11   Effect of fleet size on average wait time at two demand levels. Note: fleet size is adjusted according to demand level

- Under high demand condition, at lower fleet level, the average wait time is lower as compared to under lower demand condition. At higher fleet level, the average wait time under the two demand conditions is identical (Figure 8.11).
- Relocations under lower demand scenario are higher than under the higher demand scenario. This advantage of lower relocations holds at all levels of fleet-to-demand ratio (Figure 8.12).

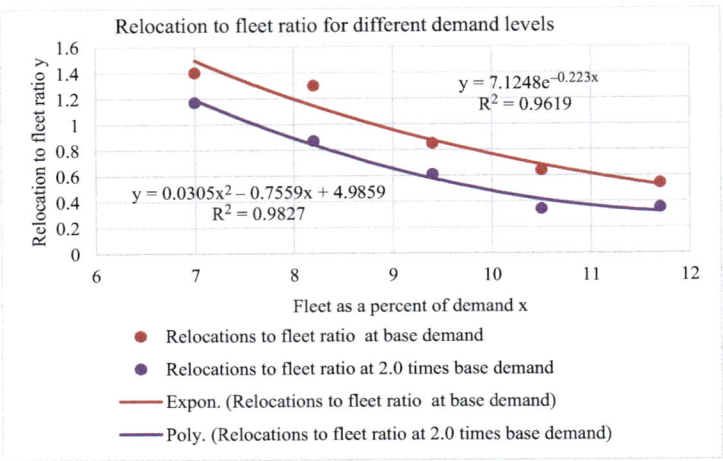

Figure 8.12   Effect of fleet size on relocations at two demand levels. Note: Fleet size is adjusted according to demand level

Taken together, the results of sensitivity analysis of the effect of higher demand on the SVS show the resilience of the system in accommodating higher demand without an adverse effect on service, provided that the fleet size is adjusted according to demand level. For this reason, within the range of values used in simulation studies, the SVS performs better under high demand condition as compared to the low demand base case, provided that the fleet size is adjusted according to demand level.

8.6   Design for ride-hailing EVs

Ride-hailing is expected to serve a part of urban transportation demand in the postpandemic era. Also, these vehicles are likely to continue to be privately owned and will serve the market as a part of a TNC fleet. The increasing popularity of ride-hailing services experienced in the prepandemic period calls for their safe and efficient accommodation regarding infrastructure and traffic management measures [26,27]. The following parts of this chapter cover the planning and design enhancement of these systems.

8.6.1   Pick-up and drop-off facilities

The ride-hailing service of nonautomated vehicles requires pick-up and drop-off locations at the curb as well as off-street sites. As the demand for these tasks will rise, there will be intense competition for space with other travel modes and urban services. According to recent experience, the operation of ride-hailing vehicles of TNCs have placed increasing demands on limited curbside space for pick-ups and drop-offs. The future growth in this category of shared mobility will require policy, planning, and design actions by urban governments [27,28].

Urban planners and designers are likely to be tasked by municipal governments to define and implement measures to enhance safety, accessibility, and efficiency of shared mobility while at the same time reducing traffic disruption, and improving curb management. In addition to the orderly accommodation of ride-hailing service on uncongested and safe curbside, off-curb facilities will be required. Looking ahead, among other options, the cities are already considering pricing and regulatory measures for efficient allocation of limited curb space, so as to expedite turnover of vehicles. Planning and design initiatives will be required for increasing access of shared mobility vehicles to delivery zones and pick-up/drop-off areas while treating all modes of travel on an equitable basis [27].

In the postpandemic period, as the curbside and off curb solutions evolve for use by ride-hailing services, these will be formally shown in the traffic network maps as well as in smartphone applications.

8.6.2 Charging facilities

Ride-hailing electric vehicles are likely to be driven more kilometers per day throughout the year than the average vehicle that is not used for shared mobility service [27]. The TNC drivers will be eager to charge batteries more quickly to avoid missing fares [29]. Research shows that strategically placed fast chargers with available charging time slot reduced travel time to access chargers by 49%. Also, it was found that charging time is the dominant component of vehicle downtime. So, for increasing shared mobility service availability, enhanced opportunity to charge a ride-hailing electric vehicle is necessary [30].

Although the business model that is based on monetizing privately owned vehicles does not call for providing parking/charging facilities, due to their highly intensive use, these electric vehicles are likely to require fast charging locations during their business hours. From a design perspective, the ride-hailing part of nonautomated vehicles requires access to fast-charging sites that are expected to be available throughout the urban area. In the presence of a network of SVS, the battery charging need of ride-hailing fleet may be met to some degree by allowing the use of a fast charger (for a fee) at any station, subject to availability of 30 min charging time windows.

8.7 Conclusions

The two forms of shared mobility provided by nonautomated electric vehicles, namely customer-driven and TNC driver-driven, are likely to continue until the automated vehicle technology will mature. The customer-driven vehicle fleet will require parking and charging infrastructure provided at stations, also called depots, that are strategically located within a well-defined market area. The major design variables include vehicle fleet and infrastructure. The infrastructure is characterized by the number, size, and location of park-and-charge stations.

The free-floating nonautomated car shared mobility, although very convenient for customers, poses operational challenges for relocating vehicles from the drop-off location to park and charge stations. However, the availability of automated vehicles will make the floating car service the same as the station-based option.

Simulation and optimization methods are available to design station-based shared vehicle systems. Uncertainties in balancing demand with supply of service and those in the traffic network can be analyzed with the Monte Carlo simulation-based and the Bayesian methods. Additional information on these methods is provided in Chapter 12. The methodological framework based on the discrete event simulation technique in association with the Bayesian method used for the Ottawa case study is suitable for shared vehicle system design, while taking into account party size as well as vehicle size.

The Ottawa case study suggests that serving 95% of trip requests is efficient with respect to the number of vehicle relocations, vehicle-to-trip ratio (i.e., an indicator of the fleet size), and therefore the number of parking stalls per station. Meeting less than 100% requests avoids a large nonproductive fleet and excessive parking stalls. Adding extra vehicles results in diminishing marginal gains in serving demand. Although percent demand served rises with increasing vehicle-to-trip ratio, the gain in capacity increases at a diminishing marginal rate.

Results show that the SVS performance is highly sensitive to vehicle-to-trip ratio as well as parking stall-to-vehicle ratio. For the case study, a vehicle-to-trip ratio of 10.5% was found to be efficient for serving travel demand in terms of satisfying an average user wait time of about 11 min and keeping the number of vehicle relocation to about 30% per operation day while avoiding high queue lengths.

The SVS can function well under higher demand, provided that the fleet size is increased at the same rate as the increase in demand.

In the case of the Ottawa case study, the trips/stations ratio in percentage terms was about 84%. The vehicle fleet size was found to be about 48% of parking stalls. These design factors compare well with other simulation and real-world case studies.

The TNC driver-based model for on-demand mobility commonly referred to as ride-hailing service, based on monetizing privately owned vehicles requires well-defined pick-up and drop-off locations on the curb and at off-street locations. Urban governments in association with TNC companies have to plan, design, and manage these facilities. Further, the electric vehicles used for ride-hailing will require access to fast chargers in the city. For this reason, adapting city infrastructure for ride-hailing service will be a challenge.

Acknowledgments

The author wishes to thank Dr. Akhtar Hossain for literature review and permission to use simulation material. Also, Mr. Seth Gatien is thanked for reviewing a draft of the chapter.

Author contribution

The author reviewed and approved the manuscript.

Declaration of conflicting interests

The author declared no potential conflict of interest with respect to publication of this chapter.

References

[1] Institute of Transportation Engineers (ITE). 'MaaS/MOD. ITE Initiative about the MaaS/MOD adapted from the June 2019 ITE Journal'. *ITE Website*. Available from https://www.ite.org/technical-resources/topics/maas-mod-ite-initiative/.

[2] U.S. Department of Transportation. *Mobility on demand, operational concept report*. Final Report – September 2017, FHWA-JPO-18-611.

[3] Shaheen, S., Cohen, A., and Zhohdy, I. *Shared mobility, current practices and guiding principles*. U.S. DOT. FHWA-HOP-16-022. April 2016

[4] Kreetzer, A. "Putting the 'M' in motion – VOLVO's new shared service." *Auto Futures*. June 12, 2019.

[5] Vulog. "Volkswagen launches world's largest electric car-sharing operation in Berlin supported by Vulog." June 27, 2019. Vulog Canada, Toronto, Canada.

[6] Ditta, S., Urban, M.C., and Johal, S. *Sharing the road, the promise and perils of shared mobility in the GTHA*. Mowat Research #124, University of Toronto, August 2016.

[7] Transportation Research Board (TRB). "Forum on preparing for automated vehicles and shared mobility, mini-workshop on the roles of government and private sector." *Transportation Research Circular No. E-C258*, November 2019.

[8] Abel, S. "Mobility and the public right of way." *ITE Journal*. 2019.

[9] Jörg, F., and Müller, M. "Free-floating electric car-sharing fleets in smart cities: The dawning of a post-private car era in urban environments?" *Environmental Science and Policy*. 2015; 45: 30–40.

[10] Volkswagen. "WeShare launched in Berlin as full-electric service." *VW Newsroom*. Available from https://www.volkswagen-newsroom.com/en/press-release, June 27, 2019. Also, see 'VW's new car sharing service aims to be a little different from the rest', posted by Trevor Mogg on June 27, 2019 & 'Volkswagen launches WeShare all-electric car sharing service', by Darrel Etherington, June 27, 2019.

[11] Feigon, S., and Murphy, C. *Shared mobility and the transformation of public transit*. Washington, DC: TCRP Research Report 188; 2016.

[12] Shared Mobility Center. *Shared mobility and the transformation of public transit. research analysis TCRPJ-11/Task 21*. 2016.

[13] Hossain, A., and Khan, A.M. "Multiple station shared vehicle systems design and operations modelling framework." *ITS World Congress*, Detroit, 2014.

[14]   Barth, M., and Todd, M. "Simulation model performance analysis of a multiple station shared vehicle system." *Transportation Research Part C: Emerging Technologies*. 1998–1999; 7(4): 237–259.

[15]   Barth, M., and Shaheen, S. "Shared-use vehicle systems framework for classifying carsharing, station cars, and combined approaches." Transportation Research Record 1791 _Paper No. 02-3854.2002.

[16]   Hermann, S., Schulte, F., and Vos, S. "Increasing acceptance of free-floating car sharing systems using smart relocation policies strategies: A survey based study of car2go Hamburg." *International Conference on Computational Logistics*, pp.151–162. Springer. 2014.

[17]   Laarabi, H.M., Boldrini, C., Bruno, R., Porter, H., and Davidson, P. "On the performance of a one-way car sharing system in suburban areas: A real-world use case." *3rd International Conference on Vehicle Technology and Intelligent Transport Systems*. Proceedings Volume: VEHITS December 2016. DOI: 10.5220/0006307901020110

[18]   Ribberink, H., Wilkens, L. Abdullah, R., McGrath, M., and Wojdan, M. "Impact of clusters of DC fast charging stations on the electricity distribution grid in Ottawa, Canada." *EVS30 Symposium*, Stuttgart, Germany, October 9–11, 2017

[19]   Nicholas, M., and Hall, D. *Lessons learned on early electric vehicle fast-charging deployments*. White paper July 2018. The International Council on Clean Transportation (ICCT), Washington, D.C.

[20]   Congressional Research Service. *Vehicle electrification: federal and state issues affecting deployment*. R45747. 2019. June 03, 2019.

[21]   US Dept. of Energy Office of Energy Efficiency and Renewable Energy. *Costs associated with non-residential electric vehicle supply equipment. Also see Enabling fast charging: A technology gap assessment*. October 2017.

[22]   Hossain, A., and Khan, A.M. "Modelling demand and supply interaction in multiple stations shared vehicle systems." *ITS World Congress*, Montreal. 2017.

[23]   Ciari, F., Schuessler, N., and Axhausen, K.W. "Estimation of carsharing demand using an activity-based microsimulation approach: model discussion and some results." *International Journal of Sustainable Transportation*. 2013; 7(1): 70–84.

[24]   Mathworks. *Direct search*. 2020. Available from http://www.mathworks.com/help/gads/direct-search.htm.

[25]   Mathworks. *Genetic algorithm and integer programming*. 2020. Available from http://www.mathworks.com/help/gads/genetic-algorithm-options.htm http://www.mathworks.com/discovery/integer-programming.htm.

[26]   Massachusetts Institute of Technology (MIT). *Insights into future mobility*. MIT Energy Initiative, Cambridge, MA. USA. 2019. Available from http://energy.mit.edu/insightsintofuturemobility.

[27]   City of Toronto. *Automated vehicles tactical plan readiness 2022. IE87 – Attachment 1*. 2019.

[28]   International Transport Forum (ITF). *Shared mobility, innovation for liveable cities*. Corporate Partnership Board (CPB), OECD. 2016.

[29]   Woods, E., Rames, C., Kontou, E., Motoaki, Y., Samrt, J., and Zhou, Z. *Analysis of fast charging station network for electrified ride-hailing services*. SAE Technical Paper 2018-01-0667, doi:10.4271/2018-01-0667. 2018.

[30]   Roni, M.S., Yi, Z., and Smart, J. "Optimal charging management and infrastructure planning for free-floating electric vehicles." Submitted to *Transportation Research Part D*. 2019.

The views expressed are those of the authors.

Chapter 9

Design of systems with automated and electric vehicles

Guoyuan Wu[1] and Matthew Barth[2]

In recent years, our contemporary transportation systems are rapidly evolving based on several influencing factors: (1) connected and automated vehicles (CAVs) are quickly emerging with the advancement of wireless communication technologies and with different levels of automation being introduced [1]; (2) vehicles are also becoming increasingly electrified with the advancement of efficient electric drivetrains and battery technology [2]; (3) new forms of shared mobility continue to emerge, including the concept and practice of Mobility-as-a-Service (MaaS) such as ride-sourcing [3] and car-sharing across different modes, including passenger cars, transit, micromobility platforms, and even freight vehicles; and (4) the roadway infrastructure (not only for traffic surveillance but also for traffic control) and charging facilities (both wired and wireless) are rapidly improving, along with the electric grid to support this vehicle electrification. The nexus among transportation, information, and energy/environment is becoming increasing important due to the rise in public awareness of efficiency, equity, and sustainability. In particular, shared electric and automated mobility (SEAM) has been considered as a key enabler to change our standard paradigm of urban transportation systems [4].

In this chapter, we review the state of the practice in shared automated mobility (SAM) as well as SEAM across the globe and emphasize some key features of automated and electric vehicles that play a role in how shared mobility systems operate. In the context of shared mobility, we detail the system level design for SAM and SEAM, followed by a specific case study using microscopic traffic simulation. Lastly, we discuss the potential challenges and opportunities of the deployment of SAM and SEAM systems.

9.1 Introduction and background

The global urbanization trend has witnessed an average annual growth rate of 1.8% for cities with more than 300,000 inhabitants over the past three decades, and a

[1]Center for Environmental Research and Technology, University of California at Riverside, Riverside, CA, USA
[2]Department of Electrical and Computer Engineering, and Center for Environmental Research and Technology, University of California at Riverside, Riverside, CA, USA

projected 68% of the world's population residing in urban areas by 2050 [5]. Although urbanization is considered to be a key driver for economic growth, poverty reduction, and human development in general, it also raises a series of potential issues on our economy, society, and environment throughout the world. For example, it is estimated that cities account for more than 70% of CO_2 emissions and around 72% of global energy use [6]. In the United States, the average daily congested duration is approaching 4 h for major metropolitan areas [7]. As a result, sustainable development has become an important topic and vital strategy for managing our urban mobility systems.

Among all the emerging technologies and services, connected and automated vehicles (CAVs), shared mobility, and transportation electrification are regarded as the most remarkable disruptions to the movements of both people and goods [8,9]. These technologies and services are enabling the urban mobility system to be more efficient, affordable, and sustainable. Much effort has been devoted to the development and deployment of these technologies and services as well as their convergence, that is, combining sharing, electrification, and automation, by the government, industry, and academia through various partnerships worldwide.

The U.S. Department of Transportation's (USDOT) Federal Transit Agency (FTA) has initiated a 5-year Strategic Transit Automation Research (STAR) Plan [10,11] to investigate the application of automation technologies to bus transit operation. In addition, the Mobility on Demand (MoD) Sandbox Program [12] and the Accessible Transportation Technologies Research Initiative (ATTRI) have incentivized a variety of projects on SAM-related technologies and services that may target for different user groups [13]. As the successor to the CityMobil project [14], the CityMobil2 project co-funded by the European Union's Seventh Framework Programme has cultivated a collaborative environment for manufacturers and demonstrated automated shuttle services in different countries such as the Netherlands, France, Germany, and Switzerland in recent years [15]. Figure 9.1 summarizes the worldwide efforts in automated shuttle testing by 2019. The CoEXist project is another ongoing project funded by the European Union's Horizon 2020 Research and Innovation Programme, where the shared automated mobility service has been examined in a more realistic urban environment involving various penetration levels of conventional vehicles and connected and automated vehicles (CAVs).

In Asia, the Chinese government and Society of Automotive Engineers of China (SAEC) have issued a roadmap for intelligent and connected vehicles (ICVs) that not only sketches out the plan for future urban mobility, but also encourages industry (e.g., Didi, Geely, and Baidu) involvement and collaboration from the aspects of shared mobility, electric vehicles, and autonomous driving, as part of the "Made in China 2025" plan [17]. Under the Innovation of Automated Driving for Universal Services (adus) plan sketched by the Cross-Ministerial Strategic Innovation Promotion (SIP) Program, Japan has been actively testing automated shuttle services for older adults in rural areas and autonomous buses in some airports to prepare for the commute between terminals during the Tokyo Olympic Games [18]. South Korea has also designated autonomous driving as one of its top 13

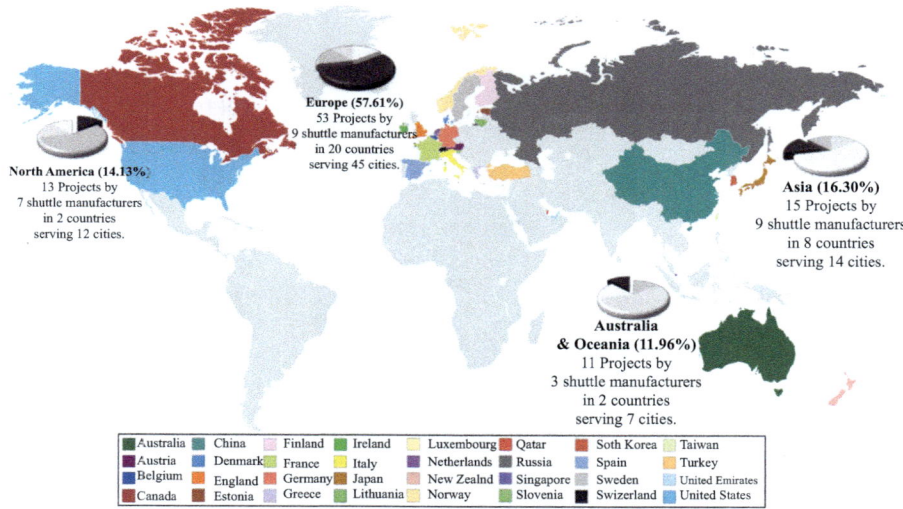

Figure 9.1   A summary of worldwide automated shuttle experiment. Adapted from [16]

Industrial Engine Projects and its Ministry of Land, Infrastructure, and Transport (MOLIT) also initiated the Connected & Automated Public TrAnsport INovation (CAPTAIN) project, which focused on connected and automated driving (CAD) for both large transit buses and smaller cut-away shuttle vehicles [19]. With the amendments to the Road Traffic Act (RTA), Singapore's Land Transport Authority (LTA) has been supervising and facilitating the nationwide development and deployment effort on automated public transportation. It plans to pilot the auton-omous bus and on-demand shuttle service in Punggol, Tengah, and the Jurong Innovation District in 2020, as part of the Singapore's Smart Nation strategy [20]. Following the National Policy Framework for Land Transport Technology, Australia just delivered the latest National Land Transport Technology Action Plan [21], which identifies new technological priorities in low and zero emissions vehicles, mobility as a service, integration of CAV operation with infrastructure and land use planning, and freight transportation. A few driverless electric bus trials have been conducted on the public roads in New South Wales (e.g., Armidale, Coffs Harbour, and Sydney) and Northern Territory (e.g., Darwin).

9.2   Features of shared automated and electric vehicles

9.2.1   Fundamentals of automated vehicles

Automated vehicles (AV), or in a broader sense, automated driving systems (ADS), are vehicles that can leverage (1) advanced on-board sensors (see Figure 9.2 regarding the state-of-the-practice sensing technologies for AVs) and/or communication technologies

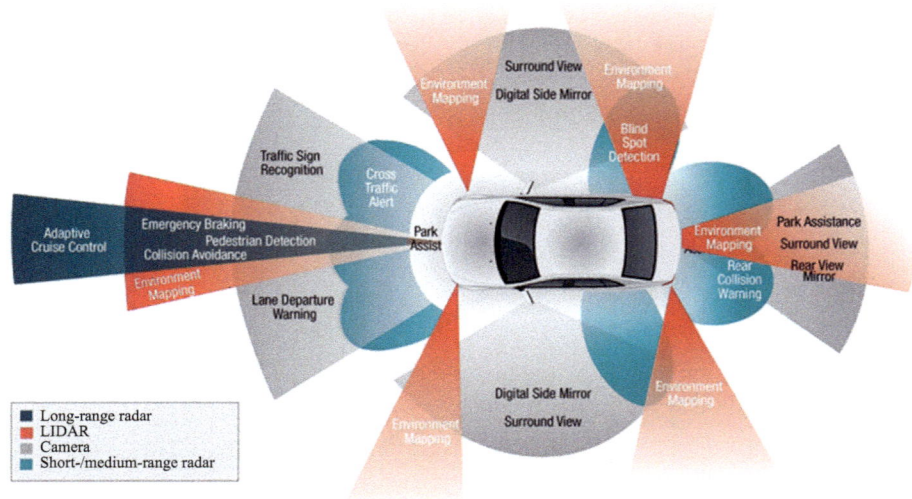

*Figure 9.2   Illustration of the state-of-the-practice suite of sensors on AVs.
            Courtesy of SAE International*

(if applicable) for situation awareness, for example, localization, in-vehicle monitoring, or surrounding perception; (2) available computing power for decision making, for example, initiation of a warning message or calculation of an energy-efficient speed trajectory; and (3) effective interface for interaction with users (e.g., driving assistance) and environment (e.g., automatic control, or communication with infrastructure and other road users), to fulfill our daily needs with various levels of automation (see Figure 9.3 for the SAE's definition on Levels of Driving Automation). Such needs may be much beyond just transporting people and goods from point A to point B, but also facilitating these trips (i.e., relocation or rebalancing), serving as a surveillance or regulation system for traffic, or even just accompanying and escorting for protective purpose.

9.2.2   Fundamentals of electric vehicles

In a broad definition, electric vehicles (EVs) refer to road vehicles whose propulsion involves electricity [22], including hybrid electric vehicles (HEVs); plug-in hybrid electric vehicles (PHEVs); battery electric vehicles (BEVs); and fuel-cell electric vehicle (FCEVs). Figure 9.4 illustrates a general EV configuration which is composed of three major subsystems: (a) electric propulsion; (b) energy source; and (c) auxiliary. As shown in the figure, the electric propulsion subsystem consists of motor(s), transmission, power converter, and electronic control units (ECUs), and may have different configurations depending on the consideration of size, compactness, weight, cost, reliability, and performance (e.g., maximum cruising

Figure 9.3 Definition on level of automation by SAE International (2018a). Adapted from [3]

speed, gradeability, and acceleration). The energy storage unit, energy management unit, and energy refueling unit comprise the energy source subsystem. In practice, the most widely adopted energy storage device for EVs is the battery, due to their characteristics in terms of relatively high energy density, compact size, and reliability. Other devices may include ultra-capacitors (UC), flywheels, and hydrogen tanks which can be utilized as an auxiliary energy source or hybrid energy source. For plug-in electric vehicles (PEVs), their battery packs can be recharged from an external source of electricity, for example, power grid. The auxiliary subsystem involves auxiliary power supply unit, power steering unit, and A/C control unit.

9.2.3 Convergence of automated and electric vehicles for shared mobility

The convergence of driving automation, transportation electrification, and shared mobility has the potential to enable a paradigm shift for urban transportation and have significant impacts for different stakeholders. Table 9.1 summarizes the key factors that determine a variety of shared mobility services, which may include sharing type, vehicle ownership, who the driver is, what the mobility service is used for, vehicle classification, and powertrain type. Vehicle automation would wipe out

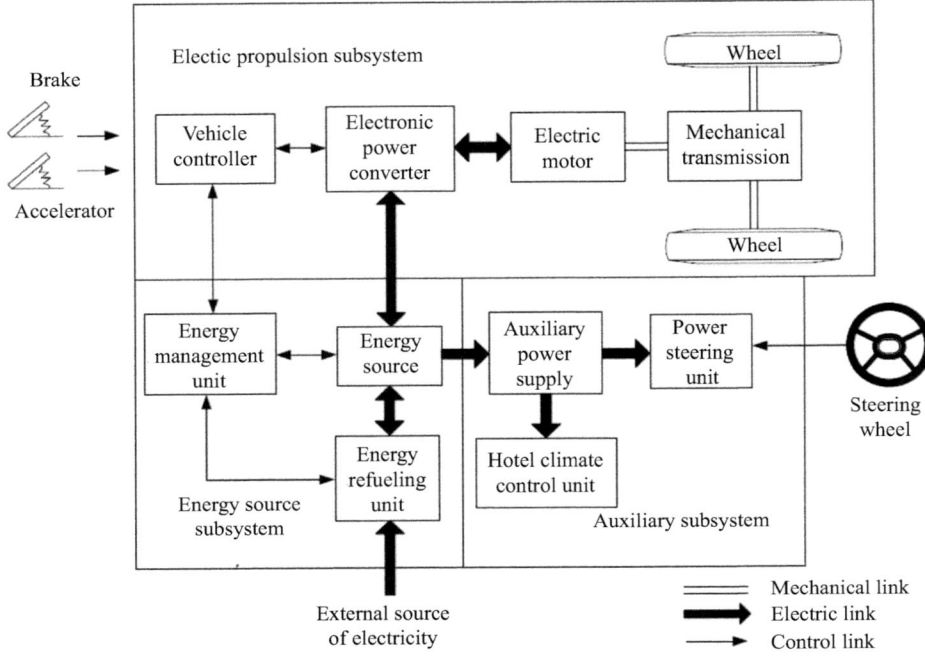

Figure 9.4   A representative configuration of EV. Adapted from [23]

the "who drives" factor but would instead shift to "who operates." This may blur the boundary between conventional car sharing and ride-sourcing services with nonautomated vehicles. Although transportation electrification would only change the dialogue of the powertrain type for individual vehicles in this table, more profound impacts would permeate into transportation and power systems as well as our daily behaviors. A more detailed explanation on the relationship between automation, electrification, and shared mobility is presented below.

9.2.3.1   Automation and shared mobility

Vehicle automation takes the driver out of the equation which can considerably improve the productivity of shared mobility service as well as reduce the labor costs from the perspective of service providers. In addition, customers may enjoy more cost-effective, convenient (24/7), privacy preservative, and pandemic-proof (due to no human driver involved) shared mobility service due to the introduction of automated vehicles. As illustrated in Figure 9.5, considering the current industry's technology readiness level in vehicle automation (we currently lie in between Level 2 and Level 3), various types of shared mobility services have been piloted all over the world by following two major paths toward the full automation services:

1. The *"revolutionary"* path or the *"everything somewhere"* strategy which consists of little human intervention to drive within some confined context and

Table 9.1   Key factors and levels for different types of shared mobility service

Factors	No. of levels	Description	Note
Sharing type	2	• Time sharing • Space sharing	Vehicle sharing vs. ride sharing (or co-loading)
Vehicle ownership	3	• Private/family • Company • Public agency	Vehicles owned by private or public agencies may provide nonprofitable shared service
Who drives?	4	• Owner • Hired • User • Machine	For SAM, machine would be the only driver. Who operates the vehicle would be important
Trip purpose	3	• For people • For goods • For both	The SAM may facilitate the integrated people and freight delivery
For profit?	2	• Yes • No	Acquaintance-based or organization-based ridesharing may be nonprofitable.
Vehicle type	Multiple	• Microvehicle • Small vehicle • Mid-size vehicle • Large vehicle • Delivery robot • Pick-up and van • Rigid truck • Truck w/ trailer	Strongly related to the trip purpose. For example, microvehicles through large vehicles are mainly for people movement, while others are mainly for goods
Powertrain Type	Multiple	• ICE • HEV/PHEV • BEV • FCEV	For SEAM, the electric motor would be the main propulsion component

gradually expands such operation to a much more complicated environment. The context herein may refer to geographical scale, operation time, roadway type, traffic state, weather condition, and many others. Novel shared automated mobility (SAM) service providers adopt this approach as they entered the arena with the expertise in vehicle automation. Many focus on the operation of low-speed autonomous shuttles or delivery vehicles within confined areas (e.g., fixed-route service, the first/last mile service). For example, EasyMile and Navya have demonstrated their automated shuttle services in many countries in Europe and North America, and Nuro [24] was just approved to test driverless

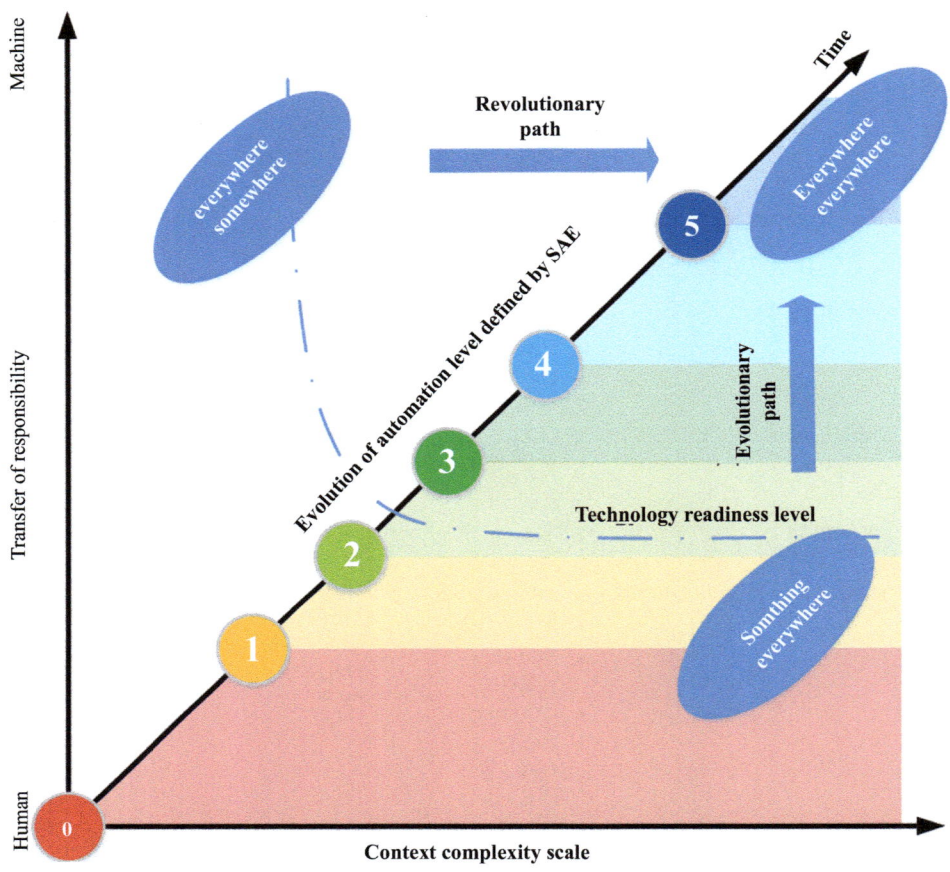

Figure 9.5   Illustration of states, strategies and paths to full automation

delivery in California. Others are targeting at automated taxi directly. Waymo One [25] is considered to be at the forefront, which has been providing real-world self-driving ride-hailing service for about a year in Chandler, Arizona. Baidu just initiated free trials of its robo-taxi service in Changsha, Hunan province, China [26]. Very limited tests have been emphasized on full-sized automated buses. For example, Daimler tested its Mercedes-Benz Future Bus for bus rapid transit service from the Schiphol airport to Haarlem in Amsterdam in July 2016 [27].

2. The "evolutionary" path or the "*something everywhere*" strategy which aims to gradually improve advanced driver assistance systems (with relatively low levels of automation) that are widely available in existing vehicle models, shift more and more dynamic driving tasks from human drivers to ADS. This is the major pathway taken by many automakers, for example, General Motors [28], who can leverage their capabilities in car manufacturing and niches to push

forward their own SAM services mainly in the form of automated taxi. As another example, Tesla recently reiterated its "Tesla Network" plan to launch its own ride-sharing app before achieving full autonomy [29].

9.2.3.2   Electrification and shared mobility

Transportation electrification, which involves major disruption in the vehicle powertrain, physical and digital infrastructure (such as charging facility, power grid, and communication network), and travel behavior, is considered as a promising solution to mitigating the carbon emissions associated with our daily transportation activities. From the perspective of shared mobility service providers, the introduction of electric vehicles (EVs) to their fleets would significantly reduce the fuel and maintenance costs. In particular, for hub/depot-based car sharing services, EVs may fit well with the fleet usage patterns (i.e., short trip distance profiles and good utilization rates) while taking advantage of the charging availability and schedule flexibility at the hub/depot if applicable. However, to the free-floating system based car sharing service providers and app-based ride-sourcing services by transportation network companies (TNCs), there are a few barriers to EV adoption in their fleets at the current stage:

1. Compared to conventional vehicles, EVs are generally more expensive to purchase (for now), and have fewer options to meet the operational requirement for shared mobility services, such as long all-electric ranges, high seat capacity, and large trunk space.
2. The number of public charging facilities, especially fast chargers, is still limited. Drivers may experience significant difficulties when searching for available chargers with long charging times, resulting in lower productivity for their time.

Over time, these challenges could be addressed through the development of battery/charging technologies (e.g., higher energy-density sources, faster and wireless charging), better plans of charging networks (to improve accessibility), and innovative user incentives.

9.2.3.3   Shared electric and automated mobility

The SEAM service would greatly strengthen the bond between mobility and energy while increasing the system complexity. In spite of some current challenges on electrifying shared mobility as described above, there are potential synergy opportunities between electrification and SAM from both the technical and operational perspectives:

1. In spite of the rapid advances in energy efficiency, on-board electronics in autonomous vehicles utilize significant power themselves, likely requiring larger energy storage;
2. Owing to the availability of more drive-by-wire components, it is easier and cheaper to implement automated driving technologies in electric vehicles (EVs);

3. Due to the high utilization rates of commercial fleets (if any) for SAM, lower operational and maintenance costs of EVs would be an attractive feature, and the requirements for larger battery packs or frequent recharging may be mitigated by fleet management optimization;

4. Compared to fossil fuel-based vehicles, EVs are considered to be much more environmentally friendly over their life cycle [30]. Therefore, the convergence of electrification and SAM would effectively alleviate the sustainability concerns resulting from the induced travel demand by SAM; and

5. Taking the Transportation-as-a-System (TaaS) approach, the integrated optimization of SAM service and power grid operation would yield significant system-wide energy benefits given the necessary digital technologies and incentives are in place. This allows the fleet service providers to have much more rooms to reduce the operational costs.

9.3   Design of shared electric and automated mobility system

9.3.1   5A+S rule for SEAM system design

Inspired by Shrestha *et al.* [31], we propose herein a 5A+S rule for the shared electric and automated mobility (SEAM) system design: *Availability, Affordability, Adaptability, Accessibility, Attractability,* and *Safety. Availability* refers to the service (both mobility and charging) location and schedule within the reach of right demands, which may be a challenge for rural areas. *Affordability* means that the service should be economically viable and socially equitable to avoid potential conflicts. *Adaptability* involves the system agility in response to variations of demands as well as the system resilience to withstand major disruptions. *Accessibility* is concerned with the quality of SEAM, associated infrastructure (i.e., pick-up and drop-off locations), and information to be accessed by the customers (including passengers with physical limitations or other different abilities). Accessibility is also concerned with the charging facilities or maintenance stations. *Attractability* focuses on the design of SEAM (both vehicle interior and exterior) and user interfaces (e.g., smartphone apps) to call for the service. *Safety* touches on not only the conventional safety performance of vehicles but also the personal safety of customers, especially for minority or seniors, without on-board supervision.

9.3.2   SEAM service type

Based on the trip purpose, SEAM may provide three types of services: *passenger-oriented, freight-oriented,* and *integrated*, as described in Figure 9.6.

9.3.2.1   Passenger-oriented service

The majority of shared mobility services target moving people. Depending on the seat capacity and operating speed, at least three sub-groups of shared automated mobility (SAM) or SEAM service can be identified.

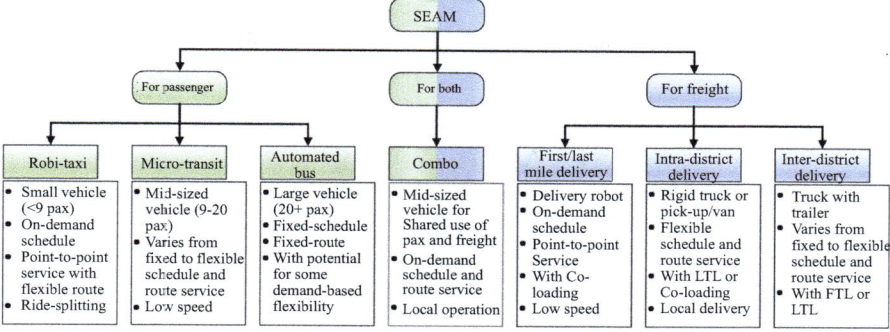

Figure 9.6   Different purposes and types of potential shared automated mobility service for people and goods

1. *Automated Taxi* which usually hosts a small number of passengers (e.g., no more than eight) and provides on-demand schedule and door-to-door service. It can be considered as the automated version of existing ride-sourcing or taxi service. And ride-splitting which enables passengers to share rides and split the cost may be an option.

2. *Microtransit* which accommodates a medium group of passengers (about 9–20) and mainly operates within a geographically confined area (e.g., within a city, university campus). Flexibility in terms of schedule and route may largely rely on the passenger demands and context complexity within the operational region. To only fulfill local travel needs, its operating speed may not be very high (say, below 50 mph) and the vehicles to be used may vary from large SUVs to vans to mid-size shuttle buses.

3. *Automated Bus* which supports long-distance (e.g., intercity) commute with relatively large capacity (say, greater than 20). Its schedule and route are not very flexible, and the majority of its trip is freeway segment and/or major arterials with high operating speed. Effective travel demand agglomeration as well as efficient multimodal connectivity play a critical role in guaranteeing its system performance.

9.3.2.2   Freight-oriented service

Compared to shared mobility for people movement, limited information or research is available for freight-oriented services. Here we propose three potential types of SAM or SEAM service for ground freight, based on the geographical scale of service.

1. *First/last mile shared automated delivery.* An autonomous delivery robot (e.g., Nuro's R2) is applicable for this type of service, which may only operate at low speed. On-demand schedule and point-to-point service are preferable to guarantee the service quality. As the robot serves the terminal of logistics and major interface with customers, it needs to be designed in a more user-friendly manner and to enable more efficient space sharing for the goods.

2. *Intra-district shared automated delivery.* The type of service is considered as local delivery (in terms of logistics management), mainly comprising the trips between warehouses and distribution centers and trips circulating within the district. For logistics between warehouses and distribution centers, the schedules and routes could be less flexible, while the delivery of large goods (e.g., furniture) to the customers should be more flexible and less than truckload (LTL) freight shipping strategies should be applied to provide more attractive service.

3. *Inter-district shared automated delivery.* Large trucks (with trailers) are usually required for this type of delivery service, featured with long-haul travel between ports and warehouses or between warehouses. Considering the relationship between supplies and demands as well as the fleet management strategies, schedules and routes may vary from fixed to flexible, and different freight shipping strategies such as full truckload (FTL) or LTL may be adopted.

9.3.2.3 Integrated service

Like existing courier network service (CNS) or crowd-shipping, integrated service aims to satisfy the mobility needs for both people and goods with the same vehicle or trip. Although there are more flexible service targets to choose and more synergy opportunities to combine the delivery trips, vehicle designs would likely have to change to adaptively share room for different situations and for the fleet management to efficiently satisfy mobility needs across different commodities. Considering the potential overlapping trip segments between people movement and goods movement as well as other common features, local operation in mid-sized vehicles with flexible schedules and routes would be preferable.

9.3.3 SEAM system framework and key elements

Similar to the existing shared electric mobility systems, a typical shared electric and automated mobility system for a large-scale on-demand passenger-oriented service consists of at least four elements as illustrated in Figure 9.7: (1) users; (2) shared electric and automated vehicles (SEAVs); (3) charging facilities; and (4) service providers with fleet management.

9.3.3.1 Customers

As the demand side for SEAM service, customers can make instant requests via dedicated smartphone apps based on their travel needs. They may be provided with different options as trade-offs between convenience and cost. These options may include different type or size of vehicles, exclusive service versus ride sharing, door-to-door service versus walking to/from alternative pick-up/drop-off spots in exchange of money saving, and accommodation of special needs (e.g., wheelchair access and large luggage). Once the requests and preferences are confirmed, customers can follow the guidance to approach the designated pick-up locations and meet with service vehicles. Certain identity verification procedures are necessary to

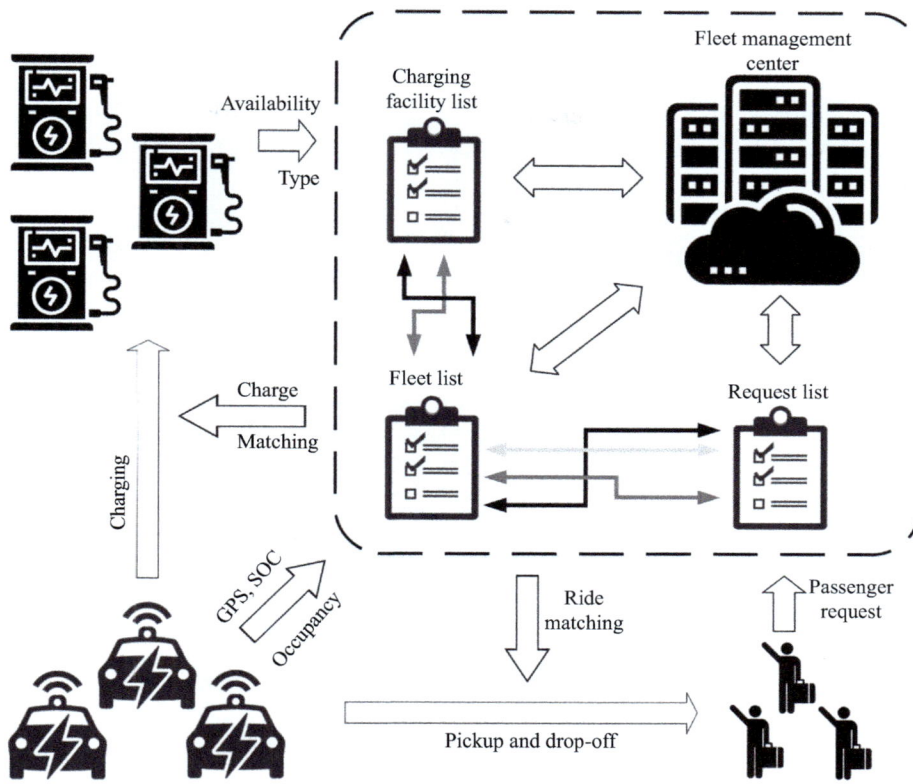

Figure 9.7   System overview of shared electric and automated mobility

authorize customers' access to the shared-use vehicles. Upon the ID verification, customers can get on-board and travel toward the designated drop-off locations. The payment methods (i.e., payment types and when to pay) must be in agreement between the customers and service providers before the initiation of service.

9.3.3.2   Shared electric and automated vehicles (SEAVs)

As the centerpiece of the SEAM service, SEAVs work as the bridge between customers and service providers. In addition, they generate the energy demands for charging activities. SEAVs must connect with the fleet management center and report on their critical states, including their instantaneous locations, speed, vehicle occupancy, state-of-charge (SOC), and confirmation of the customers' pick-up or drop-off location, in order to facilitate the optimal planning and system operation by the fleet management center. For the fleet health monitoring and predictive maintenance purpose, more detailed in-vehicle diagnostics information must be transmitted to in real-time and monitored by the management center. If the SOC of an SEAV does not satisfy the service need, it should be dispatched to an available charging station for refueling. Also, once an SEAV completes a round of service, then it may be repositioned to a nearby charging stall or a designated spot where service requests may be very likely to spawn very soon in the neighbor area (according to the predictive analytics by the fleet management center).

9.3.3.3   Charging facilities

As the energy supplier for SEAVs, the available charging facilities will significantly affect the entire system efficiency. Current capacities of charging infrastructure may not be sufficient to support the vehicle clustering effects that are likely with SEAM service. To mitigate the potential bottlenecks caused by the peak charging demands from SEAV swarms, key information of charging facilities such as the levels of charging (i.e., level-1, level-2, and level-3), charging type (i.e., wire or wireless), and availability should be constantly communicated with the fleet management center to maximize the utilization of their charging capabilities. As more investment flows into charging infrastructure (i.e., to expand the energy supply capacities) better scheduling for charger usage would help reduce the energy costs of SEAM service providers and minimize the impacts of fleet charging on the power generation at the plant and power distribution over the grid.

9.3.3.4   Fleet management centers

As the "brain" of the SEAM service system, fleet management centers need to receive a plethora of information from customers, SEAVs, charging facilities, and other sources (e.g., geographical information systems, traffic surveillance systems, and weather monitoring systems) in real-time, constantly process all this information, sketch the best operational strategies within reasonable time windows, and promptly send out confirmation or rejection messages to customers, itineraries and even routing guidance to SEAVs, and reservation or cancellation messages to charging facility managers. The fleet management centers may store massive historical data and perform predictive analytics to improve the overall system performance.

9.3.4   *Methods to support SEAM system design*

From the SEAM system perspective, the customer needs are the major driver for the service and would be impacted by the key elements mentioned earlier: SEAVs, charging facilities, and fleet management centers, all of which should be modeled within the design framework. The SEAV design model should focus on the vehicle characteristics and performance such as seat capacity, maximum operating speed, acceleration, charging level compatibility, all electric range, and energy consumption rate or factor. The charging facility design model should at least reflect the facility location, number of chargers, and type of chargers. The fleet management center model should be able to consider the real-time traffic information along each road section and charging schedule, estimate the shortest paths or least energy consumption paths, and determine the optimal fleet assignment as well as associated itineraries. The input of the design framework includes all the SEAM system constraints (e.g., investment level) and parameters (such as predictive demand for SEAM service), while the output covers various costs or objective performance indices for different stakeholders (e.g., customer waiting time and total revenue) and decision variables (such as fleet size and number of charging stations).

Compared to the vehicles and the charging facilities, the fleet operation and management itself is much more complicated. Over the past few years, researchers have proposed different strategies for optimizing SEAV fleet management and have developed various tools to model and evaluate SEAM service. Most of them relied on agent-based simulation to model the SEAM system operation using the C++ program [32] or MATSim [33]. Others have used MATLAB® for numerical simulations, attempting to solve the mixed-integer programming problem [34]. Also, the majority of existing studies have been focused on SEAM system operation at the regional or city level such as Lisbon in Portugal [35], Austin in Texas [36,37], Ann Arbor in Michigan [38], New York City [39], Tokyo in Japan [40], and Rouen Normandie in France [41]. Only a few have extended their modeling capability to a national scale [42]. These studies aim to investigate how various SEAM system parameters (including travel demand, value of travel time, fleet size, fleet composition, vehicle occupancy, all electric range, charger location and type, charging availability, and price of electricity) would interact with each other and affect the travel behaviors as well as system-wide costs and efficiencies for mobility, energy, and environment.

To improve the adoption of SEAM service and better steer its sustainability, some studies performed detailed cost and benefit analyses [43], while others conducted comprehensive literature reviews to synthesize users' perception [44] and examine potential impacts. For example, Rojas-Rueda *et al.* [45] discussed the direct and indirect impacts of SEAM service on public health in a comprehensive manner (see Figure 9.8). Nevertheless, enabling cost-effective deployment of charging infrastructure and deliberate design of relevant policies [46,47] would be indispensable to ensure synergies between electrification and SAM and tap the full potential of SEAM system.

9.4 SUMO-based modeling for demand-side cooperative shared automated mobility system: a case study

In this section, we provide a case study on the system design by modeling a novel demand-side cooperative shared automated mobility (SAM) system using the Simulation of Urban MObility tool (SUMO) [48], a microscopic traffic simulator.

9.4.1 General description

Conventional ride-hailing services (e.g., Uber and Lyft) or automated taxi operation models described in the existing literature are demand-side oriented, where customers can enjoy the utmost convenience by requesting the pick-up and drop-off locations at their own will. However, the supply sides (i.e., drivers or shared automated vehicles) usually have to adapt to satisfy the requests (e.g., making significant

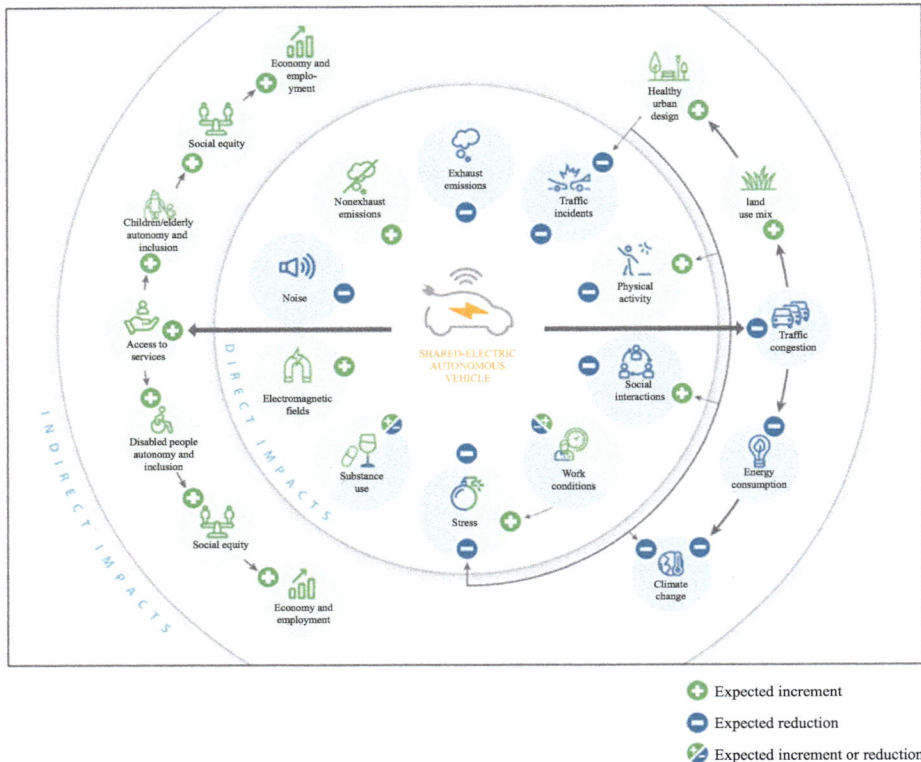

Figure 9.8   Direct and indirect impacts of SEAM on public health. Adapted from [45]

detours en-route, or incurring delays at left-turns). Without this adaptation, customers' experiences and system-wide performance would suffer. In contrast, a demand-side *cooperative* SAM system aims to improve overall system efficiency (on both demand and supply sides) in exchange for a small amount of customers' walking from origins to alternative locations to facilitate pick-ups or be dropped off at slightly different locations to enable agglomeration of drop-offs. The alternative locations are very scenario-specific which could be just the other door of a large building or the other corner of a nearby intersection. Customers can specify their walking distance limit for cooperation level negotiation (with potential incentives in terms of monetary compensation).

9.4.2   Simulation framework

The system is built in SUMO with background traffic and consists of a group of *customers*, a fleet of *shared automated vehicles* (SAVs), and a *service coordinator* as illustrated in Figure 9.9. The customers' demands for the SAM service are generated randomly over the simulation network. The requests are sent to the SAM service coordinator, including the information of time stamps, locations, group sizes, trip origins (if different from the request locations), and trip destinations. Once the customers get the confirmation of ride matching as well as the walking guidance related to alternative pick-up and drop-off locations (if any), they will proceed in the simulation to follow the guidance until the completion of their trips. At the core of the SAM system, the service coordinator keeps monitoring the states of SAVs (e.g., location and seat availability) and network traffic in real-time, receives the requests information from customers, determines the optimal ride-matching for

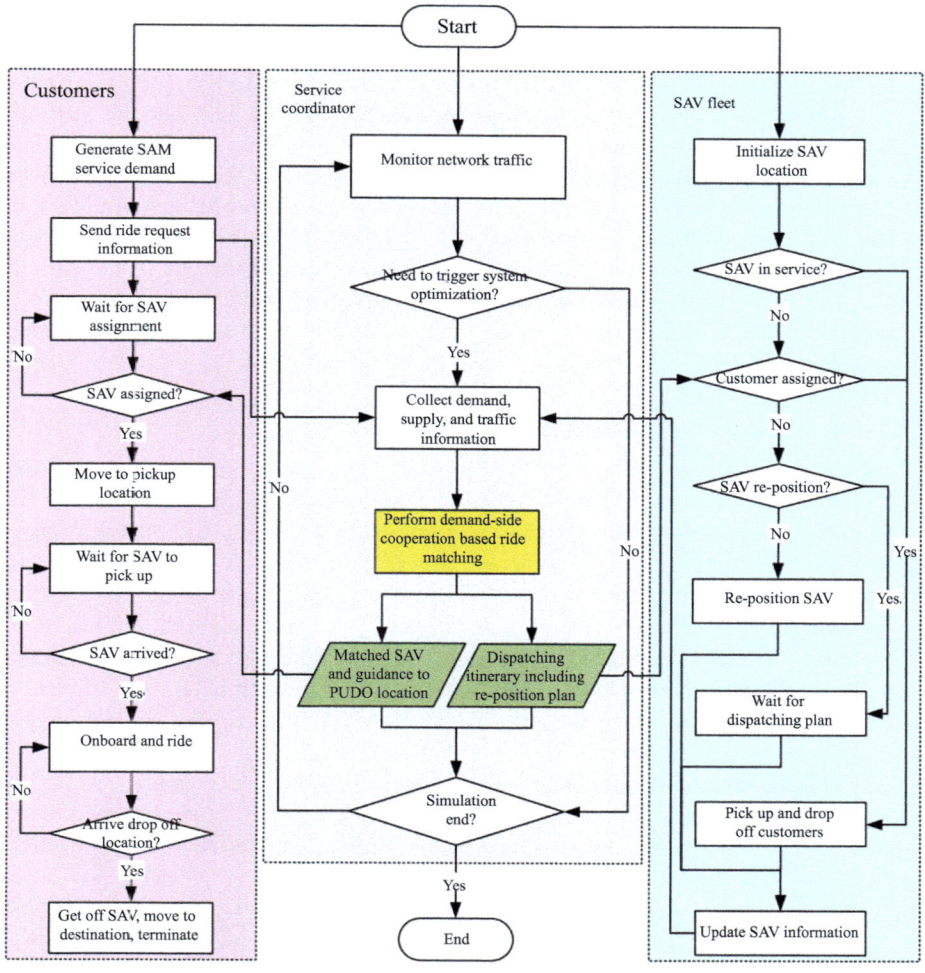

Figure 9.9   Simulation framework in SUMO. Adapted from [50]

each customer-SAV pair and itinerary as well as suggested route for each SAV, and communicates this information with the designated customer as well as the associated SAV. SAVs follow the ride-matching plan, suggested itineraries, and route guidance (for both service and re-position) from the service coordinator to provide the SAM service. If they complete their service rounds, they will reposition to the suggested locations by the service coordinator. These locations may be identified from the spatio-temporal predictive analytics of historical SAM service request data.

9.4.3   Ride matching optimization

The SAV ride matching optimization with demand-side cooperation can be modeled as a mixed integer programming (MIP) problem. The objective of ride matching is to maximize the total profit of providing the SAM service which is the sum of difference between the revenue and cost for each transaction. In the simulation, the revenue and cost for each transaction are crafted to be highly relevant to the request trip length and instant route distance between the SAV and the customer to provide enough driving force for the SAV to initiate the service. However, if there are not enough supply capacities (i.e., totally available seats of the SAV fleet) compared to the service demands, some requests (e.g., less profitable) would not be responded in a prompt manner and would be put into the waiting list for next round system optimization or even be dropped without service at the end of the simulation. A python script-based Application Programming Interface (API) is coded to apply the GUROBI Optimizer [49] to solve the MIP-based ride matching problem online in SUMO.

9.4.4   Simulation setup and results

As a case study, the New York City network provided by the Open Street Map (OSM) is imported into SUMO, which includes all the roadway geometries and default traffic signal plans. In the simulation, four shared automated vehicles (SAVs) with seat capacity of three are generated to serve 140 requested trips by customers (with group size of one). It turns out a larger scale of the optimization problem may not be handled in a computationally efficient manner by the GUROBI Optimizer or other commercial solvers. A heuristic algorithm should be developed to find the right balance between solution optimality and computational efficiency. Background trips (other traffic) are spawn uniformly (one trip per simulation second) over the time and the entire network (see Figure 9.10).

Besides the *optimal ride matching with demand-side cooperation* (ODC) algorithm, two other "door-to-door" service algorithms are coded with APIs in SUMO for comparison. One is the *spatiotemporal incremental matching* (Baseline) algorithm and the other is the *optimal ride matching without demand-side cooperation* (ONDC) algorithm. The first algorithm is a heuristic one where the ride-matching results depend on both the chronological order of customers' requests (i.e., the earlier the higher priority) and the route distances between available SAVs and customers on the waiting list (i.e., the closer the higher priority). The ONDC

Figure 9.10   Simulation network of New York City in SUMO

algorithm (also solved with the GUROBI Optimizer) is similar to the ODC algorithm but no alternative pick-up or drop-off locations for customers (from the demand-side perspective) are considered. For both the optimal ride matching algorithms (i.e., ODC and ONDC), an optimization window of 120 s is selected on a sensitivity analysis basis. In other words, the system will be reoptimized for the ride-matching between the available SAVs and customers' requests every 120 simulation seconds.

From the supply-side (SAVs) perspective, two key performance metrics: vehicle-miles traveled (VMT) and vehicle-hours traveled (VHT) are evaluated in the simulation. From the customer's perspective, trip detour factor (TDF) is concerned which represents the ratio between the customer's trip distance under the ridesharing mode and under the exclusive riding mode. The simulation results of different dispatching algorithms are presented in Table 9.2. It can be observed that both ODC and ONDC algorithms can remarkably reduce VMT and VHT of the SAV fleet, which indicates that the SAM services with optimal ride matching are much more efficient than the baseline service. In addition, compared to the SAM

Table 9.2   Comparison of performance measures for different fleet dispatching algorithms[a]

Algorithm	Performance metrics					
	VMT (veh-mile)		VHT (veh-hour)		TDF	
	Value	% change	Value	% change	Value	% change
Baseline	605.4	/	51.6	/	9.2	/
ONDC	295.1	51% ↓	29.2	43% ↓	4.5	52% ↓
ODC	283.0	53% ↓	27.9	46% ↓	4.1	56% ↓

[a]The baseline algorithm results are used as the benchmark for estimating % change.

service without demand-side cooperation (ONDC), the ODC algorithm can further improve the system efficiency by more than 4%. In addition, the optimal ride matching algorithms (both ODC and ONDC) can significantly alleviate detoured travel distances for customers, and the savings can be as much as 56% and 52%, respectively. The SAM service with the ODC algorithm can bring about an additional 9% reduction than the one without demand-side cooperation.

9.5   Conclusions and discussion

It is widely accepted that connected and automated vehicles should be largely shared and in operation with zero emissions (e.g., electricity) to bring about the deep de-carbonization of the transportation sector. In this chapter, we have presented the state-of-the practice in shared automated mobility (SAM) as well as shared electric and automated mobility (SEAM) across the globe, followed by the review of some fundamentals on vehicle automation and electric vehicle. We have proposed some guidelines and issues on the SAM or SEAM system design and used the SUMO-based demand cooperative SAM system modeling as a study case.

It is also acknowledged that there are quite a few challenges and opportunities on the design, development, and deployment of the SAM and SEAM systems. From the design perspective, most of the existing efforts have been focused on mobility for people (such as automated-taxi and low-speed autonomous shuttle), and SAM or SEAM service for freight or synergy of both passenger and goods movements should receive more attention. Due to the nature of scattered spatiotemporal demands in the rural areas, it is a major challenge to design cost-effective SAM or SEAM services for improving the transportation availability, affordability, and accessibility in these communities. Transportation innovations and collaborations among public agencies, private sectors, academia, and local communities are definitely required. In addition, the recent global outbreak of COVID-19 has had a profound impact on all aspects of transportation including shared mobility, and autonomous and electric vehicles. We have a chance to learn a big lesson from this pandemic in planning resilient, sustainable, and equitable SAM or SEAM systems for the future.

In terms of modeling and evaluation of SAM or SEAM systems, the majority of research relies on numerical simulation or agent-based modeling without consideration of the interaction between agents or capturing more realistic traffic dynamics. Innovative tools and testing approaches/platforms are desperately needed for model fidelity, which may include but are not limited to (1) multiresolution simulation models; (2) integration of models for travel behavior, vehicle/traffic dynamics, energy consumption, and charging facility or power grid operation; and (3) everything-in-the-loop (XiL) modeling and testing platforms. Furthermore, with the rolling out of some pilot programs or projects, more opportunities would be unlocked to validate the proposed models and better understand the impacts of SAM and SEAM services in the real-world implementation.

Lastly, although a definite uptake in the adoption of electric vehicles (EVs) has been witnessed over the past decade, a few roadblocks need to be cleared along the path toward mass deployment of SEAM systems:

1. *Technology improvement for EVs.* With the recent advances in battery technologies, less people have the "range anxiety" concern when using modern EVs. However, research and development on more cost-effective battery materials, faster charging capabilities, and wireless charging capability are necessary to make SEAVs more attractive.

2. *Expansion of charging capacities.* Increased public charging infrastructure investments and major grid upgrades by various governmental agencies would be valuable for the introduction of large-scale SEAM services. Installing curb-side fast charger stalls, building charging stations in public parking lots, and setting up battery swap facilities (if any) could offset the lack of charging capacities.

3. *Incentives for SEAM service providers.* To date, many incentives for EV adoption are targeted at the individual consumer purchasing a vehicle. More thinking processes are needed to create incentives or even policies that support SEAV fleet operators if the governments and municipalities are serious to reduce vehicles on the road.

Author contributions

The authors confirm contribution to this chapter. All authors reviewed and approved the manuscript.

Declaration of conflicting interests

The authors declared no potential conflict of interest with respect to publication of this chapter.

References

[1]   SAE International. *J3016 (revised): Taxonomy and definitions for terms related to driving automation systems for on-road motor vehicles*. June 2018.

[2]   Jones, P.B., Levy, J., Bosco, J., Howat, J., and Van Alst, J.W. *The future of transportation electrification: utility, industry and consumer perspectives*. FEUR Report No. 10, August 2018.

[3]   SAE International. *J3163: Taxonomy and definitions for terms related to shared mobility and enabling technologies*. September 2018.

[4]   Sperling, D. *Three revolutions: Steering automated, shared, and electric vehicles to a better future*. Washington, DC: Island Press. 2018.

[5]   United Nations, Department of Economic and Social Affairs, Population Division. *World urbanization prospects: The 2018 revision (ST/ESA/SER.A/420)*. New York: United Nations. 2019.

[6]   Seto K.C., Dhakal, S., Bigio, A., *et al.* "Human settlements, infrastructure and spatial planning." In Edenhofer O. *et al.*, (eds). *Climate change 2014: Mitigation of climate change, Contribution of Working Group III to the Fifth Assessment Report of the Intergovernmental Panel on Climate Change (IPCC, Geneva)*, Cambridge UK/New York: Cambridge University Press; 2014; pp 923–1000.

[7]   Federal Highway Administration. *The urban congestion report*. (UCR): July–September 2019.

[8]   Lazarus J., Shaheen, S., Young, S., *et al. Shared automated mobility and public transport report*. 2017. https://doi.org/10.7922/G2HQ3X3V

[9]   Stocker, A., and Shaheen, S. *Shared automated vehicles: Review of business models*. International Transport Forum. 2017.

[10]   Federal Transit Agency (FTA). *Strategic transit automation research plan. Final report, FTA Report No. 0116*, January 2018.

[11]   Federal Transit Agency (FTA). *Transit bus automation market assessment. Final report, FTA Report No. 0144*, October 2019.

[12]   Federal Transit Agency (FTA). 2020. Available from https://www.transit.dot.gov/research-innovation/mobility-demand-mod-sandbox-program.

[13]   U.S. Department of Transportation (US DOT). 2020. Available from https://www.its.dot.gov/research_areas/attri/index.htm

[14]   van Dijke, J. P., and van Schijndel, M. "CityMobil, advanced transport for the urban environment: Update." *Transportation Research Record*. 2012; 2324: 29–36.

[15]   European Commission (EC). *CityMobil2: Cities demonstrating cybernetic mobility*. Final report, FP7-TRANSPORT. 2017.

[16]   Antonialli, F. "International benchmark on experimentations with autonomous shuttles for collective transport." *27th International Colloquium of Gerpisa,* February 2019, Paris, France. ffhal-02489797f

[17]   Teece, D. "China and the reshaping of the auto industry: A dynamic capabilities perspective." *Management and Organization Review*. 2019; 15(1): 177–199.

[18] Sugimoto, Y., and Kuzumaki, S. "SIP-adus: An update on Japanese initiatives for automated driving." *Mobility Road Vehicle Automation*. 2018; 5. https://doi.org/10.1007/978-3-319-94896-6_2.

[19] SK Telecom. "Building cooperative automated public transportation system and HD Map." Presentation in the *1st Navigation Data Standard (NDS) Public Conference*, June 2019.

[20] Land Transport Authority (LTA). *Autonomous vehicles*. Available from https://www.lta.gov.sg/content/ltagov/en/industry_innovations/technologies/autonomous_vehicles.html [Accessed March 20, 2020].

[21] Transport and Infrastructure Council. *National land transport technology action plan 2020 – 2023*. August 2019.

[22] Chan, C. C. "The state of the art of electric and hybrid vehicles." *Proceedings of the IEEE*. 2002; 90(2): 247–275.

[23] Ehsani, M., Gao, Y., and Emadi, A. *Modern electric, hybrid electric and fuel cell vehicles – Fundamentals, theory, and design,* 2nd Edn. Taylor and Francis. 2010.

[24] NURO. 2020. Available from https://www.theverge.com/2020/4/7/21212719/nuro-driverless-car-test-california-dmv-delivery

[25] Waymo. 2020. Available from https://waymo.com/waymo-one/

[26] Baidu. 2020. Available from https://www.chinainternetwatch.com/30485/baidu-free-robotaxi-service/

[27] Daimler. 2016. Available from https://www.daimler.com/innovation/autonomous-driving/future-bus.html

[28] International Transport Forum (ITF). *Automated and autonomous driving: regulation under uncertainty: Corporate partnership board report*. OCED. 2015

[29] Tesla. 2020. Available from https://electrek.co/guides/tesla-network/

[30] Knobloch, F., Hanssen, S.V., Lam, A., *et al.* "Net emission reductions from electric cars and heat pumps in 59 world regions over time." *Nature Sustainability*. 2020. https://doi.org/10.1038/s41893-020-0488-7

[31] Shrestha, B.P., Millonig, A., Hounsell, N.B., and McDonald, M. "Review of public transport needs of older people in European context." *Population Ageing*. 2017; 10: 343–361.

[32] Chen, D., and Kockelman, K. "Management of a shared, autonomous, electric vehicle fleet: Implications of pricing schemes." *Transportation Research Record*. 2016; 2572: 37–46.

[33] Loeb, B., Kockelman, K.M., and Liu, J. "Shared autonomous electric vehicle (SAEV) operations across the Austin, Texas network with charging infrastructure decisions." *Transportation Research Part C*. 2018; 89: 222–233.

[34] Iacobucci, R., Mclellan, B.C., and Tezuka, T. "Modeling shared autonomous electric vehicles: Potential for transport and power grid integration." *Energy*. 2018; 158: 148–163.

[35] International Transport Forum (ITF). *Urban mobility system upgrade: How shared self-driving cars could change city traffic*. 2015.

[36] Chen, T.D., Kockelman, K., and Hanna, J. "Operations of a shared, autonomous, electric vehicle fleet: Implications of vehicle & charging infrastructure decisions." *Transportation Research Part A*. 2016; 94: 243–254.

[37] Farhan, J., and Chen, D. "Impact of ridesharing on operational efficiency of shared autonomous electric vehicle fleet." *Transportation Research Part C*. 2018; 93: 310–321.

[38] Kang, N., Feinberg, F.M., and Papalambros, P.Y. "Autonomous electric vehicle sharing system design." *Journal of Mechanical Design*. 2017; 139.

[39] Bauer, G., Greenblatt, J., and Gerke, B.F. "Cost, energy, and environmental impact of automated electric taxi fleets in Manhattan." *Environmental Science & Technology*. 2018; 52: 4920–4928.

[40] Iacobucci, R., Mclellan, B.C., and Tezuka, T. "Optimization of shared autonomous electric vehicles operations with charge scheduling and vehicle-to-grid." *Transportation Research Part C*. 2019; 100: 34–52.

[41] Vosooghi, R., Puchinger, J., and Bischoff, J. "Shared autonomous electric vehicle service performance: Assessing the impact of charging infrastructure and battery capacity." *Transportation Research Part D*. 2020.

[42] Sheppard, C.J.R., Bauer, G.S., and Gerke, B.F., "Joint optimization scheme for the planning and operations of shared autonomous electric vehicle fleets serving mobility on demand." *Transportation Research Record*. 2019; 2673(6): 579–597.

[43] Quarles, N., and Kockelman, K. "Costs and benefits of electrifying and automating U.S. bus fleets." *Proceedings of the 97th Annual Meeting of the Transportation Research Board*. Washington D.C., January 2018.

[44] Axsen, J., and Sovacool, B. "The roles of users in electric, shared and automated mobility transitions." *Transportation Research Part D*. 2019; 71: 1–21.

[45] Rojas-Rueda, D., Nieuwenhuijsen, M.J., Khreis, H., and Frumkin, H. "Autonomous vehicles and public health." *Annual Review of Public Health*. 2020; 41: 329–345.

[46] Cohen, S., and Shirazi, S. *Can we advance social equity with shared, autonomous and electric vehicles?* UC Davis, Policy Brief, February 2017.

[47] Goetz, M. *Electric vehicle charging considerations for shared, automated fleets*. UC Davis, Policy Brief, October 2017.

[48] SUMO. *Simulation of urban mobility*. 2020. Available from https://sumo.dlr.de/docs/index.html

[49] GUROBI. *GUROBI optimization*. 2020. Available from https://www.gurobi.com/

[50] Zhu, L., Zhao, Z., and Wu, G. "Shared automated mobility with demand-side cooperation: A proof-of-concept microsimulation study." *Sustainability*. 2021; 13(5): 2483.

The views expressed are those of the authors.

Chapter 10

Demand for shared mobility to complement public transportation: human-driven and automated vehicles

Shadi Djavadian[1], Bilal Farooq[2], and Seyed Mehdi Meshkani[2]

Recent advances in communication technologies and automated vehicles have opened doors for alternative mobility systems (e.g., app-based taxis, app-based carpool, demand-responsive transit, peer-to-peer ridesharing, carsharing, and shared automated vehicles/shuttles). These new mobility modes have attracted the attention of researchers, public agencies, and private sector companies. Among other potential applications, these can serve as candidates for shared mobility modes that can complement public transportation. An example is the first/last-mile transportation need in low-density urban areas where implementation of high-frequency buses is not feasible. There is also the potential to substitute a new version of shared mobility mode as a substitute for fixed transit in low-density urban areas. In this study, we investigate the effects of ride-sharing service on travel demand and welfare, as it complements public transportation, thus addressing the first/last mile problem.

Given that field studies are costly and may be impossible to implement, we developed modeling and simulation tools for analyzing different scenarios. Two types of management and vehicle types are investigated: crowdsourced human-driven vehicles (HDVs) (e.g. Uber and Lyft) and centrally operated shared automated vehicles (SAVs). The influence of fare discount on demand and mode shift is also investigated. A case study of Oakville road network in Ontario, Canada is conducted using real data. The results show that ride-sharing with commuters who live in the study area has the potential of increasing ridership by 76% and decreasing wait time by 47% if centrally operated shared automated vehicles are used and 50% fare discount is offered for the use of shared first mile/last mile mobility service.

[1]Greenfield Labs, Ford Mobility, Palo Alto, CA, USA
[2]Department of Civil Engineering, Laboratory of Innovations in Transportation (LiTrans), Ryerson University, Toronto, Canada

10.1 Introduction

Among the societal impacts that will be caused by increasing urbanization over the coming decades, the transportation sector is likely to act both as a driver and a recipient of change: improving mobility with privately owned vehicles will make cities attractive sites for living and employment for many residences, but ever-growing private mobility demand will cause capacity issues that even massive investments into new infrastructure cannot cope with. Already today, traffic congestion in urbanized regions around the world is a major problem. For example, in the European Union (EU), it costs nearly 100 billion Euros, or 1% of the GDP, per year [1]. Although high capacity public transit will continue to serve commuters in well-defined corridors, accessing fixed-route transit hubs/stations in low-density residential areas will remain as an issue. In addition to the well-known first mile/last problem, serving the transportation needs in low-density urban and rural areas efficiently will continue to be a challenge [2].

Urban regions are keen on finding solutions that will help in avoiding many well documented adverse effects of traffic congestion, including economic losses. The Greater Toronto Area serves as a good example, given that it is already the sixth most congested city in the world [3]. The forecasted cost of congestion to its economy will balloon to $7.2 billion by year 2031 [4].

One potential solution would be to provide high frequency and more accessible public transit systems to mitigate congestion and discourage the use of low occupancy private cars from some roads. However, the demand for transit, especially in low-density areas, is often hindered by the absence of well-organized and effective multimodal services for accessing high-capacity public transit service stations. As noted above, this is the well-known "first mile/last mile" problem [5].

New technologies, in particular, shared on-demand flexible transit and the future use of shared automated vehicles (SAVs) by public agencies as well as private sector transportation companies are likely to help by offering new and fundamentally more flexible mobility services as a compliment to fixed-route high capacity transit [6,7]. The effectiveness of SAVs as compared to privately owned vehicles has been studied using simulation methods for both European cities (e.g., Berlin and Lisbon) and American cities (e.g., Austin). The results show that under favorable conditions, one SAV has the potential of replacing ten privately-owned conventional vehicles [8–10].

In the pre-COVID 19 pandemic period, the transportation network companies (TNCs) used conventional human-driven vehicles with success in gaining market share at the expense of traditional taxi services. In some instances, the ridership of public transit was also affected. It is logical that many public transport operators, either alone or in public–private partnerships, have explored ways to provide new mobility services and piloted different types of so-called flexible transportation systems, including demand-responsive shuttle, ride pooling, or micro-transit services. Also, shared automated shuttle demonstrations have been initiated in a number of urban areas.

For a long list of prepandemic era public and private partnership projects on shared vehicles in the United States, the interested reader is referred to American Public Transportation Association (APTA) [11]. A recent example of such services in Ontario (Canada) is in the City of Belleville which partnered with the Toronto-Based company Pantonium [12]. In this on-going project, the city's bus service acts more like a ride-hailing service from 9 PM to 12 AM, during weeknights.

The reason behind this experiment is that the City of Belleville with a population of 50,000 cannot offer conventional bus transit service. The annual transit ridership of this city is not higher than even the daily transit ridership in Toronto. Therefore, providing a high-frequency fixed-route traditional transit service in a small city or in low-density areas of large cities is not feasible. However, flexible transit system can be a viable solution. Due to the success of the initial pilot project, the City of Bellville (Ontario) is considering changing more routes to flexible service as well as decreasing reliance on 40-foot diesel buses used on fixed routes [13].

Many modeling and simulation studies on shared mobility (e.g. ridesharing, and carsharing) have either considered demand fixed as an exogenous input for optimizing profit, fleet size, station location, and reservation policies [14], or used a fixed fleet size for studying the interaction of demand with supply [15]. However, as shown by Djavadian and Chow [16], the flexible transit service (FTS) behaves as a two-sided market due to the distinct interests of service operators and travelers. Failure to apply the adjustment process reflecting the interaction of operator as well as traveler interest leads to unrealistic estimates of demand, welfare, and fleet size. Since real-world field experiments are impossible under controlled conditions, well-designed simulation studies are necessary. A measure of welfare is the consumers' surplus, which is defined as the difference between what the travelers might be willing to pay and what they actually pay.

Djavadian and Chow [16,17] proposed a simulation-based method to investigate the market for FTS. They adopted an agent-based day-to-day adjustment process designed to reach the stochastic user equilibrium (SUE). However, in their studies, they only evaluated the dynamic FTS for an HDV single ride service. In the study reported in this chapter, we extend the work of Djavadian and Chow [16] to shared HDV and AV services.

For comparison purposes, a simulation-based case study similar to [16] is conducted using Oakville (Ontario, Canada) road network. This study aims to evaluate the demand and social welfare effects of shared HDV and AV services under different dynamic operating policies (e.g. fleet vehicle capacity, fare price, and minimum wage of drivers).

The remainder of this chapter is organized as follows. In Section 10.2, an overview of agent-based day-to-day adjustment process for evaluating dynamic FTS will be presented. Section 10.3 describes the design of our case study for evaluation of dynamic shared HDV and AV services. Section 10.4 presents the results and analysis and Section 10.5 describes the summary and future work directions.

10.2   Background

In planning transportation services for low-density areas, public policymakers and planners are often faced with questions such as the ones noted below:

- What fleet size should be employed and what dispatch algorithm should be used for the given fleet size?
- Should they offer single ride or shared ride?
- Should such a service be operated 24 h a day, or only during a well-defined other period (e.g., peak period)?
- What pricing scheme should be used?

Each of these design decisions and operating policies leads to different level of service (LOS) offered to travelers, which in turn might lead to different demand and different impacted welfare (as measured by consumers' surplus).

Literature studies show that flexible transit, specifically taxi, has received research attention. These studies have frequently focused on fleet size, pricing, and vehicle routing problems. They mostly investigated taxi service or other flexible services from the point of view of service operators where the aim is to minimize cost within the day dynamics (or maximize profit for a private sector operator). An assumption is commonly made that the system is at steady-state, meaning that tomorrow's travel condition will be the same as today and that the travelers will make the same choices the next day even under different operating conditions.

The supply-demand equilibrium considered in these studies is based on a market equilibrium where the learning behavior of travelers is not considered. However as noted by Quadrifoglio and Li [18], the demand for flexible transit varies according to the LOS experienced by travelers. Due to this and other reasons, demand may change from one day to another. Therefore, the stochastic nature of demand deserves much attention when it comes to transportation planning and is of much importance to public agencies as well as private sector companies when evaluating alternatives.

The FTS, as noted by Djavadian and Chow [16,17], is fundamentally different from general transportation systems for the following reasons:

- The system performance is dependent on the choices of the travelers as well as the policies adapted by the FTS manager serving as an additional decision-maker.
- Unlike a traffic network where the route choice exclusively depends on the traveler decision, in the case of FTS, a passenger's route is decided by the operating policy of the FTS manager. In turn, the operating policy of the FTS is also affected by the choice sets of all travelers.
- As opposed to the traffic network cost function which is monotonic, it has been shown in the literature (e.g., [19]) that demand-responsive public transit cost function can be nonmonotonic with respect to flow. The link costs for an FTS are dependent on the operating policy and may be nonmonotonic or follow discrete step functions.

Due to distinct characteristics of the FTS, the methodology for defining, analysis, and evaluation of different designs should formally treat the following requirements:

- Heterogeneity of travelers and interaction between travelers (if applicable);
- Effect of operating policy on traveler decision;
- Impact of traveler decisions on FTS level of service;
- Day-to-day learning process of both travelers and the FTS operators (in the two-sided market).

Because of the complex characteristics of the FTS as defined above, a steady-state model would not be suitable due to the sensitivities attributed to within-day dynamic operating policies. Therefore, a day-to-day dynamic model should be used.

In their study, Djavadian and Chow [16,17] proposed an agent-based day-to-day adjustment process for analyzing the effects of design parameters and operating policy of flexible transportation system on equilibrium demand and associated welfare effect. They demonstrated that the interaction between the operator and travelers was equivalent to a two-sided market made popular by companies such as Uber, where the trading platform is the spatial network. This was done so that the operating decisions of the service operators become part of the output of the modeling framework.

The proposed methodology explained in this chapter incrementally adjusts demand levels interacting with a stochastic dynamic system operator in search of the intended solution. The first mile/last mile case study results covered in this chapter, based on the use of real data from Oakville, Ontario, demonstrate the sensitivity of the day-to-day model to operating policies.

The approach proposed by Djavadian and Chow [16,17] is organic and supportive of the advanced mobility systems that are taking place or have the potential for future application in transportation planning. However, in their study, the focus was only on single-ride FTS driven by human drivers. In the case study described in this chapter, the methodological framework is extended for application to the flexible transport service using HD and AV modes, in providing ride-sharing service as opposed to single-ride service.

10.3   Methodology

In support of planning the first mile/last mile service, we developed an original agent-based transportation simulation tool framework and applied it to the study of shared mobility and automated vehicles (see Figure 10.1). In order to have the capability to answer research questions, the state-of-knowledge advances in modeling and simulation tools are incorporated in the development of all parts of the methodology and their interactions.

The user input interface is designed to receive the following data that are required for simulations:

- FTS characteristics
- Traffic/transit network
- Transport survey data and commuter population generation.

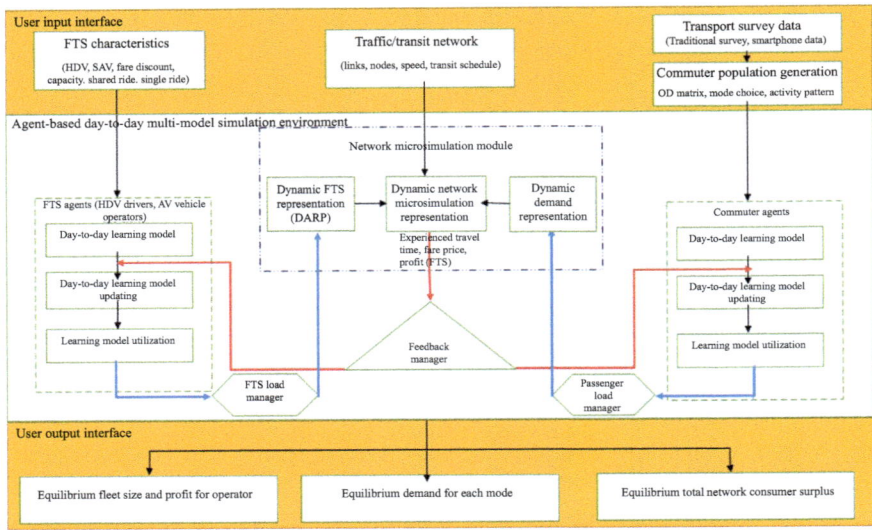

Figure 10.1   Agent-based transportation simulation tool framework

For use in the present case study, FTS characteristics are required for HDV and SAV alternative modes and associated fare option and capacity information. The choice of shared ride and single ride options is provided. Using the standard practice of coding study area traffic/transit network, the user enters data on links, nodes, speed, and transit schedule. The transport survey input can be sourced from traditional travel surveys and smartphone data. The traveler (i.e., in this case study, the commuter) population is generated to the necessary level of detail (i.e., origin-destination matrix, mode choice, and activity pattern).

The computation part of the methodology is designed to work within the agent-based day-to-day multimodal simulation environment. As shown in Figure 10.1, there are three interactive components of the computation part. These are:

- FTS agents (HD drivers/operators and AV vehicle operators).
- Network microsimulation module (includes dynamic FTS representation as a dial-a-ride problem (DARP), dynamic network microsimulation representation, and dynamic demand representation); managers for FTS load, passenger load, and feedback to agents.
- Commuter agents.

The user output interface includes equilibrium level:

- Fleet size and profit for operator;
- Demand for each mode serving commuters;
- Total network consumer surplus.

The agent-based transportation simulation tool includes a number of functions and calibrated models required for providing outputs to the user-specified inputs applicable for the last mile/first mile transportation problem (Figure 10.1). This state-of-knowledge methodology has the capability to answer present and tomorrow's questions on shared mobility and automated vehicles in the context of last mile/first mile travel needs.

10.3.1   Ride-sharing

Flexible on-demand ridesharing refers to a mode of transportation which allows users to share a vehicle. These travelers have similar itineraries and time schedules [20]. Since they live in the same general area of a city, they can potentially meet socially and need not be strangers. If they choose to use shared mobility service, they avail benefits of a reduction in travel costs. Potentially, ridesharing is a system that can combine the advantages of private cars and fixed-line transit systems, that is, flexibility and speed of private cars with the reduced cost of fixed-line systems, at the expense of convenience [20]. Saving travel cost, reducing travel time (as compared to fixed route transit), mitigating traffic congestion, conserving fuel, and reducing air pollution are a number of ridesharing benefits that can be availed by various interest groups [20]. Examples of flexible and shared use mobility systems are demand-responsive micro-transit, and special services offered by TNCs as a result of a contract with an urban government.

10.3.1.1   Dynamic dial-a-ride problem

Since the 1970s, various vehicle routing policies have been studied for a dial-a-ride-problem (DARP), focusing on both static and dynamic characteristics. What distinguishes the dynamic dial-a-ride problem from the static dial-a-ride problem is that in the dynamic dial-a-ride problem, vehicles' routes are modified in real-time in response to trip requests arriving over time. The dynamic dial-a-ride problem usually has two conflicting objectives, identified as (1) system efforts and (2) the customer's interests [21].

In this study, the dynamic DARP proposed by Hyytiä *et al.* [21] is adapted. The same model was used by Djavadian and Chow [16]; however, in their study, the capacity of fleet size was highly constrained and only single ride FTS was assumed. Our microsimulation implementation handles fleet vehicles and updates their path en-route to accommodate added pick-ups and drop-offs dynamically.

At the beginning of each day, a fixed set of v uncapacitated vehicles (i.e., they can serve the market) with constant speed is assumed to provide pick-up and delivery service for customers. The total fleet size on each day Λ_d, depends on the policy of the service provider. In this study, two types of fleet vehicles and management styles are used. These are centrally operated AV fleet and crowdsourced HDVs. The strategy for updating Λ_d for AVs is explained in Section 10.3.2, whereas the strategy for updating the latter is taken from Djavadian and Chow [16].

Within a day, the vehicle to customer assignment is as follows. When a trip request is placed, the customer is assigned to a specific vehicle immediately, and the vehicle's route plan (list of nodes/intersections to visit) is then updated to include both the pickup location and delivery location in that order, with no request ever being rejected. An example of a tour for a ride-sharing problem is $\xi = \{1\ 2\ -2\ 3\ -1\ -3\ 0\}$. The numbers represent customers' ID in the request table list, whereas positive value denotes pick-ups, negative values denote drop-offs, and 0 denotes the depot. An example for single ride service tour is $\xi = \{1\ -1\ 2\ -2\ 0\}$. Since this service is a shared-use type as a dynamic DARP, a customer may be delayed in being dropped off in favor of another customer if it minimizes total cost.

As mentioned before, ξ for each active vehicle $v \in \Lambda_d$ is updated using the model proposed by Hyytiä *et al.* [21] where they considered the nonmyopic dynamic DARP as a multiserver queue problem. The model is being nonmyopic since the decision is not only based on current information, but it also takes into account the future conditions with steady-state queue characteristics.

The model allows user to choose between myopic and nonmyopic assignments. Since the first mile/last mile problem is considered in this study and travelers are assumed to have defined departure time as opposed to random time, in this study myopic assignment is used. Here, $\kappa = 0$ refers to a myopic system whereas $\kappa > 0$ refers to a nonmyopic system.

In their study, Hyytiä *et al.* [21] assumed the process of modeling vehicle assignment to be similar to assigning a customer to a server in a multiserver queue and they developed a policy called mm1 that aims to minimize a weighted sum of the mean passengers' travel time and distance the vehicles travel (per passenger) as shown in (10.1):

$$mm1: argmin_{v,\xi}[C(v, \xi) - C(v, \xi')] \tag{10.1}$$

Equation (10.2) presents the calculation for the relative cost of vehicle-route pair (v, ξ) as a sum:

$$C(v, \xi) = \gamma T(v, \xi) + (1 - \gamma)\left(\kappa T(v, \xi)^2 + \sum_i S_i(v, \xi)\right) \tag{10.2}$$

where ξ is a new tour and ξ' is a prior tour (the tours are constructed using a Traveling Salesman Problem with Pickup and Delivery (TSPPD) heuristic; C is the cost value; $T(v, \xi)$ is the current work backlog of vehicle v (measured in time); $S_c(v, \xi)$ denotes the residual sojourn time of a customer c in the system; γ corresponds to the combination of the minimization of the system's effort and travelers' costs and can take values between 0 and 1. In this study, a value of 0.5 is used, which takes into account both system's effort and the travel time cost incurred by the traveler.

It is useful to note that the DARP implemented in this study based on Hyytiä *et al.* [21] is with centralized dispatch.

10.3.1.2   Fleet vehicle path updating

In the study conducted by Djavadian and Chow [16,17], since the purpose was to model a single ride, the microsimulation tool updated the path of fleet vehicles only at pick-up and drop-off locations and the path remained the same in between. However, to accommodate ride-sharing, changes are made to the microsimulation path finding.

For example, let's assume there are two commuters, 1 and 2. With 1 having origin/ destination pair (A-C) and 2 having origin/destination pair (D-C) (Figure 10.2). The request time of 1 is 8:00 AM. Further, let us assume a fleet size of 1 with a seat capacity one. When commuter 1 makes a request, the tour of the vehicle is {1 −1 0} and the path is {Depot, A, B, C, E, Depot}. However, let us assume that commuter 2 makes a request when vehicle carrying commuter 1 is on link connecting A and B. In the case of single ride as noted in [16,17], the tour and path of the fleet vehicle would have been updated as follows: {1 −1 2 −2 0}, {Depot, A, B, C, E, D, C, E, Depot}.

In the case of a single ride, the entire path from start to end is calculated. However, in the case of shared-ride (with the use of a vehicle with seats appropriate for the travel market), only the next intersection on the list is known to the vehicle.

For example, at 8:00 AM, the tour of the vehicle is {1 −1 0}, therefore the path is {Depot, A}. Once the vehicle arrives at node A, it updates the next node on its path to destination C which is B (based on shortest travel time), so the new path becomes {Depot, A, B}. Once traveling on the link between A and B, it is notified of new passenger pick up at D, so the tour is changed to {1 2 −1 −2}. As such when the vehicle arrives at B, it looks for the shortest path to D and the next node on its path which is E, as such the path becomes {Depot, A, B, E} and so on. This not only allows us to implement ride-sharing, but it also allows us to further extend the framework to dynamic distributed routing, similar to Djavadian and Farooq [22] where AVs are guided by a network of intelligent intersections.

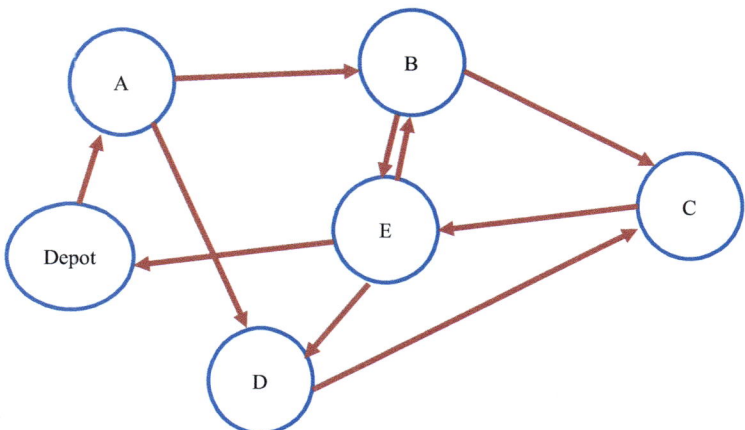

Figure 10.2   Sample network

10.3.2   Shared automated vehicles

If supported with well-studied infrastructure plans and designs, AVs offer the potential to improve transportation system operations in many ways [22]. Increased safety, enhanced efficiency, and reduced environmental impact (especially electric vehicles) are frequently noted as benefits of AVs [7]. AVs can be used as a part of car-sharing systems during and postpandemic era if suitable design and operational measures are implemented as described in Chapter 18. Further, under favorable market conditions and with public health measures in place, AVs can be used for ride-sharing. Shared automated vehicles enable people to call up distant SAVs using mobile phone applications. Moreover, they can anticipate future demand and relocate in advance to better match vehicle supply and travel demand.

An on-demand flexible transit system can be operated by a decentralized operator or a centralized operator. For the purpose of this study, it is assumed that the centralized operator is a public agency with maximum M registered automated vehicles. Unlike FTS with HDVs [16], in the case of FTS with automated drivers, the central operator is the main decision maker.

An agent-based methodology shown in Figure 10.3 has the capability to find day-to-day stochastic user equilibrium (SUE) for centrally operated AV fleet used for flexible transit services. In this methodology, the central operator decides how many vehicles should be used on each day.

The green square in Figure 10.3 highlights the role of centrally operated AVs in this study. What separates centrally operated SAV in this study from the centrally operated HDV in Djavadian and Chow [17] is that in their study, the day-to-day adjustment process of only travelers was considered. The fleet size remained the same from one day to another, whereas in this study, for realism, fleet size changes from one day to another, depending on demand level and feedback from the system.

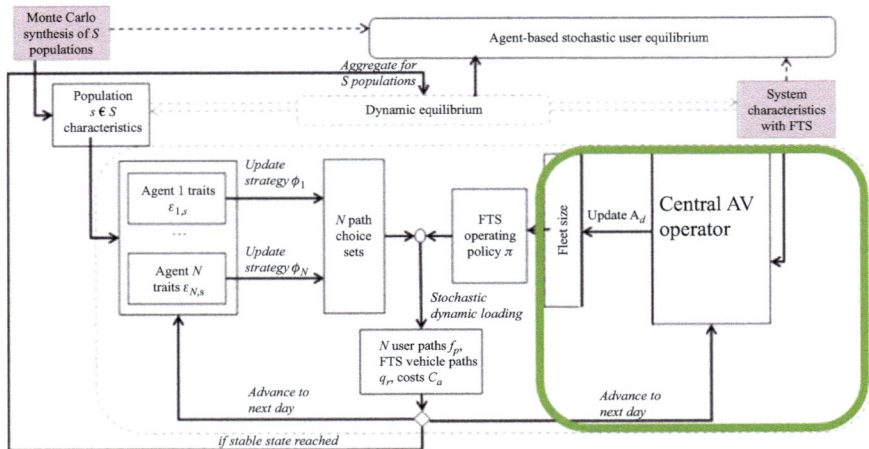

Figure 10.3   Agent-based day-to-day SUE framework for centrally operated AV fleets

As a public service provider, the aim is to serve the citizens. However, maintaining AVs is also costly, and as such the strategy of the central operator on each day d is to reach social optimality where the needs of both parties are met.

The results from Djavadian and Chow [16] showed that for each fare price, social optimality occurs when there is a perfect match between supply and demand (market equilibrium). Therefore, to reach system optimality, (10.3) is used to adjust the fleet size of AVs from one day to another:

$$\Lambda_d = \frac{\Upsilon_d}{\varpi}, 0 \leq \Lambda_d \leq M \tag{10.3}$$

where Λ_d is the AVs fleet size on day d and Υ_d is the predicted total demand on day d.

For simplicity in this study, it is assumed that $\Upsilon_d = \Theta_{d-1}$, where Θ_{d-1} is actual demand on day $d-1$. This means that the operator assumes that the demand on day d is the same as demand on day $d-1$. This is a myopic estimation of demand; however, as shown by Djavadian and Chow [16,17], the day-to-day adjustment process for evaluating FTS eventually reaches stochastic user equilibrium where demand for day d is equal to demand for $d-1$.

ϖ is the capacity of each AV vehicle.

In this study, the capacity of both HDV and AV is set to four-passengers. However, we also investigated the seven-passenger vehicles but found no major differences.

10.4   Case study

The modified agent-based day-to-day adjustment process is applied to a hypothetical ride-sharing flexible transit system using real data obtained for Oakville, Ontario (Figure 10.4). The proposed ride-sharing FTS is used as a potential feeder service solution for the first/last mile problem. It is intended to move residents from home to the GO Transit rail station. The network chosen for this case study has been previously used for first/last mile study site by Djavadian and Chow [16,17] and Alshalalfah and Shalaby [23]. The use of a common network allows us to draw comparisons with the previous studies.

10.4.1   Case study objectives

The focus of this study is on those residents of the town of Oakville who commute to downtown Toronto for work during the morning peak period by taking Government of Ontario (GO) rail transit out of the Oakville Station. There are four major differences between the case study presented in this chapter and the ones conducted by Djavadian and Chow [16,17]:

- The dynamic within-day routing policy used by Djavadian and Chow [16] only focused on single-ride service, whereas in this study ride-sharing option is introduced.

Figure 10.4   Oakville road network and location of GO Transit rail station

- In their study, the focus was only on human-driven FTS, whereas in this study we introduce shared automated vehicles.
- The fleet size in Djavadian and Chow [16] depended on the day-to-day strategy of the drivers themselves (entering the market or not) without any decision made by the operator of the FTS. However, in this study for the case of AVs, the operator decides to change the fleet size from one day to another.
- We investigate the effect of vehicle capacity on ride-sharing demand.

Since the focus of this study is on home-to-work trips, it is assumed that all the commuters are willing to share a ride with persons living in their neighborhood. These commuters can get to know each other and reach an understanding to share the first mile/last mile service. This is a reasonable assumption, as studies have shown that when it comes to work trips, commuters put more emphasis on travel time and other benefits as opposed to with whom they share a ride [24].

10.4.2   Problem definition

For comparison purposes, similar to Djavadian and Chow [16,17], the focus of this case study is on home-to-work (H-W) trips and five access modes to the Oakville (Ontario) GO rail transit station are modeled: bus, automobile, walk, fixed-route feeder bus transit, and taxi. These modes are listed in the Transportation Tomorrow Survey (TTS) [25]. During the study period (6:30–7:30 AM), 2000 commuters access Oakville GO rail station for work trips from Oakville to Toronto. Based on the statistics obtained from 2011 household survey by TTS [25], the market share for different modes used by commuters to access the Oakville GO Transit are as follows: 73% used auto as the access mode, 19% used bus, ~1% used taxi (17 commuters), 6% used bike, and 1% walked to the Oakville Go Station. As can be seen from the access mode statistics, auto is a major access mode to Oakville Go Station and because of this high dependency on auto as an access mode to the station, a significant problem facing Go Transit in Oakville is that all its park-and-ride parking lots are operating at capacity.

For the purpose of this study, we "assume" that a public transit agency would like to provide commuters better access to Oakville Go Station (as a solution to the first mile/last problem). One way for the public agency to achieve this goal is to provide door-to-Go Station flexible transit service as studied by [16,17,23]. Similar to the study conducted by Djavadian and Chow [16,17], a taxi service is treated as a ride-sourcing flexible transit mode.

For the purpose of this case study, we assume that the government agency would like to improve the level of service of current FTS by changing its design and operating policies in the following aspects:

- Offering first mile/last mile ride-sharing service;
- Offering cost sharing to reduce fare cost;
- Using shared automated vehicles (SAVs) versus crowdsourced ride-sharing service using human-driven vehicles (HDVs);

The case study calls for answers to the following questions:

- What will be the effect of ride-sharing on demand and consumer surplus?
- What will be the effects of fare discount on demand for ride-sharing?
- Will automation affect demand for ride-sharing?

The network, base case population, and calibrated Logit Model from Djavadian and Chow [16,17] are used in this study so that comparisons can be made with some of their findings. Similarly, unless it is stated otherwise, the same values are used for the agent-based day-to-day adjustment model parameters.

A maximum fleet size of 10 vehicles are assumed for illustration purposes. Table 10.1 presents test scenarios for this case study. In total, eight scenarios are tested for investigating the effect of automation, fleet vehicle capacity, and fare discount on equilibrium demand and consumer surplus (i.e., welfare). According to scenario design, shared vehicles are AVs and conventional vehicles have human drivers. The base case scenario (Day 0) is obtained from Djavadian and Chow [16].

In this study, the effect of traffic congestion is not considered due to the assumption that both HDV and AVs travel at free-flow speed within the low-density Oakville traffic network. For the information of the reader, Chapter 11 will describe the effect of traffic conditions on in-vehicle travel time.

10.4.3 Results

To illustrate the sensitivity of flexible transit demand to vehicle capacity (seats), fare discount, and operation management (i.e., crowdsourced HDVs or centrally controlled AVs), a total of eight scenarios were tested, with ten runs (days) for each scenario (Table 10.1). The "Day 0 equilibrium state" (characterized by demand, mode choice, departure time choice, and desired arrival time), sourced from Djavadian and Chow [16], is used as the base case scenario. It serves as the starting

Table 10.1   Test scenarios attribute summary

Scenario	Fleet type and available seats	Max available fleet size	Profit ($) threshold	Fixed fare price	Fare Price ($)/ additional 130 m	% fare discount	Operating cost ($)/km
Base case	HDV single ride	10	1	4.25	0.25	0	–
HDV_0%[a]	HDV four seats	10	25	4.25	0.25	0	0.51
HDV_15%	HDV four seats	10	25	4.25	0.25	15	0.51
HDV_25%	HDV four seats	10	25	4.25	0.25	25	0.51
HDV_50%	HDV four seats	10	25	4.25	0.25	50	0.51
AV_0%	AV four seats	10	–	4.25	0.25	0	0.51
AV_15%	AV four seats	10	–	4.25	0.25	15	0.51
AV_25%[b]	AV four seats	10	–	4.25	0.25	25	0.51
AV_50%	AV four seats	10	–	4.25	0.25	50	0.51

[a]HDV_0%: human-driven fleet—0% discount for sharing the ride.
[b]AV_25%: automated vehicle fleet—25% discount for sharing the ride.

point for each test scenario. Results are presented in Figures 10.5–10.9. The following observations are drawn from an examination of the results. In discussing results, cross-effects are pointed out by drawing attention to applicable figures (e.g., Figure 10.5 for demand and Figure 10.7 for fleet size).

Figure 10.5 presents shared FTS ridership demand over the range of 10 days for different fare discount levels and management/vehicle types (crowdsourced HDV and centrally operated AV). The results illustrate the effect of discount rate and management/vehicle type on ridership demand adjustment from one day to another. A couple of significant findings can be observed in Figure 10.5. First, the demand for a human-driven vehicle when the fare is discounted to 50% level drops to zero. There is no incentive for the driver to enter the market. This observation is confirmed with zero fleet (vehicle) in service for this case (see Figure 10.7).

Another notable observation is from the results as shown in Figure 10.5 is the increase in ridership due to an increase in vehicle capacity (i.e., availability of seats). In the base case scenario (based on a single seat for a single ride case), the equilibrium ridership was 17. However, by increasing capacity to four seats and offering ride-sharing, the ridership increased 76% to as high as 30 from the original 17.

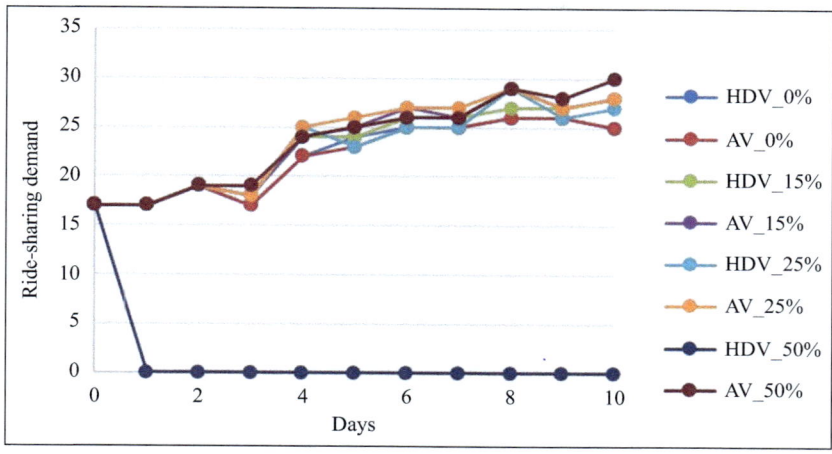

Figure 10.5   Ride-sharing demand

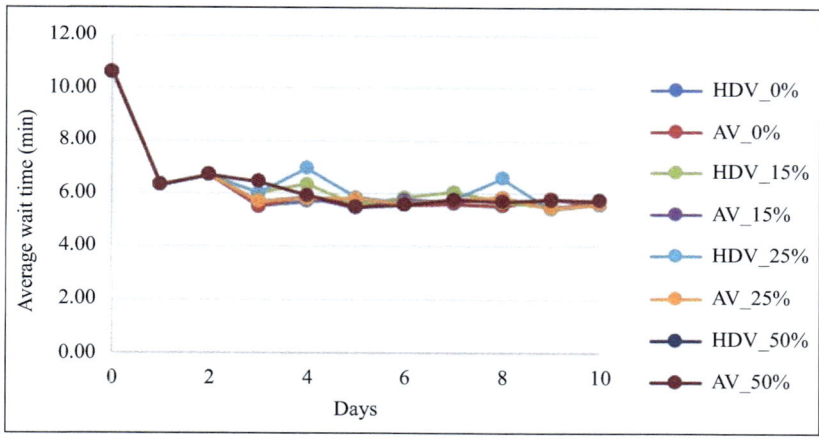

Figure 10.6   Average wait time (min)

The notable shift in demand shown in Figure 10.5 can be attributed to a substantial decrease in wait time to access service (47%) and fare discount as compared to the base case as shown in Figure 10.6. Since this is a first mile travel case and riders have a common destination (Go Transit station), with a given fleet size, it is more efficient to share a ride.

Another noteworthy finding from Figure 10.5 is that the centrally managed AVs resulted in slightly higher demand compared to crowdsourced HDVs. The reason for this occurrence, as shown in Figure 10.7, is that the fleet size is more stable under centrally operated AVs as opposed to crowdsourced HDVs. The fleet size stability can be attributed to one main factor, which is updating the strategy of

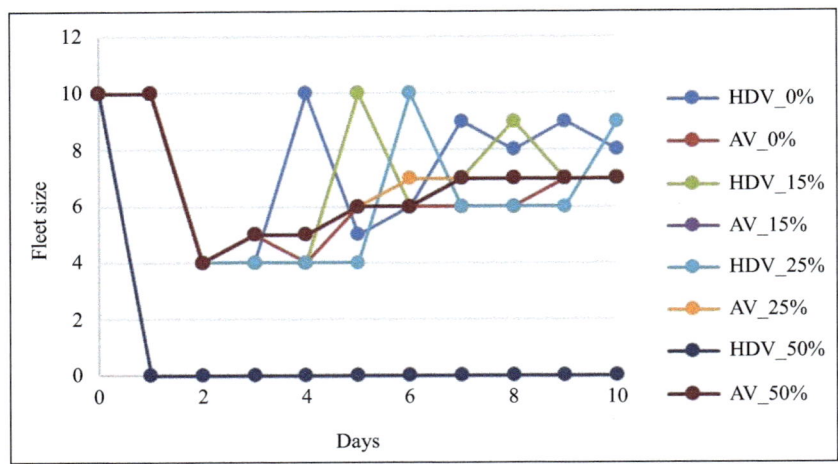

Figure 10.7   Fleet size

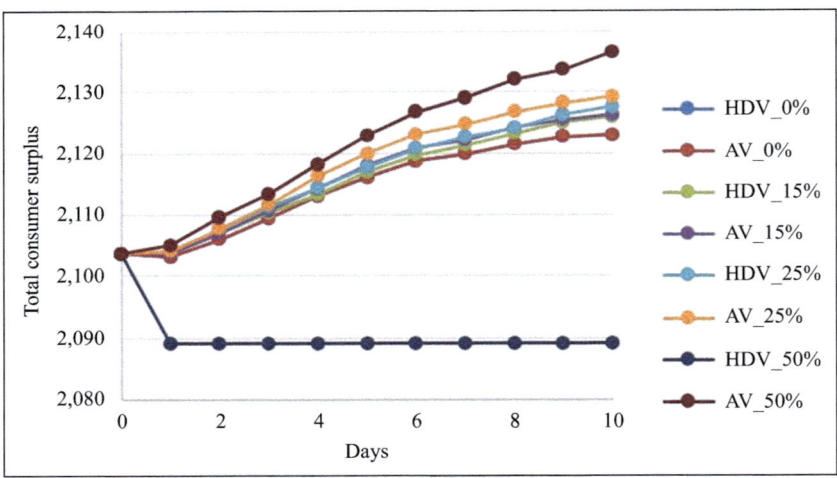

Figure 10.8   Total consumer surplus

the service provider. In the case of crowdsourced HDVs, the fleet size is determined by the number of drivers entering the market on each day, which depends on the perception of profit by drivers. If the perceived profit is high, more drivers will enter the market and vice versa.

The fluctuation in fleet size causes fluctuations in demand, which in turn will result in fluctuation in profit. In the long run, convergence to a stable point can occur.

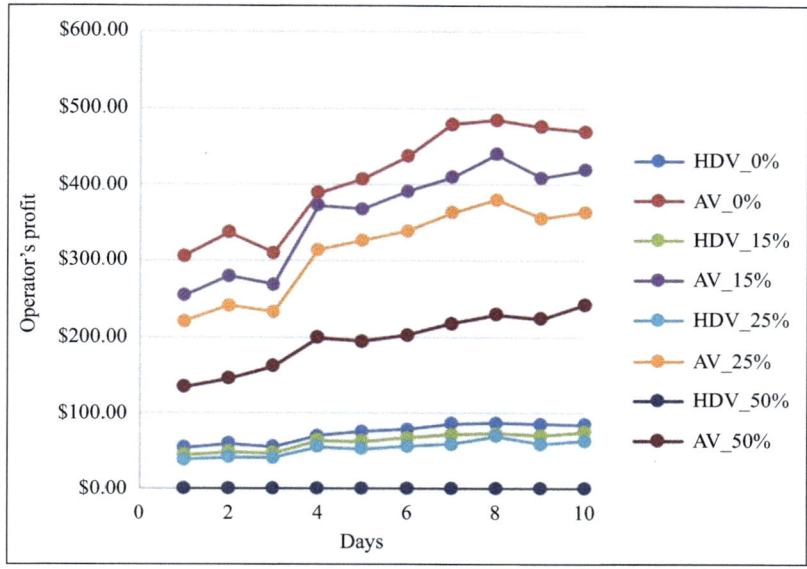

Figure 10.9   Total profit

Comparing Figures 10.5 and 10.7, it can be seen that there is a more obvious change in fleet size compared to change in demand, mostly in the case of HDVs. This shows the sensitivity of the HDV fleet to minor changes in profit. However, in the case of centrally operated AVs (operated by a public agency) as noted earlier, the aim is to reach social optimality. In this case, the fleet size is updated in such a way so that supply matches demand. Since the central operator knows demand on each day, it has the advantage of adjusting fleet size accordingly for the next day. This stability in fleet size in turn results in stability in demand.

From Figure 10.5, it can be seen that a discounted fare has resulted in increase in ridership which is an intuitive effect. However, this increase is more pronounced under centrally operated AV fleets. The reason for this phenomenon is that in the case of HDV fleet, higher discount means less profit for drivers and the number of drivers entering the market drops resulting in lower fleet size (Figure 10.7) or no drivers at all (case of 50% discount). Conversely, in the case of centrally managed AVs, the goal is to reach social optimality (i.e., matching supply to demand), the fleet size is not negatively affected by increase in discount rate.

Figure 10.8 presents the total consumer surplus under the tested eight scenarios. It can be seen that with the exception of HDV fleet size and 50% discount, the overall introduction of ride-sharing resulted in an increase in consumer surplus with the most increase being 2% under centrally operated AV and 50% discount.

For AV-operated fleet and 50% discount, seven commuters switched from auto to ride-sharing and five commuters switched from regular transit to ridesharing.

The results shown in Table 10.1 suggest a significant difference between two fleet size updating objectives, one being profit-maximizing and the other being social optimality.

Figure 10.9 illustrates the difference between owning and operating AV fleet centrally and outsourcing rides in terms of profit. As can be seen from Figure 10.9, the operator can accumulate more profit if it owns and operates the fleet itself. In the case of outsourcing, the operator will only accumulate commission from the rides.

10.5   Conclusion

In this study, the agent-based day-to-day adjustment framework for evaluating dynamic FTS proposed by Djavadian and Chow [16,17] is extended to include new and upcoming disruptive mobility services such as shared mobility and automated vehicles. A first mile/last mile mobility case study of Oakville, Ontario compared the effect of ride-sharing on demand and consumer surplus with the single ride service. In total, eight scenarios are tested with the aim of investigating the effect of fleet vehicle capacity, fare discount, and management/vehicle type (crowdsourced HDV fleet vs. centrally operated AV fleet) on travel demand and welfare.

The results obtained highlight the presence of a two-sided market where the choices of travelers are highly impacted by the choices made by the operators (level of service offered) and vice versa.

One of the main findings is that introduction of ride-sharing with a discount for the last mile/first mile travel resulted in a 76% increase in ridership and a 47% decrease in wait time and a slight increase in consumer surplus.

The comparison of centrally operated AV fleet with crowdsourced HDV fleet shows the important differences between profit-maximizing service and social optimal seeking service. The effect on fleet size, demand, and their impacted welfare is noted.

For the Oakville case study, under a centrally managed AV fleet, the optimal case is when a 50% discount is offered for sharing, which resulted in an increase in demand, a decrease in wait time, and an increase in operator profit. On the other hand, under crowdsourced HDV fleet, the optimal case is when a 25% discount is offered. The results show that fleet size is more stable under centrally operated AV cases.

A future study can explore the role of differentiated services on the travel demand and welfare. Also, the methodological framework can be used to study congested networks (e.g., downtown Toronto).

Author contributions

The authors confirm contribution to this chapter. All authors reviewed and approved the manuscript.

Declaration of conflicting interests

The authors declared no potential conflict of interest with respect to publication of this chapter.

References

[1]   E. Commission. [Online]. Available from https://ec.europa.eu/transport/themes/urban/urban_mobility_en.
[2]   EPSON. "Policy brief: Shrinking rural regions in Europe. Towards smart and innovative approaches to regional development challenges in depopulating rural regions," 2018.
[3]   Shum D. "Toronto ranked worst city for commuting in North America: Study." *Global News*. 21 June 2018.
[4]   Metrolinx. Cost of traffic congestion in GTA and Hamilton. 2008.
[5]   Li X., and Quadrifoglio L. "Feeder transit services: choosing between fixed and demand responsive policy." *Transportation Research Part C*. 2010; 18(5): 770–780.
[6]   Lekach S. "Driverless shuttles are here, outpacing personal self-driving vehicles." *Mashable*. 2019.
[7]   Fagnant D.J., and Kockelman K.M. "The travel and environmental implications of shared autonomous vehicles, using agent-based model scenarios." *Transportation Research Part C: Emerging Technologies*. 2014; 40: 1–13.
[8]   Bischoff J., and Maciejewski M. "Autonomous Taxicabs in Berlin: A spatiotemporal analysis of service performance." *Transport Research Procedia*. 2016; 19: 176–186.
[9]   Fagnant D.J., and Kockelman K.M. "Dynamic ride-sharing and fleet sizing for a system of shared autonomous vehicles in Austin, Texas." *Transportation*. 2016; 1–16.
[10] Martinez L.M., and Viegas J.M. "Assessing the impacts of deploying a shared self-driving urban mobility system: An agent-based model applied to the city of Lisbon, Portugal." *International Journal of Transportation Science and Technology*. 2017; 114(3): 462–467.
[11] APTA. American Public Transportation Association. [Online]. Available from https://www.apta.com/resources/mobility/Pages/First-Last-Mile-Solutions.aspx.
[12] Pantonium. [Online]. Available from https://pantonium.com/our-company/.
[13] McLeod J. "Belleville transit pilot project ditches fixed routes for bus-hailing system." *Financial Post*. 10 September 2018.
[14] Shen Y., Zhang H., and Zhao J. "Integrating shared autonomous vehicle in public transportation system: A supply-side simulation of the first-mile service in Singapore." *Transportation Research Part A: Policy and Practice*. 2018; 113: 125–136.

[15]   Becker H., Balać M., Ciari F., and Axhausen K.W. "Assessing the welfare impacts of shared mobility and Mobility as a Service (MaaS)." *Arbeitsberichte Verkehrs- und Raumplanung.* 2018; 1378.

[16]   Djavadian S., and Chow J.Y.J. "An agent-based day-to-day adjustment process for modeling 'Mobility as a Service' with a two-sided flexible transport market." *Transportation Research Part B: Methodological.* 2017; 104C: 36–57.

[17]   Djavadian S., and Chow J.Y.J. "Agent-based day-to-day adjustment process to evaluate dynamic flexible transport service policies." *Transportmetrica B: Transport Dynamics.* 2017; 5(3): 286–311.

[18]   Quadrifoglio L., and Li X. "A methodology to derive the critical demand density for designing and operating feeder transit services." *Transportation Research Part B: Methodological.* 2009; 43(10): 922–935.

[19]   Morlok E. "Short run supply functions with decreasing user costs." *Transportation Research Part B: Methodological.* 1979; 13(3): 183–187.

[20]   Furuhata M., Dessouky M., Ordóñez F., Brunet M.E., and Wang X., Koenig S. "Ridesharing: The state-of-the-art and future directions." *Transportation Research Part B: Methodological.* 2013; 57: 28–46.

[21]   Hyytiä E., Penttinen A., and Sulonen R. "Non-myopic vehicle and route selection in dynamic DARP with travel time and workload objectives." *Computers & Operations Research.* 2012; 39(12): 321–3030.

[22]   Djavadian S., and Farooq B. "Distributed dynamic routing using network of intelligent intersections." *Proceedings of Intelligent Transportation Systems Canada Annual General Meeting 2018 Conference.* Niagara Falls. June 2018.

[23]   Alshalalfah B., and Shalaby A. "Feasibility of flex-route as a feeder transit service to rail stations in the suburbs: Case study in Toronto." *Journal of Urban Planning and Development.* 2012; 138: 90–102.

[24]   Lavieri P.S., and Bhat C.R. "Modeling individuals' willingness to share trips with strangers in an autonomous vehicle future." University Transportation Center at The University of Texas at Austin, Austin, 2018.

[25]   DMG. *Transportation tomorrow survey.* 2011. [Online]. Available from http://dmg.utoronto.ca/.

The views expressed are those of the authors.

Chapter 11

Demand for shared mobility to replace private mobility using connected and automated vehicles

Seyed Mehdi Meshkani[1], Shadi Djavadian[2], and Bilal Farooq[1]

We examine how the introduction of shared connected and automated vehicles (SCAVs) as a new mobility mode could affect travel demand, welfare, as well as traffic congestion in the network. To do so, we adapted an agent-based day-to-day travel adjustment process supported by a SCAV fleet dispatching system, which is implemented on an in-house traffic microsimulator. A two-sided market based on the demand and supply interaction was simulated in which demand for SCAV service and the supply of required fleet size change endogenously. In dispatching SCAV fleet, the changing traffic conditions were taken into account. For realism, the effect of SCAVs on the use of privately owned connected automated vehicles (CAVs) was studied using the downtown Toronto network during peak period travel conditions.

The results based on a part of peak period traffic show that the demand for SCAVs went up by 43% over seven study days, from 670 trips on the first day to 959 trips on the seventh day. During the same period, there was a 10% reduction in the use of privately owned CAVs, from 2,807 trips to 2,518 trips. Moreover, total travel time in the network dropped by 7%, suggesting that traffic congestion was reduced in the network. In order to achieve traffic congestion reduction effect, infrastructure modifications will be necessary so that pick-up and drop-off operations will not disrupt the flow.

11.1 Introduction

Due to the increasing urbanization trend, access to sustainable mobility means has become a serious issue around the world. The pace of development of services and infrastructure in cities has traditionally been slower than the rate of

[1]Department of Civil Engineering, Laboratory of Innovations in Transportation (LiTrans), Ryerson University, Toronto, Canada
[2]Greenfield Labs, Ford Mobility, Palo Alto, CA, USA

increase in transportation demand [1]. Conventional solutions such as adding parking spaces or expanding roadways result in increased traffic which adversely impacts the environment, the livability of cities, and in almost all cases, these solutions are financially not affordable. Therefore, it is necessary to consider new and potentially transformative transportation solutions [2]. Transportation experts around the world view the emergence of CAVs with an interest in developing sustainable future mobility solutions [3]. As discussed in Chapter 10, CAVs enable carsharing and ridesharing under favorable market conditions. Of course, CAVs can be used in on-demand shared mobility systems with or without ridesharing service [3,4].

Despite the obvious advantages of ridesharing, in order to be widely adopted, it must be able to compete with one of the biggest advantages of private car usage: immediate access to door-to-door transportation [5]. Moreover, there are some barriers, including the requirement that itineraries and schedules have to be coordinated between participants and the lack of effective methods to encourage participation that have inhibited a wider adoption of ridesharing [6]. Also, in some markets, ridesharing with strangers may not be acceptable to some travelers.

Technological advances in hardware and software, nevertheless, have helped to find solutions to mobility constraints. GPS-enabled smartphones, social networks, data repositories, and the Internet enable practical dynamic or real-time ridesharing. Dynamic ridesharing refers to a system which supports an automatic ride-matching process between participants on very short notice or even en-route [7,8]. However, ultimately it is up to the traveler to accept or reject the service supplier's offer of ridesharing.

SCAVs either with or without ridesharing, enable people to schedule a ride using mobile phone applications. Moreover, the system can anticipate future short-term demand and relocate vehicles in advance to better match supply with travel demand. Well-known socio-economic and environmental benefits of using SCAVs can be achieved such as reduced vehicle ownership and parking needs as well as the potential to reduce vehicle-kilometer traveled (VKT) [9,10]. Of course, electrification delivers additional benefits.

One question in search of an answer is that how would emerging SCAVs with ridesharing capability when used as a new transportation mode affect demand for private mobility as well as users' welfare (also called utility as measured by consumers' surplus). Another question is that how they would influence traffic congestion in the network. A part of the answer to the first question is provided in Chapter 10. It was shown that a centrally operated CAV fleet in comparison with crowdsourced human-driven vehicles (HDV) (e.g., the current version of Uber and Lyft) fleet yields enhanced mobility and welfare.

In this study, as an extension of Chapter 10, we answer the aforementioned questions and evaluate the effects of SCAVs as a new mode on travel demand, users' welfare, and traffic congestion. To do so, we employ an agent-based simulation framework for SCAVs. We developed a central dispatcher as well as SCAV demand and SCAV fleet size estimators. Demand and SCAV fleet size change every day based on day-to-day adjustment process developed by Djavadian and

Chow [11,12]. The designed SCAV system is implemented and tested on an in-house agent-based traffic microsimulator developed by Djavadian and Farooq [13].

The remainder of this chapter is organized as follows. A general review of literature related to shared mobility is presented in Section 11.2. In Section 11.3, we explain the designed framework for SCAVs. Section 11.4 illustrates the performance of the designed system based on a real-world road network. Section 11.5 discusses the results and Section 11.6 provides a summary of the study reported in this chapter and advances concluding remarks on future research.

11.2 Background

Shared mobility (e.g., ride hailing and carsharing) has attracted research attention over the years. Most previous studies assessed the operations of existing or future on demand services that use HDV and in these research studies, the operator or users' objective function was modeled. Given recent advances in connected and automated vehicle technologies, researchers have mounted efforts to investigate the use of SCAV fleets in serving urban travel needs. In these projects, SCAV fleet size, wait time, financial, and environmental impacts are some indicators that have been studied. One of the most important potentials of SCAVs is the capability to rideshare in favorable markets. In theory, ridesharing is a transportation mode which benefits both users and society by saving travel cost, reducing travel time, mitigating traffic congestion (if supported by necessary infrastructure adaptations), conserving energy, and reducing emissions.

Burns *et al.* [14] examined the performance of a shared automated fleet in three distinct city environments: a mid-sized city (Ann Arbor, Michigan), a low-density suburban development (Babcock Ranch, Florida), and a large densely populated urban area (Manhattan, New York). They compared the SCAVs with privately owned vehicles in terms of operational cost and fleet size. The study found that in mid-sized urban and suburban settings, each shared vehicle could replace 6.7 privately owned vehicles. In the dense urban setting, the current taxi fleet could be downsized by 30% with the introduction of automated driving technology and the average wait times could be reduced substantially.

Fagnant and Kockelman [9] presented an agent-based system which examines the implications of SCAVs with capability of dynamic ridesharing across the Austin (Texas) network. Results reveal that dynamic ridesharing leads to a reduction in total service times (wait times plus in-vehicle travel times) and travel costs for SCAV users. In addition, overall vehicle-miles traveled can be reduced as trip-making intensity (SCAV membership) rises and/or users become more flexible in their trip timing and routing.

Zhang *et al.* [3] using an agent-based simulation model in a grid-based hypo-thetical city, estimated how SCAVs affect urban parking demand under different operations scenarios. They reported that one SCAV will be able to replace around 14 privately owned vehicles and approximately 90% of the parking demand for the participating clients can be reduced. In another research effort, Chen *et al.* [15]

examined the implications of shared, automated, and electric vehicle fleet in a discrete-time agent-based model. The simulation was conducted under various vehicle range and charging infrastructure scenarios in a 100-mile by 100-mile gridded city of Austin, Texas. Results reveal that fleet size is sensitive to battery recharge time and vehicle range. Simulation predicts a decrease in fleet empty vehicle miles (3% to 4%) and average wait times (2 to 4 min per trip), with each vehicle replacing 5 to 9 privately owned vehicles.

Levin *et al.* [16] proposed an event-based framework in which cell transmission model (CTM) was used as a realistic flow model to obtain more accurate predictions about SCAV operations. A heuristic was presented to study SCAVs with dynamic ridesharing capability. The authors compared SCAVs scenarios (including dynamic ridesharing) with personal vehicle scenarios. Results show that a smaller SCAV fleet can service all travel demand in the morning peak. However, some SCAV scenarios also increased congestion due to empty repositioning trips to reach travelers' origins. In such a situation, it is important to model congestion when studying SCAVs to attain realistic estimates of quality of service. Furthermore, SCAVs may be less effective than previously predicted for peak-hour scenarios. Nevertheless, SCAVs with dynamic ridesharing provided services comparable to personal vehicles.

Becker *et al.* [17] developed the first joint simulation of carsharing, bike-sharing, and ride hailing for a city-scale transport system with a fixed fleet size using MATSim. Their results showed that the introduction of shared modes may increase transport system efficiency by up to 7% and this efficiency may reach to 11% if shared modes were used as a substitute for public transport in lower-density areas.

Most of the above-noted studies have modeled a one-sided market. That is, either demand or the fleet size was fixed as an exogenous input. There are only a few previous studies, for example, Djavadian and Chow [12] and Chapter 10 of this book that assessed a two-sided market using an agent-based day-to-day adjustment simulation in which both demand and fleet size are determined endogenously. However, in their proposed framework for dynamic on-demand services, they did not consider traffic congestion in the network.

This study aims to gain an understanding of the extent to which shared automated vehicle-based service can compete with the use of privately owned automated vehicles. For modeling purposes, the demand and supply are considered as a two-sided market. The two sides are the interaction of user interest (as characterized by the attributes of the service) and the interest of the supplier of the service (quantified by vehicle fleet). An additional challenge in this study is that in the simulation of SCAVs operation, traffic congestion in the network is taken into account.

It should be noted that in the study described in Chapter 10, ridesharing service used HDVs and SCAVs. The potential competition between private mobility and shared mobility was not modeled. Whereas, in this study, we consider private CAVs and SCAVs as the available modes. Another difference is that the Chapter 10 study addressed the first mile/last mile problem which is known as many-to-one transportation problem while this study addresses the many-to-many problem.

11.3   Methodology

In the future, the availability of safe and affordable CAVs is expected to result in private ownership of these vehicles and also these are likely to serve the travel market as SCAVs. A number of contributors to this book have pointed out that future automobile ownership for the purpose of private mobility may not continue at the historical rate and that shared mobility may become acceptable to future travelers under favorable travel market conditions. The study described in this chapter aims to provide a better understanding of the acceptance of SCAV services versus the ownership and use of a CAV. Also, it is intended to find the effect of SCAV service on traffic congestion using the total network-level travel time changes as an indicator.

In this study, we employ an agent-based simulation framework for modeling SCAVs that compete with privately owned CAVs. As can been seen in Figure 11.1, the framework is composed of three major parts that interact with each other:

1. Dynamic demand, including demand for private CAVs and demand for SCAVs;
2. Centralized SCAV system which consists of SCAV demand, SCAV fleet size, and dispatcher; and
3. An in-house traffic microsimulator, which models traffic conditions in the network.

11.3.1   Dynamic demand

Users in the network choose their mode of transportation based on a learning process gained from their experiences from the previous days. There are two available modes in the network: private CAVs and SCAVs. To model users' mode choice, a binary logit model was developed which predicts the mode choice of users on each day. The number of users choosing each mode defines the demand for private CAVs and the demand for SCAVs. These demand levels change day-to-day, as on each day some users might switch from CAV to SCAV, and vice versa. Eventually, after several days of learning, the demand for travel would reach a steady state. For the day-to-day adjustment process, we use a method similar to the one developed by Djavadian and Chow [11,12].

The logit model belongs to a well-known family of discrete choice models based on random utility theory. Over the years, modeling travel behavior has advanced both in theory and successful practical applications. The model incorporates user attributes, characteristics of available modes of travel, and the travel environment in which the competing modes operate.

A logit model requires statistically satisfactory calibration before application as a part of the travel demand prediction modeling framework. For new modes of

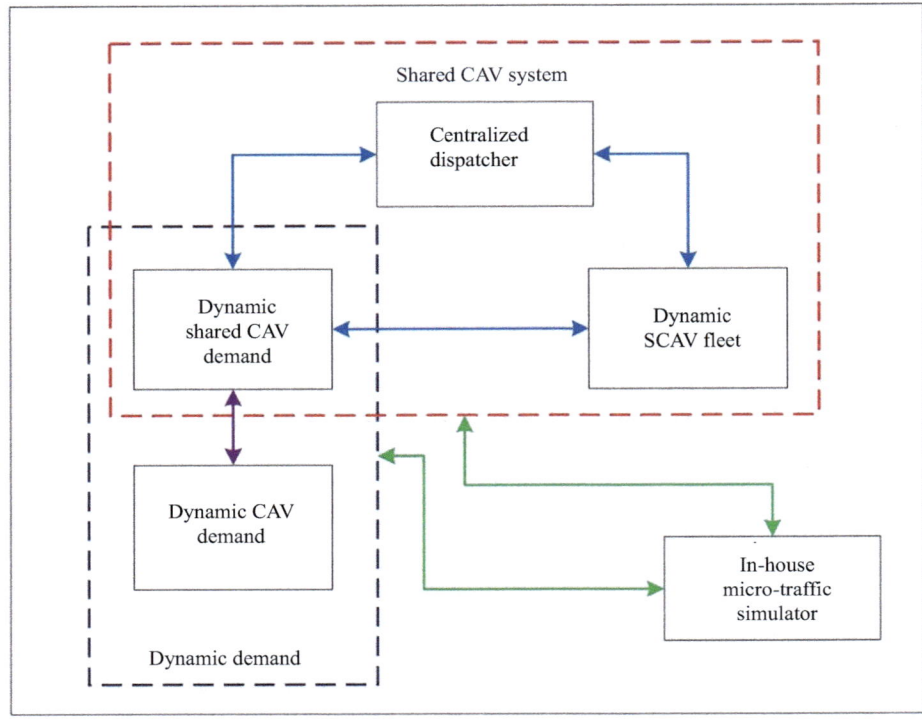

Figure 11.1   Agent-based simulation framework for SCAVs

travel such as CAVs and SCAVs, the calibration constants and coefficients of a logit model that represent existing modes (e.g., private car, taxi, ride hailing service, and public transit) can be used, provided that care is exercised in finding similarities in modal attributes of existing modes and their replacements. For example, the traveler perception of the cost of owning a nonautomated car and a CAV need not be different, even if a CAV may cost more to buy. Likewise, the traveler perception of waiting and in-vehicle times for public transit can be assumed to be similar to these modal attributes for a future SCAV service.

The fleet size for the SCAVs during simulations varies from one day to another. It is determined using the model explained in Chapter 10. These are vehicles that serve travelers. The day-to-day adjustment process for determining fleet size eventually reaches a stochastic user equilibrium.

For routing vehicles, a system is used based on a network of intelligent intersections and mature technology-based CAVs and SCAVs [13]. In this system, when CAVs and SCAVs arrive at each intersection, they announce their destination to the intelligent intersection. Then, the intelligent intersection based on the current status of the traffic congestion in the network guides these vehicles to the direction they should take and thus routes them to the next intersection in such a way that it minimizes individual vehicle as well as network-level travel time and maximizes capacity. For more information about the end-to-end distributed routing system, refer to [13].

11.3.2   Centralized SCAV dispatcher system

The centralized SCAV system (Figure 11.2) proposed in this study consists of two network layers and three types of agents. The two network layers are the communication network where the information exchange between vehicles, links, and intersections takes place, and the physical road network used for modeled vehicle travel. Physical road network is represented by a network $G(I, L)$ which consists of I intersections (nodes/vertices) and L links (edges).

Three agents in this system are SCAV agents $(f \in F)$, passenger agents $(p \in P)$, and infrastructure agents. There is a central dispatcher (D) in the system, which is responsible for receiving requests from travelers, matching them to

```
j=1;
while # of passengers < total demand do
    if (j mod Δ j)= 0 then
        ... Dispatcher
        Passenger (p) sends a request for scav (f) to the dispatcher;
        while p < = total number of requests do
            Dispatcher creates a request-table based on passengers' itinerary;
            Dispatcher checks to see if there is available SCAV at intersection depot;
            if SCAV (f) is available and has enough capacity then
                Dispatcher assigns to passenger (p);
                Dispatcher updates request table, including passenger status to Assigned and SCAV
                    status to busy;
                SCAV (f) creates an itinerary for on board passengers;
                if SCAV (f) does not have any capacity then
                    leaves intersection I_i;
            
            else end
                Passengers (p) waits on request-table for a passing SCAV from intersection (I_i) or a
                    SCAV coming back to intersection I_i;
            end
            ...vehicle trajectory update
            while f < = F do
                When SCAV arrives at I_i Checks to see if any passenger needs to be dropped off from
                    customer on board itinerary or picked up from request table;
                If picking up or dropping off, dispatcher and f update their request and itinerary
                    tables;
                If all passengers of f have been dropped off, it goes back to the depot; f=f+1;
            end
        end
    end
    j=j+1;
end
```

Figure 11.2   Pseudo-code for centralized SCAV dispatcher system

SCAVs and ensuring all travelers are served. However, no central depot exists and depots are distributed throughout the network. Δ is dispatch update cycle, which is set to 1 min in this study, and j is a time whose unit is second.

This system is based on several assumptions:

1. Only intersections where demand is produced have a depot;
2. SCAVs have a capacity for four passengers; and
3. There is no threshold for waiting time and passengers always wait on the waiting list.

Once a passenger needs a SCAV, they make a request to the dispatcher. The dispatcher checks to see if there is any SCAV available at the intersection depot where the passenger makes a request. If available, it is assigned to the passenger. If a SCAV is not available, the passenger request would place on the waiting list.

The dispatcher is aware of the status of all SCAVs, including location as well as capacity. When any SCAV arrives at this intersection, the dispatcher checks to see if there is enough capacity and then assigns it to the passenger waiting at this intersection. If there are multiple requests at the intersection, passengers are served in a first-come-first-serve (FCFS) order.

When a SCAV is assigned to a passenger, SCAV picks the passenger up from the intersection en route to the passengers' destination. While arriving at each intersection, first a check is made to see if any passenger needs to be dropped-off then another check is made to see if the dispatcher assigns the SCAV any new passenger(s). The general policy for dropping-off is first-in first-out (FIFO). Each SCAV after dropping off the last passenger goes back to its own depot if en route to the depot there is no new assignment. The pseudo-code for the proposed centralized system is shown in Figure 11.2.

Similar to the SCAV demand discussed in the previous part, the SCAV dispatcher has a learning process and is updated day-to-day using Equation (10.3) from Chapter 10.

11.3.3   *Traffic microsimulator*

The proposed centralized SCAV system is implemented on an in-house traffic microsimulator which was developed by Djavadian and Farooq [13]. In this simulator, a traffic management system using network of intelligent intersections was proposed that is capable of dynamically routing intelligent vehicles (that are connected and automated) from origin-to-destination in dense urban areas. In this system, real-time traffic information is collected by links using sensors and is frequently exchanged among intelligent intersections using Infrastructure-to-Infrastructure (I2I) communication.

A traffic microsimulator advances vehicle states on a second-by-second (or even shorter time interval) basis. The vehicles are driven according to car following, route choice, and other applicable algorithms. Characteristics of driver (human or automated), road and traffic control system, and driving environment are coded in the simulator. The stochastic traffic simulators are verified so as to ensure that traffic flow results are realistic. The microsimulator used in this study is adapted for simulating AVs and SCAVs.

11.4   Case study

For obtaining answers to questions on the role of SCAVs in competing with private mobility, replacing some private cars, and effect on traffic congestion, the developed methodology was applied to the downtown Toronto street network (Figure 11.3). Because of the high current demand levels and congestion in the downtown Toronto network, it is a good test case for the methodology.

The test network consists of 76 nodes, 223 links, and 26 centroids (matched to nearest nodes). As noted in the methodology section, depots are distributed throughout the network. The study period for our simulations is a part of the morning peak period and only vehicles whose both origin and destination are in downtown Toronto are modeled.

The demand information used in this study is time-dependent type. It is obtained from the origin-destination (OD) matrices based on 5 min intervals for a part of the morning peak period, reported by the 2011 Transportation Tomorrow Survey (TTS) [18]. We used a growth factor to convert this data to current demand level. During the study period, the total demand is 3477 travelers using the test network, which is distributed randomly within 5 min using a Poisson probability distribution model. The simulation is run for seven consecutive days for the demand and supply to come to a steady state.

The objectives of this case study are to evaluate how introducing a shared transport mode using SCAVs affects users' mode choice as well as traffic congestion. To model users' mode choice, we developed a binary logit model which is represented in Tables 11.1 and 11.2. We synthesize a population such that all observable traits are captured from survey data and all unobservable variables ε are randomly drawn for each user *n* to fit the observed choices from sample data [11].

As noted in Table 11.1, the binary logit model is based on two competing modes, namely the privately owned connected and automated vehicle (CAV) and the shared mobility mode vehicle that is a specially designed connected and automated vehicle.

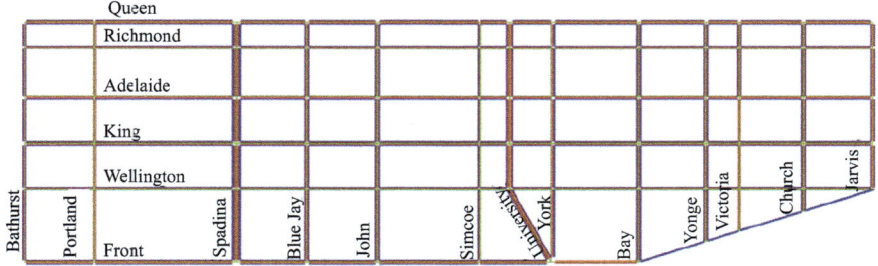

Figure 11.3   Downtown Toronto street network. Adapted from [13]

Table 11.1   Utility function

#	Mode	Utility
1	CAV (car)	$U_{(CAV)} = ASC_{CAV} + \beta_1(T_{CAV} - T_{SCAV}) + \beta_2(\text{ratio}^*) + \varepsilon$
2	SCAV	$U_{(SCAV)} = 0$

*Ratio of (number of cars in household/number of licence holders in the household)

Table 11.2   Systematic utility parameters

Name	Value	Std err	t-test
ASC_{car}	−1.91	1.22	−1.56
β_1	−0.153	0.0798	−1.92
β_2	2.24	0.776	2.89
Initial log-likelihood		−42.975	
Final log-likelihood		−21.038	
Likelihood ratio test		43.874	
Adjusted rho-square		0.441	

The utility functions are specified for calibrating the binary logit model by anchoring the U(SCAV) at zero and the U(CAV) is varied so as to achieve calibration. The ASC_{CAV} is the alternative-specific constant and T represents total travel time, including wait time and in-vehicle time. The ratio variable defined below captures user attributes and ε accounts for random effects. The effect of relative cost and convenience of modes is captured by the modal constant. In the model, β_1 and β_2 denote the decision-maker's sensitivity to changes in the values of the variables.

In the model, T_{CAV} is the total travel time of private CAV, and T_{SCAV} is the total travel time of SCAV, including wait time and in-vehicle time.

The values of utility parameters and calibration statistics are presented in Table 11.2. The signs of parameter values result from the logit model calibration. The calibration statistics indicate a reasonable quality binary logit model. The standard error (std err) of the estimate and the t-test values for the model parameters are acceptable. Likewise, the log-likelihood and adjusted rho-square results suggest a good quality model.

For the study of traveler decision to select the privately owned CAV versus SCAV, the utility values are to be computed and used in the binary choice logit model. The model specification suggests that traveler attributes (i.e., availability of a car, model constant for the private car) and the relative efficiency of the competing modes will influence the choice. Since the cost of automation is likely to affect both CAVs and SCAVs approximately uniformly, the cost variable does not appear as a distinct variable.

Two scenarios are considered in this study. The base case scenario represents the demand for private CAV and current public transit as two available modes. In scenario # 1, we replace the public transit with on-demand SCAV and run it for seven consecutive days and then compare the demand of each day with the base case scenario. In the base case, from the overall 3,477 travelers, 2,807 choose CAV and 670 choose public transit. The current demand for public transit is used for the first day. In scenario # 1, we set the initial number of SCAV fleet at 200 and the maximum number at 300. A learning process is used for updating demand and fleet size each day based on Djavadian and Chow [11,12].

11.5   Results

In this study, different indicators, including CAV and SCAV demand, SCAV fleet size, welfare level (based on consumers' surplus), and network total travel time for seven consecutive days were measured. Figure 11.4 illustrates the demand for CAV and SCAV over these 7 days and Figure 11.5 shows the SCAV fleet size (i.e., number of vehicles that were used).

As can be seen in Figure 11.4, the demand of public transit in the base case is 670 trips which are taken over by the SCAV service. From there on, more users switch to SCAV. The SCAV demand increases to 959 trips on the seventh day, showing 43% growth. In contrast, CAV demand decreases from 2,807 trips to 2,518, which shows a 10% reduction due to travelers' mode shift from private CAVs to SCAVs. The switch to SCAV leads to a 19% increase in SCAV fleet size from 200 vehicles on the first day to 238 vehicles on the seventh day, which is illustrated in Figure 11.5.

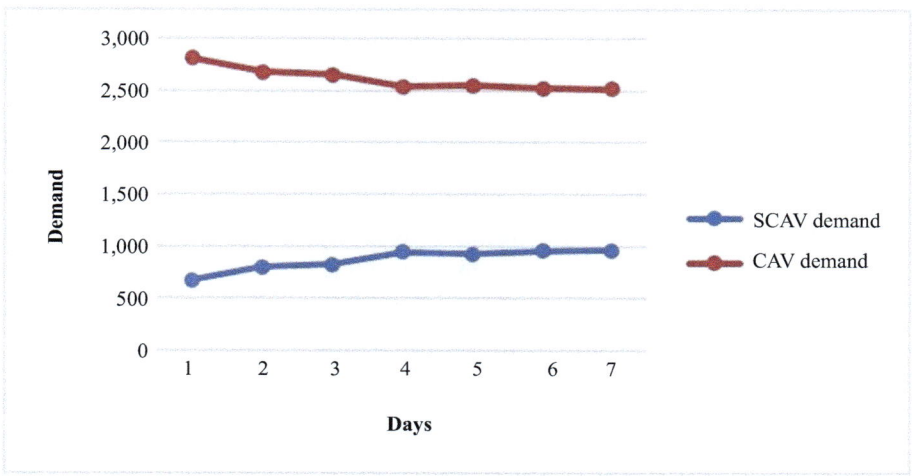

Figure 11.4   CAV and SCAV demand

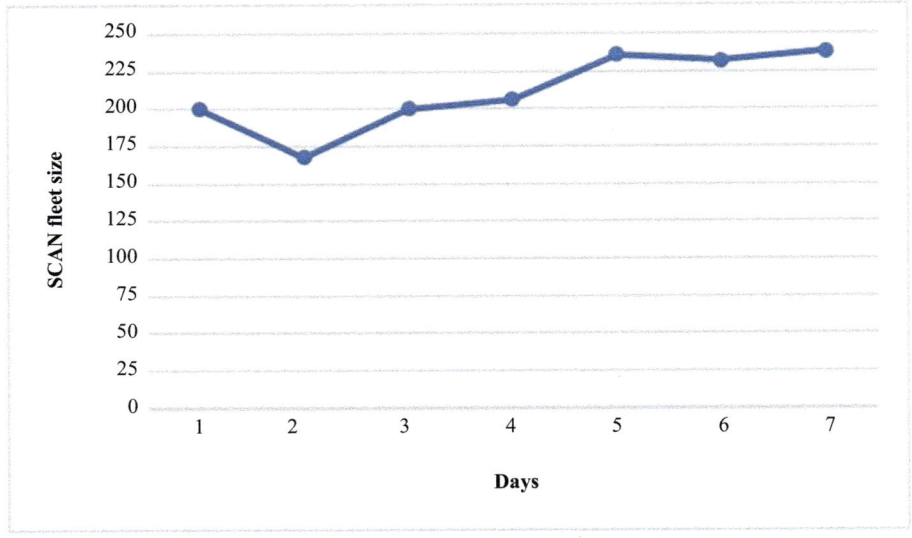

Figure 11.5   SCAV fleet size

As shown in Figure 11.4, for the second day, SCAV demand increases whereas CAV demand reduces. This is because, on the first day, SCAV users get a good experience of using SCAVs in terms of total travel time. So, on the second day, more users are willing to use SCAVs. On the other hand, since SCAV fleet size is determined based on SCAV demand from the previous day, on the second day, there is a reduction in the fleet size, which is shown in Figure 11.5, whereas SCAV demand grows. Therefore, SCAV users on the second day experience some inconvenience in terms of total travel time. This is why for the third day shown in Figure 11.4, there is a reduction in SCAV demand and a rise in CAV demand.

This interaction between demand and SCAV fleet size is simulated for all other days to reach an equilibrium condition in which users do not change their transport mode and the demand of each mode for two consecutive days is the same.

Figure 11.6 presents the total utility for users who used privately owned CAVs (cars). As can be seen, by introducing the SCAV service, the total utility of the CAV users decreases. The inference to be drawn is that in the presence of shared CAVs, private CAV use is less attractive to users.

Another important measure that reflects on the congestion level in the network is the total travel time. As shown in Figure 11.7, the introduction of a SCAV mode with ridesharing results in a decrease in total travel time from 35,675 (veh.min) on the first day to 33,220 (veh.min) on the seventh day which shows a 7% reduction in total travel time and consequently indicates that traffic congestion decreases in the network. It is assumed that owing to infrastructure adaptations to accommodate the SCAV service requirements, the pick-up and drop-off operations will not disrupt traffic flow.

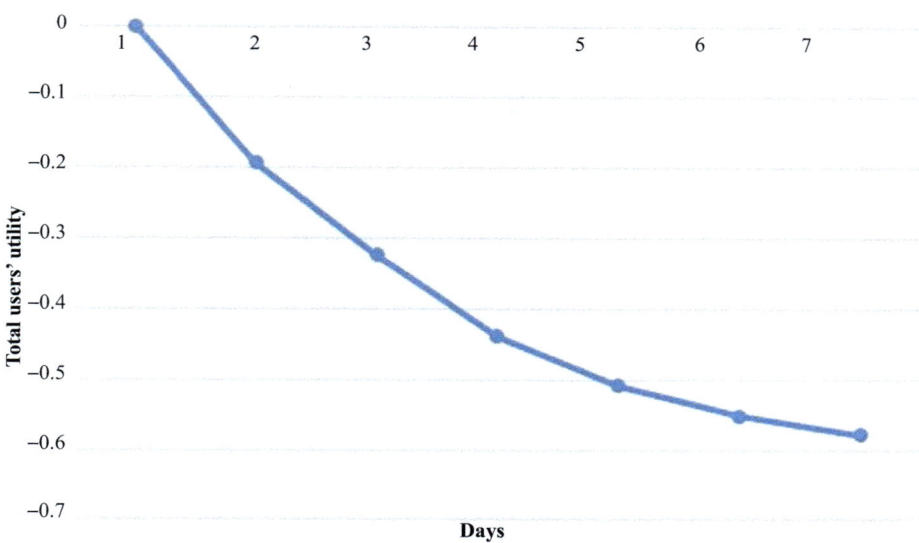

Figure 11.6 Normalized total users' utility for CAVs

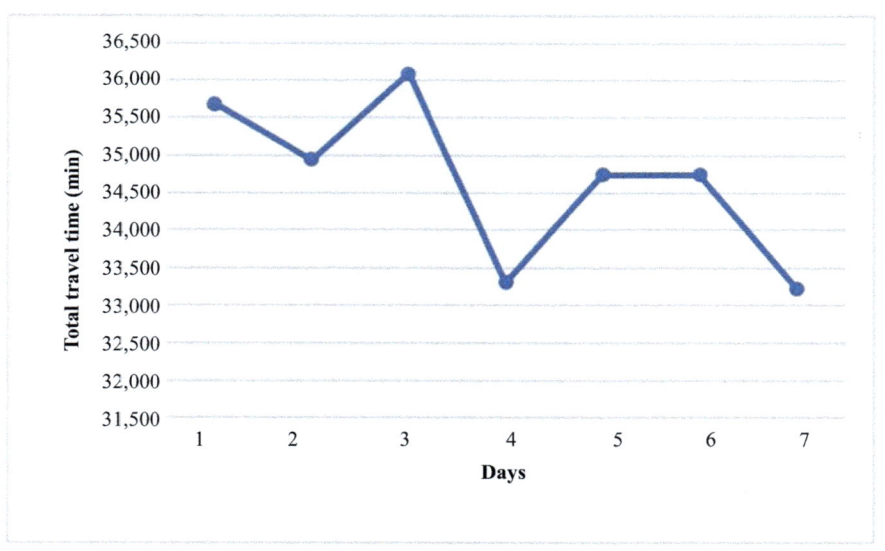

Figure 11.7 Network total travel time

11.6 Summary and conclusion

In this study, we analyzed SCAVs as a new mode to compete with and replace some private vehicles in the future travel network. To do so, we employed an agent-based simulation framework, which is composed of dynamic demand, SCAV system, and an in-house traffic microsimulator. For SCAV system, we developed a centralized dispatcher for managing SCAVs as well as passenger requests for service in the network. Dynamic demand and SCAV fleet size are updated every day using a day-to-day adjustment process in a two-sided market [11,12]. The in-house traffic microsimulator [13] routes CAVs in the network based on dynamic traffic conditions. In our methodological framework, demand and SCAV fleet size are determined each day endogenously. For simulating shared CAVs, we considered the real-world traffic congestion in the network.

Two research questions are answered in this study: How the use of SCAVs in a new shared mode service affects the demand for private cars and how it influences traffic congestion in the network. Each day, people choose their mode of transport based on their experience from the previous day. To capture an individual's mode choice decisions, a binary logit model was developed. Downtown Toronto road network was chosen to implement the proposed system with real data and travel was simulated for seven consecutive days.

Results show that:

- The demand for SCAV travel goes up from 670 trips on the first day to 959 trips on the seventh day, showing 43% growth over these 7 days whereas the demand for private CAV drops from 2,807 trips to 2,518, indicating a 10% decrease. This change in the travel market reflects the efficiency of shared mobility mode. The use of SCAVs (i.e., fleet size) goes up by 19% (i.e., from 200 vehicles to 238 vehicles) over these 7 days in order to meet the extra demand of those who change their mode of travel and demand SCAV service.
- Total network travel time is also measured and shows a 7% reduction over the 7 days, which indicates that using SCAVs as a shared mode has a positive effect on traffic congestion of the network. In order to achieve this positive effect, infrastructure modifications will be necessary so that pick-up and drop-off activities will not disrupt traffic flow.

A number of directions are suggested for future research.

- This study focuses on centralized dispatching systems, while in the future, distributed dispatching systems can be researched.
- In the dispatching system used in this study, passengers are served based on first-in first-out (FIFO) concept while in the future, an operational policy can be defined based on advanced optimization models that take into account differentiated service attributes.
- The methodological framework can be updated and applied to investigate the effect of using electric SCAV fleets for carsharing on the demand for travel based on privately owned CAVs. The SCAV fleets will operate out of a number of parking and charging stations/depots located in the urban areas.
- Another direction is that relocation can be taken into account since it reduces the vehicle kilometer traveled in the system.

Author contributions

The authors confirm contribution to this chapter. All authors reviewed and approved the manuscript.

Declaration of conflicting interests

The authors declared no potential conflict of interest with respect to publication of this chapter.

References

[1]   Spieser K., Treleaven K., Zhang R., Frazzoli E., Morton D., and Pavone M. "Toward a systematic approach to the design and evaluation of automated mobility-on-demand systems: A case study in Singapore." *Road Vehicle Automation*. 2014: 229–245.

[2]   Spieser K., Samaranayake S., Gruel W., and Frazzoli E. "Shared-vehicle mobility-on-demand systems: A fleet operators guide to rebalancing empty vehicles." *Transportation Research Board 95th Annual Meeting*. Washington, DC, USA, 2015.

[3]   Zhang W., Guhathakurta S., Fang J., and Zhang G. "Exploring the impact of shared autonomous vehicles on urban parking demand: An agent-based simulation approach." *Sustainable Cities and Society*. 2015; 19: 34–45.

[4]   Kornhauser A. "Uncongested mobility for all: New Jersey's area-wide aTaxi system." *Operations Research and Financial Engineering*. 2013.

[5]   Agatz N., Erera A. L., Savelsbergh M. W., and Wang X. "Dynamic ridesharing: A simulation study in metro Atlanta." *Procedia-Social and Behavioral Sciences*. 2011; 17: 532–550.

[6]   Furuhata M., Dessouky M., Ordóñez F., Brunet M. E., Wang X., and Koenig S. "Ridesharing: The state-of-the-art and future directions." *Transportation Research Part B: Methodological*. 2013; 57: 28–46.

[7]   Agatz N., Erera A. L., Savelsbergh M. W., and Wang X, "Optimization for dynamic ridesharing: A review." *European Journal of Operational Research*. 2012; 223(2): 295–303.

[8]   Nourinejad M., and Roorda M.J. "Agent based model for dynamic ridesharing." *Transportation Research Part C: Emerging Technologies*. 2016; 64: 117–132.

[9]   Fagnant D. J., and Kockelman K. M. "Dynamic ridesharing and fleet sizing for a system of shared autonomous vehicles in Austin, Texas." *Transportation*. 2016; 1–16.

[10] Fagnant D. J., Kockelman K. M., and Bansal P. "Operations of shared autonomous vehicle fleet for Austin, Texas, market." *Transportation*

Research Record: Journal of the Transportation Research Board. 2015; 2536: 98–106.

[11]   Djavadian S., and Chow J. "Agent-based day-to-day adjustment process to evaluate dynamic flexible transport service policies." *Transportmetrica B: Transport Dynamics*. 2017; 5(3): 286–31.

[12]   Djavadian S., and Chow J. "An agent-based day-to-day adjustment process for modeling 'Mobility as a Service' with a two-sided flexible transport market." *Transportation Research Part B: Methodological*. 2017; 104C: 36–57.

[13]   Djavadian S., and Farooq B. "Distributed dynamic routing using network of intelligent intersections." *Intelligent Transportation Systems Canada Annual General Meeting 2018 Conference*, Niagara Falls, Canada, 2018.

[14]   Burns L. D. "Sustainable mobility: A vision of our transport future." *Nature*. 2013; 497.

[15]   Chen T. D., Kockelman K. M., and Hanna J. P. "Operations of a shared, autonomous, electric vehicle fleet: Implications of vehicle and charging infrastructure decisions." *Transportation Research Part A: Policy and Practice*. 2016; 94: 243–254.

[16]   Levin M. W., Kockelman K. M., Boyles S. D., and Li T., "A general framework for modeling shared autonomous vehicles with dynamic network-loading and dynamic ridesharing application,' *Computers, Environment and Urban Systems*. 2017; 64: 373–383.

[17]   Becker H., Balac M., Ciari F., and Axhausen KW. "Assessing the welfare impacts of shared mobility and mobility as a service (MaaS)," *Transportation Research Part A: Policy and Practice*. 2020; 131:228–43.

[18]   DMG. *"Transportation Tomorrow Survey."* 2011. [Online]. Available from http://dmg.utoronto.ca/.

The views expressed are those of the authors.

Chapter 12

Matching demand and supply under uncertainty

Ata M. Khan[1]

For successfully meeting the service expectations of shared mobility users in various market niches, the supply of service has to be defined that is dynamically in balance with demand. The demand and supply interaction is to be modeled by treating the stochastic nature of demand as well as uncertainties on the supply side, including those in the traffic environment. Efficiency consideration requires that while meeting the service criteria, the supply of service is optimized. The demand and supply balance can be examined within the simulation and optimization methodological frameworks.

This chapter addresses matching demand with the supply of shared mobility enabled by electric automated vehicles. The shared automated vehicle (SAV) system provides mobility on demand (MOD) service to customers in an urban transportation environment. The origin-destination service is demanded in the form of variable party size that reflects the market being served. The supply system consists of electric vehicles and a number of stations are used for battery charging and parking. For the customer, the methodology enables the study of the probability of accessing the vehicle with the required seats. The methodology can be used to estimate the probability of charger availability when needed. For the use of the SAV system operator, a method is defined for directing an automated vehicle with state of charge (SOC) deficiency to the optimal location for charging taking into account uncertain traffic congestion states.

12.1 Introduction

The shared mobility road transportation system is intended to serve selected market niches in a technologically advanced environment and it should also meet the sustainability objective [1–3]. For successful implementation of a shared automated vehicle (SAV) system, the following factors should be taken into account:

- Spatio-temporal variability of demand
- Variable party size

[1]Department of Civil and Environmental Engineering, Carleton University, Ottawa, Canada

- Congestion in the traffic network
- Battery re-charging requirement.

In the SAV service environment, a fleet of self-driving electric vehicles meets the travel demand for individuals as well as groups. The request for a vehicle with required seats is sent to the operating system with the use of a cell phone. Matching such demands with supply is a probabilistic process. If the request cannot be fulfilled at that time, the request is added to the spatially localized queue. The wait time to access a vehicle or a seat in a vehicle is based on a user-defined level of service.

The SAV system design enables one-way or two-way on-demand vehicle sharing. When the vehicles are not in service, these can be charged and/or parked at stations located throughout a well-defined service area in the city [4]. When the traveler or the group are delivered to their destination, the vehicle becomes available for the next call for service, provided that the battery state of charge (SOC) is favorable. If the SOC is deficient, the vehicle requests a charging slot at nearby stations equipped with quick chargers. If a number of charging stations have available slots, the SAV system operator makes the choice on the basis of minimum expected travel time found with the use of a Bayesian method for decision-making under uncertainty of traffic congestion in the network.

The automated connected vehicles can re-balance themselves by moving to stations with chargers and parking stalls. This technological capability enables system-wide coordination for enhancing the availability of vehicles with required seats.

The performance of the demand-supply balance is assessed by the probability of vehicle (seats) availability upon request's arrival. If the probability is low, that implies waiting for service. If queues build up at a number of locations, this implies that the demand cannot be served by the fleet as an on-demand service. However, the routing of vehicles to locations of higher demand can be assisted by the dynamic traffic assignment [5].

This chapter on balancing demand with supply starts with the rationale for the study of this subject. The follow-up topics covered are accessing a vehicle with required seats, accessing battery charging slot, and deciding to find the optimal location for battery charging under uncertain traffic conditions. Here, the demand and supply factors mainly apply to a multistation shared mobility system and ride hailing is not covered. In the automated shared mobility context, users access a vehicle by subscribing to a shared SAV fleet of vehicles.

12.2   Purpose of demand-supply balance study

The SAV system design has many requirements to meet. First, the desired vehicle availability within an acceptable time (i.e., the service supply) depends on characteristics of demand as well as supply. The demand is influenced by the level of service as well as the price, which in part is defined by the fee to join the pool of subscribing members. Therefore, the SAV system design methodology should treat the demand-supply, specifically the dynamic balance of demand and supply within a short time period during a 24 h cycle of SAV service.

Second, the design methodology should be able to handle the system consisting of a network of multistations, rather than analysis of a single station. The third requirement is the recognition and treatment of uncertainty in demand for mobility on demand (MOD) service. Given that the SAV system will materialize in the future when automation in driving will mature, there is no real-world experience on the behavior of the traveling public in the use of SAV services. A fourth requirement is to address uncertainties on the supply side as well as uncertain traffic congestion states which affect travel times. A fifth requirement is to study demand, supply, performance measures, costs, and revenues for a multiyear analysis period (e.g., 10 years) rather than for a particular target year. This chapter describes methods that address requirements one to four, and the fifth requirement is addressed in Chapter 16.

The methods for designing the SAV system are not well researched. A number of authors have reported demand-supply interaction at the microscopic level for controlling the logistics of ride hailing [6,7]. These require data which are generally not available for real-world complex networks. The mesoscopic level demand-supply dynamic balance described in this chapter is suitable for SAV system design.

12.2.1   Methodological framework

The methodology advanced in this chapter is intended to address SAV system design requirements noted above. It is supported by methods with the capability to analyze the performance of components and the overall multistation system (Figure 12.1). In addition to analysis and evaluation of service functions, a system economic viability study can be supported by the methodology. One other notable attribute of the methodology is that it formally takes into consideration the uncertainties involved in the study of dynamic demand-supply balance.

Applicable system design parameters are selected for the study that contributes to SAV system efficiency [8]. These include technological and operational variables, vehicle availability, and wait time. The demand for service is analyzed on the basis of short intervals within the daily cycle. The methodology applied in this chapter can form the basis of demand estimates and when data on subscribed members and their use of SAV service becomes available, refinements can be made. The request for accessing a vehicle of required size incorporates the party size variable, which is essential for characterizing demand.

For the study of demand-supply balance within short periods of time within the service period (e.g., 24 h), a common set of design and service variables is used. The level of service factor for a system state is represented by the acceptable probability of accessing a vehicle with required seats, and the acceptable probability for a vehicle to access a fast charger when needed. If the service factors do not meet the service objectives and performance criteria, a search is initiated to identify new values of variables that will improve the efficiency of system design.

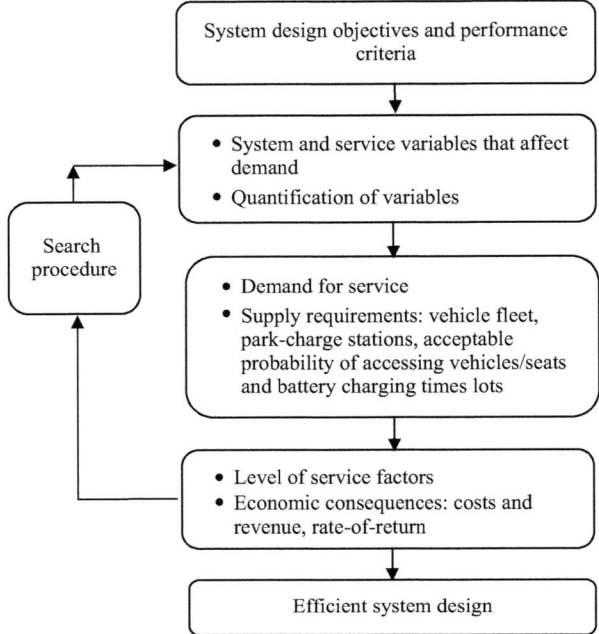

Figure 12.1   Demand-supply balancing methodology

12.2.2   Treating demand as well as supply as uncertain

The demand for the SAV service will in part depend upon the perception of travelers of its service characteristics. Since there is no experience with in-service SAVs at present, this perception will be difficult to model with certainty. In the future, even with some real-world experience, the spatial and temporal characteristics of demand will continue to be uncertain, it is, therefore, desirable to treat the travel demand as uncertain.

Two approaches can be considered to incorporate uncertainty in the analysis of system performance. The first calls for applying probability distributions for the perception coefficients involved in the applicable demand functions. If these distributions are randomly sampled, a drawn set of values of such coefficients will define a probable demand state and an associated probability of occurrence. In the second approach, a probable range of variation in the "service request rate by a specific party size" is specified. Associated with each "demand state" is a probability of occurrence. The distribution of probabilities over this range may be treated as rectangular, triangular, or as bell-shaped.

The first approach implies a rather large number of probability distributions if demand models are of the disaggregate type. There is the additional issue of data availability, which may lead to the choice of distributions without a scientific base. Also, this approach is time-consuming. The second approach, which is recommended for use, is simpler and requires no specific data. Demand can be treated as uncertain in estimating the expected value of a performance variable such as waiting time or delay in accessing a vehicle with required seats. Also, demand uncertainty can be taken to the level of an economic feasibility study.

In addition to uncertain demand states, the values of some supply variables may also be uncertain. Examples are the availability of a vehicle of required size, and the availability of a fast-charging slot within the multistation system. Uncertain demand as well as uncertain supply variables cause much complexity in modeling the demand-supply balance in analytical terms.

12.2.3 Planning infrastructure, fleet, and control system

The analyst may be interested in finding out fleet composition and size for acceptable quality of service. In the context of an optimization study, the optimal fleet size may be of interest. The variables that are used to define fleet size are arrival rate of service request, average origin-destination (O-D) distance, demand distributions, and average speed. Owing to the random nature of demand within short intervals of time, it can be assumed to be uniformly distributed. At the level of minimum fleet size, the customer waiting time may be unacceptably high. Therefore, the fleet size should be increased so as to ensure acceptable quality of service. For example, in a Singapore study, during peak times with a high fleet size, the peak wait time was found to be less than 15 min [9].

12.2.4 Demand-supply balance in operations

In demand factors for various market niches, three variables are of interest: specific time for requested service, specific origin-destination combination, and party size. Combinations of these define the demand packets that will be served.

Methods for demand prediction are covered in Chapters 10 and 11 of this book. However, here, a brief note is presented. SAV demand can be estimated on the basis of trip productions for a zone or even a sub-zone. Such information can be sought from the city transportation planning department. Another source of information is the dynamic demand tables that are compiled from taxi data or other means. Should such information be available, the SAVs can preemptively relocate in order to serve the demand.

Travel between stations located near major generators is challenging to model. These trips are taken for a variety of purposes and can be planned to be carried out at desired departure time. If travelers do have a requirement to arrive at the destination at a specific time, they are likely to request service taking into account potential delay in accessing vehicles and uncertain travel times due to congestion.

Serving demand for travel in high-density corridors is of interest to suppliers in order to become a part of mobility-as-a-service (*MaaS*) [2,10–12]. On the demand side, trip generators frequently locate along such corridors for ease of access to multimodal travel services. However, travelers who cannot use public transit directly along such corridors can request SAV service. First mile/last mile services are recognized as worthy of planning attention. Users of high-capacity public transit systems can benefit from the availability of SAV service to and from stations.

The SAV shuttle services generally follow a predefined route in order to serve such market niches as travel between an airport and a parking location or between a rapid transit station and a medical or a university campus. Of course, other shuttle markets can also benefit from SAV technology. Ride hailing market niches are generally not structured as other niches defined above. However, these demands can be served by SAVs that are in circulation or located at a nearby station.

12.3   Example applications

According to performance criteria, the successful operation of a multistation system based on shared electric automated vehicles requires the availability to a customer (i.e., an individual or a group) of a vehicle with required seats and sufficient SOC level for the trip. Another requirement is the availability of a time slot for charging battery (Figure 12.2).

12.3.1   Modeling availability of vehicles

A Monte Carlo simulation-based model was developed to estimate the availability of a vehicle with required seats to a subscribing customer or a group of customers. See Reference [14] for an introduction to simulation and Monte Carlo methods. During a given session (e.g., a time slot of 30 min) within an operating cycle such as 24 h, the

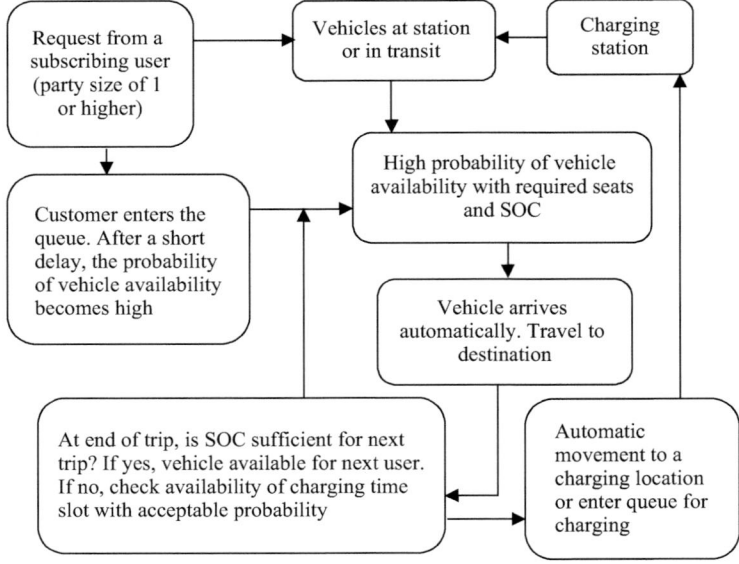

Figure 12.2   Vehicle and charger time slot availability. Adapted from [13]

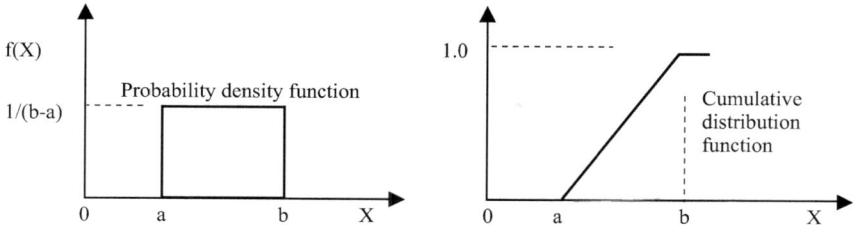

Figure 12.3 Uniform probability density function applied to demand for vehicle and charger time during short time intervals

probability of demand for a vehicle with required seats is represented by the uniform probability distribution (Figure 12.3). The choice of this probability density function is based on the highly uncertain characteristic of the real-time demand.

This continuous uniform distribution, also called the rectangular distribution, is a family of symmetric probability density functions with the characteristic that all intervals of demand for service are equally probable. In this application, the demand is defined by two parameters, a, and b, which represent the range set by the analyst. This probability distribution is the maximum entropy probability function for a random variable X. The applicable constraint is the range of values (i.e., a and b) [15].

Notable characteristics of the uniform probability distribution are noted next.
Probability density function: $P(X) = 1/(b - a)$ for $X \in |a,b|$ and 0 otherwise.
Mean: $\frac{1}{2}(a+b)$
Median: $\frac{1}{2}(a+b)$
Mode: any value in (a,b)

The supply for service is represented by the triangular probability distribution function (Figure 12.4). It is a continuous probability distribution function, defined by three values: the minimum value a, the maximum value b, and the peak (i.e., the mode or most likely) value c. This probability density function is widely used for the reason that in real-life problem-solving conditions, the analyst can often estimate the maximum and minimum values, and also the most likely outcome. The assignment of these values can be done without knowing the mean and standard deviation. It enables the analyst to avoid unnecessary extreme values due to definite upper and lower limits. Another desirable feature is that it is a good model for skewed distributions [16].

The notable statistics for the triangular probability density function are shown below:

$$P(X) = 2(X - a)/[(b - a)(c - a)]$$
for $a \leq X \leq c$ and
$$2(b - X)/[(b - a)(b - c)] \text{ for } c \leq X \leq b$$
Also, $P(X) = 0$ for $X < a$ and $X > b$

where $c \in |a,b|$ is the mode.
The mean is $1/3[(a+b+c)]$

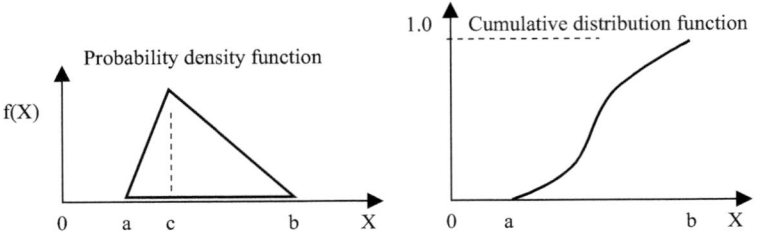

Figure 12.4   Triangular probability distribution function applied to supply of vehicle and charger during short intervals of time

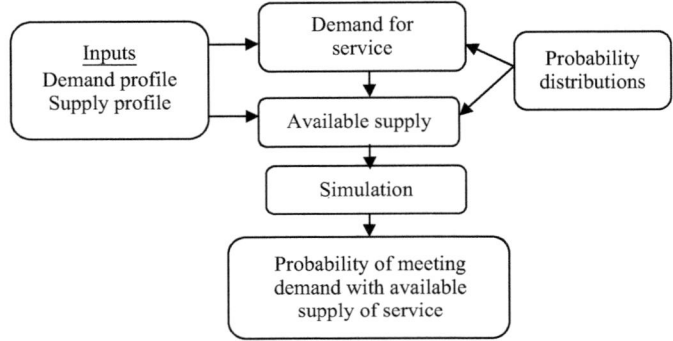

Figure 12.5   Simulation of meeting demand with available supply

A Monte Carlo simulation-based methodology was developed and applied for evaluating the effect of stochastic variables (defined through probability distribution) on the adequacy of supply factor in meeting demand (Figure 12.5). In each simulation, the demand and supply functions are compared and the adequacy or deficiency information is computed. Following the completion of specified runs, the probability of meeting the demand is computed. The methodology has the capability to treat constant, uniform, normal, and triangular distribution functions. The results include the distribution and probability of meeting the demand. Different random number streams can be defined.

In the present application, as previously noted, the demand for access to a vehicle of required size is represented by the uniform probability density function, which reflects a higher degree of uncertainty than other probability functions. However, the triangular probability density function is best suited for the supply-side factors, namely the availability of vehicles and charger time slots.

The simulation method was applied to test the extreme cases as well as to study a large number of other demand-supply balance cases in order to obtain results that are useful for designing the SAV system. In this application, the simulated system has 100 vehicles and five stations for charging and parking. The vehicle fleet can be assumed to consist of various sizes. Each station has four fast-charging stalls that can also be used to park vehicles when these are not used for charging batteries. Also, there are 16 parking stalls in each station that are not equipped with battery chargers. In total, there are 100 parking stalls (including 20 equipped with chargers) that can accommodate 100 automated vehicles when these are not serving customers. It should be noted that in the case of nonautomated vehicles, approximately two parking stalls per vehicle are required due to an imbalance of vehicles requiring temporary parking before relocation can take place.

The first test of the simulation method is to find out at what demand level, the probability of accessing a vehicle is 1.0 under the following supply condition: $a = 0$, $b = 100$, and $c = 100$. That is, the probability of accessing zero vehicles is zero and it increases with the increasing number of vehicles. The most likely value of vehicle availability is 100. Under this supply condition, the probability of accessing a vehicle is 1.0, provided that up to 50 vehicle access requests are received during a time interval of 30 min. On the other hand, if a highly constrained supply side is represented by $a=0$, $b=100$, and $c=0$, the probability of 1.0 for meeting the demand can occur for 0 to 8 vehicles within a time interval of 30 min. An observation of these extreme demand-supply balance conditions suggests that the results are logical.

The developed simulation model can be applied at a station level or to the entire system consisting of five stations. A station with a cluster of four chargers can offer eight slots of 30 min for battery charging. Due to the randomness of demand and service factors, the probability of charger access for up to ten vehicles is about 0.4. At a probability of 1.0 (i.e., without waiting), three vehicles can be offered access during the 30 min time slot at one station.

The model can be applied to fleets consisting of a single size (e.g., a sedan) or a mixture of different vehicle sizes. If vehicles of different sizes are required in a market and the overall fleet consists of vehicle sizes that are in demand, the model can be applied for on-demand service requests that arrive within short time durations (e.g., 30 min) as noted next. A demand for one seater solo can be met by solo or a larger vehicle. A demand for a two-seater mini can be met with a mini or a larger vehicle. Likewise, a demand for a sedan can be met by a sedan or a larger vehicle. However, if a van is needed, smaller vehicles cannot be assigned to the customer.

For use in the design of SAV system and operation, four scenarios of service availability during a session with differing demand and service characteristics were simulated.

Scenario 1: High constant demand (up to 100 vehicles) and changing service ($a = 0$, $b = 100$, c ranges from 50 to 100 vehicles).

Scenario 2: Lowered constant demand (80 vehicles) and changing service ($a = 0$, $b = 100$, c ranges from 50 to 100 vehicles).

Scenario 3: Changing demand (from 40 to 100 vehicles) and service held constant ($a = 0$, $b = 100$, $c = 50$).

Scenario 4: Changing demand (from 55 to 100 vehicles) and service held constant at an increased level ($a=0$, $b=100$, $c=75$).

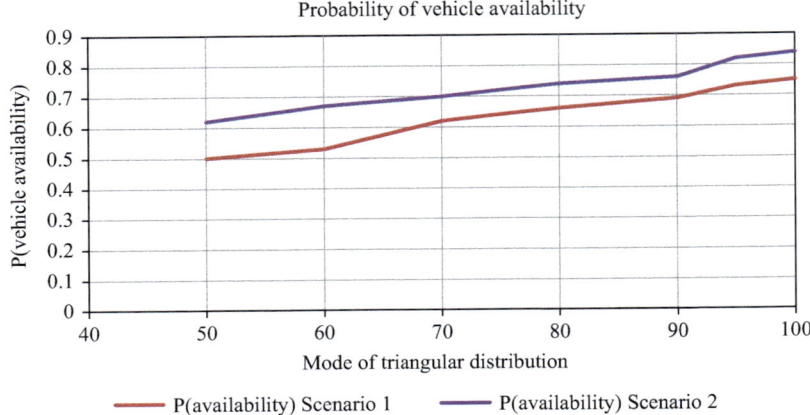

Figure 12.6   Probability of vehicle availability (Scenarios 1 and 2)

These scenarios are based on demand and service features for the five-station SAV system. For illustration purposes, all 100 vehicles are of the same size. However, as noted earlier, the model can be applied to fleets consisting of a single size (e.g., a sedan) or a mixture of different vehicle sizes.

Results for scenarios 1 and 2 in terms of the probability of requested service availability are presented in Figure 12.6. These results suggest that the probability of vehicle availability increases as the most likely number of vehicles ready for service increases. However, according to Scenarios 1 and 2, under specified demand and supply condition, the probability of accessing a vehicle is less than 1.0, which implies that the customer request has to join the queue and wait for a while for service.

Figure 12.7 shows results of Scenarios 3 and 4. In these cases, at moderate demand levels in association with improved supply, the probability of accessing a vehicle on-demand reaches 1.0 and there is no need to wait for service. However, as expected, should the demand increase while the supply remains constant, the probability of accessing a vehicle decreases.

The simulation results imply that the SAV system operator can enhance the probability of vehicle availability by managing high peaks in demand through pricing incentives and at the same time enhancing the availability of vehicles ready for serving customers with reliable and durable vehicles coupled with the efficient operation of the fleet, including optimal routing for travel to customer pick-up and drop-off locations as well as for reaching charger location.

12.3.2   Modeling availability of charging time

As previously noted there are a total of 20 fast chargers installed in five stations and each charger can offer two 30 min charging slots per hour. The 30 min include 15 min for accessing and aligning with robotic charger and 15 min for fast

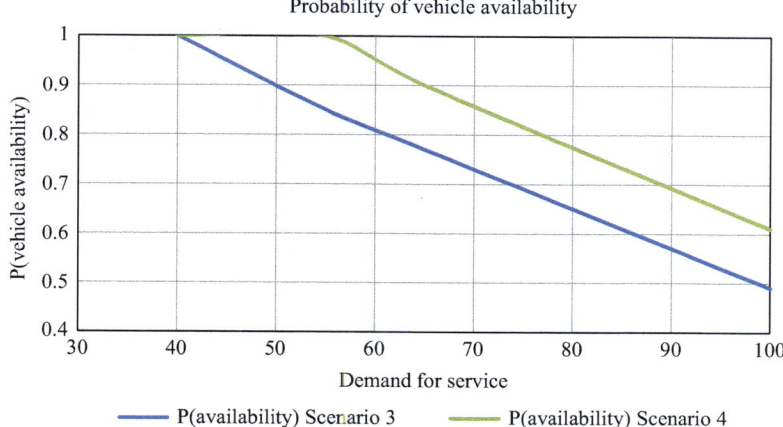

Figure 12.7   Probability of vehicle availability (Scenarios 3 and 4)

charging. So, the total charging time slot in the SAV system amounts to 40 on-demand charging requests. Four scenarios are defined for the availability of charger time slots in the SAV system. A uniform probability distribution for demand is used and the triangular probability distribution represents the supply of charging time slots.

Scenario 1: Demand varies from 5 to 35 charger time requests. The supply is held constant as defined by the following values of variables for the triangular probability distribution function: $a=0$, $b=40$, $c=8$.

Scenario 2: Demand is varied from 5 to 35 requests. Supply is increased, but it is not varied from case to case. The values of supply variables are $a = 0$, $b = 40$, and $c = 16$.

Scenario 3: Demand is varied from up to 5 to 35. The supply is increased but held constant from case to case. The values of supply variables are $a = 0$, $b = 40$, and $c = 20$.

Scenario 4: Demand is varied from up to 5 to 35. The supply is increased from case to case. The values of variables are $a = 0$, $b = 40$, and $c = 24$.

The simulation results on the probability of charger availability are presented in Figure 12.8. These results suggest that under favorable demand and supply conditions, the probability of charger availability reaches 1.0 and there is no need to wait in a queue. However, the charger availability becomes scarce as the number of vehicles requesting charger time increases. In relative terms, as expected, the ranking of scenarios in terms of probability of charger availability is as follows (from high to low): Scenarios 4, 3, 2, and 1.

The implications of results for SAV system design and operations are as follows: (1) Reduce surges of demand for fast charging with the policy of charging batteries to full capacity when chargers are not fully occupied. This will reduce the nonproductive time for chargers and reduce the need for frequent requests for

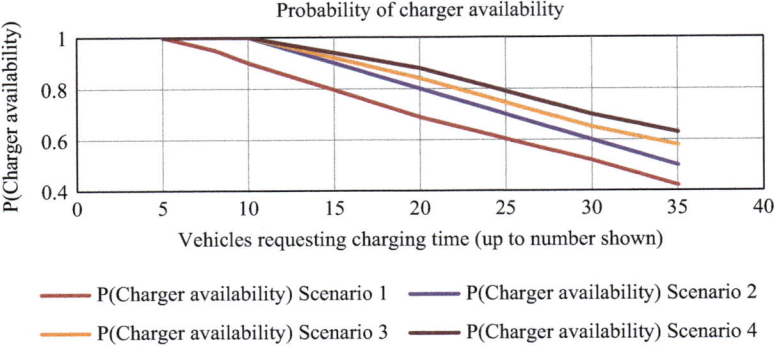

Figure 12.8 Probability of charger availability

charger time during high-demand periods. (2) Direct vehicles to charger locations that can be accessed with minimum expected travel time. This action will reduce the time for a low SOC vehicle to arrive at the charger location. (3) Increase the most likely number of available chargers by efficiency measures such as assigning higher priority to requests from SAV system vehicles as opposed to other vehicles that may be allowed to use chargers for a fee.

12.3.3 Decision to select the charging location

As illustrated in Figure 12.2, following the completion of a trip, if the SOC is sufficient for another trip, the vehicle can accept another on-demand call for service and will travel to the location of the customer for pick-up. On the other hand, the SOC may be at a level that will require the vehicle to reserve a charging slot at a station nearby. If a number of stations have charger slot availability, the SAV system will choose a station on the basis of the minimum expected time to reach the charger location, taking into account the uncertainty of travel time due to congestion.

The Bayesian method is best suited to compare the charger station locations on the basis of expected travel time. Since a shared mobility vehicle should reduce nonrevenue time, and also the vehicle should reach the charger location at the reserved time, decision-making on the basis of minimum expected travel time is the logical approach. In this application, the Bayesian method will quantify the trade-offs between various decisions using probabilities and travel time that accompany such decisions. Also, in the era of connected automated mobility, this method will enable the control system to take into account new data/information on travel conditions in modifying the initial estimates obtained with the use of prior probabilities.

The basic data required to solve the charger station choice decision is as follows [17–20]:

1. A set of alternative charging stations: $S_1, S_2, \ldots \varepsilon$ s, wherein the decision is to be made by selecting a single station S, which is a part of the set of stations.

2. The possible states of congestion (states of nature in decision theory language): C_1, C_2, ... ε c. The states of traffic congestion are probabilistic and probabilities have to be assigned to each state in the set of states of congestion. The decision maker or the decision-making system follows the philosophy that the selection of an alternative S from the set s is dependent upon the actual state of congestion which is not known with certainty.

3. In order to improve the knowledge of travel time, additional information can be acquired, provided that the value of such information is positive (as explained later in this section). In the present application, there are two basic information acquisition alternatives that could be analyzed: do not seek new information that could take time to obtain (e_0) and obtain new information (e_1) from a city traffic management center or a commercial entity, if available. Using the decision theory terminology: e_0, $e_1 \varepsilon$ E, the decision maker (or the system) may decide to obtain more information about the actual state of traffic prior to selecting a station for battery charging.

4. The set of results of information acquisition are r_0, r_1.....ε R. For each information acquisition option, e, there is a space of possible outcomes for that action. For no additional information, e_0, the outcome is r_0. The space of possible outcomes R is defined to encompass any outcome of any e in E.

5. The travel time function or disutility DU(e, r, S, C) represents the decision maker's preferences for all combinations of e, r, S, C. The sequence of the course of action is as follows. The decision maker selects the information acquisition means e, observes a result r, selects a particular S, and then a particular state of traffic, C, occurs. Due to the probabilistic nature of the r and C, the space of all possible combinations of (e, r, S, C) is of interest.

6. The disutility function DU(e, r, S, C) consists of two additive components. That is, in accordance with the principles of utility/value theory, the disutility DU(e, r) and DU(S, C) are additive. The disutility in this application is "time to obtain information" and the "travel time." The DU functions reflect the value structure of the decision maker (i.e., the SAV control system). Although such functions may exhibit diminishing marginal gain of reducing travel time or step functions (with well-defined threshold values) may be relevant, in this application, a linear function is assumed. In research conditions where preference data on travel time reduction are available, a disutility function can be estimated. Here, due to lack of data, no attempt is made to define applicable disutility values.

7. The probability distributions of $P'(C)$ and $P(r|C, e)$ define a joint probability measure $P(C, r|e)$ over the space for each information acquisition option. The $P'(C)$ (a prior probability) represents the decision maker's judgment about the relative likelihood of values of C, and $P(r|C, e)$ (a conditional probability) characterizes each information acquisition option. It is the probability that the outcome r will be observed if additional information acquisition activity e is initiated, and C is the value of the state of traffic congestion variable. Further information on probability measures is provided next.

For solving the Bayesian decision problem, a joint probability $P(C,r|e)$ is to be found for the joint distribution of C and r, over the space c x R for each information acquisition option e. This implies that the decision maker (or the decision system) should define the reliability of each possible information outcome r in predicting the true state of traffic C, for each information acquisition option e.

The joint probability measure relates to four other probability measures defined below.

1. The prior measure $P'(C)$ on the states of traffic congestion that the decision system would assign to C prior to observing the outcome r of additional information acquisition option e.
2. The conditional measure $P(r|C, e)$ on the space R—the probability that the outcome r will be observed if e is initiated and C is the true value of the state of traffic congestion.
3. The marginal measure $P(r|e)$ on the space R for all C is the probability of observing outcome r from e. It is computed as follows:

$$P(r|e) = \Sigma P'(C)P(r|C,e)$$

The posterior measure $P''(C|r, e)$ on the space c represents the likelihood of different traffic states C, given r and e. The decision maker assigns this probability measure to the space c after knowing the outcome r of e. It is computed using the following equation:

$$P''(C|r,e) = \frac{P(r|C,e)\ P'(C)}{P(r|e)}$$

According to the Bayesian method, each e can be characterized by a conditional probability distribution $P(r|C, e)$, such that the relationship between the prior and posterior distribution is given by the Bayes Theorem shown above.

In the present Bayesian method application, the (e, r, S, C) sequence is studied in the form of DU(e, r, S, C) for the purpose of minimizing time to reach the charging location. The decision maker wishes to choose an information acquisition option and therefore has to evaluate all the possible sequences or courses of action.

The method of analysis for evaluation of alternative information acquisition means and determining the most desirable e is known as the preposterior analysis. In this application, to determine if additional information should be obtained, the preposterior analysis method is used. However, if the purpose of the analysis is to compare results under prior and posterior probability distributions, the mode of analysis is termed posterior analysis.

In this application, both preposterior analysis and posterior analyses are used to find out if additional information on the uncertain states (i.e., level of congestion) will be useful and also to identify the optimal S. Following the acquisition of additional information, the selection of a charger station can be made. For preposterior analysis, the following sequence of operations is required [17]:

1. The likelihood of different states of traffic congestion, C, is expressed in the form of prior probability distribution $P'(C)$.
2. The conditional probability characteristics $P(r|C, e)$ are determined by the control part of the SAV system on the basis of the reliability of e, inferred from actual data or simulations.
3. The marginal probabilities $P(r|e)$ for e are computed as noted above, wherein for the null alternative (e_0), the probability is equal to 1: $P(r_0|e_0) = 1.0$
4. The posterior probability distribution $P''(C|r, e)$ is computed from prior and conditional probability distributions.
5. For each combination of e, r, S, C, travel time is found: DU (e, r, S, C).
6. The expected travel time for each station S, for each (e, r) combination is as follows:

   The expected travel time for the posterior part of the analysis is found by the equation:

   $$DU^*(S, r, e) = \sum_C P''(C|r, e)DU(e, r, S, C)$$

   However, for the prior branch, where no new information is acquired,

   $$DU^*(S, r_0, e_0) = \sum_C P'(C)DU(e_0, r_0, S, C)$$

7. For each (e, r) combination, the optimal charging location is determined as follows:

   $$DU^*(r, e) = \text{Min}_S DU^*(S, r, e)$$

8. For each information acquisition option e, the expected travel time is computed:

   $$DU^*(e) = \sum_r P(r|e)DU^*(r, e)$$

9. The optimal e^* is that e for which $DU^*(e)$ is a minimum. That is:

   $$DU^*(e^*) = \text{Min}_e DU^*(e)$$

10. In the preposterior analysis, if the "null" option is not the optimal option, the value of additional information is calculated as follows.
    (i)   Find optimal S by calculating and comparing $DU(S,r,e)$ for each r. Call this Sr.
    (ii) Find optimal S for r_0 by calculating and comparing $DU(S,e_0)$. Call this S'.
    (iii) For each r, find $DU(S'-Sr)$.
    (iv) Calculate value of information $Vt^*(e) = \sum_r P(r|e)[DU(S'-Sr)]$.

According to theory, the variable $V_t^*(e)$ can be considered as the expected reduction in risk. In this application, it is the maximum amount of time that can be spent on additional information acquisition. If $V_t^*(e)$ is equal to zero or it is found to be negative, then e_0 is the best course of action (i.e., it is not desirable to acquire additional information in support of decision-making).

11.   Next step is to find expected disutility $DU^*(S)$ for each S,
$\sum_r P(r/e)[DU(S,r)]$, and identify the optimal S with minimum expected DU^*(i.e., the S with the minimum expected travel time).

12.3.4   Application of the Bayesian method

Following the delivery of a customer or group of customers, the SOC value may not be sufficient to accept a new call for service and therefore the control function will send a request for a charging slot to stations in the SAV system. On the assumption that more than one station can accept the request, a decision will be required to direct the vehicle to one station on the basis of minimum travel time. Given that the travel time will depend upon traffic congestion in various parts of the traffic network and it is not possible to know the level of congestion with certainty, the decision to choose a station will be made under uncertainty. The Bayesian method is best suited for analyzing the decision problem.

For an illustration of the Bayesian method application, travel time estimates to three stations (S_1, S_2, S_3) are available under three traffic congestion states (C_1, C_2, C_3). The congestion level increases from C_1 to C_2 and C_3 (Figure 12.9). An examination of travel time estimates indicates that under C_1 and C_2, station 1 is the choice. On the other hand, travel to station 3 minimizes travel time if C_3 becomes the true state of traffic congestion. Since states of congestion are uncertain, the travel times are uncertain. Therefore, the use of the Bayesian method becomes necessary for the analysis of charger station alternatives.

Two scenarios of the application of the Bayesian method are presented next. For Scenario 1, prior probabilities and conditional probabilities are noted in Table 12.1. The prior probabilities do not appear to imply the decision maker's concern that travel will take place in the congested condition. The conditional probabilities suggest a reasonable level of reliability of additional information if acquired from the traffic control center. Past experience with the reliability of additional information can be used for assigning conditional probabilities.

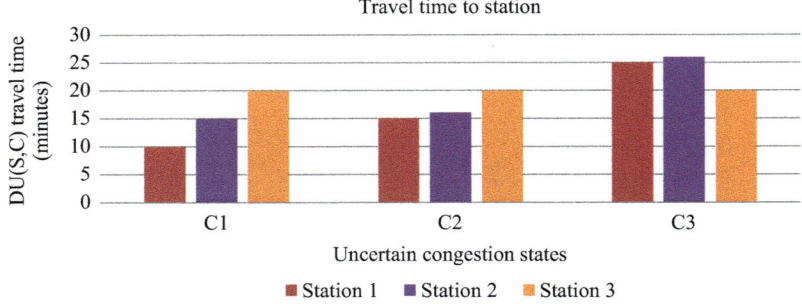

Figure 12.9   Travel time to stations

Table 12.1 Scenario 1 input probabilities

Uncertain states of congestion C	C1	C2	C3	Sum
Prior probabilities $P'(C)$	0.3	0.4	0.3	1.0
Conditional probabilities $P(r|C,e)$				
r1	0.6	0.2	0.1	
r2	0.3	0.6	0.3	
r3	0.1	0.2	0.6	
Sum	1	1	1	

Table 12.2 Scenario 1 outputs

Posterior probabilities $P''(C|r,e)$					Marginal probabilities $P(r|e)$	
	C_1	C_2	C_3	Sum		
r_1	0.62	0.28	0.10	1.0	r1	0.29
r_2	0.21	0.58	0.21	1.0	r2	0.42
r_3	0.10	0.27	0.62	1.0	r3	0.29
					Sum	1.0

The outputs of preposterior analysis of Scenario 1, namely posterior probabilities, marginal probabilities, expected travel time for applicable combinations of e,r,S, and value of information are shown in Table 12.2. The application of marginal probabilities to the expected travel time estimates for the e,r,S combinations results in the expected travel time for the candidate charger stations. Results show that the expected value of information is positive, which suggests that the initiation of additional information acquisition is beneficial. Also, under Scenario 1, the travel time will be minimized by directing the electric automated vehicle to charger station S_1.

Expected travel time (minutes)		Value of information	r_1	r_2	r_3
$DU(S_1,r_1,e)$	12.93 (min)	S_r	12.93	16.07	20.00
$DU(S_2,r_1,e)$	16.41	S'	16.50	16.50	16.50
$DU(S_3,r_1,e)$	20.00	$S'-S_r$	3.57	0.43	−3.50
$DU(S_1,r_2,e)$	16.07 (min)	$P(r|e)$	0.29	0.42	0.29
$DU(S_2,r_2,e)$	17.93		$V^*_t = 0.2$		
$DU(S_3,r_2,e)$	20.00		Exp. Travel time $(S_1) = 16.5$ min		
$DU(S_1,r_3,e)$	20.69		Exp. Travel time $(S_2) = 18.7$min		
$DU(S_2,r_3,e)$	22.10		Exp. Travel time $(S_3) = 20.0$ min		
$DU(S_3,r_3,e)$	20.00 (min)				

Table 12.3   Scenario 2: input probabilities

Uncertain states of congestion C	C_1	C_2	C_3	Sum
Prior probabilities $P'(C)$	0.1	0.2	0.7	1.0
Conditional probabilities $P(r|C,e)$				
r_1	0.6	0.2	0.1	
r_2	0.3	0.6	0.3	
r_3	0.1	0.2	0.6	
Sum	1	1	1	

Table 12.4   Scenario 2 outputs

Posterior probabilities $P''(C|r,e)$						Marginal probabilities $P(r|e)$
	C_1	C_2	C_3	Sum		
r_1	0.35	0.24	0.41	1.0	r_1	0.17
r_2	0.08	0.34	0.58	1.0	r_2	0.36
r_3	0.02	0.09	0.89	1.0	r_3	0.47
					Sum	1.0

The Scenario 2 differs from Scenario 1 in terms of prior probabilities that are assigned with the expectation that travel is likely to be impacted due to traffic congestion. (Table 12.3). A relatively high probability assigned to congestion level C_3 is based on the controller's (i.e., decision maker's) prior knowledge of traffic conditions at the applicable time of day and network congestion pattern. The conditional probabilities regarding the reliability of additional information are the same as in Scenario 1.

The outputs of Scenario 2 analysis shown in Table 12.4 reflect the effect of high prior probability for the occurrence of S_3. On the basis of minimization of travel time, charger station S_3 is the choice in this scenario. The expected value of information is positive, which suggests that the additional information obtained about traffic conditions is beneficial prior to selecting the charger station. It appears that the travel time to charger station S_3 is not affected by travel conditions as compared to other chargers due to its location within the city.

Expected travel time (minutes)	Posteriors	Value of information	r_1	r_2	r_3
$DU(S_1,r_1,e)$	17.35 (min)	S_r	17.35	20.00	20.00
$DU(S_2,r_1,e)$	19.76	S'	20.00	20.00	20.00
$DU(S_3,r_1,e)$	20.00	$S'-S_r$	2.65	0	0
$DU(S_1,r_2,e)$	20.42	$P(r/e)$	0.17	0.36	0.47
$DU(S_2,r_2,e)$	21.75	$V^*_t = 0.45$			

(Continues)

(Continued)

Expected travel time (minutes)	Posteriors	Value of information	r_1	r_2	r_3
$DU(S_3,r_2,e)$	20.00 (min)		Exp. travel time ($S1$) = 21.5 min		
$DU(S_1,r_3,e)$	23.83		Exp. travel time ($S2$) = 22.9 min		
$DU(S_2,r_3,e)$	24.91		Exp. Travel time ($S3$) = 20.0 min		
$DU(S_3,r_3,e)$	20.00 (min)				

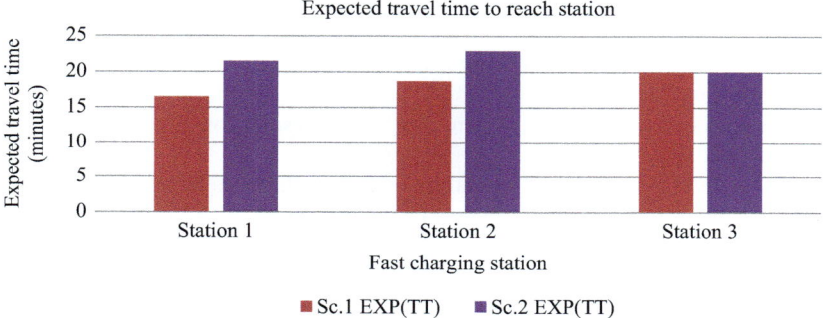

Figure 12.10   Expected travel time and choice of charging station

Results of the two scenarios illustrated in Figure 12.10 suggest that in Scenario 1, the expected travel time to charger station 1 is the lowest due to almost uniform prior probabilities of the occurrence of various levels of congestion in the network. On the other hand, according to Scenario 2 results, the relatively high prior probability assigned to congestion level C_3 in association with a reasonably high level of conditional probability results in changes in expected travel time and causes station 3 to be the choice. As in Scenario 1, the expected value of travel time is positive in Scenario 2, which suggests that new information plays a role in decision making.

12.4   Conclusions

1. Given that at present there is no real-world experience with demand for services of SAVs and also there is a lack of experience on the supply side, it is necessary to approach the matching of demand for vehicle access (and seats) and supply under uncertainty.
2. Although demand-supply balancing methods exist in other fields, a new methodological framework became necessary for the SAVs due to the lack of real-world data and uncertainties on both the demand as well as supply sides.

3. For the simulation of demand for service and corresponding supply of service for short time intervals such as 30 min, the probability density functions chosen for use in the developed methodology are suitable. The uniform probability density function for on-demand access to a vehicle for a short time interval reflects a high level of uncertainty. The triangular probability function for characterizing the availability of vehicles enables the setting of the range and the most likely value of vehicle availability.

4. Likewise, for charger slot demand, the uniform probability density function is well suited and for the availability of charger slot, the triangular distribution function is a logical choice.

5. The simulation model based on the Monte Carlo method meets the requirements in terms of treating both the demand and supply variables as uncertain and enables sampling from the uniform as well as the triangular probability distribution functions.

6. The implications of SAV system simulation results for design and operations are noted next.

   (i) System design and operations can be guided in terms of defining conditions under which the probability of vehicle and seat availability can be enhanced, thus improving the level of service.

   (ii) The probability of vehicle availability can be improved by managing high peaks in demand (i.e., a smaller number of requests for access within a short time period) through pricing incentives.

   (iii) The availability of vehicles ready for serving customers with reliable and durable vehicles is necessary and so is the efficient operation of the fleet.

   (iv) Optimal routes can be identified for travel to customer pick-up and drop-off locations.

   (v) Surges of demand for fast charging can be reduced by using the policy of charging batteries when chargers are not fully occupied. This will reduce nonproductive time for chargers and reduce the need for frequent charger time during service periods.

   (vi) Direct vehicles to charger locations that can be accessed with minimum expected travel time. This action will reduce the time for a low SOC vehicle to arrive at the charger location.

   (vii) Increase the most likely number of available chargers by efficiency measures such as assigning higher priority to requests from the SAV system vehicles as opposed to other vehicles that will use the chargers for a fee.

7. The Bayesian method is suitable for selecting a charger location (station) under the uncertainty of travel time in the urban transportation network.

8. The Bayesian method can also be applied to evaluate alternate routes for an automated vehicle to reach a customer or drop-off a customer.

Acknowledgments

The author wishes to thank Seth Gatien for reviewing a draft of the chapter.

Author contribution

The author reviewed and approved the manuscript.

Declaration of conflicting interests

The author(s) declared no potential conflict of interest with respect to publication of this chapter.

References

[1] Massachusetts Institute of Technology. *Insights into future mobility.* Cambridge, MA: MIT Energy Initiative. 2019. http://energy.mit.edu/insig

[2] City of Toronto. *Automated vehicles tactical plan. IE87 – Attachment.* 1. 2019.

[3] Transportation Research Board (TRB). *Forum on preparing for automated vehicles and shared mobility (E000SA).* Lectern Session 1679. Transportation Research Board (TRB) Annual Meeting 2019.

[4] Mitchell, W.J., Borroni-Bird, C.E., and Burns, L.D. *Reinventing the automobile: Personal urban mobility for the 21st century.* Cambridge, MA: The MIT Press; 2010.

[5] Chiu, Y-C., Bottom, J., Mahut, M., et al. "*Dynamic traffic assignment: A primer.*" The Transportation Network Modeling Committee, Transportation Research Board, Washington, DC; 2011. www.TRB.org

[6] Agatz, N. A., Erera, A. L., Savelsbergh, M. W., and Wang, X. 'Dy-namic ride-sharing: A simulation study in metro Atlanta'. *Transportation Research Part B: Methodological.* 2011; 45(9): 1450–1464.

[7] Fagnant, D., and Kockelman, K. *Preparing a nation for autonomous vehicles: Opportunities, barriers and policy recommendations.* Eno Center for Transportation. October 2013.

[8] Kang, N., Feinberg, F.M., and Papalambros, P.Y. "Automated electric vehicle sharing system design." *Journal of Mechanical Design.* 2017.

[9] Pavone, M. "Autonomous mobility-on-demand systems for future urban mobility." Chapter 19 of M. Maurer *et al.* (eds). *Autonomous driving.* 2016. DOI 10.1007/978-3-662-48847-8_19.

[10] Abel, S. "Mobility and public right of way." *ITE Journal.* 2019.

[11] Institute of Transportation (ITE). "MaaS/MOD, ITE initiative about the MaaS/MOD adapted from the June 2019 ITE Journal." ITE Website. Available from https://www.ite.org/technical-resources/topics/maas-mod-ite-initiative/.

[12] World Economic Forum and the Boston Consulting Group. *Reshaping urban mobility with autonomous vehicles: Lessons from the City of Boston.* Boston, USA. 2018.

[13] Cepolina, E.M., and Farina, A. "A methodology for planning a new urban car sharing car sharing system with fully automated personal vehicles." *European Transportation Research Review*. 2014; 6: 191–2014.

[14] Rubinstein, R.Y., and Kroese, D.P. *Simulation and the Monte Carlo method*. Wiley. 2016.

[15] Walpole, R.E., Myers, R.H., Myers, S.L., and Ye, K. *Probability and statistics for engineers and scientists*. Boston, MA: Prentice Hall. 2012. pp. 171–172.

[16] Evans, M., Hastings, N., and Peacock, B. 'Triangular Distribution'. Chapter 40 in *Statistical distributions*, 3rd edn. New York: Wiley; 2000. pp. 187–188.

[17] Raiffa, H. and Schlaifer, R. *Applied statistical decision theory*. The MIT Press. 1968.

[18] Carlin, B.P. and Louis, T.A. *Bayes and empirical bayes methods for data analysis*. New York: Chapman & Hall; 1996.

[19] Korb, K.B., and Nicholson, A.E. *Bayesian artificial intelligence*. UK: Chapman & Hall/CRC; 2004.

[20] Congdon, P. *Bayesian statistical modelling*, 2nd edn. John Wiley and Sons, Ltd. 2006.

The views expressed are those of the authors.

Chapter 13

Operations and management

Shams Tanvir[1], Kanok Boriboonsomsin[2], and Matthew Barth[3]

The essence of transportation is moving people or goods from point A to point B. Through this process, society creates the demand to move, and vehicles of all sorts supply this demand. Shared mobility emerged as a special variety in this range of demand–supply interaction which enables sharing the vehicles among multiple demand entities to complete a trip [1]. In recent years, shared mobility services have expanded to almost all corners of the world catering to every class of mobility purpose. This rapid expansion is attributed in part to the information technology revolution of the twenty-first century; particular focus is on five major technological innovations—the Internet, wireless connectivity, Global Navigation Satellite Systems (e.g., GPS), electronic payment systems, and smartphones. Even at the current state of technology, new modalities of shared mobility are evolving every year ranging from motorcycle ride-hailing to luxury car sharing. These new modalities offer new functionalities to address varying mobility needs; however, the core functionalities of shared mobility services remain the same—the ability to pair mobility demands with a variety of supplies, executing the operation, and completing the economic transactions. Operational steps additional to core functionalities often arise from the need to "scale-up" the shared mobility operations, improve operational efficiency, reduce costs, comply with regulations, and increase competitiveness. In this chapter, we review the core operational steps of shared mobility activity along with the auxiliary operational modules.

Recent developments in automated vehicles have added to the possibilities of shared mobility applications (see Chapter 9). One of the major drawbacks of shared mobility is that the profitability and operational efficiency tend to be lower at lower demand and supply densities. Automated vehicles (AVs) significantly reduce the constraints on the supply side. For example, higher-level automated vehicles (Levels 4–5) can remove human drivers [2]; therefore, can be made available to shared mobility service subscribers

[1]Department of Civil and Environmental Engineering, California Polytechnic State University, San Luis Obispo, CA, USA
[2]Center for Environmental Research & Technology, University of California at Riverside, Riverside, CA, USA
[3]Department of Electrical and Computer Engineering, and Center for Environmental Research and Technology, University of California at Riverside, Riverside, CA, USA

over a wide geographic area and periods. These anticipated spatial and temporal coverage enhancements have the potential to extend and, in some cases, simplify shared mobility operations. Therefore, AVs will likely increase auxiliary operational steps to accommodate scaling up of services and, at the same time, reduce complexities of several core functionalities of existing shared mobility offerings. We review each operational step in the light of potential transformations enabled by automated vehicles.

13.1   Introduction and concepts

The history of ridesharing dates back to 1942 when the US Office of Civilian Defense developed a system to match riders with similar trip origin and destination through workplace bulletin boards [3]. In 1948, carsharing started in Zurich, Switzerland as part of a self-drive club, SEFAGE [4]. In both cases, the motivation was to increase the utilization rate of vehicles, thus, reducing the operating expense. In these early years, all of the shared mobility operations were managed manually—increasing the overhead expenses for managing shared mobility services. Hence, the spatial and temporal scope of shared mobility operation was limited and almost all available modes that offered shared mobility options were auto-based. With the advent of computing and information technology, the marginal overhead expenses to manage substantially large shared mobility operations were reduced drastically. Consequently, an exponential increase of shared mobility services ensued in the last two decades; particularly in low-value modes such as scooters and bikes and nonwork-based trip purposes such as recreation and package delivery.

Technology also enabled different business models to operate shared mobility. In the Business to Customer (B2C) business model, vehicles are owned by the shared mobility fleet managers. As owners of vehicles, businesses are required to maintain, store, reorganize, prove roadworthiness, and assume liabilities of the vehicles. Whereas in Peer-to-Peer (P2P) business models, vehicles are privately owned and shared mobility operators work as a broker between the vehicle owner and the users. Modern P2P service operators maintain an online platform that enables the communication between the two parties. A common complement to the online platform is smartphone applications that are capable of identifying vehicles through geolocation capabilities, reserving vehicles through two-way cellular or wireless communications, providing customer service and troubleshooting, and handing payments [5]. Smartphone applications ensured the ubiquity of shared mobility services and as a result, the scale of operation increased significantly. Transportation network companies (TNC) seized this opportunity to increase the efficiencies and flexibilities in the core functionalities provided by the existing shared mobility services to further increase the scale of operation. For example, with the same app-based concept, Uber now operates in more than 900 cities worldwide providing a range of shared mobility offerings [6]. Due to technological enhancements, the cost of operation for a similar trip can be lowered either by increasing the scale of operation or by improving the efficiency of operation. Although the core operating principles remained the same, the cost reduction measures engendered several auxiliary operations—often varying by the type of operation and by the business models. The effectiveness of auxiliary operations often dictates the cost efficiency of shared mobility offerings; hence, these operations determine the competitiveness of emerging mobility products with traditional travel modes and existing shared mobility platforms.

Automated vehicles (AV) are considered a symbiotic addition to vehicle technologies in shared mobility operations. Researchers anticipate AVs will help the reach of shared mobility services [7]. Conversely, shared mobility business models can help in recovering high purchase prices of AVs [8]. Even though the actual way of shared automated vehicles (SAV) operation is still unknown, educated guesses can be made on the types of SAV business models that may be adopted in the future. Stocker and Shaheen [9] predicted six possible business models for SAVs: (1) B2C with a single owner-operator, (2) B2C with different entities operating, (3) P2P with third party operator, (4) P2P with decentralized operations, (5) hybrid ownership with the same entity operating, and (6) hybrid ownership with third-party operators. Like previous technologies, only the SAV business models that can improve the cost-effectiveness of shared mobility offerings will be adopted widely. Similarly, SAV will not change the core operating principles of shared mobility; however, some auxiliary operations may be very different from AVs. To understand operational changes brought on by AVs, we will explore different components of cost-effectiveness for SAV operation.

13.1.1 Cost-effectiveness for SAV operation

The business decision to operationalize a new mobility technology is often driven by the unit cost or average cost per user per distance traveled. In the case of shared mobility services, the average cost is the ratio of total costs and the scale of the operation. Total cost is the combination of overhead expenses, fixed costs, and variable costs. The scale of operation has both temporal and spatial aspects. The duration of service represents the temporal scale and the number of users is a good indicator of the spatial scale. Therefore, new services have the option to drive down the total cost or increase the scale of operation to minimize the average cost of operation.

$$\text{Average cost} = \text{total cost}/(\text{service time} \times \text{number of users})$$

$$\text{Total cost} = \text{overhead expenses} + \text{fixed costs} + \text{variable costs}$$

Overhead expenses include the cost of computing and IT systems and the cost of complying with government regulation. The cost of computing and IT infrastructure will be much higher for AVs, especially in the initial stages. The cost of complying with government regulation might also increase due to the expected increase in vehicle miles traveled (VMT) with SAVs [10,11]. Expenses might be added to monitor or supervise an SAV fleet in the early days of automated vehicle technology. For example, California adopted a regulation to administer automated vehicle tests [12] that requires a "remote operator" to continuously supervise the operations of an automated vehicle. Additionally, manufacturers are required to submit "disengagement reports" to share how frequently and why test vehicles had to disengage from autonomous mode during tests.

Fixed costs include costs for vehicles leased or purchased, and potential costs associated with parking. The base price for SAVs will be much higher than the traditional shared mobility vehicles. Fagnant and Kockelman [13] suggested SAV capital cost of $70,000 and an operating cost of $0.50 per mile. Bosch *et al.* [14] assumed an approximately 20% increase in fixed cost due to automation considering four different vehicle sizes (i.e., solo, midsize, van, and minibus). Fixed cost components such as depreciation, interest, and taxes are mostly unaffected by the operations and management protocols. The level of sophistication required in sensing technology and software will likely drive the cost of the vehicles. In most cases, technology choices available to SAV operators are governed by safety requirements. However, total fixed cost to the fleet owner can be minimized through "vehicle right-sizing," a concept that uses optimum vehicle size according to the characteristics of the shared vehicle trips.

Variable costs are operational costs such as fuel cost, maintenance cost, insurance cost, and labor. Stephens *et al.* [15] estimated that fuel cost could be reduced by approximately 10% due to more balanced driving in automated systems. Insurance costs will be significantly lowered because of the likely decrease in incident rates with automated systems. The maintenance cost might be decreased in the longer term as the vehicle automation features can be used in predictive maintenance of different vehicle components. However, the sensors installed in the vehicles will need periodic maintenance, adding to the variable cost of operating an SAV fleet. Vehicle cleaning will be vital for an SAV fleet as there is no driver available to conduct the cleaning procedure. It is expected that a vehicle in an SAV fleet will make empty trips to cleaning and refueling facilities. Bosch *et al.* [14] estimated that cleaning costs will be 29% of total cost per passenger-km in an SAV fleet. The cost of labor is the item where SAVs have the potential for maximum savings. Currently, more than 80% of the total cost is spent on driver salaries for conventional taxi services. With SAVs, only a fraction of salaries will be spent on remote operators and fleet managers.

On average the capacity utilization ratio is 30% higher for transportation network companies (TNC) than traditional taxi services [16]. However, in the current TNC economy, drivers are voluntary participants in the system. Therefore, a portion of the TNC vehicle service hours is unutilized moving between work and home locations of the drivers. In contrast, vehicles in an SAV fleet will be operated throughout the day except for maintenance periods. This extended period of serviceability provides a better trip economy for SAV fleets than any other shared mobility option. Figure 13.1 shows a conceptual diagram of changes in the scale of operation and anticipated average costs for varying levels of technological sophistication in shared mobility technologies.

13.1.2   Core operating principles

Any shared mobility service, manual or automated, needs to guarantee that a trip from origin to a destination can be completed successfully. Like any other service, there needs to be a way to make requests, a way to find available providers, a way

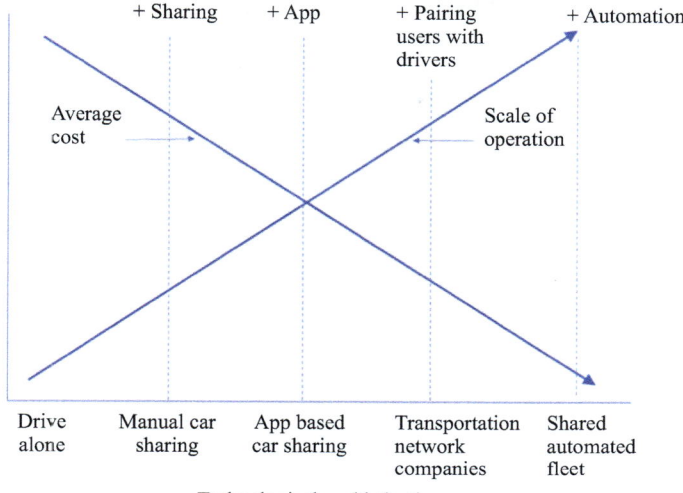

Figure 13.1 Improved cost-effectiveness of shared mobility services with enhanced technologies. Schematic shows a larger scale of operation enabled by technologies is the main driver for average cost reduction

to match requests with providers, and a way to pay for the service. Therefore, the core functionalities of shared mobility operations can be listed as follows (corresponding operations are shown in parentheses):

1. Ability to request for a trip and the ability to declare availability (*Reservation and access*).
2. Ability to match requested demand with available supplies in a way that matches the requirements of both the users and drivers (*Dispatching and ride-matching*).
3. Ability to travel from the origin to destination within a reasonable time and ensuring an acceptable level of comfort and convenience (*Routing and fleet monitoring*).
4. Ability to handle payments and user feedbacks (*Payment and pricing*).

Operationally, the trip reservation process and the ride-matching processes as listed in the first two points happen prior to the actual trip. The payment and trip evaluation component occurs at the end of the service.

13.1.3 Auxiliary operations with and without AV

As described previously, the cost-effectiveness of a new shared mobility service is often determined by the scale of its operation. The technological revolutions brought on by the Internet, wireless communication, global navigation satellite systems, and smartphones have added many possibilities for minimization of cost and increasing the scale of operation. In a way, many auxiliary functionalities are already operationalized through existing shared mobility services. Some example operations are:

- Forecasting demand and supplies at a given location and time period;
- Attracting users and drivers to participate in the service;
- Rating users and drivers for more personalized matching in the future;
- Observing fleet operations through a centralized platform;
- Detecting anomalies in the system and troubleshoot accordingly;
- Developing performance metrics, availing data to monitor performance over time, and taking actions to improve performance;
- Addressing regulatory concerns and adapting the service paradigm within those policies;
- Storing data generated during the service and automating decision making using stored data such as fraud detection and driver onboarding;
- Repositioning vehicles within the service area to maximize service availability and reduce the waiting period; and
- Refueling and maintaining vehicles for a single owner-operator fleet.

These auxiliary operations occur over a longer time period before and after the trip compared to the core functionalities. Figure 13.2 shows the timeframe for auxiliary operations relative to core operations.

With SAVs, the potential for more nuanced auxiliary operations increases tremendously. Since automated vehicles can be utilized in all three major modes of shared mobility, that is, carsharing, ridesharing, and microtransit, a flexible SAV fleet can be developed. An emerging mobility concept, namely, Mobility-as-a-Service (MaaS) is particularly suited for such flexible operations [17,18]. Through MaaS, a centralized subscription-based system can be developed that can meet the mobility needs of a variety of users. For example, a user may reserve an automated single-occupant vehicle to connect to a dynamically routed on-demand automated microtransit. Although MaaS does not inherently change the core functionalities of the individual shared mobility service segments, it does allow for increasing the

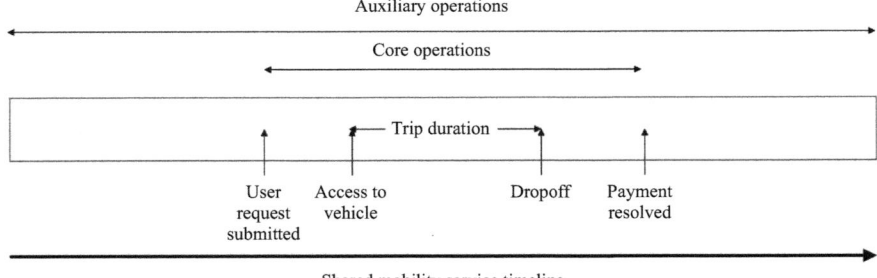

Figure 13.2   Schematic of the operations timeframe for shared mobility services showing the relative duration of auxiliary operations

scale of overall operations. The vehicle reservation and access process is much more streamlined in a unified mobility framework provided by MaaS. With vehicle automation, there are more opportunities for optimization in vehicle dispatching, routing, and repositioning.

The innovations in vehicle automation can be broadly categorized into three types of technologies—sensing, computing, and communication. Several new operations are feasible with advanced in-vehicle and environmental sensors deployed in an automated vehicle. Apart from the safety-critical applications, environment-friendly routing and operations applications can be developed using the high-performance onboard computers. Enhanced computing and communication capabilities in automated vehicles also enable on-board and edge data processing, service optimization, fleet management, and user tracking.

Management of an SAV fleet has a two-fold meaning: (1) regular operations management to maximize profit and user benefits; and (2) maximizing societal benefits or minimizing the unintended consequences. For regular operations, fleets are managed in a way that the most number of users can access the shared vehicle at minimum waiting and trip times, at minimum routing costs, or/and at a maximum profit for the fleet operating companies. However, user-benefit maximization may not guarantee maximum societal benefits as users may choose to not share their rides (increases number of vehicles on road) or travel longer distances and run empty trips (increases the distance traveled). Hyland and Mahmassani [19] described the SAV fleet optimization problem in three aspects—(a) routing, (b) scheduling, and (c) fleet management. They formulated each subproblem as dynamic multivehicle pickup and delivery problems with explicit or implicit time-window constraints.

In Section 13.2, we review the system components of the shared mobility services. Then, we discuss the core operations of shared mobility in light of possible and anticipated changes from vehicle automation. Within the realm of auxiliary operations, fleet performance optimization, vehicle repositioning, and fleet refueling and maintenance are discussed in detail.

13.2   System components

Modern shared mobility services rely on information technology and advanced computing to make a connection between demand and supply and execute the operation. The main system components are client applications, communication channels, request aggregators, cloud databases, and data processing systems. Figure 13.3 shows a schematic of the processes involved in a typical shared mobility service. In the following paragraphs, different components and processes in a modern shared mobility service system will be discussed, with a focus on potential changes due to vehicle automation.

Client interfaces: The most commonly used approach to communicate with the users and drivers is through smartphone applications. Depending on the need, the service providers design applications for other clients such as mobility aggregators,

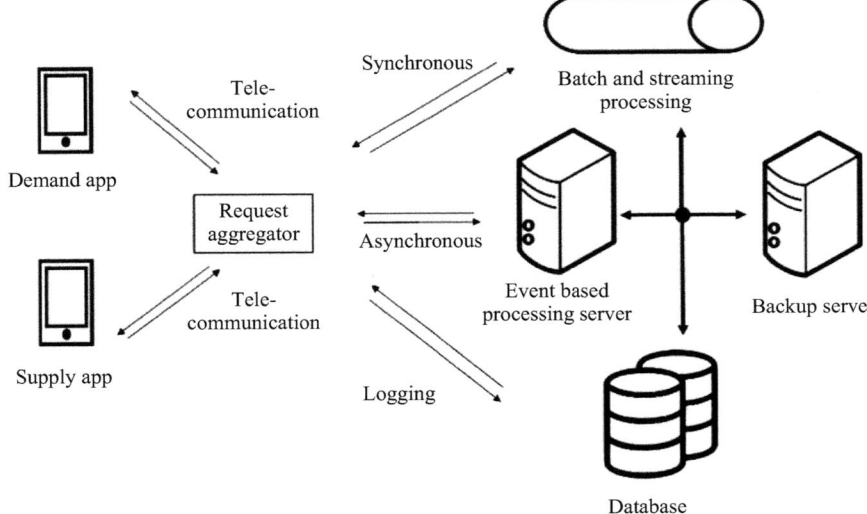

Figure 13.3   System architecture of a typical modern shared mobility service

third-party providers, and infrastructure managers. Additionally, dashboards are designed for fleet administrators to enable monitoring. Typically, these applications communicate with the main server through application programming interfaces (API).

The demand-side application captures riders' geolocation once a trip is requested. In addition, different request details are captured. Once the ride is reserved, the demand app is used to display the state of the vehicle such as the location, number of occupants, level of fuel, etc.

The supply-side application with human drivers shows the request for rides, location of the riders, navigation, payment information, user ratings, and estimated discounts. With automated vehicles, no display interface will be needed on the supply side. The server will communicate directly with the onboard vehicle computer. The location and status of the SAV will be communicated in real-time with the remote server. Similarly, any routing and relocation decision for the vehicle will be sent to the operating system of the vehicle.

Communication and telematics: Existing shared mobility services mostly rely on cellular communication to transmit information to and from the vehicle. Many safety-oriented applications have been proposed with vehicle to vehicle communication (V2V) and vehicle to infrastructure (V2I) communication [20]. Historically, the development of digital short-range communication (DSRC) has been geared to accommodate low-latency requirements for safety-critical applications [21]. In recent years, 5G cellular communication has opened up new opportunities for even lower latency applications such as sensor information sharing, trajectory sharing, and coordinated driving [22]. However, there is still scope for traditional 4G/LTE communication in nonsafety critical applications such as energy-saving at intersections [23]. The communication requirements for reservation, dispatching, routing, and payment have similar flexibility.

Telematics systems refer to installed tracking devices in the vehicles that transmit, receive, and store telemetry data. Usually, the tracking device is connected to the vehicle onboard computer through the onboard diagnostics (OBDII) port and is connected to the cellular network through a built-in SIM card. The system combines GNSS locations and other relevant in-vehicle sensor information, and sends it as general packet radio service (GPRS), 4G mobile, or satellite signal. The server receiving the data makes decisions and displays it at the client interfaces. In an automated vehicle environment, telematics will be an in-built feature of the vehicle hardware. Sensors used for perception such as radar, LIDAR, and infrared cameras will generate part of the data sent via telemetry [24]. Most of this detailed and high-volume information will be consumed locally for low-level operations. Therefore, DSRC and 5G communications will be most suited for these applications. Broadly, telematics functionalities in helping automated driving can be classified into four categories [25]:

1. Perception—sharing high throughput-sensor data and real-world models;
2. Trajectory planning—intention and trajectory sharing for faster and safer maneuvers;
3. Real-time local updates—real-time sharing of local data with infrastructure and other vehicles such as high-resolution 3D maps; and
4. Coordinated driving—intention and sensor data sharing for more predictable and smoother driving.

Request aggregators: The request aggregators are interim services and infrastructure that connect to the front-facing applications on one side and the backend infrastructure on another. These services receive requests from client applications, filter suspicious requests, and assign requests to the appropriate servers and databases.

Filtering the requests coming from client applications is performed through web application firewalls (WAF). Over-reliance on shared information to make critical maneuvering and routing decisions may induce additional vulnerabilities from cyberattacks. Kong *et al.* [26] presented an attack tree analysis approach to analyze threats and vulnerabilities for various cyberattacks in a vehicular ad hoc network (VANET). For a shared automated vehicle fleet, spoofing and jamming of GNSS signals may cause a breakdown of the dispatching and routing process. Payment information and personal information of the system users are especially vulnerable to cyberattacks. Fleet operators usually adopt stringent communication and database security protocols to encrypt some of this sensitive information. Even with a deidentified and anonymized database, risks of inference attacks remain.

Load balancers are hardware or software assigned at the entry points of the web servers. There are several load balancing algorithms available depending on the needs and types of the service such as the round-robin method, least connection method, least bandwidth method, least response time method, and hashing method.

The various purposes for load balancing include distributing the service request to the appropriate server, detecting and monitoring backend server health, and redirecting service requests to backup servers in the event of failure. Load balancing will be one of the most critical features in SAV fleet operation as automated vehicles will generate high-volume and high-frequency data; therefore, the possibility of failure at the backend infrastructure will be much higher.

Database: Databases store historic information about the trips, users, and drivers. Also, databases store automatic logs from client interfaces, administrative dashboards, and interactions among different services within the backend infrastructure. Historic data are used to debug different applications, train artificial intelligence models, and optimize different operations. In an automated vehicle environment, database system response time needs to be reduced and the available disk storage space needs to be increased to accommodate various advanced applications such as teleoperations, scenario simulation, and network optimization. Large-scale TNC operators have moved to nonrelational databases for applications requiring speed and performance. For complex data and analytics, operators are moving from a single central relational database management system (RDBMS) to distributed computing and storage using technologies such as Hadoop Distributed File System (HDFS).

Servers: Main web servers compute the dispatching, ride-matching, and routing algorithms. Usually, communication between client applications with the server is maintained through web sockets. Web sockets are the preferred method of connections for asynchronous event-based computation. In the web socket connection, client applications subscribe to the server and when a trigger event happens such as the driver accepts a ride or the passenger unlocks a vehicle, a real-time notification is sent among all the subscribed client applications. Generally, servers are managed as data centers. Shared mobility service providers arrange data centers according to the size of their operations and geographical proximity. To increase operational reliability, some fleet operators use a hybrid cloud model where the entire operation is distributed among multiple cloud providers and multiple active data centers.

Some applications within the shared mobility service require real-time, fault-tolerant, and streaming processing. For example, the calculation of the estimated time of arrival (ETA) involves the use of real-time streaming data.

13.3 Reservation and access

Reservations are one of the core functionalities of shared mobility. It is often the first step in the overall process of a shared mobility trip. Generally, a transportation service reservation method includes the following components [27]:

1. Receiving a ride request specifying an origin and a destination with or without a preferred arrival time;
2. Generating a list of possible combinations of vehicles and service types that is capable of meeting the ride request;

3. Computing the probability of choosing a service or combination of service by the rider, followed by computing the expected number of ride requests in the future;
4. Selecting an assortment of services to be displayed to the user requesting the service based on choice probability and future demand; and
5. Transmitting and displaying the options at client interfaces, followed by capturing the final selection from the user and reserving the service accordingly.

The above approach is particularly suited for E-hailing, whereby a smartphone app or networked computer interface is available to complete the reservation. Reservation can be done in advance through a telephone-based dispatching system or a website or app. In contrast, mobility-on-demand (MOD) systems allow for no advanced notification from the users [28]. The most primitive approach for a shared mobility reservation system is a street hail where users can form sharing by raising hands on the street, standing at a taxi stand, or a specified loading zone.

With MOD and interactive client interfaces, users now have the option to make a more informed decision. Traditionally, travel time and user costs have been the two most important aspects of a trip for making reservations. However, an increasing number of users are now concerned with environment and negative externalities associated with their travel. As a result, fleet operators are providing more flexible service options and eco-friendly choices. For example, Incentrip is an app used in the Washington DC-Baltimore area that displays energy consumption and emissions associated with different trip choices [29]. Green Mobility, a one-way electric car-sharing service in Demark, shows available cars to the users along with information about vehicle battery status.

Accessing the vehicle or the service is the next logical step after reservation. In a service paradigm with human drivers and operators, verification can be made manually. Barth *et al.* [30] mentioned five methods for shared vehicle access: (1) lockbox, (2) common key, (3) common smart card, (4) personalized smart card, and (5) personalized smart card with PIN confirmation.

13.3.1   *Potential changes due to vehicle automation*

Automated vehicles have much more flexibility and availability compared to human driven vehicles. This added flexibility translates to potential new options for operations. Dandl and Beogenberger [31] presented two different operational paradigms with SAVs: automated car-sharing (aCS) and automated taxis (aTaxi). In the aCS approach, users have the option to drive manually by overriding the vehicle automation, withhold the information of the destination, have more flexibility in choosing routes, stops, and duration of the reservation. In contrast, in the aTaxi approach, the fleet operator will have more control of the vehicle operation and requires the users to specify a fixed origin and destination.

Vehicle automation also increases the precision of the vehicle time of arrival at the destination and other facilities along the route. This enhanced level of precision allows for the reservation of facilities in exchange for priority or security of service. Mashayekhi and List [32] proposed a multiagent auction-based system at signalized intersections that can prioritize the highest bidders to use the green time. Similar prioritization approaches have been proposed for parking [33], toll lanes [34], and roundabouts [35]. The automated vehicle can decide to reserve the facilities based on users' preference for flexibility and willingness to pay. In addition, automated vehicles allow users to input a daily schedule for pick-ups and drop-offs for the nondrivers. The reservation system will assign an appropriate vehicle at the desired time and location according to the needs and preferences of the users.

The reservation mechanism with automated vehicles offers the chance to address some system-level concerns. The use of dynamic pricing for SAV loading and unloading activities has been proposed to ensure the serviceability of the curbside [36]. Advanced reservation for an automated vehicle fleet has been proposed as a means for demand management [37].

The process for accessing the vehicle fundamentally changes with no drivers on board. In addition to smart cards and software keys using near-field-communication (NFC) protocols, riders can be authenticated using a biometric vehicle access systems such as fingerprint readers and eye scanners. Another fundamental difference to the access protocol with automated vehicles is that the vehicle can drive to a preferred stop specified by the users.

13.4   Dispatching and ride matching

Dispatching is the process of assigning a specific vehicle to the travelers. In the reservation step, the ride requests are mostly specifying different trip requirements in a broad sense, such as the number of passengers, type of vehicle (cargo, luxury, etc.), expected time of arrival, destination, preference for ridesharing, child-seats, and wheel chair accessibility. Based on these requirements, a human dispatcher or a computerized dispatching system assigns a specific vehicle and the corresponding driver to complete a single trip or a sequence of trips. Cummings *et al.* [38] listed seven main functions of a human dispatcher: communication, resource allocation, navigation support, contingency management, monitoring, logging, and training.

The decision variables involved in a dispatching process include (1) the number of vehicles and size of the vehicles to operationalize during a given time; (2) the distance of the vehicle from the request origin location; (3) whether the vehicle is completing another trip; (4) the availability of the driver during the entire duration of the trip; (5) the willingness of the driver to make a trip to the destination; and (6) the availability of empty seats. Dispatching is usually a more efficient and cost-effective process for in-advance requests compared to on-demand services [39].

Conventional vehicle dispatching problems are treated as vehicle routing problem with pick-up and delivery (VRPPD) or vehicle routing problem with time windows (VRPTW). Jung *et al.* [40] added three considerations for the dynamic shared-taxi dispatch problem: (1) the complete vehicle schedule may not be known at the beginning of the trip; for example, schedules may need to be revised in real-time with updated passenger requests, traffic information, and a potential vehicle breakdown; (2) fast response time, and (3) change in the objective function to minimize response time. Additionally, three types of taxi dispatch algorithms have been tested by Jung *et al.* [40]: (1) nearest vehicle dispatch (NVD); (2) insertion heuristics (IS); and (3) hybrid simulated annealing (HSA). The performance of the algorithms depends on the objective functions used in optimization. HSA significantly reduced waiting time for passenger cost minimization framework, whereas HSA rejected many trip requests for a maximized profit.

In addition to matching riders with drivers, dispatchers also need to match multiple riders in a ridesharing system. Wen *et al.* [39] found such sharing brings down the average trip costs by reducing the fleet size; however, adversely affect travel experiences such as higher detour distance, longer travel time, and higher uncertainty in travel time.

13.4.1 *Potential changes due to vehicle automation*

Automated vehicles are expected to be operated within an automated dispatching system in which trip requests and schedules will be collected and sorted by the server and vehicles will be dynamically allocated. However, Cummings *et al.* [38] predicted three new automated vehicle supervisory control and monitoring functions, as follows:

1. *Remote control or teleoperation:* At the early stages of automated vehicle development, situations might frequently arise where human oversight might be required. For example, Nissan designed a *Seamless Autonomous Mobility* system that included a call center with human operators in case the vehicle is deemed inoperable [41]. Teleoperation is similar to "Drive-by-Wire" in which remote operators take full control of the vehicle in the event of an emergency.
2. *Communication with passengers:* In the absence of a human driver, providing instructions and information to the passengers during unexpected situations.
3. *Fleet management:* Communicating important weather and roadway information to a fleet of vehicles at once. Redirecting vehicles in case of emergencies such as bridge collapse, road flooding, may be necessary.

Madrigal [42] mentioned three different dispatching functions for Waymo's robo-taxi operation in Phoenix, AZ: traditional dispatch, fleet response, and rider support.

Another future application area for SAV dispatching will be mode selection of fleet operation—such as single occupancy mode, ridesharing mode, and door-to-door mode. Enoch [43] envisioned that buses, cars, and conventional taxis would converge to a universal automated taxi system. In such a system, the fleet composition will be needed to be adjusted to accommodate more trips during peak hours and more rider flexibility during the off-peak hours.

Automated vehicles allow for various operational paradigms for ride-matching. Zhu *et al.* [44] modeled two different operations for SAV in the traffic microsimulation software called SUMO: on-demand door-to-door ridesharing, and on-demand fixed route-ridesharing. Moreover, automated vehicles allow for additional customization of the in-vehicle features such as music, air conditioning temperature, and color themes to match fellow riders' preferences. Zhang and Zhao [45] proposed such ride-matching based on preference implementing maximal cardinality matching instead of the traditional approach for maximal system efficiency.

13.5   Routing and fleet monitoring

In a typical routing application, the road network is modeled as a graph with nodes representing intersections, edges representing road segments, and edge weights representing distance, travel time, or other costs. A range of routing algorithms are available to solve the problem of routing the vehicles on the network; the simplest of them being the Dijkstra's algorithm. However, current large-scale shared mobility operations typically deal with thousands of vehicles at one time and therefore requires faster methods and algorithms. One major bottleneck for faster routing method is updating edge weights representing estimated travel time under current traffic conditions. Algorithms that use heuristics such as A* [46] are used commonly to complete the computation of routing within a reasonable timeframe.

Initiation of ride-matching and routing requires knowledge of the state of all the vehicles in a fleet—location of the vehicles, the status of the vehicles (idle, en-route, out of service), number of occupants, etc. This information is usually gathered from vehicle telematics. Fleet monitoring systems, in general, collect four types of information: (1) real-time location, (2) driver behavior metrics, (3) vehicle health data, and (4) in-vehicle, and vehicle local environment data [47,48]. Freight operators use fleet monitoring to provide feedback to drivers that improve energy efficiency and safety of operation.

13.5.1   Operations change with automated vehicles

The advent of SAVs allows a feasible path to operationalizing system optimal routing applications. Liang *et al.* [49] proposed a profit maximization algorithm for routing automated taxi trips using dynamic travel times. Liang [50] also formulated optimization schemes for automated taxi routing with real-time demand with efficient algorithms. Horvath *et al.* [51] devised a routing method for automated vehicles that utilize predefined stops dynamically for more demand-responsive time.

Monitoring is not optional in the automated vehicle era as the human component is almost removed from the control of the vehicle. This continuous monitoring will cause a high volume of vehicle and sensor data to be sent to the centralized servers for processing. Additional operational measures need to be taken to ensure the appropriate level of data filtering and extraction of key performance indicators.

Telematics will be an important component for troubleshooting automated driving algorithms and gathering training data for offline simulation software. Furthermore, telematics can help in predictive maintenance of the vehicle including monitoring the need for charging in an electric SAV fleet.

13.6 Payment and pricing

Shared mobility payment models can be broadly categorized into two models—membership-based, on-demand, or some combination. Membership payments are collected generally as prepaid subscription fees. Conversely, on-demand payment models establish a new transaction each time a user accesses the service. The on-demand or "pay-as-you-go" model is quite popular with the flexibility of choosing different services and mobile payment services.

Integrated, seamless, and transparent payment service is essential for modern shared mobility services. The overall payment process, although a crucial component for the sustainability of the service, is often regarded as a barrier for new customer entry and existing customer retention. Users are deterred in using a service if there are multiple steps in completing a transaction. Most app-based services use payment gateways that accept multiple payment methods such as credit cards, debit cards, and bank transfers. These gateways are called payment service providers (PSP). Usually, PSPs collect fees as a percentage of each transaction or a fixed cost per transaction. The service providers need to follow payment card industry data security standards (PCI DSS) guidelines to ensure the security of user information.

Pricing for per unit service depends on the revenue generation and average cost as delineated in Section 13.1. There are three accepted models for revenue generation for shared mobility companies: user charge, public subsidies, and private subsidies. The user charge is levied either through direct payment from customers or by indirect revenue generated by selling users a complimentary product (often in the form of advertisement). Public subsidies are common for transits; however, vanpools and carpools obtain subsidies from the transportation agencies to help manage the demand. In recent years, public–private partnerships (PPP) have emerged to provide integrated solutions. For example, the City of Monrovia, CA recently launched a partnership with Lyft that will allow users connecting to the LA Metro Gold Line discounted price of $0.50 per ride. Similar PPP arrangements have been made to improve paratransit service coverage and first-and-last mile transit access. Private subsidies allow shared mobility services to stay competitive and capture a greater market share.

13.6.1 Operations change with automated vehicles

Pricing has been used as a tool to manage demand and assure service availability [52]. With automated vehicles, the need for pricing to adjust the demand to the level of available supply is even greater as fleet owners will try to minimize SAV fleet size (due to the high acquisition cost of vehicles). Chen and Kockelman [53] simulated four pricing strategies to manage an electric SAV fleet: distance-based pricing, origin-based pricing, destination-based pricing, and combination pricing. They found that the SAV business model struggles to recoup capital if the pricing is too low and a large proportion of low value of time (VOTT) travelers cannot afford the service if the pricing is too high. To balance profitability and service completeness, Chen and Kockelman [53] suggested two classes of service—refined, work enhancing environments at higher fares for high VOTT users; and discounted, sufficiently basic service for low VOTT users.

The choice of vehicle powertrain technology in the SAV fleet is a determining factor in pricing. Loeb and Kockelman [54] compared gasoline hybrid-electric vehicles (HEV) with battery electric vehicles (BEV) for an SAV-based solution in Austin, TX. The HEV fleet was found more profitable than the BEV fleet up until the cost of gasoline exceeds $10 per gallon or the cost of BEV fell under $16,000. Such an unfavorable business proposition for BEVs necessitates public or private subsidies given the long-term environmental benefits of BEVs [55].

In addition to dynamic pricing or surge pricing to manage demand, cities around the world are considering congestion pricing in the form of cordon pricing or other geofencing methods. The automated vehicle technology is highly suited for such pricing. Most likely, the payment interface will be integrated into the bio-metric verification system in automated vehicles. Any toll, parking, and congestion pricing can be processed in the backend and a combined fee can be charged to the users. Predictive models can be developed to inform travelers of the expected price of a service before the trip starts. Furthermore, an integrated payment system allows for the in-vehicle purchase of different amenities such as music, use of the Internet, and e-commerce sites.

13.7   Performance optimization

Most auxiliary functionalities are related to the performance optimization step. To optimize a shared mobility business, operators need to identify the key performance indicators (KPI). Large-scale mobility service providers create their analytics pipeline to get the user data, process and contextualize information, and monitor their KPIs consistently. New services specifically need to monitor two sets of KPIs: (1) demand and supply: how many potential users are available at a given time, how many rides are requested, how many get fulfilled, and where are the rides requested; (2) utilization and revenue: driver wait time, deadheading or empty time, occupancy ratio, fees collected, and discounts and rebates.

Another avenue to improve service performance is through regular subjective and objective feedback from the users, drivers, and operators. Most TNC applications have driver and rider rating services. However, the subjective basis for these ratings may generate discrimination against racial minorities [56].

Automated vehicles permit a higher level of flexibility to plan and operationalize sophisticated performance optimization. The fleet size and the composition of vehicles in the fleet can be adjusted at any given time without repercussions of driver unrest and human rights violation. Centralized control also allows fleet operators to coordinate with the agency transportation management centers to manage demand, change scheduling and routing policy to comply with regulation and optimize revenue, and manage the charging scheme for electric vehicles to avoid peak unit electricity costs. Figure 13.4 shows a multiobjective management scheme for the SAV fleet according to a typical city demand pattern. Although such

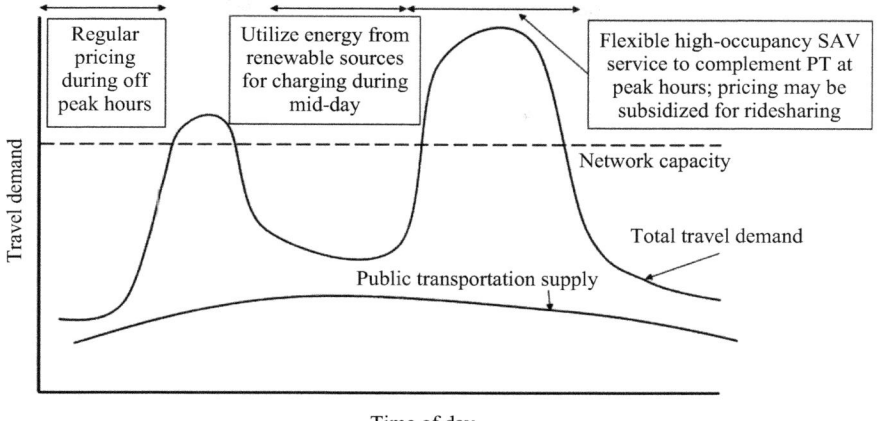

Figure 13.4   *Multiobjective optimization scheme for shared and automated vehicle (SAV) fleet management*

comprehensive optimization may seem utopic, a higher level of transparency among public and private entities enables many mutually beneficial applications.

13.8   Vehicle repositioning

Vehicle repositioning, also referred to as vehicle rebalancing or vehicle relocation, is an important aspect of shared mobility system operations and management as it can significantly affect the level of service and the operational efficiency of the system. This is especially true for one-way shared mobility systems that, while more attractive for users than round-trip systems, may lead to an imbalance between user demand and vehicle supply. In station-based one-way carsharing or bikesharing systems, an insufficient number of vehicles at a station can lead to a degraded level of service and lost revenues. On the other hand, an excess of vehicles at a station can cause underutilization of assets and result in parking issues if there are not enough parking spaces at the station [57].

In general, vehicle repositioning strategies aim to match the spatial distribution of supply (vehicles, bicycles, scooters, etc.) at any given time with that of demand (riders, users, etc.) to ensure that the demand can be served without significant delay or wait time and that the supply or asset is well utilized. Vehicle repositioning strategies can be designed to achieve this goal efficiently by taking advantage of historical data on the spatial pattern of user demand to predict where and how much vehicle supply would be needed at a given time.

13.8.1   Nonautomated vehicles

As shown in Figure 13.5, vehicle repositioning strategies in shared mobility systems operated with nonautomated vehicles depend largely on the business model of

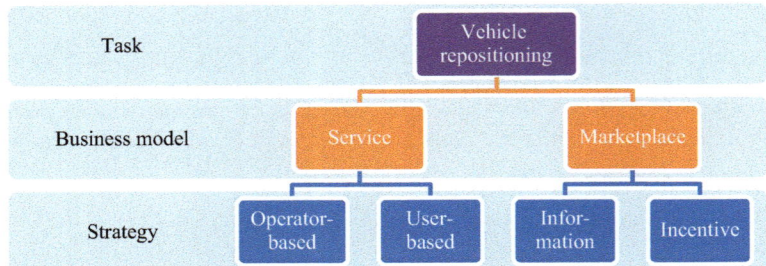

Figure 13.5   Strategies for vehicle repositioning under different business models of shared mobility systems with nonautomated vehicles

the system, especially as related to whether the system operator has control over its vehicle supply. In the market place model where system operators act as a facilitator of the transactions between individual buyers (e.g., riders) and sellers (e.g., drivers) [58], vehicle repositioning cannot be controlled or enforced by the operators. Instead, the operators may provide information and/or incentives to encourage sellers to relocate to where there is higher buyer demand. For example, Uber's Driver app provides a heat map of demand for rides in the city, indicating areas with low, high, and surging demand [59]. In addition, the app may occasionally alert the drivers to a High Demand Area nearby. Drivers can earn bonus payments for providing rides in high-demand areas, which incentivizes drivers looking for rides to relocate to those areas. The effectiveness of information and incentives in relocating vehicle supply according to the rider demand depends on the incentive structure and the behavior of drivers in response to the information and incentives.

In contrast, system operators in the service model own and operate the vehicle supply themselves [58], and thus, can decide how the vehicles in the system should be distributed at the beginning of the day and rebalanced throughout the day. The task of repositioning the vehicles can be performed by staff employees of the operator or by the system users.

Common techniques for operator-based vehicle repositioning include towing and ridesharing, either among employees performing the vehicle repositioning or with system users [60]. The vehicle repositioning task is often formulated as a mixed-integer programming optimization problem to determine the number of staff employees needed to perform the necessary vehicle relocations in a least-cost manner while maintaining a certain level of service [61,62]. When multiple staff employees or rebalancing drivers are used to relocating vehicles, their distribution across the stations may become imbalanced as well. To address this problem, a joint optimization framework for balancing both vehicles and rebalancing drivers at the same time has been proposed [63,64]. In such a framework, a user may be driven to her destination station by a rebalancing driver if the system needs to move one or more rebalancing drivers to that station.

Studies have shown the effectiveness of the optimization approach through simulation. In practice, the number of vehicles that can be realistically relocated by staff employees in a period is limited, depending on the size of the service area, transportation mode for returning to the origin station (e.g., car, scooter, or public transit), and traffic condition, among others. For example, in Munich, Germany, it was measured to be 0.7 to 1.1 relocations per hour per staff employee [65]. Santos and Correia [66] estimated this number to be 2 to 3 in Lisbon, Portugal. At this rate, they found that vehicle relocations performed by staff provide only a small improvement in the revenues, which may not be sufficient to overcome the costs of the staff itself.

For user-based vehicle relocation, examples include trip joining and trip splitting [60]. Trip joining occurs when multiple users who plan to travel from a station with low vehicle supply to a station with high vehicle supply share a ride in a single vehicle. This reduces the number of vehicles that will need to be relocated back to the low-supply station. Conversely, trip splitting can be utilized when a group of users intends to share a ride from a station with a high vehicle supply to a station with a low vehicle supply. The system operator can ask them to drive separate vehicles to help balance the number of vehicles in the system. Simulation studies have demonstrated the effectiveness of user-based vehicle relocation techniques [60,67], but these studies assume 100% user participation. However, there are barriers to the implementation of these techniques in practice. For example, trip splitting would only work when there are enough trips made by groups of ridesharing users rather than by a single driver [68]. In the case of trip joining, users may not be willing to share rides with strangers due to safety and privacy concerns [3].

To encourage users to accept user-based vehicle repositioning requests, a system operator may offer the users incentives such as discounted fares as necessary. The operator can extend such an approach by pricing trips differently depending on the number of vehicles at the origin and destination stations to keep the vehicle stock at the different stations in balance [69]. In free-floating one-way shared mobility systems where vehicles can be returned anywhere within a geographic area [58], the idea of using incentives to encourage users to help relocate vehicles can also be applied to destination choice where the system operator may offer incentives for users to end their trips in zones with a shortage of vehicles rather than the intended destination zones [70].

In contrast to the use of incentives, it has also been proposed that the system operator may selectively decline trip reservations that would contribute to an excessive imbalance of the vehicle supply in the network [71,72]. While this approach is effective in maintaining the balance of vehicle supply in the network, the users whose trip reservations are declined would experience the same rejection as when there are not enough vehicles to serve user demand, which could lead to user dissatisfaction with the system in the long run.

13.8.2 Automated vehicles

While the current shared mobility systems that utilize nonautomated vehicles operate in either the market place model or the service model, shared mobility systems of the future that utilize automated vehicles would be more likely to operate in the service model. Over the last decade, several companies have been working on various forms of Automated Mobility on Demand (AMoD) or Robo-Taxi system, and a number of field operational tests have been initiated since 2016 [73]. The operators of these AMoD systems would have full control of the fleet management scheme, allowing them to optimize system operations to achieve maximum efficiency [74].

Vehicle repositioning will still be a major task in AMoD system operations as long as there is spatiotemporal variation in user demand. Some of the operator-based and user-based vehicle repositioning techniques developed for shared mobility systems utilizing nonautomated vehicles could potentially be adapted for use in AMoD systems [75]. On the other hand, other techniques such as towing and using rebalancing drivers to move vehicles would no longer be relevant. The ability of shared automated vehicles to relocate themselves will reduce the costs associated with the vehicle repositioning task, and allow system operators to rely less on user-based vehicle repositioning techniques to make the systems economically viable. Thus, the operators can focus on devising strategies for relocating the vehicle supply to meet the user demand in a centralized fashion. Example strategies include relocating vehicles based on expected user demand and relocating vehicles to balance vehicle stock based on predicted supply [74]. These strategies can be formulated as a mixed-integer programming optimization problem. Since there will be no need for rebalancing drivers, the objective could then be to minimize the number of rebalancing trips or the total travel distance, travel time, or monetary cost of the vehicle repositioning operation.

The effectiveness of vehicle repositioning strategies in an AMoD system depends on many factors such as fleet size, user demand pattern, parking constraint, and traffic condition, among others. On the other hand, vehicle repositioning strategies can have an influence on those factors as well. For instance, Zhu and Kornhauser [76] investigated the effects of vehicle repositioning strategies on the required fleet size and the level of service of an AMoD system in New Jersey, assuming that all nonwalking travel demand is served by the AMoD system. They found that with local vehicle repositioning, all of the travel demand can be served by the AMoD system with a fleet size of about half the existing number of vehicles operating in New Jersey. That fleet size would result in 8% of the total vehicle miles traveled attributable to the repositioning of empty vehicles.

It is expected that the vehicles used in AMoD systems would be electric vehicles (EVs). During the time that EVs still have a shorter driving range and take longer time to refuel than their gasoline or diesel counterparts, EV charging requirements need to be considered in all aspects of AMoD system operations including vehicle repositioning [77]. For example, there may be a need for an EV to be relocated specifically for the purpose of getting charged. And the decision on where an excess EV at one station should be relocated to may depend on not only the user demand at the destination station but also whether the destination station has EV chargers available.

13.9   Vehicle refueling and maintenance

Vehicle refueling and maintenance are other key aspects of shared mobility system operations and management. They are necessary tasks that need to be included as part of system operation routines. During refueling and maintenance, the vehicles are pulled from service and cannot be used to serve the user demand. Thus, consideration needs to be given to when, where, and how refueling and maintenance activities should be performed so that they have minimal impacts on other aspects of system operations. The discussion in this section will focus on refueling activities as they occur more frequently than maintenance activities.

One approach is for the system operators to refuel the vehicles at the end of the day or when the user demand is usually lowest. Station-based system operators with a large number of vehicles may set up their refueling pumps at one or more stations. They may also use mobile refueling services where a refueling truck carries a large amount of fuel to the station and refuel most or all of the vehicles at that station. However, these refueling techniques may not be cost-effective for small station-based system operators and operators of free-floating one-way systems. Another refueling approach is to rely on the system users to perform the refueling task. The operators may leave a fuel card in each vehicle so that the users can refuel the vehicles if needed. Instead of using a fuel card, the operators can also choose to reimburse the cost of fuel to the users after they refuel the vehicles. As in the case of vehicle repositioning, incentives may be offered to encourage this type of user-based refueling technique.

If a shared mobility system is operated with EVs, then the refueling or charging of these EVs becomes more complex and requires extra consideration. For station-based systems, it is intuitive that EV charging equipment should be placed at the system stations. But the system operators need to decide how many chargers should be installed at each station. It would be convenient to have a dedicated charger for each parking space at the stations so that the EVs can be charged every time they are not in use. However, this approach will incur a large amount of charging infrastructure costs and, depending on the usage pattern of the EVs, may not be economically justified.

In a free-floating shared mobility system, the placement of EV charging stations needs to be carefully designed concerning the transportation network the EVs will operate in and the spatial pattern of the travel demand they will serve [78]. Also, the type of EV chargers (e.g., Level 1, Level 2, or DC fast charge) and the driving range of EVs are important variables for determining the fleet size required and the economic competitiveness of the system [79]. In addition, as the share of EVs in the vehicle fleet grows, the scope of EV charging management in shared mobility systems should be expanded to include consideration of its interaction with the electricity grid [80].

13.10   Summary

Operations and management of a shared automated vehicle (SAV) fleet involve multiple stakeholders. On the one side, fleet operators want to maximize profits by minimizing costs and increasing the user base. On the other side, the city managers and the government need to ensure that overall transportation demands are met in an equitable and environmentally sustainable manner. Further, customer needs also put specific constraints on the overall operation.

In this chapter, we have analyzed the fundamentals for operationalizing new functionalities in a shared automated vehicle fleet. Most anticipated operational changes due to vehicle automation are incremental and designed to increase the cost-effectiveness of the new mobility solution being marketed. Although academics and futurists propose many utopic solutions with an SAV fleet, the path for integrating these solutions in practical operation splits in two ways: by reducing the average cost of new services or, by subsiding the operations to optimize some societal objectives.

At the minimum, a viable SAV fleet needs to satisfy the core functionalities of a sharing system. Most changes in the SAV fleet operations are transformative, yet straightforward to interpret—transition from driver to algorithmic decision making, availability of more on-board sensors, superior computing and communication capabilities, among others. However, with enhanced capabilities and flexibilities come concerns of unintended consequences such as induced demands and risk of cyberattacks. Thus, the overarching operational theme for an SAV fleet will not be a discrete one, rather integrated with other stakeholders in the mobility ecosystem.

Key operational and management changes in shared mobility operations due to automation are listed below:

- *System components*: No driver in the vehicles means no need for a driver interface. Operational decisions will be communicated directly with the onboard computer. SAVs allow more on-board, on-the-edge computing capabilities permitting a greater degree of decentralized decision making.
- *Reservation*: Users of an SAV fleet will be able to reserve single trips or according to a trip schedule. Also, road infrastructures such as parking spots and toll roads can be reserved in the same system.
- *Access*: Biometric user authentication and personalization will be more prevalent in an SAV fleet.
- *Dispatching*: In the early days of SAV, manual dispatchers will be needed to monitor fleet operation in case of emergencies and unexpected situations. Unless these services are provided by the OEM, this overhead cost will be a significant burden on the fleet operator.
- *Ride matching*: More preference-based ride matching options will be available in an SAV fleet along with a higher level of vehicle personalization.
- *Routing*: Unlike humans, SAVs can be programmed to comply with routing instructions; hence, different routing and operational schemes can be implemented in an SAV fleet.

- *Fleet monitoring*: Predictive maintenance of vehicles will be more common.
- *Payment*: Mobility-as-a-Service (MaaS) and AV paradigms will need to converge, resulting in an integrated payment method.

Author contributions

The authors confirm contribution to this chapter. All authors reviewed and approved the manuscript.

Declaration of conflicting interests

The authors declared no potential conflict of interest with respect to publication of this chapter.

References

[1] Shaheen, S., Chan, N., Bansal, A., and Cohen, A. "Shared mobility: Definitions, industry development, and early understanding." 2015. URL: http://innovativemobility.org/wp-content/uploads/2015/11/SharedMobility_ WhitePaper_FINAL.pdf.
[2] SAE International. "On-Road Automated Driving Committee." *SAE J3016. Taxonomy and Definitions for Terms Related to Driving Automation Systems for On-Road Motor Vehicles*. tech. rep., SAE International, 2016.
[3] Chan, N.D. and Shaheen, S.A. "Ridesharing in North America: Past, present, and future." *Transport Reviews* 32.1; 2012. 93–112.
[4] Rain Magazine. The CarSharing Handbook (Part 1). Archived from the original on 20 July 2007. Retrieved 5 July 2020.
[5] Berglund, E. Z., Monroe, J. G., Ahmed, I., *et al.* "Smart Infrastructure: A Vision for the Role of the Civil Engineering Profession in Smart Cities." *Journal of Infrastructure Systems*, 26(2); 2020. 03120001.
[6] Uber. "When and where are the most riders?" 2020. https://help.uber.com/ driving-and-delivering/article/when-and-where-are-the-most-riders? nodeId=456fcc51-39ad-4b7d-999d-6c78c3a388bf, accessed April 2020.
[7] Thomas, M. and Deepti, T. "Reinventing carsharing as a modern and profitable service." 2018. URL: https://ridecell.com/wp-content/uploads/White-Paper-Presentation_Reinventing-Carsharing-As-A-Modern-And-Profitable-Serv ice.pdf.
[8] Gurumurthy, K.M. and Kockelman, K.M. "Analyzing the dynamic ride-sharing potential for shared autonomous vehicle fleets using cellphone data from Orlando, Florida." *Computers, Environment and Urban Systems* 71; 2018. 177–185.
[9] Stocker, A. and Shaheen, A. "Shared automated vehicle (SAV) pilots and automated vehicle policy in the US: Current and future developments." *Road Vehicle Automation 5*. Springer, Cham. 2019. 131–147.

[10]   Alama, Md. J. and Habib, M.A. "Investigation of the impacts of shared autonomous vehicle operation in Halifax, Canada using a dynamic traffic microsimulation model." *Procedia Computer Science*; 2018. 130: 496–503.

[11]   Zhang, W. and Guhathakurta, S., "Residential location choice in the era of shared autonomous vehicles." *Journal of Planning Education and Research*; 2018: 0739456X18776062.

[12]   DMV. 2019. https://www.dmv.ca.gov/portal/uploads/2020/06/Adopted-Regula tory-Text-2019.pdf

[13]   Fagnant, Daniel J., and Kara M. Kockelman. *Dynamic ride-sharing and optimal fleet sizing for a system of shared autonomous vehicles*. No. 15-1962. 2015.

[14]   Bösch, P. M., Becker, F., Becker, H., and Axhausen, K. W. "Cost-based analysis of autonomous mobility services." *Transport Policy*. 2018; 64: 76–91.

[15]   Stephens, T. S., Gonder, J., Chen, Y., Lin, Z., Liu, C., and Gohlke, D. *Estimated bounds and important factors for fuel use and consumer costs of connected and automated vehicles* (No. NREL/TP-5400-67216). 2016. National Renewable Energy Lab.(NREL), Golden, CO (United States).

[16]   Cramer, J., Krueger, A. B. "Disruptive change in the taxi business: The case of Uber." *American Economic Review*, 106(5); 2016. 177–182.

[17]   Li, Y. and Voege, T. "Mobility as a service (MaaS): Challenges of implementation and policy required." *Journal of Transportation Technologies*, 7(2); 2017. 95–106.

[18]   Shaheen, S. and Cohen, A. "Mobility on demand (MOD) and mobility as a service (MaaS): Early understanding of shared mobility impacts and public transit partnerships." *Demand for Emerging Transportation Systems*. 2020; 37–59.

[19]   Hyland, M. F. and Mahmassani, H. S. Taxonomy of shared autonomous vehicle fleet management problems to inform future transportation mobility. *Transportation Research Record*, 2653(1). 2017. 26–34.

[20]   Dong, C., Wang, H., Li, Y., Liu, Y. and Chen, Q. "Economic comparison between vehicle-to-vehicle (V2V) and vehicle-to-infrastructure (V2I) at freeway on-ramps based on microscopic simulations." *IET Intelligent Transport Systems*, 13(11). 2019. 1726–1735.

[21]   Kenney, J. B. "Dedicated short-range communications (DSRC) standards in the United States." *Proceedings of the IEEE*, 99(7). 2011. 1162–1182.

[22]   Shah, S. A. A., Ahmed, E., Imran, M. and Zeadally, S. "5G for vehicular communications." *IEEE Communications Magazine*, 56(1). 2018. 111–117.

[23]   Hao, P., Wu, G., Boriboonsomsin, K. and Barth, M. J. "Eco-approach and departure (EAD) application for actuated signals in real-world traffic." *IEEE Transactions on Intelligent Transportation Systems*, 20(1). 2018. 30–40.

[24]   Avr, A., Tanvir, S., Rouphail, N. M. and Gupta, R. "Capturing vehicular space headway using low-cost LIDAR and processing through ARIMA prediction modeling." arXiv preprint arXiv:1907. 2019. 11648.

[25]   Misener, J. "Smart Transportation." 2020. Accessed from https://www.qualcomm.com/media/documents/files/smart-transportation-presentation.pdf. Accessed on September 5; 2020.

[26] Kong, H. K., Hong, M. K. and Kim, T. S. "Security risk assessment framework for smart car using the attack tree analysis." *Journal of Ambient Intelligence and Humanized Computing*, 9(3). 2018. 531–551.

[27] Ikeda, T., Ben-Akiva, M. E. and Atasoy, B. *U.S. Patent No. 10,628,758.* 2020. Washington, DC: U.S. Patent and Trademark Office.

[28] Mitchell, W. J., Borroni-Bird, C. E. and Burns, L. D. *Reinventing the automobile: Personal urban mobility for the 21st century.* 2010. MIT press.

[29] Mohan, S., Yan, F., Bellotti, V., Elbery, A., Rakha, H. and Klenk, M. "On Influencing Individual Behavior for Reducing Transportation Energy Expenditure in a Large Population." In *Proceedings of the 2019 AAAI/ACM Conference on AI, Ethics, and Society.* 2019. pp. 461–467.

[30] Barth, M., Todd, M., and Shaheen, S. "Intelligent transportation technology elements and operational methodologies for shared-use vehicle systems." *Transportation research record*, 1841(1); 2003. 99–108.

[31] Dandl, F. and Bogenberger, K. "Booking processes in autonomous carsharing and taxi systems." In *Proceedings of 7th Transport Research Arena, Vienna.* April 2018.

[32] Mashayekhi, M. and List, G. "A multiagent auction-based approach for modeling of signalized intersections." *IJCAI Workshops on Synergies Between Multiagent Systems, Machine Learning and Complex Systems.* July 2015. pp. 13–24.

[33] Huang, C., Lu, R., Lin, X., and Shen, X. "Secure automated valet parking: A privacy-preserving reservation scheme for autonomous vehicles." *IEEE Transactions on Vehicular Technology*, 67(11). 2018. 11169–11180.

[34] Liu, Z. and Song, Z. "Strategic planning of dedicated autonomous vehicle lanes and autonomous vehicle/toll lanes in transportation networks." *Transportation Research Part C: Emerging Technologies*, 106. 2019. 381–403.

[35] Martin-Gasulla, M., and Elefteriadou, L. "Single-lane roundabout manager under fully automated vehicle environment." *Transportation research record*, 2673(8). 2019. 439–449.

[36] Mitman, Meghan, A.I.C.P., and Steve Davis P.E. "Introducing ITE's New Curbside Management Practitioners Guide." Institute of Transportation Engineers. *ITE Journal*. 2019; 35–41. ProQuest, https://search.proquest.com/docview/2197278089?accountid=6724.

[37] Lamotte, R., De Palma, A., and Geroliminis, N. "On the use of reservation-based autonomous vehicles for demand management." *Transportation Research Part B: Methodological*, 99. 2017. 205–227.

[38] Cummings, M., Li, S., Seth, D., and Seong, M. *HAL2020-01: Concepts of Operations for Autonomous Vehicle Dispatch Operations.* Humans and Autonomy Laboratory. Duke University NC. 2020. Accessed from https://hal.pratt.duke.edu/sites/hal.pratt.duke.edu/files/u36/CONOPS%20dispatch_Final_compressed.pdf

[39] Wen, J., Chen, Y. X., Nassir, N., and Zhao, J. Transit-oriented autonomous vehicle operation with integrated demand-supply interaction. *Transportation Research Part C: Emerging Technologies*, 97. 2018. 216–234.

[40]   Jung, J., Jayakrishnan, R., and Park, J. Y. "Dynamic shared-taxi dispatch algorithm with hybrid-simulated annealing." *Computer-Aided Civil and Infrastructure Engineering*, 31(4). 2016. 275–291.

[41]   Nissan. *Seamless Autonomous Mobility: The Ultimate Nissan Intelligent Integration*. Mountain View, CA. 2017.

[42]   Madrigal, A. C. "Waymo's Robot Cars, and the Humans Who Tend to Them." 2018. Retrieved September 6, 2020, from https://www.theatlantic.com/technology/archive/2018/08/waymos-robot-cars-and-the-humans-who-tend-tothem/568051/

[43]   Enoch, M. P. "How a rapid modal convergence into a universal automated taxi service could be the future for local passenger transport." *Technology Analysis & Strategic Management*, 27(8). 2015. 910–924.

[44]   Zhu, L., Wang, J., Garikapati, V., and Young, S. "Decision Support Tool for Planning Neighborhood-Scale Deployment of Low-Speed Shared Automated Shuttles." *Transportation Research Record*. 2020. 0361198120925273.

[45]   Zhang, H., and Zhao, J. "Mobility sharing as a preference matching problem." *IEEE Transactions on Intelligent Transportation Systems*, 20(7). 2018. 2584–2592.

[46]   Hart, P. E., Nilsson, N. J., and Raphael, B. "A formal basis for the heuristic determination of minimum cost paths." *IEEE transactions on Systems Science and Cybernetics*. 4(2). 1968. 100–107.

[47]   Tanvir, S., Frey, H. C., and Rouphail, N. M. "Effect of light duty vehicle performance on a driving style metric." *Transportation Research Record*, 2672. 2018; 25: 67–78.

[48]   Tanvir, S., Chase, R. T., and Roupahil, N. M. "Development and analysis of eco-driving metrics for naturalistic instrumented vehicles." *Journal of Intelligent Transportation Systems*. 2019; 1–14.

[49]   Liang, X., de Almeida Correia, G. H., and van Arem, B. "An optimization model for vehicle routing of automated taxi trips with dynamic travel times." *Transportation Research Procedia*, 2017; 27: 736–743.

[50]   Liang, X. *Planning and Operation of Automated Taxi Systems*. Delft University of Technology. 2019.

[51]   Horváth, M. T., Tettamanti, T., and Varga, I. "Multiobjective dynamic routing with predefined stops for automated vehicles." *International Journal of Computer Integrated Manufacturing*. 2019; 32(4–5), 396–405.

[52]   Castillo, J. C., Knoepfle, D., and Weyl, G. "Surge pricing solves the wild goose chase." *Proceedings of the 2017 ACM Conference on Economics and Computation*. June 2017; pp. 241–242.

[53]   Chen, T. D., and Kockelman, K. M. "Management of a shared autonomous electric vehicle fleet: Implications of pricing schemes." *Transportation Research Record*, 2572. 2016; (1), 37–46.

[54]   Loeb, B., and Kockelman, K. M. "Fleet performance and cost evaluation of a shared autonomous electric vehicle (SAEV) fleet: A case study for Austin, Texas." *Transportation Research Part A: Policy and Practice*. 2019; 121: 374–385.

[55] Tanvir, S., Hao, P., and Boriboonsomsin, K. "Emerging transportation technologies and implications for traffic-related emissions, air pollution exposure, and health." *Traffic-Related Air Pollution*. 2020; pp. 511–530. Elsevier.

[56] Rogers, B. "The social costs of Uber." *U. Chi. L. Rev. Dialogue*. 2015; 82: 85.

[57] Bruglieri, M., Colorni, A., and Luè, A. "The relocation problem for the one-way electric vehicle sharing. *Networks*." 2014; 64(4): 292–305.

[58] SAE International. "J3163™: Taxonomy and definitions for terms related to shared mobility and enabling technologies." 2018. Issued September 2018.

[59] Uber. https://www.uber.com/us/en/about/uber-offerings/. 2020. [Accessed 05 July 2020].

[60] Barth, M., Todd, M., and Xue, L. "User-based vehicle relocation techniques for multiple-station shared-use vehicle systems." *Proceedings of the 83rd Annual Meeting of the Transportation Research Board*. 2004. Washington, DC.

[61] Kek, A., Cheu, R., Meng, Q., and Fung, C. "A decision support system for vehicle relocation operations in carsharing systems." *Transportation Research Part E: Logistics and Transportation Review*. 2009; 45(1): 149–158.

[62] Nair, R., and Miller-Hooks, E. "Fleet management for vehicle sharing operations." *Transportation Science*. 2011; 45(4): 105–112.

[63] Smith, S., Pavone, M., Schwager, M., Frazzoli, E., and Rus, D. "Rebalancing the rebalancers: Optimally routing vehicles and drivers in mobility-on-demand systems." *Proceedings of the 2013 American Control Conference*. 2013. Washington, DC.

[64] Nourinejad, M., Zhu, S., Bahrami, S., and Roorda, M. J. "Vehicle relocation and staff rebalancing in one-way carsharing systems. *Transportation Research Part E: Logistics and Transportation Review*. 2015; 81: 98–113.

[65] Weikl, S., and Bogenberger, K. "Integrated relocation model for free-floating carsharing systems: Field trial results." *Transportation Research Record*, 2536. 2015a; 19–27.

[66] Santos, G. G., and Correia, G. H. "Finding the relevance of staff-based vehicle relocations in one-way carsharing systems through the use of a simulation-based optimization tool." *Journal of Intelligent Transportation Systems*. 2019; 23(6): 583–604.

[67] Uesugi, K., Mukai, N., and Watanabe, T. "Optimization of vehicle assignment for car sharing system." *Knowledge-Based Intelligent Information and Engineering Systems*. 2007; 4693: 1105–1111.

[68] Jorge, D., and Correia, G. "Carsharing systems demand estimation and defined operations: A literature review." *European Journal of Transport and Infrastructure Research*. 2013; 13(3): 201–220.

[69] Jorge, D., Molnar, G., and Correia, G. H. "Trip pricing of one-way station-based carsharing networks with zone and time of day price variations." *Transportation Research Part B: Methodological*. 2015; 81(2): 461–482.

[70] Febbraro, A., Sacco, N., and Saeednia, M. "One-way carsharing: Solving the relocation problem." *Proceedings of the 91st Annual Meeting of the Transportation Research Board*. 2012. Washington, DC.

[71]   Fan, W., Machemehl, R., and Lownes, N. "Carsharing: Dynamic decision-making problem for vehicle allocation." *Transportation Research Record.* 2008; 2063: 97–104.

[72]   Fan, W. "Management of dynamic vehicle allocation for carsharing systems: Stochastic programming approach." *Transportation Research Record.* 2013; 2359: 51–58.

[73]   Vincent, J. "World's first self-driving taxi trial begins in Singapore." The Verge. 2016. https://www.theverge.com/2016/8/25/12637822/self-driving-taxi-first-public-trial-singapore-nutonomy, accessed April 2020.

[74]   Fagnant, D. J., and Kockelman, K. M. "The travel and environmental implications of shared autonomous vehicles, using agent-based model scenarios." *Transportation Research Part C: Emerging Technologies.* 2014; 40: 1–13.

[75]   Zhao, L., and Malikopoulos, A. A. "Enhanced mobility with connectivity and automation: A review of shared autonomous vehicle systems." *IEEE Intelligent Transportation Systems Magazine.* 2020. DOI: 10.1109/MITS.2019.2953526.

[76]   Zhu, S., and Kornhauser, A. L. "The interplay between fleet size, level of service and empty vehicle repositioning strategies in largescale, shared-ride autonomous taxi mobility-on-demand scenarios." *Proceedings of the 96th Annual Meeting of the Transportation Research Board.* 2017. Washington, DC.

[77]   Weikl, S., and Bogenberger, K. "A practice-ready relocation model for free-floating carsharing systems with electric vehicles – mesoscopic approach and field trial results." *Transportation Research Part C: Emerging Technologies.* 2015b; 57: 206–223.

[78]   Kang, N., Feinberg, F. M., and Papalambros, P. Y. "Autonomous electric vehicle sharing system design." *Journal of Mechanical Design.* 2017; 139:1.

[79]   Chen, T. D., Kockelman, K. M., and Hanna, J.P. "Operations of a shared, autonomous, electric vehicle fleet: Implications of vehicle & charging infrastructure decisions." *Transportation Research Part A: Policy and Practice.* 2016; 94: 243–254.

[80]   Iacobucci, R., McLellan, B., and Tezuka, T. "Modeling shared autonomous electric vehicles: Potential for transport and power grid integration." *Energy.* 2018; 158: 148–163.

The views expressed are those of the authors.

Chapter 14

Impacts on the public realm: understanding the context

Gerry Tierney[1]

Chapters 14 and 15 will examine the potential impacts of automated vehicles on the public realm, primarily within both an urban and suburban context, as well as on land use and suburban sprawl. Incorporated into this will be an examination of external forces, such as vehicle fleet ownership and deployment, that could lead to either positive or negative outcomes.

But first we need to establish a shared understanding of the physical, economic, and social contexts into which automated vehicles or indeed this entire new emerging mobility ecology is being situated. We also need to acknowledge the viewpoints of multiple stakeholders by, for instance, understanding where mobility ranks as a priority in people's daily lives; for most people, it is only one among many competing priorities. As planners and engineers working in this field, we need to understand that automated vehicles are simply a tool which ideally will work to improve both people's lives as well as the environment. We also need to acknowledge that there is a difference between a tool (something that enables us) and a crutch (something that we come to rely upon in the absence of a complete system). And if today's automobile has become the crutch for our incomplete mobility system, it is worth remembering that we have been at this nexus before.

14.1 Introduction

A little over 100 years ago, our industrialized societies could be forgiven the decisions (or lack thereof) they made regarding the arrival of the private automobile, and its ultimately negative impacts upon the public realm, due in large part to their lack of awareness of what was about to overwhelm them. However, as we begin this new mobility era, posterity will not be so kind to us if we fail to take a holistic view of the potential impacts upon our cities, towns, and suburbs of not just automated vehicles, but the whole emerging mobility ecology which includes the emergence of the smart

[1]Perkins&Will, San Francisco, USA

phone enabled e-commerce and the shared economy. We need to take the lessons learned from the previous impacts of the private automobile on our cities and towns and project these forward so that this time we get our priorities and outcomes right. We cannot let this new tool to again become a crutch.

In examining the potential impacts of this emerging mobility scenario on street design, land use or parking, it is important that we understand the priorities regarding public space and land use which underlie Chapters 14 and 15. First and foremost, we need to make clear that our priority for the public realm of cities and towns is about people and their daily interactions. Streets are not simply conduits for vehicles, but are key elements of a city's public life. In the foreword to the UN-Habitat's 2013 report titled "Streets as Public Spaces and Drivers of Urban Prosperity" Dr. Joan Clos, Under-Secretary-General and Executive Director, UN-Habitat notes that while in the history of cities successful urban development has not been possible without an organized layout and system of streets, these same "streets, plazas and designed public spaces have contributed to define the cultural, social, economic and political function of cities" and observes that they are "the first element to mark the status of a place." This public realm is where the casual interactions of life occur, such as serendipitous encounters with our fellow residents, our browsing of storefront windows, the celebration of major cultural events and the assembly of populations to register their approval, or disapproval, of their leader's policies. In using the word "casual" it is not meant to demean the importance of such activities, since they frame how we feel about an urban space or city. Positive associations regarding these activities are what we have in mind when we refer to a city or town's "quality of life," or the "attractiveness" of a space beyond its aesthetic and functional qualities; they help define what Dr. Clos notes above, "the status of a place."

Just as important, we need to understand that the examination of the potential impacts covered in Chapters 14 and 15 fall into the category of "wicked problems" described by Professors Horst Rittel and Melvin Weber's in their "Dilemmas in a General Theory of Planning"[1]. In fact, the characteristics for a wicked problem can be applied to most of what follows. And in the interest of full disclosure, the author acknowledges at the start a bias in the discussions to follow toward a public realm which is designed to serve the larger goals of society and does not privilege one group of users at the expense of all others.

Chapters 14 and 15 are intended to work together: Chapter 14 sets out the typological contexts and high-level spatial implications of these emerging mobility scenarios as well as examining a key determinant that could have a major impact upon our being able to realize the potential benefits that automated vehicles and the emerging mobility ecology could bring us; Chapter 15 will look at current development as well as study precedents for street design, parking and the future of retail. Chapter 15 will also summarize and set out conclusions which should form the basis of the next steps required to see that the outcomes of this "disruptive" force are for the good of all and not just for the few. That is, as a society, let's not make the same mistakes as we did in the early twentieth century.

14.2   Precedents and the understanding of positive space

To better understand the potential impacts that shared mobility and automated vehicles could have on our streets and land use, it is useful to understand the context in which this implementation will occur. The following is a high-level summary from the urban design and planning point of view of both the pre- and post-automobile urban and suburban public realms in which shared mobility and automated vehicles will operate.

14.2.1   *Urban neighborhood typology examples (or the status of place vs. the geography of nowhere and why this matters)*

To start with, let us compare the public realms of two pre- and post-automobile neighborhoods in Paris and Shanghai and see the "status of place" that the residents of these cities assign these realms. Most people would recognize that experientially there is a qualitative difference between, for example, the pedestrian experience on the sidewalk outside the café "Les Deux Magots," on the corner of Boulevard Sant Germain and Place Saint Germain des Pres, in Paris's 6th arrondissement, and the pedestrian experience on the sidewalk at Place Jean Millier at the La Defense development in the Department Haute-de-Seine. Similarly, in Shanghai, there is experientially a qualitative difference between for example the pedestrian experience on the sidewalk at Fumin Street or Huaihai Middle Road in the Luwan District versus the pedestrian experience on the sidewalk at Lujiazui East Road at Yingcheng Middle Road in the Pudong New District. The intersections of Blvd. St. Germain with Pl. Ste. Germain de Pres and Huaihai Middle Rd. with Donghu Rd. each have a noticeable "status of place" as in "let's meet at …." These two pre- and post-automobile neighborhood locations are cited because they represent similar urban developments, built at similar times in two world class cities (Figures 14.1 and 14.2).

The Boulevard Saint Germain, with its six floors of residential-over-ground floor commercial uses, was constructed during the 1850s as part of Haussmann's

Figure 14.1   Blvd. St. Germain

Figure 14.2   Luwan District

reconstruction of Paris, while La Defense is a new business and finance district 5½ miles to the west, constructed primarily during the 1970s through the 1980s. Whereas the population density of La Defense is slightly greater than that of Blvd. St. Germain at Pl. St. Germain des Pres, and la Defense has a network of pedestrian concourses and walkways such that objectively it meets and incorporates many key urban design metrics and goals, the two areas are experientially totally different from each other. In contrast with the 6th arrondissement's pedestrian friendly streets and boulevards, the La Defense development chose to vertically segregate the pedestrian realm from the vehicular so that a visitor may find oneself being dropped off into a severely degraded pedestrian experience at the ground or vehicular level of the neighborhood. This vertical segregation has resulted in the pedestrian precinct above lacking an authentic 24/7 vibrancy due to it being a facsimile of an authentic street, remote from the mix of storefronts and activities, vehicular as well as pedestrian, that give vibrancy to the traditional street favored by most people.

Similarly, Shanghai's Luwan district's Fumin Street or Huaihai Middle Road and adjoining streets were constructed at the turn of the twentieth century as part of the former French Concession. The urban form and typology are similar in many respects to Paris's 6th arrondissement in that there was a mixture of residential- or commercial-above-ground level retail. By contrast the Pudong District's development, 3½ miles away to the east, was constructed during the 1990s–2000s and similar to La Defense, its pedestrian traffic is often vertically segregated from the vehicular. And for the same reasons that come into play at La Defense, it also suffers from a degraded street level pedestrian experience. Objectively both districts may have similar urban design metrics regarding density, pedestrian walkability and a mix of uses, however experientially the Pudong district is very different to the Luwan district (Figures 14.3 and 14.4).

Both Paris's 6th arrondissement and Shanghai's Luwan neighborhood have vibrant street life and are places that people chose to go to as destinations in and of themselves. They may not have a defined reason to visit, such as an appointment or a business meeting, but simply enjoy the activity of the neighborhood with its potential for serendipitous encounters. When we examine these neighborhoods, we notice two things. First, while both were constructed in the mid to late nineteenth century, they both have mixed-use building typologies which are typically located

Figure 14.3   La Defense

Figure 14.4   Pudong District

on the principal streets, usually with commercial or residential activities above street-level retail. They also allowed for a generous public realm designed to enhance the daily activities of its residents. Second, both were planned and constructed before the arrival of the private automobile, resulting in a closer equilibrium between street modalities and sidewalk activities.

Whereas both the Boulevard Saint Germain and Huaihai Middle Road were designed for wheeled traffic, the relative speed differential between that traffic and the pedestrian was not as pronounced as that which would later occur in the La Defense or Pudong neighborhoods. For safety reasons, the increased average speed of the automobile required more space per vehicle and so, as automobiles became more numerous, they required more space. And since space in an urban setting is finite, this became a simple exercise in geometry; the more space required by automobiles came at the expense of other modalities. This usually came at the expense of the pedestrian. When that space is filled, one must either increase the space horizontally (make wider streets) or vertically (provide segregated levels for autos and pedestrians).

What we learn from these two examples is that the simple application of objective urban design metrics and parameters relating to population density, mixed-use land-planning and the provision of copious pedestrian walkways does not necessarily create the quality "subjective" experience of a space that people are drawn to, or which contribute to a city's "quality of life" or "attractiveness." And this is an important distinction. If we are to consider that cities and their public spaces should be primarily about people and their daily activities then we need to be conscious of these subjective qualities; successful design seeks to strike a balance between the objective criteria of program and the subjective spatial and material qualities of the space in which the program occurs. It needs to convey a "status of place."

From a traffic engineering point of view, the two options of either increasing the space horizontally (wider streets) or going vertical (segregated levels for autos and pedestrians) are perfectly rational. And if we look at the work of two of the twentieth century's leading architects, Frank Lloyd Wright and Le Corbusier, we see illustrations of a future where transportation modalities and land-uses are clearly segregated. Whereas the traffic engineer addresses the issue from an engineering point of view, one would like to think that Wright's Broadacre City and Le Corbusier's Ville Radieuse were also addressing the problem from an experiential point of view (Figures 14.5 and 14.6).

Figure 14.5   Broadacre City

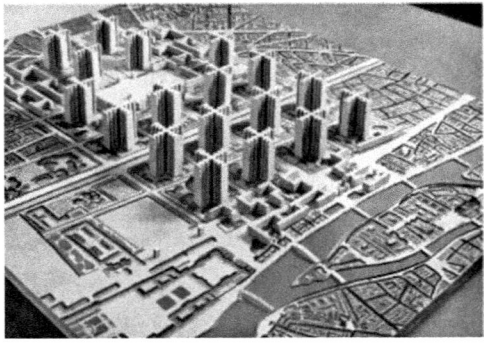

Figure 14.6   Ville Radieuse

However, "point-of-view" takes on a new meaning when one considers that the illustrations for each of these were primarily "birds-eye" or aerial views, with few if any views illustrating the ground level user experience or even considering the vibrancy of the public realm. While each of these theoretical designs for new cities contained ideas of merit, when their relationship between buildings and the street were copied elsewhere it typically resulted in a degraded public realm lacking vibrancy or meaning. Lewis Mumford, the architectural historian and critic, is quoted in Kunstler's book "The Geography of Nowhere" [2], as observing in his critique of Ville Radieuse that "the space between the high-rise floating in a superblock became instant wastelands, shunned by the public." Unfortunately, Mumford's observation has been validated in many twentieth century contemporary urban developments around the world whose conceptual design was based upon the Ville Radieuse model, with their high-rise buildings disengaged from the street.

So why does any of this matter? It matters because we need to ask ourselves if the values and understanding of urban space that gave us Paris's 6th arrondissement or Shanghai's Luwan neighborhood still exists. A contemporary indication of what we now consider to be important might be inferred in the editorial decision by the publishers of the children's book "The Ultimate Book of CITIES" [3] whereby it chooses to open a book introducing children to cities and urbanism with a double-page spread titled "Traffic." In all likelihood this was a subconscious decision; however, it is an indication that even when thoughtful people think of a City, their first choice of an image, before buildings, parks, museums, cultural buildings, or urban spaces, is that of traffic and the thoroughfares that accommodate it.

Is this children's book an indication that we have, perhaps subconsciously, accepted the La Defense or Pudong type of neighborhood to such an extent that we are now telling our children that in order to understand a city you must first start with traffic? If so, then we need to reconsider our values and biases. A City is more than simply a utilitarian construct; it is just as much a social construct that relies upon social interaction. As validated in the UN-Habitat's "Streets as Public Spaces and Drivers of Urban Prosperity" referenced earlier in this chapter, successful cities have a public realm that supports and reinforces social interaction.

By contrasting the experiential qualities of the pre-automobile public spaces encountered in Paris's 6th arrondissement, or Shanghai's Luwan District, with those of the auto-centric La Defense or Pudong Districts, we can see that the former creates more walkable, vibrant, and ultimately more adaptable and sustainable urban public realm. However, if we look at the editorial decision for opening the "The Ultimate Book of CITIES" with traffic, we need to ask if we can take it for granted that society will choose in the future to develop walkable vibrant neighborhoods instead of auto-centric ones. Is this the urban design equivalent of the Canary in the coalmine? If we really believe that cities are first and foremost about people, then the space occupied by the public realm can no longer be privileged by one group of users over another. The city's public realm is for everyone and the monopolization of that space by one user group or modality is neither sustainable nor in our long-term interest.

While the above are examples of pre- and post-automobile urban public realms located in both developed and emerging markets, there are other development typologies, the suburb, and the smaller city or rural town that we need to understand. The suburban typology discussed in Section 14.2.2 is sub-divided into two categories: late nineteenth century/early twentieth century "garden" or "streetcar" suburb and the mid-twentieth century to the present, auto-centric suburb. In Section 14.7, we will take a brief look at rural towns or smaller cities, which are often County administrative seats, with populations of less than 100,000. The choice of this population figure is discussed in the section and is derived largely from current minimum population figures required to support a diverse retail and "main street" environment.

14.2.2   Suburban typologies

For the purposes of this discussion, suburbia is being divided into two subheadings:

The late nineteenth century/early twentieth century "garden" or "streetcar" suburbs of major urban centers, which are often referred to as "inner suburbs" due to their relative adjacency to the urban core.

The mid-twentieth century to the present day, auto-centric suburb, which due to its increasing distance from the urban core, and along with its relative lack of public transport, this is often referred to as "outer-suburbia."

The defining differences are typically adjacency to either the urban core or an established neighborhood center, density, age of dwellings, and access to public transportation (usually commuter rail or light-rail/streetcar).

Originally conceived of by Sir Ebenezer Howard in the late nineteenth century as a means to provide an alternative for the working class to escape the unsanitary living conditions and related health issues resulting from residential overcrowding in the inner cities of the rapidly industrializing United Kingdom. These "Garden City" or "streetcar" suburban developments were constructed before the arrival of the private automobile and were often developed by railway companies, for example, The Metropolitan Railway's Harrow and Wembley suburbs northwest of London. Sometimes, the railway company formed a partnership with a real-estate company, for example, The Key System partnering with Mason McDuffie to develop various Berkeley and Oakland, California neighborhoods. In so relocating people to the edge of the City, away from their places of employment which were typically in or around the inner-city, the success of these developments relied on an efficient and accessible commuter rail or streetcar network, with any portion of a residential development being no more than a 10-min walk from a transit stop.

In the case of the Mason McDuffie–Key System's Berkeley Hills developments, this resulted in the construction of an intricate network of attractive staircases and paths running perpendicular to the hillside contours, along which the principal streets and their related streetcar lines were laid out. Convenient neighborhood retail centers allowed for a walkable neighborhood suburb which was essential in the pre-automobile era of the late nineteenth century. This layout followed the guidelines developed by The Hillside Club, founded in the late nineteenth century to promote good design practices in the Berkeley hills, as outlined by Charles Keeler in his book "The Simple Home" [4].

Beginning at the start of the twentieth century with Parker-Unwin's master plan for Letchworth, Hertfordshire, UK, there was an evolution of the "Garden City" suburban typology that began to adapt to the automobile. Whereas Letchworth, with its tighter grain and terraces of houses predated the mass adoption of the automobile, by 1920 Ebenezer Howard and Louis de Soissons Welwyn Garden City, also in Hertfordshire, was already beginning to accommodate the car. With wider streets and enough space between the typically semidetached houses to accommodate automobile garages, one can see the first glimpses of what suburbia was to become. More explicitly, the never fully completed Radburn development in Bergen County, New Jersey, by Stein, Wright and Cautley advertised itself as "A Town for the Motor Age" [5]. This development not only acknowledged but made a design feature out of the different demands and expectations of the car and the pedestrian. With a housing typology that was designed to be "dual-fronted," there were streets and courts for the automobile on one side, while on the other side there were landscaped "fairways" and open spaces onto which the principal living rooms opened; pedestrian footpaths linked the courts to the fairways. However, the onset of the Great Depression meant that the overall development plan was never completed. These examples are cited as they each illustrate the centrality of the pedestrian experience to the overall development plan, with gradual accommodation being made for the automobile.

Whereas the "Garden City" movement initiated by Sir Ebenezer Howard had an explicitly visionary and reformist viewpoint [6], the next evolution of suburbia was more pragmatic. The need for housing following World War II was intense, with returning service men and women looking to re-build their lives, usually away from the inner cities or rural communities in which they had grown up in before the war. Recognizing this and seeing how much they would be able to afford with the aid of the GI Bill, William Levitt, starting in 1947, created his prototype affordable starter home community in Nassau County, New York [7]. Unashamedly utilitarian, it set out to give "everyone" [8] the opportunity to buy an affordable house in which to raise their family. In so far as this was out in Nassau County, and was not explicitly related to transit with the Long Island Railroad's Hicksville and Bethpage stations being 2 to 3 miles away, the use of the private car was assumed as being part of everyday life. No longer was there to be convenient retail accessible by human-scaled streets or along paths running through open parkland. The public realm was designed in a very matter-of-fact manner to accommodate the needs of the car and the servicing of the houses. Levitt & Sons went on to replicate this model in New Jersey, Pennsylvania, Maryland and Puerto Rico [9]. By common consensus, this model is seen as the foundation of contemporary suburbia.

In contrast to Radburn, New Jersey's "A Town for the Motor Age," this model continues today, relying as it does upon an auto-centric public realm with its underlying assumption of the primacy of the car over the pedestrian. The street now becomes simply a means for the conveyance of automobiles and storm water, whose width is determined by a public works department's requirement to allow cars to pass stopped garbage trucks or for on-coming emergency vehicles to pass each other. The notion of an attractive pedestrian or public realm that links neighborhood to retail or community activities is now entirely absent, having been made redundant by the automobile.

14.3   Determinants of potential outcomes

Before we broadly examine the potential spatial, land-use and parking implications of automated vehicles on the urban and suburban typologies outlined above, we first need to consider how these vehicles will be deployed. Specifically, since the ownership model will impact the size of fleet required to serve a given population, as well as how this fleet is parked or stored when not in use, it is worth looking at the discussions around private versus fleet ownership and its deployment as this deployment is likely to have significant impacts upon both on- and off-street parking, curb demand and potentially street capacity and design, each of which will be looked at in more detail in Chapter 15.

In October 2017, the State of Victoria, Australia, commissioned Infrastructure Victoria, an independent advisory body to the Victorian Government, to prepare an analysis of what the infrastructure impacts would be of highly automated zero emissions vehicles [10]. This analysis included the modeling of two scenarios titled "Private Drive" and "Fleet Street" which also happen to match the models examined in various scenario planning workshops that the author has participated in and which are discussed below in the section dealing with Land-use and Planning [11].

One of the conclusions of the Infrastructure Victoria analysis is that a significant determinant of the potential impact outcomes for automated vehicles is the ownership model, or more precisely, how we access and use automated vehicles. This conclusion matches what was observed in the scenario planning workshops referenced above. Specifically, there is a difference in the fleet size between the private ownership model and that accessed through a subscription-based "ownership" model. This has a significant impact upon both on-street and on-site (off-street) parking demands, both of which have a significant impact on street curb access and utilization. Also, if zero-occupant or "zombie" automated vehicles are taken into consideration, the number of vehicles utilizing the right-of-way at any given time is impacted. For instance, in a subscription-based fleet the utilization of individual vehicles will significantly increase such that zero-occupant vehicles will likely be limited to short-distance logistical repositioning. Additionally, the Infrastructure Victoria study found that there would likely be an increase in transit, or active mobility usage, in a fleet-accessed scenario.

The private ownership model for automated vehicles is the same as the current ownership structure and would essentially be a "hands-free" version of today, with shared mobility occurring only to the extent that TNCs or other forms of ridesharing provide. Currently, the utilization rate for privately owned vehicles in the

United States is 5% [12] and 4% in the United Kingdom [13]. That is, vehicles are used for making trips 4%–5% of the time; parked the other 95%–96% of the time. For the 268.8 million registered vehicles (2016) in the United States, this means that at any given time there could be up to 13.4 million vehicles on the road, and 255.4 million parked. Under the private ownership model, there is no reason to believe that individual vehicle utilization will change from its current rates; however, there is a real possibility for additional vehicles being on the road in a zero-occupant or "zombie" state as they either search for parking/storage, or run errands for their owners. It is difficult to see how the potential benefits of an automated environment will be realized in the private ownership scenario, and it is against this baseline that the size and ownership model of the future vehicle fleet needs to be made.

By way of contrast, a subscription-based model, similar to what is currently being tested by several major original equipment manufacturers (OEMs) such as General Motors, the Volkswagen Group, and Volvo would provide an on-demand "autonomous-valet" enabled service. This would be accessed in a way similar to how one access's a TNC ride today. Thus, an automated vehicle provider, who may or may not also be a manufacturer, would have subscribers access their vehicle fleet on an on-call/on-demand basis, with vehicles being matched to the needs of the requested trip. Currently in San Francisco, the Volkswagen Group, for example, has been testing a subscription access model for their Audi brand under the advertising banner of "Own the Experience, Not the Car," in which they present options to access different vehicles sized appropriately for different activities (Figure 14.7).

Figure 14.7   Audi advertisement "Own the Experience, Not the Car"

But what is the likelihood of OEMs in general adopting a subscription-based fleet model? And why would the public switch over from private to subscription-based ownership or simply not just rely on ride-hail companies? The impetus for this will be from the OEMs as they realize the benefits of amortizing their vehicle costs over higher utilization rates, as well as seeing that a subscription-based model could potentially create a stronger and more reliable revenue stream, with the resulting attraction that brings for institutional investors.

The June 9th, 2018, edition of Economist magazine noted there are benefits of scale for OEMs in owning their own fleet when they observed that "Scale will yield ... benefits. Though some people will want their own driverless car, the market is likely to favor AV-based ride-hailing services. Driverless cars will not come cheap ...." They also observed that "... cars used in ride-hailing services will cost less per mile than personal vehicles which spend much of their time sitting idle" [14]. When the current cost of the LIDAR sensors needed for an automated vehicle is greater than the cost of the vehicle itself, bringing the retail cost of an automated car down so as to be affordable to the average consumer is a long way off, if at all achievable. However, the future value of a car to an OEM does not lie in its retail value, but in its ability to generate continuous revenue over time on a per mile or per trip basis. With global sales of automobiles exceeding 78 million units, realizing a market worth over $1.6 trillion for the top ten OEMs, Morgan Stanley currently estimates the global market for personal transportation could be worth up to $10 trillion per year [15]. This value can only be tapped into in a fleet-based model.

The large gap between the value to the OEM of the current retail model versus the potential revenue that could be realized by their product in a fleet-based model presents them with a major opportunity to extract additional revenue from each vehicle. Currently, the value to an OEM of a vehicle exists primarily in the profit that they share with the dealer at the point of sale for a new car being sold for the first time. However, a subscription-accessed fleet allows each vehicle to generate a predictable and reliable income stream for the OEM, which being attractive to institutional investors allows an OEM potential access to a wider range of financing for ongoing product research and development. Additionally, as OEMs are now self-identifying as "mobility companies," we are seeing them expand beyond their traditional product area and becoming involved in nontraditional mobility services such as TNCs, bike, and scooter-share. Whereas in 2018, the combined shared mobility market in the United States, China, and Europe was worth over $53 billion. Strategy Analytics, an international consulting firm, estimates that globally the "passenger economy" created by the convergence of automated vehicles and ride-hailing will, by itself, be worth $7 trillion per year by 2050 [16]. In the September 2018 issue of Automated Vehicle Technology, John Challen notes that Ford is positioning itself, like many others, as a mobility provider and quotes Sheryl Connelly, Ford Global Consumer Trends and Futuring Manager as promising that Ford plans to "reach people who don't plan to ever own a car." As the Economist observes "In an autonomous future where ownership is optional, they (the OEMs') need to be selling rides, not cars" [17].

Further, the Economist magazine's "Reinventing Wheels" Special Report on Autonomous Vehicles [18] noted that whereas current ride-hail costs average around $2.50 per mile as against $1.20 per mile for private vehicles, when the driver, currently representing approximately 60% of the cost of the ride-hail is subtracted, along with competition and electrification, the cost per mile will drop to approximately $0.70 per mile as estimated by USB, an investment bank [19]. USB goes on to observe that once a car becomes automated, the relevance of car ownership drops. The consulting firm BCG estimates that by 2030 up to 25% of all passenger-miles traveled in the United States could be in shared, self-driving electric vehicles, leading to a fleet reduction of 60% on city streets. OEMs, either through their investment in existing or new TNCs, or acting as their own fleet managers, will essentially be charging for their product by the mile. USB has observed that this will create a level of predictability of financial yields which will be attractive to institutional investors, thus potentially changing an OEMs primary business model from one of being a vehicle manufacturer to that of being an asset manager [20].

Based upon the above, and even allowing for some overly optimistic predictions, it is fair to say that the likelihood of the current private ownership model prevailing is unlikely. With the cost of the AV sensor equipment being more than the cost of the vehicle itself, the baseline "showroom" cost to the consumer will become unaffordable, not to mention the cost of maintaining and insuring such a sophisticated piece of equipment. But more significantly, given the large discrepancy in an OEMs potential income yield between the private "for-sale" model versus a fleet model, the major global OEMs, except perhaps for all but the most expensive luxury car manufacturers, simply cannot afford to forgo the predictable revenue stream and related institutional share investment that a fleet-based model will provide. Simple economics tell us that the age of the "private automobile" is coming to an end for all but the very wealthy.

Reinforcing the likelihood of this trend occurring, it is worth noting that the agenda for the Silicon Valley AV19 Conference, a pre-eminent conference located in the heart of automated vehicle/artificial intelligence development in Santa Clara, California, included a workshop titled "Planning for the Shift from Private Vehicle Ownership to Fleet-based AV's as Deployment Accelerates" [21]. Another indicator of the seriousness being given to this scenario is the editorial message titled "Figuring out the Future" from the publisher of the industry's trade paper "Autonomous Vehicle Technology" in the January 2020 edition in which Harper Henderson, the publisher, observes that like TV companies competing with Netflix or Hulu, or media companies competing with social media, "the automotive world is … not the only industry trying to figure out what the future will look like." He goes on to inquire "will customers still own cars" or "will they just pay each time they need one to drive them somewhere?" Further, he asks "Should automakers continue to drive sales to the end customer? Or should they switch their focus to selling to a handful of ridesharing/-hailing companies and let them deal with the end customer?" Or can the OEMs exit strategy be seen from the remarks made at the launch in January 2020 of the Cruise "Origin" autonomous shuttle-like vehicle?

With General Motors and Honda being major shareholders, Cruise's CEO Dan Ammann announced that this joint GM and Honda product will only be accessed through Cruise's own ridesharing service smartphone application. Could this be an indication that GM and Honda are willing to take on the TNCs by launching, through a subsidiary, their own ridesharing service? Although the OEMs and AV developers are not announcing to the current car-buying public this trend toward a new deployment structure, one can see that preparations are being made behind the scenes for a change to the current ownership model.

In looking at a subscription or a ride-hail or TNC accessed fleet model, we see that there will be an increase in individual vehicle mileage due to an inversion of the current utilization rate of 5% to an estimated 50% or more [22]. However, this will also allow for a significant reduction in the size of the vehicle fleet required to serve a given population. A 2014 MIT study into the potential impacts of ride-sharing on the overall number of taxis in New York City, as outlined in the *Proceedings of the National Academy of Sciences* [23], showed a 78% reduction in required fleet size. Whereas this relied on close to 100% shared rides, which is probably not achievable, it is worth noting that in 2018, up to 50% of Lyft's rides in San Francisco were shared "Lyftline" rides [24]. Further, in the 2017 Infrastructure Victoria report referenced at the opening of this section, their "Fleet Street" fleet accessed on-demand scenario model showed the City of Melbourne's vehicle fleet being reduced by 93%, with a 15% reduction in car travel "as people switch from cars to public transport due to the higher perceived cost of using on-demand vehicles." As the concept of shared rides becomes more culturally acceptable, and just as importantly the vehicle's interior design reflects this (for instance Cruise's "Origin"), the numbers of shared rides will likely increase above Lyft's current 50% "Lyftline" rate and get closer to the MIT study's 100% figure, with any exact number being speculative at this time due to lack of shared data.

The Taxi/TNC scenarios noted above are one part of what is referred to as Mobility-as-a-Service (MaaS) [25]. MaaS is a rider-focused mobility scenario that combines multiple modes of integrated transportation accessed through a single portal and with a single payment platform. In other words, mobility is consumed as a service in a manner closer to how we access data or utilities today. And given the role that mobility plays in the successful daily operation of a metropolitan area, it is perhaps time that we see it as an essential utility accessed the same way as other municipal utilities. In a report published by Fujitsu America in November 2017, and separate from what has been described above, they are predicting that with the increased implementation and adoption by consumers of MaaS, the vehicle fleet utilization will increase from a current figure of 5% to 50%, while the Infrastructure Victoria analysis referenced above for their "Fleet Street" scenarios predicted between 91% and 93% fewer cars being needed to serve a given population [26]. Whereas that is a high number and may be considered by some as an outlier, from what has been described above the fleet size reduction could plausibly be somewhere between a low of around 50%, as noted by Cruise at the launch of their "Origin" autonomous shuttle-like vehicle, to either 82% based upon a 2017 report by RethinkX, a Silicon Valley and London based research group [27] or the

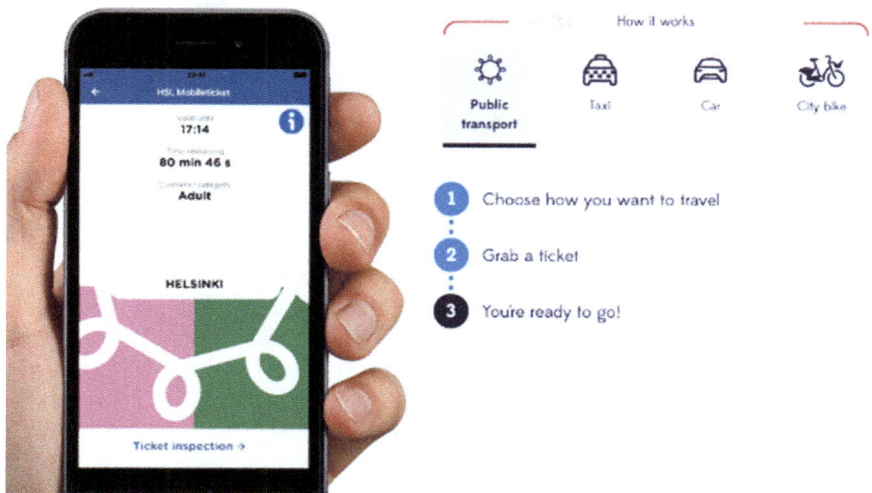

Figure 14.8   MaaS Global's Whim, Helsinki: Mobility-as-a-Service

93% as referenced in the Infrastructure Victoria analysis. Starting from the current low vehicle-utilization rate of 5% and with potential increased utilization rates resulting from an on-demand subscription-based ownership model, it is reasonable to say that there will be a significant reduction, perhaps by as much as two thirds, in the size of the fleet required to serve a given population. This will have an impact on both parking and curb demand with the knock-on effect of impacting both street and building design (Figure 14.8).

Just as important as the fleet size reduction is the fact that in a subscription or fleet scenario, the size of vehicles will be scaled to match the appropriateness of the requested trip. Whereas the majority of current automobile trips involve only one person traveling for a distance of under ten miles, the physical area taken up by a typical four-seat sedan can now be significantly reduced. When compounded, the potential reductions in fleet size (estimated at 67% pursuant to the above), the appropriate size of vehicle and its ability to utilize narrower travel lanes, along with reduced headway through "platooning" amongst other factors, the overall physical space required by an optimized fleet serving a given population can be reduced. This space reduction number becomes critical for transportation and roadway infrastructure planning as well as urban design. While this is addressed in Chapter 15, it is worth noting here that this is an area that will require further study so as to validate the current study figures based upon hypothetical scenarios. However, a pattern is already being observed which is that in order to realize the potential spatial benefits of a shared automated vehicle environment, along with the knock-on or secondary benefits this will have for the public realm, it will only occur within the context of an on-demand subscription accessed or shared fleet mobility scenario.

14.4   Spatial implications

But why, from an urban design or implementation perspective, is what is described above important? At its most basic, planning for roadway infrastructure is really a simple exercise in geometry; you have a prescribed amount of space into which a given program must fit. In cities at least, this space is prescribed by existing public rights-of-way, the edge of which in built-up urban areas are usually defined by private buildings. Into this right-of-way must go all the competing demands of a contemporary city. For millennia, this was decided in a laissez-faire manner involving size, speed, and authority. Big, faster, and more important in the middle; small, slower, and less important along the sides. And this essentially is how it was until the mid- nineteenth to early twentieth century. This worked in large part because the speed differential between the competing modalities was not so great that on a speed basis, at least, one modality dominated the other. With average speeds ranging from pedestrians at 3 mph, to horse drawn wagons at up to 10 mph, to streetcars at up to 12 mph, there was an ability for each modality to accommodate the other. And besides the sidewalk, designed to allow pedestrians to walk on a hard and level drained surface, there was no monopoly over who used the space between these sidewalk curbs, where provided. But with the advent of the internal combustion engine (ICE) powered automobile, omnibus, and freight truck, speeds increased such that for safety purposes alone low-mass, slow-speed pedestrians needed to be segregated from large-mass, relatively higher speed vehicles. This, along with a desire over the last 100 years to prioritize these new ICE-powered modalities, has led to the evolution of a public realm that has ended up privileging one group of users over all others.

If we consider current street design, typically with a paved impervious surface extending the fullwidth of the right-of-way, its two primary functions are the conveyance of powered vehicles and storm water. Pedestrians are confined to narrow strips on each side and cyclists are left to make their own way amongst the faster and heavier powered vehicles. The raised sidewalk is essentially a widened curb whose function is to guide storm water to drain inlets. Whereas contemporary street design may represent efficient infrastructure engineering, it cannot be said to encourage great urbanism. Traditionally, in cities and towns at least, the space between buildings was an integral part of the public realm, accommodating all stakeholders. As automobiles came to dominate this space, cities and towns needed to adapt the public right of way to accommodate this spatial and modality monopoly. Where once there was a heterogenous public-realm modality, now there is a modality monoculture. This efficiency comes at the expense of a vibrant street scape with a de-emphasis of street level pedestrian activity. The apotheosis of this was the grade separation or elevation of the pedestrian realm above the vehicular which we saw in many mid-twentieth century urban redevelopment schemes. Developments such as Paris's 1970s era La Defence, noted earlier, London's Barbican and San Francisco's Golden Gateway/Maritime Plaza developments are prime examples, with the latter being San Francisco Redevelopment

Agency's replacement in 1967 of the vibrant wholesale produce-market area that existed on this site.

Over the intervening 50 years City planners have come to realize the importance of vibrant street level environments and their contribution to successful cities. When one combines the potential reduction of the vehicle fleet size required to serve a given population, the elimination of on-street parking and the reduction in vehicle lane widths, there exists the opportunity to reduce the impact on the public realm of the current modality mono-culture. The "Urbanizing" issue of Technology/Architecture+Design describes the street as a "place" [28] whose primary role is human interaction and exchange, which optimally will allow for social interactions far exceeding today's experience "where people sit in vehicles close to each other, but sealed off in their own mechanical bubbles" [29]. In the 2013 UN-Habitat report cited at the opening of this chapter, the authors note that the "report is not only about the measurement of street elements, but about how streets, as public places, are associated with urban prosperity" and how "streets play a key role in productivity, infrastructure, environmental sustainability, quality of life and equity/social inclusion." If intelligently planned for, the potential benefits of a shared, electric, automated mobility environment present us with the opportunity to reclaim much of our public realm so that once again the "place" of the street will accommodate all stakeholder activities and not be simply a conveyance for vehicles (single-occupant, more often than not) and storm water run-off. Chapter 15 will examine this in more detail.

14.5   Land use: parking and retail

Changes in urban land use, as well as the potential for increased suburban sprawl, may well be two of the most important secondary impacts resulting from the introduction of a fully automated vehicle fleet. When combined with the current revolution in e-commerce, with online retailing such as Amazon's Prime Now on-demand service, the impacts will compound. However, these impacts will not be the same throughout the built environment. Just as a high-level understanding of the major urban realms was outlined above, so we need to have a similar understanding of impact zones in which the introduction of a fully automated fleet will play out on land use.

The following typological breakdown of impact zones is based upon work conducted by the University of Oregon's Sustainable Cities Initiative's Urbanism Next program [30]. Through a series of workshops [31] in 2017, 2018, and 2020, they looked at the potential impacts on downtown urban centers, prewar street-car suburb development centers, and postwar suburban strip-mall and big-box retail center.

14.5.1   Parking

The most obvious impact within the downtown core of a metropolitan area is the elimination of on-street parking with a significant reduction of off-street parking. In their study titled "Autonomous Vehicles and the Future of Parking" [32], Nelson

Nygaard quote studies by the OECD that parking could reduce by 80% in a 100% shared automated future and such parking as may still be required could either be located remotely or handled by high-density automated parking structures. On-street parking would be replaced by a combination of distributed pick-up and drop-off zones which, along with widened sidewalks, could include bikeshare programs as well as parklets serving the general public or adjacent ground-floor retail businesses. In a joint study by Arup and Perkins + Will's transportation and urban design studios [33] of a mixed-use downtown San Francisco city block, less than 50% of the space taken up by the former on-street parking spaces was required to accommodate the new TNC pick-up and drop-off zones (this study is gone into in more detail in the following chapter). Depending upon the building sizes and program of a given city's block, this will of course change; however, it would appear from initial studies that the entire city block length of sidewalk will not be required to accommodate TNC pick-up and drop-off zones.

From a development point of view, the elimination of surface parking lots, along with reductions in off-street parking required by planning codes such as the introduction of parking maximums, will both free up new land for future development as well as increase the financial feasibility of new affordable housing development options. For instance, to illustrate the impact of surface parking on a City, a study by the American Planning Association estimated that in 2010 there were approximately 18.6 million parking spaces in Los Angeles County, which in 2016 was translated into a diagram by Better Institutions showing that if consolidated, this parking footprint translates into a 200 square mile "parking crater" 16 miles in diameter. This would be capable of accommodating up to 2.3 million residents, or 22.5% of Los Angeles County's 2018 population, in 900,000 new homes, plus providing space to accommodate over 1 million workers.

Going forward cities need to give serious consideration to the introduction of parking maximums instead of the typical required parking minimums, at least in areas served by quality transit. For instance, the City of San Francisco revised its parking requirements for both commercial and residential development in its "C-3" zoned downtown core, as well as in the adjacent Transbay and Rincon Hill Development plan areas. These revised codes translated to zero spaces required for commercial, the demand for office parking being market driven instead. For residential developments, the code allows up to 0.25 spaces maximum per residential unit (0.5/unit with a conditional use permit—CUP). Additionally, in residential developments, all parking spaces have to be "unbundled" from the sales price of an individual unit such that the parking spaces must be sold separately to the unit buyers so as to externalize the real cost of parking (and by extension, of owning a car in a city). The revised San Francisco "C-3" zoning parking maximums has shown that zero parking has had no impact on the construction of new commercial or residential developments which continue at a record rate. Further, where parking has been provided at the CUP 0.5 space/unit rate, there has been difficulty in selling all of the available parking spaces.

With the combination of adjacent transit, car and bike-share programs, OEM subscription-based car access programs (see Audi's "Own the experience, Not the

Car" program referenced earlier), improved walkability created by sidewalk and public realm improvements, as well as convenient ride-hail services, there has been no negative impact on the demand for these residential units. Further, just south of downtown, an aggressive Transport Demand Management Plan (TDMP) by the University of California's San Francisco Mission Bay campus resulted in a reduction in parking demand such that their development cap, which was limited by the amount of parking initially provided, was raised by 1 million square feet [34]. These last two examples illustrate that even before the arrival of fleet-shared automated vehicles, cities can be reducing off-street parking in transit-rich neighborhoods, with residential developments requiring only 0.25 spaces per unit. Therefore, if the full arsenal of mobility options is brought into play through an aggressive TDMP, the goal of reaching 0.2 space per unit for residential development, or an 80% reduction in off-street parking, as cited at the start of this section by Nelson\Nygaard, is achievable.

While this demonstrates the scale of potential development that could be realized by the elimination of off-street parking, we also need to understand that the provision of off-street parking can be a significant cost burden on development. In the case of residential development, this can often make the difference between being able to provide affordable housing for those earning less than the median income of a metropolitan market. And our ability to be able to provide housing for all income groups within a relatively dense area impacts the viability of transit. Starting in the 1950s, we saw the beginning of what is termed "white-flight" from the urban cores to newly developed outlying suburbs, resulting in the predominantly auto-centric commute patterns we now see in metropolitan areas. However, with the return of many relatively affluent residents back to the core urban areas, we are now witnessing displacement of lower income residents to outer suburbia due to a lack of affordable housing in the urban core [35]. Since lower income families are typically more reliant upon transit, their move to the outer suburbs with its typically poor transit infrastructure requires the purchase of a car to access stores, services, and what little existing transit there is. As well as contributing to the auto-dominated commute pattern, this becomes another cost burden to a demographic that can least afford it thus furthering the economic inequality we are witnessing in our metropolitan regions.

To get an idea of the potential cost burden of providing structured parking (that is parking contained within the building either at, above, or below-grade), it is worth considering the following figures. A very efficient parking layout, often difficult to achieve in an urban infill site, requires between 330 and 350 gross square feet (gsf) per parking space which includes circulation. When we compare this to the average size of between 650 and 850 gsf for a one-bedroom unit in a multiunit development, we see that each required parking space is between 41% and 50% of the area of the unit which it is serving. Although the per square foot cost of the stall is less than the cost of the unit, it can still represent a significant added cost burden which, as mentioned previously, can determine the financial viability of providing affordable housing in a particular market. Typically, the only way for a developer to recapture these additional costs is by positioning a unit as

Living space versus parking space

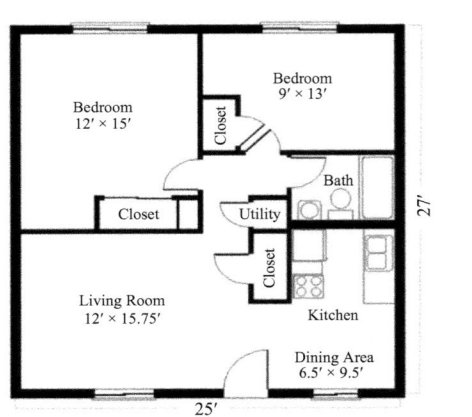 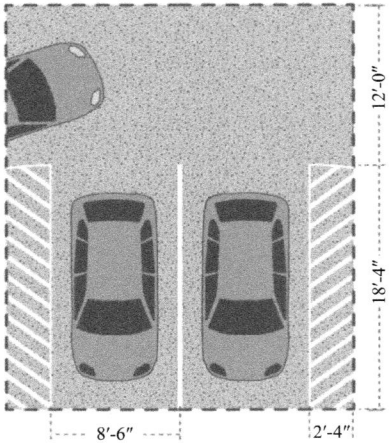

| Size for two bedroom apartment: 675 FT² | Size for two parking spaces: 650 FT² |

Sources: Transportation Cost and Benefit Analysis II - Parking Costs Victoria Transport Policy Institute (www.vtpi.org)
Graphic Adapted from Graphing Parking (https://graphingparking.com/2013/07/23/parking-across-cascadia/)

Figure 14.9   Living space versus parking space comparison (VTPI)

either market-rate for sale or for rent. This in turn makes the unit, unless subsidized, unaffordable to those earning less than the median income (Figure 14.9).

14.5.2   Retail: inner suburbs

While the major land-use impacts for downtown cores relate to the provision of parking—and the related potential viability of mixed-use affordable infill-development—the impacts on what we refer to as "streetcar suburbs" may be more benign. As noted earlier in the introduction, these suburbs were built around the concept of having an efficient and accessible commuter rail or streetcar network. In addition, they were laid out as attractive walkable neighborhoods built around conveniently accessed neighborhood retail centers which were essential in the pre-automobile era. As we look at them now, we can see that in many respects these are prototypical post-private-car-ownership shared-automated-vehicle neighborhoods, having most, if not all, of the attributes of a multimodal environment. They are walkable, usually with a grid layout that can accommodate bicycle-friendly streets parallel to major arterial roads, convenient retail centers which are not reliant upon automobile access and are usually served by mass transit. To the extent that the retail cores of these neighborhoods suffered in the postwar era, it was often due to lack of investment, with some buildings being torn down to make way for surface parking and an overall lack of building density.

In the University of Oregon's Urbanism Next workshops [36] noted earlier, the scenario-planning studies showed these types of suburb to be potentially the most resilient (these studies are gone into in more detail in the following chapter). Where there was existing surface parking (typically well in excess of demand), this could be developed as mixed-use multistory residential or professional-office style commercial over retail. Further, the density of the existing structures could be increased by the simple expedient of developing to the lot lines at the fronts and sides, as well as going up an additional couple of stories. Using these infill strategies and going from single story to three story structures on the "retail" street, a doubling or trebling of the neighborhood core density can be achieved without impacting the character of these suburbs. This increase in density is important as it can both provide the full-time residential population required to ensure a vibrant neighborhood-serving retail base (which is critical to the long-term viability of this neighborhood typology) as well as ensuring the continuing viability of any transit serving the neighborhood.

In addition to the neighborhood-serving retail model, if we look at the example of Fourth Street in Berkeley, California, we can see that the small footprint retail provided by this typology is attractive to major national tenants as they move from a "bricks and mortar" scenario to an on-line "e-commerce" environment augmented by "experiential" retail locations. While it needs to be recognized that this latter retail format cannot come at the expense of the former, the viability of these neighborhoods is enhanced by their capability of being either neighborhood-serving or regional-serving or both. And this is resulting from a neighborhood typology created in a pre-automobile era, that is well served by mass transit. At those locations where the original commuter rail or inter-urban streetcar lines have been removed, they should be replaced with either one of these modalities or with bus-rapid transit.

14.5.3 Retail: outer suburbia

Whereas the "streetcar" suburb discussed above was identified in the University of Oregon's Urbanism Next scenario planning workshops as being resilient enough to successfully adapt to a postprivate ownership shared automated vehicle environment, the same cannot be said for the auto-centric postwar strip mall or big box retail dominated suburb. These postwar suburbs, typically identified by what is sometimes referred to as a "loops and lollipops" street layout, were laid out explicitly to accommodate the automobile as well as the turning radii of city utility vehicles such as fire engines and garbage trucks. The dominance of two- and three-car garage doors facing onto the street, typical of most suburban houses, only serves to reinforce the auto-centric image.

Much like Levittown which was described earlier, when these suburbs were being developed it was understood that community serving activities such as retail, schools, local government, and religious would be accessed by the car. This understanding was embedded in the design to such an extent that in many subdivisions sidewalks were not even provided. This disbursed, low-density type of development worked against the provision of effective public transit, and to the extent that public transit was provided, it typically required some modality other than walking to access it. This gave rise to what is commonly referred to as transit's "first-mile/last-mile" problem. In effect, these suburbs are the antithesis of the "streetcar" suburb.

Compounded with this, the Urbanism Next workshops identified the impact that online e-commerce retail was having on the "big-box" retail malls. A 2017 Credit Suisse report [37] estimated that between 20% and 25% of all major big-box retail stores will close by 2022, resulting in both the downgrading and hollowing out of what little community-space vibrancy may have existed before, as well as reducing the amount of sales tax revenues for the impacted communities. Confirming this trend, the real estate services firm CoStar Group were quoted in a CNBC report that 2018 was on target to see the highest rate of store closures on record.

So, what are the options for the future of these communities in a post private ownership shared automated vehicle environment? Clearly, these suburbs are not going to disappear; however, they will likely undergo an adaptation at the level of both the individual residence and an evolution on the retail and community services areas.

With regards to the individual residence if, as appears likely, the OEMs move from a private ownership to a subscription-based model, the need for storing your private car goes away. The obvious impact of this is that two and three car garages will no longer in all instances be required, thus freeing up anywhere from 250 to 750 gross square feet, or the equivalent of a studio or one-bedroom apartment's worth of space per dwelling. Therefore, this could be converted into secondary in-law units allowing for multigenerational families where typically this had not been the case, due to either size or zoning restrictions. This would also have the added effect of doubling the unit density per acre. There is currently a significant trend in California's "single-family" housing market whereby increasingly there is a desire on the part of buyers to seek out homes which can accommodate multigenerational households—what is sometimes referred to as "one house/two mortgages." Further illustrating the changing demographic profile of potential new home buyers, an article in Builder Magazine published in October 2018, noted that based upon the most recent census data the share of households headed by married couples fell from 74% to 48% between 1960 and 2018.

This indicates that our planning assumptions around suburban development typologies and demographics, which are largely based upon a 1950s or 1960s "single-family/single breadwinner" scenario, needs to evolve to address both current demographic trends as well as the potential impacts that a shared AV environment might bring. Added to this, when we look to the future growth of e-commerce retailers, there will be the need to accommodate terrestrial and aerial AV delivery drones and where these drones can securely leave goods purchased online. The traditional mail will no longer adequately accommodate this, so perhaps a portion of the surplus car storage space can be given over to a docking or reception area that securely receives these delivery drones and their purchased contents.

Examining the retail and community core areas of existing postwar suburbs, we can identify two potential areas of evolution. Where these suburbs were built up around, or absorbed, existing smaller prewar communities the original "Main" street is often still there, all be it in a much-reduced form. But just like the "streetcar" suburbs that were discussed earlier, these have the potential to re-establish themselves as the "experiential" retail and community core that they were prior to the arrival of the auto-centric shopping malls located on the outskirts of town. What retail remains in the declining malls could then be relocated within a densified version of the "Main" street, with the vacated mall property being redeveloped at a higher density to include affordable housing and shared "WeWork" style office space, both of which are usually lacking in suburban communities. The latter would allow suburban residents to work at least one day a week from their home community, potentially reducing the demand on the transit infrastructure by up to 20%.

As an example of the potential densities that could be achieved for even a relatively modest retail mall, the 2020 Urbanism Next Workshop referenced above examined a failing strip mall in the City of Gresham, Oregon, a suburb east of Portland served by one of TriMet's streetcar lines. When overlaid with the typical 200' × 200' downtown Portland street grid, it yielded the equivalent of 16 downtown blocks for an overall development potential of over 3,000,000 gross square feet based upon six story, 85 feet high buildings conforming with the International Conference of Building Official's (ICBO) Uniform Building Code's Type 3-A construction type.

In many instances, however, these suburbs were built on "greenfield" sites with no prior communities to act as a nucleus around which to regroup, with the result that the retail big-box or strip mall often ending up acting as the center of these communities. With the accelerating rate of mall closures, these communities need to begin looking at ways to transform them into centers of daily life which will contribute to the richness of their surrounding community.

A successful example of this is the City of Westminster's redevelopment of the former Westminster Mall, located nine miles northwest of downtown Denver, Colorado. Constructed in 1977 as a regional shopping center and located adjacent to the Colorado Route 36, Denver-Boulder Turnpike, this 1.2 million leasable-square feet mall had all of the major national name-brand retailers, but by 2011, it had closed due to the loss of its six anchor tenants. The City of Westminster's Urban Renewal Authority acquired the site and have developed a new higher density urban center master plan for up to 2,300 residential units, over 2 million square feet of office space, 750,000 square feet of retail and restaurant space and 300 hotel rooms, a 5,000 square feet central plaza for community events and all located adjacent to the Denver RTD's bus station, with completion of the first residential units in mid-2019 [38]. The extension of the RTD's commuter-rail B-line is also planned to serve this new town center.

Whereas the mall had been the center of community life during the 1970s and 1980s, this waned as the mall declined. "That sense of community has been lost," the City's Development Manager was quoted in the Denver Post as saying and that what the residents wanted "was that sense of community again." Given its access to existing and planned quality transit, this development could act as a template for other cities and suburbs showing them how to take pre-emptive action in developing a master plan for their failing retail mall and create a new core around which they can develop a meaningful center for their community.

14.6   Suburbia and sprawl

The potential impacts do not just stop with the changes to different suburban typologies. There is, perhaps, more importantly, the potential impact of increased sprawl. Sprawl is a function of two key drivers, cheap available land with minimal entitlement costs and the prerequisite of many home buyers having to "drive 'til you qualify." People "choose" the outer limits of suburbia as much because of affordability as any other reason [39]. However, this perceived affordability comes at a cost, namely the increased costs of getting to and from work, school, stores, and community activities in a transit-poor environment. If the conventional wisdom holds that the cost of housing should not exceed 30% of one's income, and combined with transportation should not exceed 45%, in metropolitan regions such as the San Francisco Bay Area, this, according to a report issued in 2010 by the Centre for Neighborhood Technology, showed that these figures were now at an average of 32% and 48%, respectively, with the newer outer suburbs at 44% and 66% (or above), respectively. Significantly, where quality rapid transit, such as Bay Area Rapid Transit (BART), was provided the combined costs dropped to between 36% and 54%. From this, we can infer that the "appeal" of suburbia is in part due to the housing being affordable, but this is financially impacted by the need to rely on the private automobile for most of one's daily activities, often including significant commute distances and the costs related to that.

With this in mind, the case is being made that with lower operating costs per mile, combined with automated vehicle's elimination of the friction caused by sitting behind the steering wheel of a car stuck in endless traffic, there will be no downside on the part of the AV occupant to living in a car-dependent outer suburb or having to endure long commutes. In theory, the time spent commuting can now be put to other uses instead of being dedicated solely to the act of driving, thus inadvertently incentivizing continued suburban sprawl, as well as making it easier to locate job centers away from transit-rich environments.

Whereas we have seen that there is a qualitative difference between a car-dependent and a walkable suburb, there could also be a qualitative difference between a relatively short (sub-30 min) commute on a quality transit service, and a longer "hands-free" all-amenities commute. What if we use today's "hands free" and nominally "productive" tech-shuttle reverse commutes (i.e., from downtown or inner "streetcar" suburbs to traditional or outer suburbia) in the San Francisco Bay Area as an avatar for potential long-range AV commutes? In "off-the-record" conversations with tech company transit providers, there appears to be a limit, on the part of some employees at least, as to how much "commute" time, hands-free, and fully Wi-Fi enabled notwithstanding, they can take. These companies realize that their "luxury" shuttles are not a sustainable solution for long-term employee retention. While individual tech companies do not consider it appropriate to share information regarding reasons for employee departures, the development plans of several large Bay Area tech companies such as Google, Facebook, Twitter, LinkedIn, and Sales Force, all of whom now have significant development in downtown San Francisco indicates that this is a major concern. They are addressing this by building new offices closer to where employees want to live without requiring long commutes, even if these commutes are free, "hands free" and are in Wi-Fi enabled vehicles designed to create a "productive" environment.

For confidentiality reasons, these observations must remain as anecdotal for now. However, they fall into line with the research findings of Marchetti [40], Zahavi [41] and Anas [42]—what is often referred to as "Marchetti's Constant"—as well as commute times outlined in the US Census Bureau's most recent findings [43]. Starting with the work performed by Yacov Zahavi, he observed that with taking into account all the factors, including economic status, the global constant for mean commute travel (or "exposure") time was 1 hour per day, or a 30-min commute each way. This ties in with the US Census Bureau's finding that the average one-way commute is 26.1 min and with the top ten longest commute averages ranging from between 30.8 and 38.6 min, while the top 10 shortest range from between 15.5 and 16.9 min. And for the SF Bay Area, where the tech shuttles noted above operate, the average one-way commute time is 32.1 min. Building upon Zahavi's work, Marchetti's analysis of the evolution of a city's size based upon the speed of transportation is well illustrated by his "daily radius" diagram for the City of Berlin, which is based upon the 30-min one-way travel time noted above. Whereas Berlin, similar to the majority of other cities, had up until the early 1800s a radius of 2.5 km (based upon walking speed), with the advent of readily accessible and faster means of transport, this radius increased to around 10 km with electric trams, and again to 15 km with subways and currently 20 km with the private automobile. What this shows is "that the 'daily radius' depends on the speed of transportation ..." [44]. In comparing what he refers to as the commuting fields of eleven US cities, Marchetti illustrated that the commute distance of "driving man" is eight times that of "walking man" and that "driving man" has a potential accessible area that is sixty times larger than "walking man." While this study graphically illustrates the impact of transport modalities on the growth of city or metropolitan land areas, the variable in this equation was speed, with time being a constant; and it is this constant time factor that may ultimately define the boundaries of sprawl.

While the actual travel speed of automated vehicles may not differ too much from the current vehicle fleet, their average speed will most likely increase due to superior headway and lane control, as well as reductions in vehicle bunching, or stop-and-go traffic pulses, which serve to reduce overall average speeds. Thus, our sprawl boundaries may increase by the delta between today's average commute speeds and the average speeds attainable in an automated environment. Although the time delta will likely be less than those of earlier shifts between the pedestrian and tram, subway, or car, even a relatively small increase in average speed could translate into a large area of potential additional development area because of the logarithmic relationship between distance and area. This will need to be factored into regional land use and transportation planning. Whereas the notion of unlimited sprawl may not be borne out by real-world experiential preferences for each-way commute times in the range of ± 25–35 min, this typically applies to a mono-centric metropolitan typology. In poly-centric metropolitan regions such as the greater Los Angeles or San Francisco Bay Area, one may see a series of over-lapping 30-min commute radii resulting in an extended conurbation hemmed in only by the geography of sea, mountain, or desert. Whereas the Metropolitan Planning Organizations (MPOs) for each of these areas [45] are currently exam-ining these potential impacts, enforceable growth boundaries such as those insti-tuted by the City of Portland [46] may need to be developed for each of these regions as well as for other metropolitan regions faced with the same potential problem (Figure 14.10).

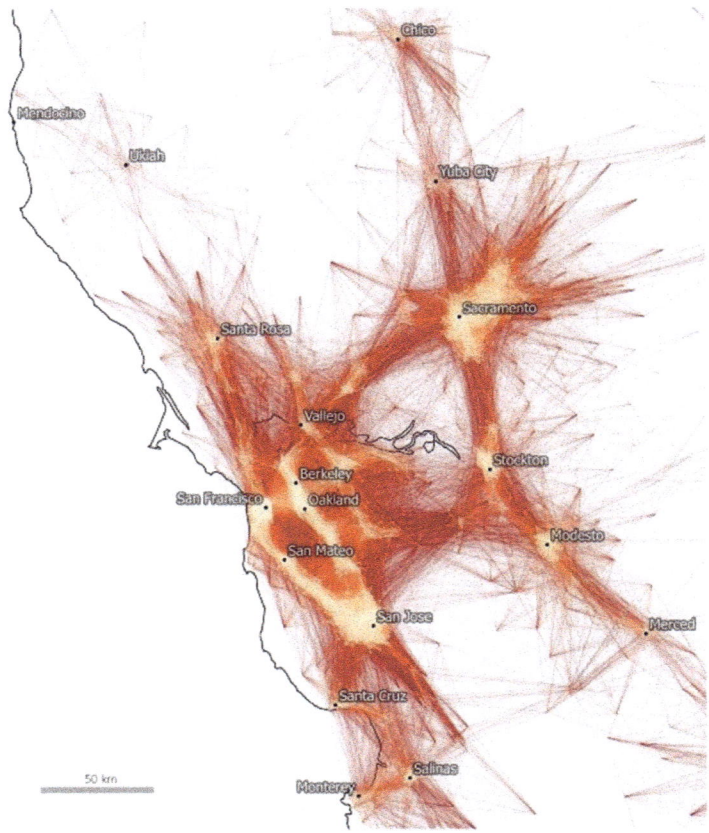

Figure 14.10   Greater San Francisco Bay Area commute patterns

The above "off-the-record" observations may be an indicator that there is a limit to the extent to which people are prepared to endure long commutes, even when the friction of driving is removed and extensive onboard amenities are provided. However, there is the simple fact that being trapped in a box, regardless of how luxurious it might be, for a significant period of one's day, is not to everyone's liking. And with companies currently headquartered in suburbia developing significant amounts of office space in the urban core, it calls into question the notion of major knowledge economy employment centers being located in outer suburbia. However, as noted earlier there may exist the potential for WeWork-like co-working spaces in suburbia, allowing workers to spend at least one day a week working in a semi-structured environment with the related positive impacts that could have on alleviating the current demands on overburdened transit. Even a modest change like this could potentially lead to a 20% reduction in commute travel.

It is also worth noting what Richard Florida wrote in an article for City Lab in 2017 titled "The New Suburban Crisis," "… today, clustering, not dispersal, powers innovation and economic growth. Many people still like living in suburbs, of course, but suburban growth has fallen out of sync with the demands of the urbanized knowledge economy …. The suburbs aren't going away, but they are no longer the apotheosis of the American Dream and the engine of economic growth." Despite that, the desire for people to raise their children in a big house with a big yard is deeply embedded in our culture and will remain so for the foreseeable future. Therefore, we need to recognize that the suburbs are not going to simply go away. Instead, we need to investigate the challenges and true economic costs related to suburbia and develop ways to guide its adaptation to a changed world where e-commerce and fleet deployed automated vehicles are the norm. We need to direct development away from continued sprawl based upon cheap land and easy entitlements. Instead, it should be directed toward a more sustainable infilling of the existing suburban cores, similar to that described above for Westminster, Colorado, and Gresham, Oregon, or towards infilling "streetcar" suburbs with their walkable neighborhood-serving commercial core and where a rich transit-based and walkable environment either exists or can be reinstalled.

Plan Bay Area 2040s "Connected Neighborhoods Scenario" [47] emphasized San Francisco Bay Area regional growth occurring in medium-sized poly-centric Priority Development Areas (PDAs with access to the region's major rail services. Anticipating regional growth between 2013 and 2040 of up to 3,000,000 jobs with 4,000,000 residents requiring an additional 820,000 dwelling units, this scenario planned for infill development and no growth outside the urban footprint on currently undeveloped land. Regional planning models such as this will ultimately be the best antidote to sprawl, regardless of the modalities or their degree of automation.

14.7   Smaller cities and rural areas

Finally, we need to consider rural areas and smaller cities with populations of up to around 100,000. The choice of 100,000 for the size of city could be considered as arbitrary, except that typically cities of this size and up can support modest main street retail with peripheral larger stores, and are often the administrative and service center of a surrounding rural agricultural community or county. Retailers in the United States looking at the viability of new stores or developments use the figure of 50,000 residents within 4–6 miles required to support what is referred to as "community" retail center, while 150,000 residents within 10–12 miles are required to support a "regional" retail center [48]. To give a further example, the population criteria required by popular restaurant chains in the United States such as Olive Garden, Red Lobster, or Longhorn Steak House are 125,000 residents within a 15-min drive [49]. Therefore, a population choice of 100,000, the mean between a "regional" and "community" retail center is, we feel, a reasonable choice. Examples of such cities in the United States are Cheyenne, WO (pop. 59,500), Champaign, IL (pop. 81,000), Redding, CA (pop. 89,000), and Ann Arbor, MI (pop. 120,000). Larger cities, typically with a population of above 250,000, such as Fresno, CA (pop. 495,000) or Wichita, KS (pop. 390,500), tend to have a discrete "downtown" district with related pre- and postwar suburbs, components of which are described earlier in this chapter and, therefore, are not as applicable for this discussion.

Regarding the implementation of automated vehicles, this is already happening in off-road situations whereby agri-business [50] and mining [51] are already adopting AV-enabled equipment. In agriculture, we already have automated equipment such as tractors or combine harvesters operate in tandem with manned vehicles, while in mining we have monitored automated haul trucks.

Overall the impacts upon rural towns and cities, at least in North America, will probably be mixed due to the fact that they will typically continue to serve a large rural hinterland where Level 5 autonomy would be required for public road use. Although there will likely be a very high level of autonomy by the agriculture and mining industries surrounding these towns, the size of their potential market may impact the economic feasibility of a shard fleet application. This will likely result in a blended fleet-deployed AV fleet alongside an AV-adaptable private fleet leased by those living in the outlying surrounding communities or areas. However, there are three interesting case studies which may offer an insight into the impacts on smaller towns and cities. These studies are the "Transforming Personal Mobility" study conducted in 2013 by Burns, Jordan, and Scarborough for Columbia University's Earth Island Institute [52], the GoMonrovia transit augmentation program administered by the City of Monrovia, CA [53], and the Livermore Amador Valley Transit Authority's (LAVTA) GoDublin! geo-fenced rideshare subsidy program in the City of Dublin, CA.

As reported in the OECD's International Transport Forum's (ITF) white paper, "Urban Mobility System Upgrade; How Shared Self-driving Cars Could Change City Traffic" [54], they cite one of the three regional case studies investigated by Columbia University which focused on the city of Ann Arbor. In this particular instance, their model found that for residents who travel less than 70 miles each day, "the shared fleet could provide near instantaneous access to a vehicle servicing their request, but with only 15% of the vehicles currently needed to carry out these trips." However, they also observed that overall VMT may increase due to the repositioning of vehicles (dead-head or "zombie" trips). In looking at the costs to the consumer, it was found that whereas the cost per mile (including ownership, operating, parking, and time-value costs) of a conventional personally owned vehicle was $1.60/mile that dropped to $0.41/mile for a shared self-driving conventional sedan vehicle and dropped further to $0.15/mile for a smaller 1-2 occupancy (Smart Car-like) shared automated vehicle. Whereas a potential reduction of at least 75% on the cost per mile makes for a compelling case, some may worry that this is based upon a shared scenario, and therefore may wonder as to how likely that will happen. This is where the smaller size of a city works to make a shared scenario more, rather than less likely to happen as we see when we look at the GoMonrovia and GoDublin! programs.

The GoMonrovia program was established by the City of Monrovia, CA (pop. 37,000—2017) in March 2018 to augment their existing Dial-a-Ride program and provide first-last mile connectivity with their LA Metro Gold Line light rail station. Other than the LA Metro Gold Line, their entire transit service was comprised of two Foothill Transit lines and the city's own on-demand Dial-a-Ride program, which cost over $19/rider but collected only $3.00 at the farebox; monthly ridership in early 2018 was about 3,000 riders. Partnering with Lyft, a TNC provider and Lime Bikes, they introduced a geo-fenced (rides only within the City of Monrovia boundary) transit augmentation program which is accessed through either the GoMonrovia or Lyft's smart-phone applications, or through a City of Monrovia "concierge" number to accommodate patrons without smart-phones or credit cards. Lime Bike access costs 50-cents/ride and each geo-fenced Lyft ride costs $3.00 or 50-cents if shared.

From an initial ridership of 5,000 in April 2018, it had reached 75,800 riders per month by January 2019, by which time fares had increased to $3.50/ride ($1.00 if shared) [55]. This represented a 25-fold increase in people using a service other than their own private vehicle. Of these rides, only 30% were first-last mile journeys to the light rail station, with the remaining 70% all being within the City boundary. And overall, shared rides represented 90% of all journeys [56]. When queried about this very high number the City observed that there were several factors at play, not least of which was the 50-cent fare, now increased to $1.00 but without appearing to have a negative impact. Whereas shared TNC rides often have a hard time attracting patrons, due in large part to the perception that this will add significant time to one's trip, in the geo-fenced area of Monrovia a shared ride adds a relatively small amount of time (in the order of 1-3 min) and there is a higher likelihood of being acquainted with your fellow rider. Whereas those who either cannot go online to access the program, or who may not have a credit card or bank account, are served by the City's "concierge" program, patron's with mobility issues which may prevent them using a conventional TNC vehicle are now served by the freed up "dial-a-ride" vans. Before the GoMonrovia program, a patron needing an accessible vehicle typically had to make an appointment 24-h in advance, thus eliminating any spontaneous access to transit. With the current accessible vehicles now dedicated solely to those with mobility needs, rides are typically available within the hour and frequently within 20 min from the time that the patron places a call [57]—this now more closely resembles a useful transit system for those who need it.

Superficially similar GoDublin! is different from GoMonrovia in two key respects: first, it is a fully integrated part of the Livermore Amador Valley Transit Authority (LAVTA); second, it is a fare subsidy program that pays 50% of the fare up to $5, for trips within a geofenced area using Uber, Lyft, or DeSoto Cab Company, and is open to any TNC, taxi company, or other mobility provider who wishes to register with LAVTA to participate in this program. With a population of over 63,000 and two BART rail stations, the GoDublin! ridership has been more transit supplemental, serving as a first/last-mile connection to BART or regional transit buses. As an integrated part of LAVTA's transit planning, GoDublin! has allowed them to reallocate vehicles and drivers from low-ridership/low-frequency routes to higher ridership lines within the district. This reallocation has allowed for shorter headway/higher frequency on their trunk lines with the result that overall ridership for 2019 is up on that for 2018. In the current US transit environment where overall ridership numbers are down, it is instructive to see a transit agency, integrating TNC/ride-share providers into their service, being able to free up resources for higher capacity/higher need lines and increase their ridership at the same time.

If we use the GoMonrovia and GoDublin! programs as a beta-test for Columbia University's Earth Island Institute study-scenario referenced above for Ann Arbor, we can see that the shared ridership assumptions of their model may not be unreasonable. The lessons being learned from both the GoMonrovia and GoDublin! programs could be directly applicable to smaller population cities which currently lack a meaningful transit system. Having a relatively small geographic area, the perceived penalty for a shared ride is likely reduced in these cities. And with a public/private partnership, the City is not required to take on additional staff to "run a transit service" [58].

Besides these two US examples, the German Federal Ministry of Transport in conjunction with the Neustadt Inner City Advisory Board is supporting the deployment of Mobility-on-Demand, a micro-transit service using ride-hailing-like cars to augment their less than robust public transit system in a city considered as being too small to attract the larger TNC companies. Again, this should be considered as a beta-test for city/transit/TNC collaboration; however, it demonstrates that we can and must think beyond the current zero-sum narrative around this topic. Ironically, whereas we tend to think of major metropolitan areas as being among the first to implement these new mobility scenarios and technologies, because of their scale the smaller rural and suburban cities currently lacking any meaningful transit are becoming the first adopters. These programs merit additional studies for addressing the needs of smaller cities and towns.

Whereas the short-term subcontracting out of an augmented transit program to TNC "contract" drivers brings up equity issues which are outside the scope of this chapter, one hopes that when the vehicles in question become automated, this becomes moot. However, GoDublin's inclusion of a "traditional" cab or taxi company into the mix illustrates that it is not just drivers operating within what is referred to as the "gig-economy" who can have a place within this evolving mobility scenario. Further, the GoDublin! program shows that a transit service can become more efficient as a result of this partnership, with transit ridership growth ensuring long-term job security for its bus drivers.

The key take-aways are that traditional transit and TNCs should not be viewed as zero-sum options, but as potential complimentary partners with each modality playing to its strength; people living in what are seen as traditional single-occupant vehicle cities are willing to switch to shared rides and transit when programs are set up that address their needs. This last point is worth more study by others as it indicates that whereas we often think of transit as a single one-size-fits-all entity, and then wonder why the majority of people do not use it, we should instead be looking at more nuanced, bespoke approaches or systems tailored to the specific needs of the populations they are designed to serve. As was noted GoMonrovia and GoDublin! are superficially the same, but with key differences that address the needs and legislative structures of their service areas. Maybe the path to success lies in integrating many smaller "organic" systems in such a way that they can effectively operate in unison. Is this a potential new role for the MPOs such as SCAG or MTC mentioned earlier in this chapter? Should we be looking at the Swiss Federal Office of Transport as a model for bringing together multiple mobility providers so as to work in unison, be that SBB/CFF/FFS (Swiss rail), Zurich's ZVV S-Bahn, or the humble PostBus waiting to pick you up at your destination? Are these signs of new mobility's maturing?

14.8   Summary of Chapter 14

As noted above, Chapters 14 and 15 are intended to work together. In Chapter 14, we have looked at the typological contexts and high-level spatial implications of automated vehicles as well as other emerging mobility scenarios. In addition, we looked at a key determinant that could have a major impact upon the fleet size required to serve a given population, and why this matters, when applied to spatial criteria impacting street design, parking, and land use. Additionally, we have seen that new mobility programs in smaller cities are leading the way in integrating TNC/ride-sharing into a structured transit environment in such a way that traditional single-occupant car users are willing to switch to transit or shared rides when the program is tailored to their needs. In Chapter 15, we will examine current precedents under development and studies into street design, parking, and the future of retail. Finally, we will summarize both chapters and draw some conclusions that inform our moving forward.

Author contribution

The author reviewed and approved the manuscript.

Declaration of conflicting interests

The author declared no potential conflict of interest with respect to publication of this chapter.

References

[1]   Rittel, H. W. J., and Webber, M. M. "Dilemmas in a general theory of planning." *Policy Sciences.* 1973;4(2): 155–169.

[2]   Kunstler J. H. *The geography of nowhere: The rise and decline of America's man-made landscape.* Simon & Schuster; 1993.

[3]   Baumann A.-S., and Baticevic D. *The ultimate book of CITIES.* Chronicle Books; 2018

[4]   Keeler C. *The simple home.* San Francisco: Smith Elder & Co.; 1904

[5]   The Radburn Association. A town for the motor age (online). 2018. Available from https://www.radburn.org/index.php/about [Accessed March 2019].

[6]   Howard E. *Garden cities of to-morrow.* London: Faber and Faber; 1902. Reprinted, edited with a Preface by Osborn F. J. and an Introductory Essay by Mumford L.; 1946. pp. 50–57; pp. 138–147.

[7]   Galyean C. Levittown. *US history scene* (online). Levittown Public Library; 2015. Available from https://www.levittownpl.org/research-history [Accessed March 2019]

[8]   Sabo K. *The Levittown legacy: Segregation in suburbia?* (online). Available from https://people.hofstra.edu/alan_j_singer/294%20Course%20Pack/x10.%20Civil%20Rights/129.pdf [Accessed March 2020]. *"Levittown was an embodiment of the American dream, with one important catch. The promise Levittown offered for the future was a racially exclusive one. African Americans were denied access to this suburban dream. Every deed signed by new homeowners contained a clause that bound them "not to permit the premises to be used or occupied by any other person than members of the Caucasian race. . ."*

[9]   *The Levittowns* (online). Available from http://levittownbeyond.com/index.html [Accessed March 2019]

[10]   Infrastructure Victoria (Australia). *Advice on automated and zero emissions vehicles infrastructure*: Infrastructure Victoria; 2018.

[11]   Larco N., and Riggs W. (moderators). *Scenario planning workshops*: *University of Oregon SCI Urbanism Next*; 2017, AUVSI San Francisco; 2018 and University of San Francisco; 2018.

[12]   Shoup D. *The high cost of free parking.* Chicago: Planners Press; 2005. Available from https://www.reinventingparking.org/2013/02/cars-are-parked-95-of-time-lets-check.html [Accessed March 2019]

[13]   Royal Automobile Club. *Spaced-out: Perspectives on parking policy* (A report). London: RAC Foundation; 2012.

[14]   "Free exchange: Road hogs." *The Economist Magazine*, June 9 2018; p. 6.

[15] "Selling Rides, Not Cars: Special report on autonomous vehicles." *The Economist Magazine*, March 2018. p. 7.

[16] "Reinventing Wheels: Special report on autonomous vehicles." *The Economist Magazine*. March 2018. p. 4.

[17] "Selling Rides, Not Cars: Special report on autonomous vehicles." *The Economist Magazine*. March 2018. p. 7.

[18] "Reinventing Wheels; Special Report on Autonomous Vehicles." *The Economist Magazine*. March 2018. p. 4.

[19] "Reinventing wheels: Special report on autonomous vehicles." *The Economist Magazine*. March 2018. p. 4.

[20] "Selling Rides, Not Cars; Special Report on Autonomous Vehicles." *The Economist Magazine*. March 2018. p. 7.

[21] Stocker A. (moderator). "Planning for the shift from private vehicle ownership to fleet-based AV's as deployment accelerates." *AV19 Conference Pre-Conference Workshop B.* February 2019, Also see https://www. AutonomousVehicles.iqpc [Accessed February 2019]

[22] "Selling rides, not cars: Special report on autonomous vehicles." *The Economist Magazine*. March 2018. p. 7.

[23] Alonso-Mora J., Samaranayake S., Wallar A., Frazzoli E., and Rus D., and Goodchild M. F. (eds.). "On-demand high-capacity ridesharing via dynamic trip-vehicle assignment." *Proceedings of the National Academy of Sciences*. 2017; 114(3): 462–467. Available from https://doi.org/10.1073/pnas.1611675114 [Accessed January 2019].

[24] Rocky Mountain Institute. *Understanding TNC's impact on cities: Looking at the VMT efficiency metric* (A report). Rocky Mountain Institute; 2018.

[25] *Future of mobility: What is MaaS?* Available from https://mobility.here.com/blog/what-does-future-mobility-look [Accessed January 2019] and MaaS alliance. Available from https://maas-alliance.eu/homepage/what-is-maas/ [Accessed January 2019].

[26] Infrastructure Victoria (Australia). Appendix A: Summary of impacts by scenario. *Advice on automated and zero emissions vehicles infrastructure*: Infrastructure Victoria; 2018

[27] RethinkX. *Rethinking transportation 2020–2030: The disruption of transportation and the collapse of the ICE vehicle and oil industries* (A report). RethinkX; 2017

[28] Favour J., Hang J., Nyamjav J., and Fisher T. *The street as place; autonomous vehicles and their impact on urban and suburban roads.* Technology/Architecture + Design; 2018

[29] Favour J., Hang J., Nyamjav J., and Fisher T. *The street as place: Autonomous vehicles and their impact on urban and suburban roads.* Technology/Architecture + Design; 2018

[30] University of Oregon. *Sustainable cities initiative: Urbanism next.* Available from https://sci.uoregon.edu/urbanism-next-0 [Accessed March 2019].

[31] University of Oregon. *Sustainable cities initiative: Urbanism next.* Available from https://sci.uoregon.edu/urbanism-next-0 [Accessed March 2019].

[32]   Nelson Nygaard. *The future of parking.* Nelson Nygaard; 2017. Available from http://nelsonnygaard.com/our-views-on-autonomous-vehicles-and-the-future-of-parking/ [Accessed January 2019].

[33]   Arup, Perkins+Will Mobility Lab. "A study of 4th Street, San Francisco." *Poster for AUVSI Symposium 2017.* Arup and Perkins+Will Mobility Lab; 2017]. Available from https://higherlogicdownload.s3.amazonaws.com/ AUVSI/14c12c18-fde1-4c1d-8548-035ad166c766/UploadedImages/2017/ PDFs/Proceedings/Posters/Tuesday_Poster%2023.pdf [Accessed January 2019]

[34]   University of California San Francisco. "Review of TDMP impacts and development cap." *Long Range Development Plan Environmental Impact Report (EIR):* UCSF; 2014. Available from https://campusplanning.ucsf.edu/ sites/campusplanning.ucsf.edu/files/reports/UCSF%202014%20LRDP %20Final%20EIR.pdf [Accessed March 2019].

[35]   Kneebone E., and Garr E. *The suburbanization of poverty: Trends in metropolitan America, 2000 to 2008* (A paper). Brookings Institute; 2010. Available from https://www.brookings.edu/research/the-suburbanization-of-poverty-trends-in-metropolitan-america-2000-to-2008/]. Badger E. "The suburbanization of poverty." City Lab; 2013. Available from https://www. citylab.com/life/2013/05/suburbanization-poverty/5633/   [Accessed February 2019].

[36]   University of Oregon. *Sustainable cities initiative: Urbanism next.* Available from https://sci.uoregon.edu/urbanism-next-0 [Accessed March 2019].

[37]   Bakashi P., Buss C., Lee S., and Shuler S. *Apparel retail & brands: eCommerce share capture accelerates and continued store & mall closings* (A report). Credit Suisse; 2017. Available from https://research-doc.credit-suisse.com/docView?language=ENG&format=PDF&sourceid=csplusrese archcp&document_id=1075851631&serialid=0H35FD75wQHBUjm3x5lk vUaUWAN03QsDSMQWFvss5x4%3D [Accessed March 2019]. Peterson H. "Wall street bank says a quarter of shopping malls will close in 5 years." Business Insider; 2017. Available from https://www.businessinsider.com/a-quarter-of-shopping-malls-will-close-according-to-credit-suisse-2017-5 [Accessed March 2019]

[38]   Westminster Economic Development Authority; 2015. Available from www.downtownwestminster.us and www.cityofwestminster.us

[39]   Kneebone E., and Garr E. *The suburbanization of poverty: Trends in metropolitan America, 2000 to 2008* (A paper). Brookings Institute; 2010. Available from https://www.brookings.edu/research/the-suburbanization-of-poverty-trends-in-metropolitan-america-2000-to-2008/]

[40]   Marchetti C. "Anthropological invariants in travel behavior." *Technological Forecasting and Social Change.* 1994;47; 75–88

[41]   Zahavi Y. "Project no. DOT-RSPA-DPB, 2-79-3." *The UMOT Project* (1979) and "Report No. DOT-RSPA-DPB-10/7." *The UMOT-Urban Interactions.* 1981. US Department of Transportation: Washington DC; 1979 and 1981

[42] Anas A. "Why are urban travel times so stable?' *Journal of Regional Science*. 2015; 55(2): 230–261. Available from https://ssrn.com/abstract=2573377 or http://dx.doi.org/10.1111/jors.12142 [Accessed February 2019]

[43] US Census Bureau (Washington, DC). Average one-way *commuting time by metropolitan areas*. US Census Bureau; December 2017. Available from https://www.census.gov/library/visualizations/interactive/travel-time.html [Accessed February 2019]

[44] Marchetti C. "Figure 2 - City dimensions and speed of transport: The case of Berlin." *Anthropological Invariants in Travel Behavior, Technological Forecasting and Social Change*. 1994; 47: 75–88

[45] Southern California Association of Governments (SCAG), Los Angeles Region Metropolitan Planning Organization (MPO). Available from http://www.scag.ca.gov/about/Pages/Home.aspx [Accessed February 2019]. MTC, San Francisco Bay Area Metropolitan Planning Organization (Metropolitan Transportation Commission MPO). Available from https://mtc.ca.gov/our-work/plans-projects/plan-bay-area-2050 [Accessed February 2019].

[46] Portland Metro. *Urban growth boundary*. Portland Metro Metropolitan Planning Organization (MPO). Available from https://www.oregonmetro.gov/urban-growth-boundary].

[47] Metropolitan Transportation Commission (MTC). *Plan bay area: 2040 connected neighborhood scenario*. MTC; 2018. Available at https://www.planbayarea.org/sites/default/files/station_4-b_-_connected_neighborhoods_scenario.pdf [Accessed March 2019]

[48] Gibbs R. J. "Public square." *CNU Journal*; 2007. Easton G., Owen J. "Supporting walkable neighborhood business districts." *APA Brown Bag Series*. American Planning Association; 2009.

[49] Retail Sites International. *Site criteria*. RSI; 2018. Available from http://www.retailsitesinc.com/site-criteria/ [Accessed March 2019]

[50] Barne W. *Setting the stage for the future of farming: Drones, autonomous vehicles and more*. IoT Tech News; 2017. Available from https://www.iottechnews.com/news/2017/jul/14/setting-stage-future-farming-drones-autonomous-vehicles-and-more/ and https://www.asirobots.com/farming/ [Accessed February 2019]

[51] Jamasmie C. *Rio Tinto autonomous trucks now hauling a quarter of Pilbara material*. Mining.com; 2018]. Available from http://www.mining.com/rio-tinto-autonomous-trucks-now-hauling-quarter-pilbara-material/ [Accessed February 2019]. Atak A. *Electric vehicles and autonomous vehicles in mining 2018–2028 technologies, challenges, benefits, markets, forecasts, key players and opportunities*. ID TechEx; 2018. Available from https://www.idtechex.com/research/reports/electric-vehicles-and-autonomous-vehicles-in-mining-2018-2028-000597.asp [Accessed February 2019]

[52] Burns L. D., Jordan W. C., and Scarborough B.A. *Transforming personal mobility*. Columbia University Earth Island Institute; 2013.

[53] GoMonrovia. Available from https://www.cityofmonrovia.org/your-government/public-works/transportation/gomonrovia and https://www.cityofmonrovia.org/your-government/public-works/transportation/monrovia-transit. [Accessed April 2019].

[54] OECD International Transport Forum. "Urban mobility system upgrade: How shared self-driving cars could change city traffic." *OECD Corporate Partnership Board Report*. OECD; 2013].

[55] Chi O., Davis P., and Tierney G. "City and TNC collaboration – A case study (moderated panel), *Urbanism Next 2019*." University of Oregon Sustainable Cities Initiative; 2019. Available from https://www.urbanismnext.com/schedule-2019 [Accessed June 2019]

[56] Interviews by the author of Oliver Chi, City Manager, City of Monrovia, December 2018–April 2019.

[57] Interviews by the author of Oliver Chi, City Manager, City of Monrovia, December 2018–April 2019.

[58] Interviews by the author of Oliver Chi, City Manager, City of Monrovia, December 2018–April 2019.

The views expressed are those of the authors.

Chapter 15

Impacts on the public realm: the potential for a positive outcome

Gerry Tierney[1]

In this chapter, we will be examining precedents which include current projects under development as well as studies into the potential impacts on street design and on-street parking. In addition, we will be looking at the potential impacts of e-commerce on the future of retail as well as the potential impacts of an AV future on building design. Finally, we will summarize both chapters and draw some conclusions that inform our next steps.

15.1 Creating positive impacts on streets, parking, and the public realm

In the September 29th, 2018, edition of The Economist magazine, they reviewed a book *"AI Superpowers: China, Silicon Valley and the New World Order"* [1]. In it, the reviewer notes that while American cities are restricting self-driving cars, "... the district of Xiong'an, 60 miles south of Beijing, is being built from scratch to accommodate them ...." A slightly more detailed description of this work is reported on by Ning Hong of CGTN.com in a May 28, 2018, article titled *"Self-Driving Cars: Road tests start in Xiong'an New Area,"* which observes that "... Xiong'an is striving to build a city of the future, and a home of innovative technology companies. Starting from scratch, city planners are introducing more innovative solutions in technology ...," and goes on to quotes Baidu's Shang Guobin Intelligent Driving Group as stating that Xiong'an's "... City Planning has adopted many designs for self-driving technology. For example, Vehicle to Everything and 5G in basic infrastructure could help us to develop a new line of technology." This summary, along with other reports noting that this could result in the elimination of traffic signals at intersections, is what many expect to be the extent of city adaptations to accommodate the automated vehicle.

If this expectation is in fact the case, we, as a society, are missing a great opportunity to reclaim our cities from the impacts of the automobile. Just as what

[1]Perkins&Will, San Francisco, USA

happened in the early part of the twentieth century, this will have again occurred due to our lack of a holistic view as to how our public realm works. For the second time, we will have allowed a single stakeholder, the private automobile (automated or not), to dominate this public space. What is being planned for Xiong'an is in fact only a small portion of the adaptation required to realize, to the fullest extent possible, the potential positive impacts of automated vehicles. Two key components of realizing this are: an understanding of the type of public realm that appeals to the majority of people while facilitating the social interaction required for a successful city; and the automated vehicle ownership model which will inform the design criteria for the type and size of the vehicle fleet that our public realm will need to accommodate.

In Chapter 14, we looked at the qualitative differences between the pre- and post-automobile public realms, as well as the differences between transit- and auto-centric suburbs. We also looked at the determinants of potential outcomes, such as the likelihood of private ownership versus subscription-accessed fleet-owned automated vehicles and the impact this will have on fleet size and street design. In contrast to the silo'd adaptation design strategies noted above for the City of Xiong'an, this chapter will examine through a more holistic viewpoint the potential positive impacts upon the urban and suburban public realm. This viewpoint considers the automated vehicle as being only one of many stakeholders and modalities competing for space in the public right-of-way. With its privilege removed, the vehicle, automated or not, is simply one of many infrastructure components of a reclaimed city, town, or neighborhood.

15.2 Precedents

Before examining some recent studies looking at potential changes to the street, it is worth taking a moment to discuss two major public–private partnership master-planned developments of publicly owned properties currently under construction in California's San Francisco Bay Area. Each of these developments chose from the very outset to adopt the holistic strategies which are outlined in this chapter and Chapter 14. And each of these developments is now in a position to show to both regulatory and financial decision makers that the holistic approaches being proposed, even in a preshared automated world, are feasible under current market conditions, and that their qualities will only be enhanced in a shared automated mobility environment. What both of these developments demonstrate is that by learning from the mistakes we have made over the last century in the fields of planning, urban design, and architecture, we can start to implement the public realm scenarios described in these chapters, regardless of the status of automated vehicle implementation

15.2.1 Treasure Island/Yerba Buena Island

The Treasure Island/Yerba Buena Island development is the first of these two master plan developments that has taken an explicit design stance to move the automobile lower down the modality hierarchy. Located on the former Naval Station Treasure Island, the 400-acre island will have 8,000 new homes, along with approximately 450,000 square feet of new retail and commercial space, and 300 acres of urban farm, parks, and wetlands, and the construction of which is being overseen by The City and County of San Francisco's Treasure Island Development Authority (TIDA) [2]. This new community of at least 20,000 residents will be housed in three distinct neighborhoods located such that they are all within a 10-minute walk of a new ferry terminal and neighborhood commercial retail center. Whereas there are existing vehicular on- and off-ramps connecting the island to the San Francisco-Oakland Bay Bridge, their capacity, as well as that of the bridge itself, are limited. Recognizing this capacity limitation, a new ferry service accommodating both pedestrians and cyclists is designed as the principal means of accessing the island, being only a 10-min boat trip from downtown San Francisco's Ferry Building (Figure 15.1).

During the master planning phase, a conscious decision was made that the hierarchy of modalities would place active transportation (pedestrian and bicycles) at the top, followed by on-island neighborhood shuttles in the middle, with private automobiles being at the bottom. The design of the residential neighborhoods is such that the pedestrian and bicycle infrastructure is paramount, with a variety of routes linking them to the new ferry terminal. These routes vary from open park-like settings with views to more sheltered intimate mews-like routes, each of which considers the unique micro-climate of the island.

As part of the de-emphasis of the automobile, the regulatory parking require-ments for the individual development parcels have been reduced to less than one space per unit, with each space being sold separately from the purchase of the residential unit itself. Located close to the center of the island will be parking structures which are designed to account for both short- and long-term parking needs that the island retail, commercial, or residential market may require. These central parking garages will be served by the on-island neighborhood shuttles. Testing of an automated on-island shuttle linking the residential neighborhoods to the new ferry terminal is planned with the intent that when the residential devel-opments are ready for the market in 2021, the on-island shuttle will be in a position to be fully automated, California Department of Motor Vehicle (DMV) approvals permitting.

15.2.2   Mission Rock

The second master-planned development is Mission Rock, a 28-acre brownfield development of former railroad yards currently used as a surface parking lot for the nearby San Francisco Giants baseball park [3]. This Port of San Francisco-owned site will accommodate between 1.3 and 1.7 million square feet of commercial development, between 150,000 and 200,000 square feet of retail/makerspace, and approximately 1,500 new residential units, housing over 3,000 new residents. Included in the development is an 850,000 square foot parking structure which is planned to serve, on a shared basis, the needs of the adjoining ballpark as well as those of the development's commercial and residential users.

Figure 15.1   Treasure Island/Yerba Buena Island illustrative master plan

As we saw in Chapter 14 with regards to the neighboring UCSF Mission Bay Campus, should parking demand decrease over time, the parking structure could potentially be dismantled and its site developed for a higher and better use. Due to the proximity of existing light-rail, bus, and water-taxi transit, combined with an infrastructure designed to enhance active transportation, none of the individual development parcels are required to have on-site parking. Building occupants instead having the option to purchase parking in the parking structure which is located at the south end of the development adjacent to 3rd Street, and its Mission Rock MUNI light-rail station. The design intent is that persons arriving by private automobile park in this structure and walk to their destination. Similar to Treasure Island, the hierarchy of modalities is active transportation (walking and bicycle) being primary, with on-site vehicle use designed primarily for taxi/ride-hail and package delivery only; the street and curb designs reflect this hierarchy (Figure 15.2).

What do these two developments teach us? These two developments are being developed by private developers whose charge was to create the maximum rate of return for their public partners, the City and Port of San Francisco, respectively, who remain owners of the land being developed. In other words, in a preshared automated mobility environment, in one of the most competitive development markets in North America, the private market has decided to invert the traditional modality hierarchy by placing active transportation at the top, with transit just below, and the private automobile at the bottom. Having a combined valuation of over $8 billion, a decision like this is not made lightly, yet in both instances, the developers are predicting a market whereby the quality of the public realm is paramount and that such quality cannot be achieved in the traditional automobile dominated public realm.

Figure 15.2   Mission Rock Waterfront (Steelblue)

If this is where the real estate development market is trending now, it reinforces and validates the observations made in Chapter 14 regarding the importance of providing a quality public realm. Further, this trend is occurring in a San Francisco Bay Area mobility environment where only 18% of daily commuters use transit [4]. With increasing shared mobility through shuttles, ride-hails, micromobility, and active transportation, and with plans to provide enhanced transit, the viability of these types of developments will only increase. Therefore, using these master plans as models for future development is both an appropriate and realistic proposition.

15.3   Studies

Returning to recent studies examining the potential changes that a shared automated mobility environment could have on the public realm, the following includes two collaborative studies conducted in 2017, which examine potential street designs as well as three separate scenario planning workshops investigating the potential impacts on neighborhood typologies. The street studies involving both cross-section and curb/sidewalk design were separate collaborative studies carried out in 2017, with one involving Perkins+Will and Arup [5], and another involving Perkins+Will, Nelson\Nygaard and Lyft [6]. The collaborative studies examined the potential impacts of a shared automated scenario on curb-side use and parking as well as the required travel lanes of existing urban arterial streets. The scenario planning workshops in 2017 included two led by the University of Oregon's Urbanism Next program [7], and a third led by The University of San Francisco (USF) at the 2018 AVS Symposium [8]. The author participated in each of the collaborative studies, as well as being a facilitator at each of the scenario planning workshops.

The University of Oregon's scenario planning workshops involved the study of prototypical neighborhood environments which were based upon actual neighborhood within the City of Portland. Current land-use, infrastructure, and transit patterns were used as the starting point from which future scenarios could be developed. The workshops looked at three neighborhood typologies consisting of the urban core, a "streetcar" suburb, and outer suburban strip-mall locations. Similarly, USF's 2018 AVS Symposium workshop [9] focused upon three street typologies: Commercial Shopping Corridor, Mixed Urban Arterial Corridor, and Suburban Neighborhood Street. For each of these street typologies, the workshop participants were asked to sketch out their assumptions, develop conceptual street designs, and list policies and strategies to achieve these visions. More detailed descriptions of the findings and recommendations of each of these scenario planning workshops are contained in separate white papers and reports, whose lead authors are Professor Nico Larco of the University of Oregon's Urbanism Next program and Professor William Riggs of the University of San Francisco, respectively. Links to each are shown in [7–9] (Figure 15.3).

15.3.1   Street study: San Francisco

Having presented together on a panel organized by SPUR [10] in May 2014, titled "Designing for the Driverless City," Perkins+Will and Arup decided to collaborate on a study [11] to examine the potential impacts of shared automated vehicles on the travel lane and curb requirements of a city street. The chosen street was Fourth Street, in the recently up-zoned South of Market Area (SOMA) district of San Francisco [12].

Originally laid out as a light-industrial/warehouse district, the SOMA neighborhood, located south of the traditional commercial core of San Francisco, is characterized by large blocks (800′ × 550′) and wide streets (80′ right-of-way typical). These large blocks are sometimes trisected in the long dimension by 40′ wide alleys and sometimes bisected by similar alleys in the short dimension. Over the past 20 years this neighborhood has served as the incubator space for numerous software technology start-up companies, and with the extension of a light-rail line through this neighborhood, the City of San Francisco took this as an opportunity to up-zone the area from light-industrial to office, commercial, and multifamily residential. Traditionally, this neighborhood had been poorly served by transit.

Running from the commercial/retail land-use of Market Street at the north end to the transit-oriented development surrounding the CalTrain depot at King Street to the south, the block of Fourth Street between Folsom and Harrison was chosen as being representative, with up-zoned residential, commercial, and a large neighborhood-serving retailer. It is currently served by two transit lines, has unprotected bike lane infrastructure and serves the southbound on-ramp to U.S. Route 101 (Highway 101) at 4th and Harrison. The new T-line light-rail "Central Subway" runs below the street with the Yerba Buena/Moscone station being located at the north end of the block at 4th and Folsom.

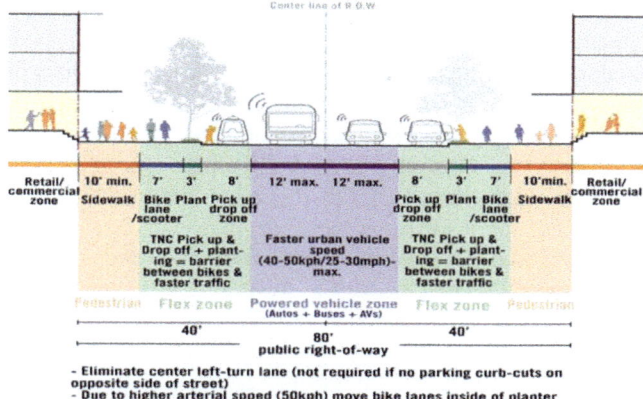

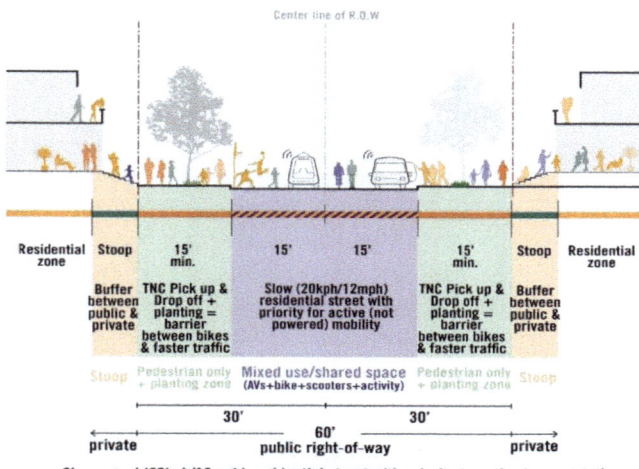

Figure 15.3   (a) Mixed urban arterial corridor (AV future) and (b) suburban neighborhood corridor (AV future)

In its current configuration, the 550′ north-south length of this block has four one-way south-bound travel lanes, two driveway curb-cuts on the east side, and two 20′ curb-to-curb alleys, Shipley and Clara, on the west side. The curbs have a total of six red-striped fire-hydrant zones (three on each side), one 15 foot wide mid-block signal-controlled crosswalk, one 75′ long dedicated bus-stop zone, one accessible on-street parking space with curb ramp, and a total of 22 on-street metered parking spaces, seven metered motorcycle parking spaces, six public bicycle racks located in a former on-street parking space and an eighty feet long on-street Ford "GoBike" bike-share dock facility along with one 22 foot long on-street "seating parklette" occupying a former parking space outside a coffee retailer. The current sidewalks are 9′-6″ wide on the east side and 12′ wide on the west side.

In addition to the mid-block crosswalk, there are traffic signal controlled cross-walks at each end of the block; however, the freeway on-ramp prevents southerly continuation of the west side sidewalk. Currently, the block has limited sidewalk activity from neighboring properties, this being confined largely to the retail super-market at 4th and Harrison, a coffee retailer across the street at 4th and Clara, a private K-8 school located at 4th and Folsom along with a neighborhood-serving senior social center between Shipley and Clara Streets. The remainder of the parcels are either multistory office, two-storey light-industrial supply businesses, or surface parking lots.

The alternative scenario examined the potential land-use impacts of the significant up-zoning that the recently approved Central SOMA Plan anticipates for this area. Where previously it has been predominantly light-industrial and ware-housing with 40′-60′ height limits, mixed-use office and residential is now allowed with heights along the entire block going up to 85′ (eight stories), or higher with conditional use permits. Based upon this up-zoning information, Perkins+Will developed massing plans and floor area yield studies for consolidated parcels. From this information, Arup developed trip generation and curb-side demand traffic models from which they created three modality scenarios: existing, transitional, and shared automated vehicle (AV). Using the shared AV modality scenario, Arup determined the travel lane and curb requirements, the latter being primarily for passenger pick-up and drop-off; it was from these requirements that Perkins+Will generated the following revised street cross-section [13].

Current street cross section:

80′-0″ ROW with 58′-6″ curb-to-curb; includes 4 × 11′ travel lanes plus 7′ parking lanes at each curb

Base-line modality split for 4th Street is:

48% legacy automobiles; 32% active mobility; 11% light-rail, and 9% legacy bus

Shared AV modality scenario, based upon NACTO's Blueprint for Automated Urbanism [14], amongst other data, is:

31% AV's (up to 75% of which are shared); 40% active mobility (pedestrians, bikes and scooters); 15% light-rail, and 15% AV bus.

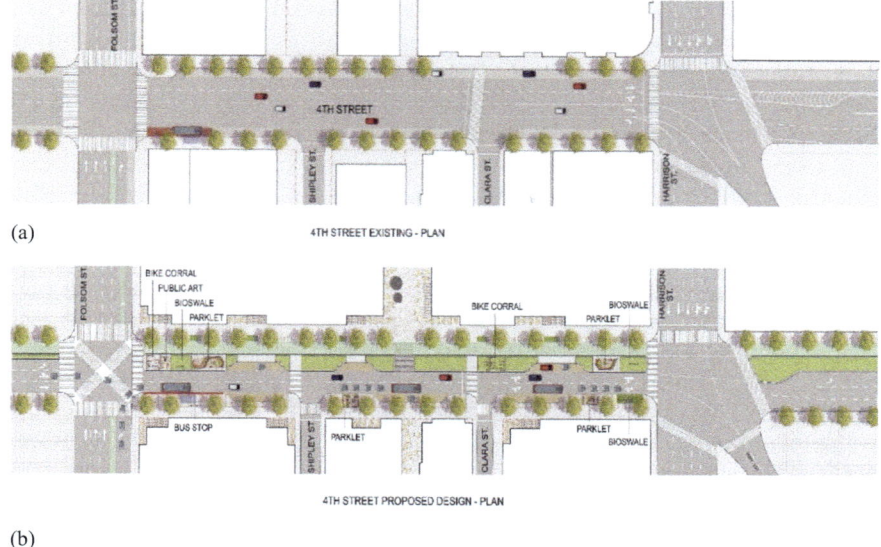

(a)

4TH STREET EXISTING - PLAN

(b)

4TH STREET PROPOSED DESIGN - PLAN

Figure 15.4   Fourth Street: current (top) and revised (bottom)

Resulting shared AV street cross-section:

- 2 × 10′ travel lanes plus 8′ bus and TNC/e-commerce drop-off zones on each side of the street
- Area reduction of road pavement (i.e., space between curbs) from 73% to 45% of overall street cross-section (i.e., right-of-way).
- Relinquished pavement is primarily given over to protected two-way bike lanes, widened sidewalks, and stormwater treatment gardens (see Figure 15.4).

Regarding the curb demand, there was a similar reduction to that seen for travel lanes. Whereas approximately 80% of the current curb is dedicated to parking, bus-stop, and loading zones, in the shared AV modality scenario, the need for on-street parking has been eliminated, with the result that passenger pick-up and drop-off zones drop to less than 40% of the overall curb length, and with freight or e-commerce drop-off accounting for an additional 10% of curb. The resulting cross-section and curb allocations allow for the creation of a significantly enhanced public realm which reclaims up to 38.5% of the curb-to-curb right-of-way from the automobile for redistribution to active- or micromobility.

15.3.2   Street study: Los Angeles

While the previous study focused primarily on travel lane and curb requirements, a subsequent study carried out by Perkins+Will and Nelson Nygaard in 2017 at the request of Lyft [15], a TNC ride-hail company, looked at the potential impacts that a shared AV mobility environment could have upon both a major arterial boulevard located in the Los Angeles metropolitan area as well as a typical suburban residential street. The locations chosen for study were Wilshire Boulevard at Veteran Avenue in the Westwood district of Los Angeles and Dos Rios Drive above Hillhaven Avenue in the Tujunga Canyon neighborhood of Los Angeles.

Wilshire Boulevard at Veteran is a major east-west arterial boulevard, adjacent to the University of California's Los Angeles campus, that connects Beverley Hills and West Hollywood to the east with Santa Monica to the west. The portion at Veteran is also one-block from the on- and off-ramps serving Interstate 405. At this location, the right-of-way width of Wilshire is 125', with four travel lanes in each direction, two dedicated left-turn lanes serving Veteran and 7' sidewalks on each side; Veteran has a 100' right-of-way with two travel lanes in each direction, one dedicated left-turn lane serving Wilshire and similar narrow sidewalks on each side. This specific location was chosen as it has both high-density mixed-use development potential on its northeast corner and is adjacent to the walkable Westwood Village on the south side of the UCLA campus.

Using NACTO capacity figures for both the current condition, as well as for a future shared AV environment (based upon their "Blueprint for Automated Urbanism" referenced earlier), it was felt that in order to properly reflect the benefits of a shared mobility environment, the capacity metrics needed to be changed to reflect people-per-hour throughput instead of the more traditional vehicles-per-hour throughput; the latter privileging single occupant vehicles at the expense of higher capacity shared vehicles or bus rapid transit. To establish a baseline, the metropolitan-wide Los Angeles modality split data from 2015 was used. The 2015 LA metro modality-split is as follows:

67.9% drive alone; 10.6% transit; 9.3% car-pool; 3.6% walk; 1.6% taxi/ motorcycle/other; 1.2% bike, while 5.7% work from home.

For 2035, the modality-split for Wilshire Boulevard is based upon the following assumptions that there will be high-capacity AV bus rapid transit, up to 50% shared TNC (similar to the 2018, 49% Lyftline usage in San Francisco and 2019 GoMonrovia shared ridership noted in Chapter 14), dedicated and protected bike lanes and an enhanced sidewalk public realm that encourages bike and pedestrian traffic. While these 2035 modality-split numbers are by necessity generalized, they are reasonable for scenario planning purposes. The 2035 Wilshire Boulevard modality-split is as follows:

50% transit/bike/walk (up from 15.4%); 25% drive alone (down from 67.9%); 20% car-pool/ride hail/taxi (up from 10.9%), and 5% work from home (down 5.7%)—in our opinion, these last two numbers are probably conservative.

Using these modality split data, and our performance metrics which focuses on people per hour instead of vehicles, the nominal Wilshire Boulevard throughputs were 29,600 people-per-hour (pph) for its current configuration and 77,000 people-per-hour (pph) for its 2035 and beyond scenario (see Figure 15.4). These figures broke down as follows:

Transit:

2015 = 3,600 pph total in both directions based upon traditional city bus.

2035 = 35,000 pph total in both directions based upon high-frequency (60 s headway) low-platform automated bus-rapid transit (BRT) with dedicated lanes and stations including preboarding fare purchase.

Cars:

2015 = 11,600 pph, with 7,400 pph in single occupancy vehicles

2035 = 3,300 – 13,200 pph, including shard rides

Bikes:

2015 = n/a (there is currently no dedicated bike infrastructure at this location)

2035 = 7,500 pph due to new dedicated and separated bikes lanes

Pedestrians:

2015 = 18,000 pph (potential capacity only; existing foot traffic is nominal)

2035 = 18,000–22,000 pph (meets or exceeds capacity due to widened sidewalks, enhanced public realm and ground floor activation of new mixed-use developments on both sides of the street).

Similar to what was discovered in the 4th Street, San Francisco study, there was a significant reallocation of space from an auto-centric modality to transit and active- or micro-mobility. Whereas in its current layout, Wilshire Boulevard with an overall potential throughput of 29,000 people-per-hour dedicates 88%, or 110', of its overall 125' right-of-way to automobiles and buses, the shared automated high-capacity BRT scenario with a potential throughput of 77,000 people-per-hour, dedicates 51.2%, or 64' of its overall 125' right-of-way to powered vehicles, including left turn lanes and BRT station platforms. This returns 36.8% of the street's curb-to-curb cross section to active- or micro-mobility with dedicated and protected lanes, wider more attractive sidewalks and storm-water filtration gardens (see Figures 15.5 and 15.6).

This is comparable to the 38.5% return noted earlier for the 4th Street, San Francisco study (Figure 15.4).

What both of these studies are telling us is that the potential exists to reclaim between 37%–39% of our current right-of-way from powered vehicles for use by active- or micro-mobility in addition to enhancing the experiential appeal of what currently can be characterized as "traffic sewers." When we look at the results of the two street studies cited above, we can see that the tyranny of the single modality can be broken; we can identify a pathway to achieving a street typology and usage that does not privilege one group of users at the expense of all others.

15.3.3   *Street study: Suburbia*

Whereas the previous two studies were for urban streets, Dos Rios Drive above Hillhaven Avenue in the Tujunga Canyon neighborhood of Los Angeles was chosen to examine the potential impacts on a typical 1960s–1970s era single-family suburban sub-division. The nearest transit is 1¾ miles away horizontal and 500' vertical, located at Summitrose Street and Tinker Avenue and consists of three LA Metro bus lines. Up to 1/3 of the walk from Dos Rios Drive to this transit stop is on streets with no sidewalk or provisions for pedestrian safety. Dos Rios Drive has a

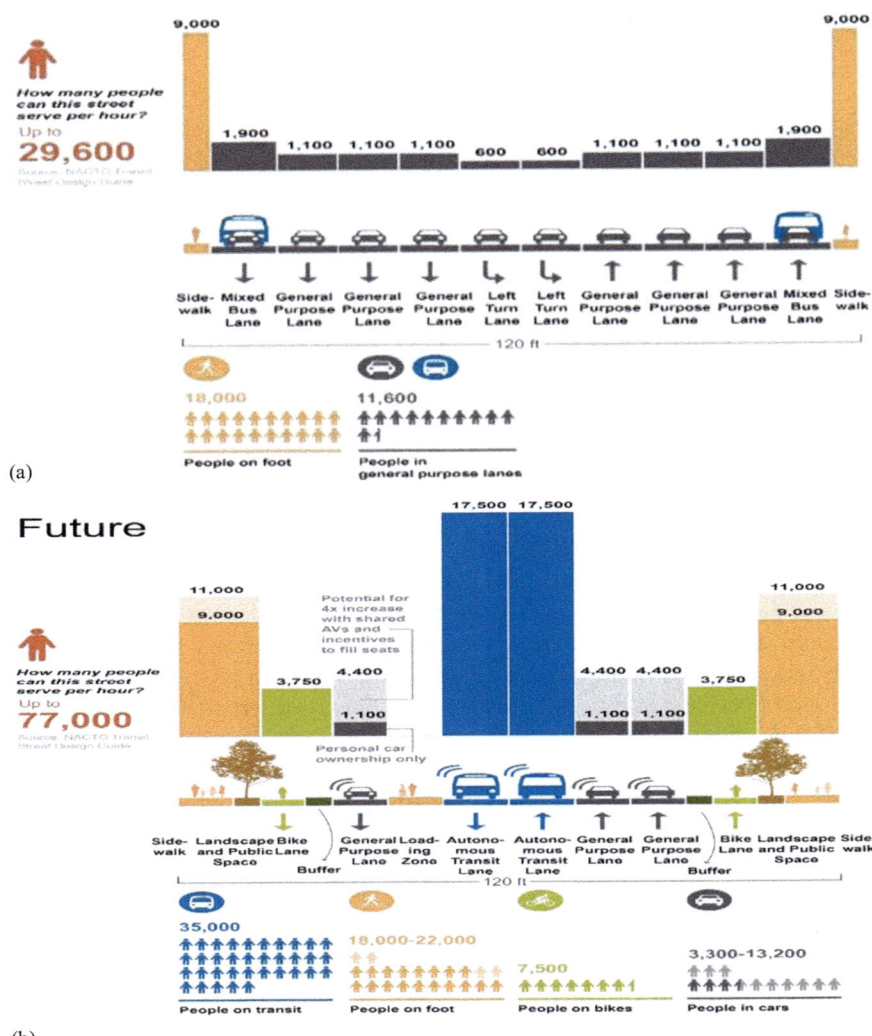

Figure 15.5 (a) Wilshire Blvd. capacity: present and (b) Wilshire Blvd. capacity: future

Figure 15.6   Wilshire Blvd. illustration of potential future (Perkins&Will)

45′ right-of-way with 5′ wide sidewalks; the turn-around shown in the background of the illustration below has a 45′ radius. All individual residences have either two- or three-car garages with driveways and full-width curb cuts; on-street parking is presumably for visitors, or for residents with surplus cars beyond the capacity of their garages and driveways. There is no travel-lane striping or marked-out on-street parking spaces.

Due to the singular nature of this residential loop road, the usual metrics that we apply to streets do not make as much sense. Typical of many loop or cul-de-sac (dead-end) streets within residential sub-divisions (sometimes referred to as a "loops-and-lollipops" typology), there is basically no through traffic or transit. Vehicles using these streets are those of the residents, their guests, delivery vehicles or city utility vehicles such as garbage trucks or emergency vehicles. Therefore, once we have provided sufficient space for these two city utility vehicles, the remainder of the space can be considered differently than one would for a typical through street, perhaps similar to a woonerf or "complete" type street. Rather than view the street as simply a conduit for vehicles, why not look at it the way that some of its residents already do, that is, the children. We are familiar with cul-de-sacs being commandeered by neighborhood children for use as impromptu play or sports areas. This tells us something about both the volume of traffic as well as the wants and needs of some of those living alongside this space. They are de-facto stakeholders so why not accommodate them? A low-speed traffic-calmed shared space could achieve this (Figures 15.7 and 15.8).

When looking at this street typology in the context of a shared and automated mobility ecology, and using NACTO's residential street guidelines, the design can start to look considerably different from its current layout of a 35′ carriageway within a 45′ right-of-way. And if one adds slow-speed residential area speed limits, such as the European Union's 20 kph (12 mph) "20 is plenty" guideline, the travel

Figure 15.7   Dos Rios's illustration of potential future (Perkins&Will)

SIDEWALK LOADING BIKE   AV   SIDEWALK
AREA   LANE LANE

DOS RIOS CROSS SECTION - EVOLVED

Figure 15.8   Dos Rios's illustrative cross section (Perkins&Will)

lane widths can be reduced to the range of 9′ each. Further, so as to encourage active mobility such as cycling, a dedicated 6′ bike lane can be provided for younger riders to use. And with the relatively low traffic volume of this street typology, there is no reason that the vehicle travel lanes themselves cannot be "sharrows" used by more experienced or confident riders. And since we are looking at this in the context of automated vehicles deployed in a shared or fleet environment, we should provide dedicated loading/unloading (pick-up/drop-off) lanes, instead of having the AV's relying on individual driveways. This arrangement within a 45′ right-of-way could allow for 2 × 9′ travel lanes + 1 × 6′ bike lane + 1× 11′ loading or pick-up/drop-off lane all within the current 35′ carriageway while retaining the existing curbs and gutters to manage storm-water flow and drainage.

Within the context of a 20 kph (12 mph) slow-street environment, this carriageway mix could be treated as a woonerf or shared street such that the 5′ sidewalks are no longer seen as pedestrian unfriendly, but simply as another protected surface that is available to pedestrians seeking additional safety or by small children on bikes. The point being that with a woonerf or slow street design, and with signage to drivers or vehicles entering a residential neighborhood requiring that vehicles yield to pedestrians, the modality delineations noted above can be handled in a variety of ways such that each street or block to be tailored to the specific needs or wants of a specific block or sub-division. Thus, the bleak vehicle dominated public realm of suburban sub-divisions described in Chapter 14 has the opportunity to be transformed into spaces that encourage active transportation, serendipitous interaction between, or occupation by, residents and which are no longer dominated by a single modality or stakeholder.

15.4   Parking

So far, we have been looking at the potential impacts on the street; however, implicit in both of the above studies is the understanding that in a shared AV mobility environment, curb utilization would no longer be dedicated primarily to on-street parking. We need to examine what is the likely future for parking as well as the curb in general. In Chapter 14, we identified that the likelihood of subscription-based access to automated vehicles is more likely than a direct ownership model. A subscription model will require maximizing vehicle utilization such that parking, or nonrevenue earning periods will be minimized, being confined to cleaning, battery re-charging or routine maintenance, all of which will take place at designated off-street facilities. Simply put, the demand for on-street parking is likely to disappear to be replaced by curb-side pick-up and drop-off. However, the street studies referenced above illustrate that not all of the curb currently assigned to on-street parking will be required, thus freeing it up for other uses. With the potential loss of revenues from parking meters and related fines, these will need to be replaced perhaps by curb-use charges for TNC's, private employer-shuttles, and e-commerce goods delivery.

In addition to the elimination of on-street parking, there will likely be a significant reduction in off-street parking for many of the same reasons. As noted in Chapter 14, we are currently seeing a significant reduction in the need for off-street parking where developments are located in transit-rich environments that have multiple mobility options; in such locations, parking demand could potentially be reduced by between 80% and 90% [16]. Also, as the cost of providing parking is externalized through "unbundled" pricing, it is causing the consumer to consider alternative options to owning a private car; the annual operating costs of which, in the United States, is typically in the range of $9,000 [17]. The result is that where possible, developers are willing to eliminate the cost burden of constructing structured parking in urban areas, which can range from between $30,000 and $60,000 or more per space in major metropolitan areas.

However, the further we move away from the urban core, the strategic storage of subscription-based vehicles may be required so as to reduce both the number and distance of dead-head journeys for "zombie" vehicles (i.e. vehicles operating between customers with no occupants on-board). In this instance, while dedicated off-street parking demand could potentially be reduced by between 50% and 60%

[16], there will be a need for the reconceptualization of off-street vehicle parking away from dedicated on-grade or structured parking to centralized vehicle charging, maintenance, and storage facilities. OEM subscription services and others may combine to construct shared facilities which could be overseen by fleet management and operation companies, similar to what we currently see at many major airports for rental car companies. An alternative to this could be provided by start-up companies such as ApParkingSpot [18], a Swedish company who are to surplus off-street parking spaces what Airbnb is to spare residential bedrooms, whereby they could allow OEM's to disaggregate their parking to a very localized or micro-level, thus significantly reducing "zombie" or dead-head VMT. This indicates that there is a potential new market which allows fleet car owners to minimize their "zombie" vehicle charges by utilizing a "micro-parking" logistical strategy. Moving further out from the urban core, these centralized strategically located facilities could reduce in size, perhaps taking over the role of the former neighborhood gas station. And as we move into ex-urban and rural locations, there may still remain a role for traditional off-street parking.

From what we see, both in the study literature published to date as well as the scenario planning exercises, there is no one-size-fits-all solution. However, studies are showing that a subscription-based vehicle access model will eliminate the need for on-street parking, replacing it with curbside pick-up and drop-off at between 40% and 50% of current curbside demand based upon the street studies referenced earlier in this chapter. For structured parking, the reductions will be dependent upon density and the richness of alternative mobility options. In transit-rich neighborhoods with multiple mobility options, the reduction could be between 80% and 90% as noted earlier; in the less dense urban periphery, this reduction could be between 50% and 60%, while for exurbia and rural communities these reductions will most likely be more modest but should be adequately accommodated within the densification of existing parking infrastructure whose space requirements could be significantly reduced by automated, or robot, parking systems.

From a land-use point of view, the reduction or elimination of off-street parking will have impacts throughout the urban-suburban-exurban spectrum. In urban cores, it will allow for both the infill of existing surface parking lots and the replacement of existing parking structures with new higher and better use development. A current prototypical example, the Hive parking structure, has been constructed in Oakland, CA, where a developer, Signature Development Group, and forward-thinking parking provider, CityLift parking, teamed up to install a fully automated, enclosed, seven-story parking tower for 39 vehicles with a footprint of approximately 1,800 gross square feet which formerly would have required at least 14,000 gross square feet of a surface lot [19].

This 7:1 ratio of parking footprint densification in today's mobility world indicates that a 10:1 reduction is entirely feasible, especially if the next generation of robotic parking systems is considered, such as Volley Automation's [20], which could increase the current rack-and-rail densities by up to 23%. The increase in building density, resulting from the elimination, or repurposing of the space formerly required by vehicle parking, also has the knock-on effect of making transit more viable and economical. Plus, with the elimination of the costs related to structured parking, the viability of providing housing and commercial spaces at a lower price point will be increased, thus enabling new developments serving a broader economic spectrum than current market conditions allow for in most major metropolitan areas.

As we move out from the core the potential exists to recapture the acres of existing on-grade parking to allow for buildings that can be brought forward to the existing right-of-way, instead of having buildings being located behind a parking-lot foreground as we see with current strip-mall typologies, thus creating a new denser and vibrant street experience. An example of this as reviewed in Chapter 14 with the City of Westminster's repurposing of a former regional mall. Additionally, the Urbanism Next scenario planning workshop addressing the former mall in Gresham, Oregon, looked at in Chapter 14, illustrates the potential area that can be reclaimed when overlaid by the downtown Portland, Oregon, street grid—these parking areas could translate into major infill development opportunities and if strategically located could help to reinforce transit routes. And for traditional suburban housing developments, the opportunity exists for the capturing of the typical two- or three-car garage square footage for use as additional living space, perhaps as either stand-alone or in-law units for multigenerational housing as described in Chapter 14.

15.5 E-commerce and its potential impacts

Whereas we noted in Chapter 14 the impacts of e-commerce, or online shopping, on traditional bricks and mortar retail stores, we need to also understand the potential impacts this could have on our streets, sidewalks, and building entries. Up until very recently, mail-order notwithstanding, most people paid for and took home their purchased merchandise from a real physical bricks-and-mortar store. And while some stores offered home delivery services, this came with a surcharge and was thus the exception. As of 2018, online retail represented 10% of all purchases and experienced a 17% annual growth rate in 2017 [21], which many analysts are predicting may grow at an even greater annualized rate.

Traditionally package delivery was confined to mail or private delivery companies with a delivery frequency that could be accommodated by the existing curb space and porch (or lobby) of a building. However, with the growth noted above, we are already seeing the impact of bulk online deliveries overwhelming what had been the "mail-room" in multifamily apartment housing, as well as adding to competition for the curb space outside of buildings receiving deliveries.

Recognizing the inefficiencies of using relatively large delivery trucks making multiple stops to deliver smaller packages, both delivery companies and online retailers are experimenting with the drone or robotic delivery (both terrestrial as well as aerial). Leaving aside the aerial drones we can see that the widespread introduction of terrestrial delivery robots will place more demand on the space in our finite rights-of-way. Further, the classification of these delivery robots as motor vehicles, which would allow them to run on the street pavement, is not yet established. If due to size or weight they are not regulated as commercial motor vehicles, will they be confined to the sidewalk? And if so, how will the pedestrian public react to having to share or compete with them for sidewalk space? The almost literally overnight introduction in 2018 of electric scooters to the streets (and sidewalks) of San Francisco may offer us a lesson as to how we need to pay attention to how we design for this. In a workshop at the Urbanism Next 2019 Conference [22] studying the potential impacts of E-commerce on our built environment, the most negative impact identified by the workshop participants was the use of the sidewalk for delivery robots. It was felt that under current conditions, there was already too much competition for sidewalk space, such that in the future either sidewalks need to be significantly widened or these delivery robots need to share a dedicated lane with other nontraditional modalities. We should therefore recognize that, along with other micro-mobility modalities, such as electric scooters and bikes, we need to create an entirely new category of street lane, similar to or sometimes in conjunction with, what was created for protected bike-lanes.

Further, we will need to account for the point of origination of these delivery robots as well as their destination. The point of origination is where goods would be transferred from larger delivery vehicles to the delivery robots. Given their relatively small size and short range, these transfer points will need to be highly distributed, occurring every several blocks or so. This could be a new use for former neighborhood gas stations now made redundant by the elimination of internal combustion engines. The pilot work that Amazon undertook in Bellevue, WA, will have generated data informing both the geographical distribution of these transfer points as well as to the sidewalk (or pavement) infrastructure required to meet their needs, as well as points of conflict with other users. Like the data generated by TNC's, cities should require the sharing of these data as a condition of e-commerce delivery entities being able to use streets and sidewalks, thus reinforcing the notion that all stakeholders need to be considered as new mobility users arrive in this space.

Regarding their termination point, we will need to consider the liability issues surrounding delivery robots traversing the private property. Do they access the front door or porch of a house or do we create a new type of docking port that allows them to discharge their cargo in a safe, secure, and weather protected environment—could this be a potential new use for the traditional residential garage, which in Chapter 14, we identified as being potentially made redundant? A similar situation would occur at both multifamily apartment housing as well as at commercial buildings; however, since the latter often have designated loading docks, or areas which could be redesigned to accommodate these micro-mobility delivery robots, the potential problem may be confined to those buildings currently without delivery or loading docks areas.

And what about the demand for the curb-space outside both individual dwellings? We are seeing that many online retailers are finding it more economical to use TNC's to deliver their customer's online purchases. Do we now have a situation of e-commerce deliveries competing for the same curb space as passenger carrying TNC rides? Earlier in this chapter, we identified that passenger TNC pick-up and drop-off would probably require around 50% of a given block's curb space. With the addition of TNC e-commerce deliveries, does this require more curb and if so at what point do the potential benefits of reclaiming the public realm get eroded to the point of being meaningless? This should cause us to think about how we want to handle the delivery of online purchases. Do we have a "right" to have purchases delivered to our front door or can we say that in the greater interest of society these deliveries will be made to a more centralized location, which itself could become a neighborhood focal point?

And pulling out further, the increase in deliveries will require more and larger distribution centers on the outskirts of towns or metropolitan areas. How these new centers are served as well as, their locations, will need to be considered such that potential social equity issues do not arise from their locations being confined to low-income neighborhoods.

What could be the best ways to address these questions and integrate the scenarios described above into the public realm? Just as we noted in Chapter 14, how planning for the integration of all forms of mobility into the urban fabric needs to be approached in a holistic manner, so too will planning for merchandise delivery. An understanding of delivery logistics will need to become an integral part of how Urban Designers and Planners design our towns and cities. Just as they currently need to understand the role of traffic and utility infrastructure, going forward an understanding of freight and delivery logistics will become as essential as their understanding of traffic and transit.

15.6 Building design

In responding to the needs described above of accommodating e-commerce, the architecture profession will also need to account for the temporal pick-up and drop-off impacts of an on-demand fleet-accessed shared automated vehicles. While the requirement for virtually all on-site parking could be eliminated, providing instead for a discretionary 10% or so of previous parking demand through high-density mechanical parking, how are the remaining 90% of building occupants going to access their modality of choice?

If we eliminate up to 90% or more of often weather-protected on-site parking, from either residential or commercial buildings, how will building users access their on-demand vehicle in a safe, secure, and weather-protected manner? For instance, in the case of multifamily housing, relying simply upon a building lobby with 50' of adjacent white-painted pick-up and drop-off street curb outside will probably not work for families having to pack children off to school or unload a week's worth of groceries. And in the case of multistory office buildings, a similar condition of lobby and white-painted street curb outside will not be able to handle the temporal pressures of morning drop-off and evening pick-up of employees.

These examples are brought up so that we can start now to build into our building design guidelines or planning codes, the means by which building occupants can be served by this changed mobility reality in a way that is not det-rimental to the public realm. In Chapter 14, we discussed the conditions required for a vibrant street and how part of this vibrancy requires an active ground floor. This condition cannot be achieved if sidewalks are continuously interrupted by curb-cuts into porte-cochere's or vehicle pull-outs serving individual buildings. Solutions could include mid-block alleys serving either protected and secure building lobbies at the rear of buildings or short-term staging garages.

And as for the impacts of e-commerce, architects will need to accommodate into their building designs both the means by which merchandise is delivered and temporarily stored. Whereas many larger commercial and residential buildings are often required by planning codes to incorporate loading docks, that is the case for only a minority of buildings. So, what is to become of building lobbies and the adjacent sidewalk and curb outside as they accommodate an ever-increasing volume of package deliveries. And how will delivery robots overcome the tradi-tional walk-up entry steps of many older buildings? If we are already seeing building lobbies being overwhelmed by deliveries, is there a case for either a shared but secure block, or neighborhood, drop-off/pick-up point (staffed or un-staffed)? Or maybe the drop-off/pick-up point is itself a mobile automated robot vehicle that circulates the neighborhood, stopping on-demand outside individual buildings so as to allow the occupants to come out and retrieve their deliveries. Again, the impacts of online retail deliveries to individual buildings will need to be considered in such a way that they do not have negative impacts upon our public right-of-way, be that the sidewalk, curb, or street.

As was discussed in Chapter 14, when addressing the design implications for both building pick-up/drop-off and e-commerce, we cannot let one activity or sta-keholder dominate the public realm at the expense of all others. Potential solutions to these design implications have precedents from an earlier time which could revitalize both a street and retail typology. Just as we saw the resilience and adaptability of "street-car" or "inner-city" suburbs, perhaps we need to look at the reinvention of the "alley" and "post office" and acknowledge the "back-to-the-future" aspects of the roles they originally played and how their time may have come back.

With the potential elimination of 90% of weather-protected on-site parking and acknowledging the physical limitations of white-curb loading and unloading zones in front of building lobbies, maybe the service alley's time has returned? Typically, with only one mid-block curb-cut on the secondary, or perpendicular, cross-street, this allows vehicular access to the formal "rear" of individual buildings without disrupting the vibrancy of the formal building "front" facing out onto the main street. These service alleys, if designed per the NACTO Commercial Alley guidelines, could be both serviceable and create a secondary series of "places." Being able to stack slow-moving vehicles accessing an individual building, weather-protected secure pull-outs could be provided where building occupants could safely load or unload their automated vehicle, and do so in a manner that does not disrupt the continuity and vibrancy of their building's main "address" street. The humble rear alley, or mews, could be reinvented as an integral part of our

Figure 15.9   St. Vincent's Court mid-block passage, Los Angeles, CA

mobility circulation network taking some of the workload off the principal street network, while at the same time, if well designed, providing a more intimate secondary pedestrian network. Precedents for this secondary network already exist, from the "Mews" of London to the "Lanes" of old Savannah, from the "Passatge" of Barcelona to the often gated "Lilong" or "Longtang" alleys of Shanghai's 19th and early twentieth century residential developments. Another example of how this could be addressed is the St. Vincent's Court mid-block passage between 7th and 8th Streets parallel to Broadway in downtown Los Angeles (Figure 15.9).

With regards to the potential impacts of e-commerce delivery on building access and package storage on lobby design, maybe we need to look beyond the building itself. What if, in order to address these physical limitations, we looked instead at a neighborhood drop-off/pick-up point? This neighborhood nexus could potentially reinforce neighborhood sociability, and in doing so we will have returned to a model familiar to those who may have grown up in rural areas without individual postal delivery. This is a reference to the role of the post-office boxes in many rural small towns throughout the United States. As the postmaster loads the individual boxes each morning, the box holders usually start gathering in an adjacent coffee shop about 30 min before the post office lobby opens, allowing neighbors to exchange news and share a community-reinforcing cup of coffee with each other, before heading across the street to retrieve whatever the outside world has brought them and then go on about their daily chores. It would be nice to think that this (almost) daily community-building ritual could reappear in our cities and suburbs, thus transforming the anonymous act of purchasing online into an activity where one gets to meet with, and hopefully share time with, one's neighbors.

It is in this complete understanding and investigation of these secondary impacts that we will hopefully be able to integrate automated vehicle, new mobility, and e-commerce into the public realm in a manner that both mitigates potential disruptive impacts while also bringing new meaning and life to declining typologies. In the case of addressing how building occupants could access vehicles in a weather-protected and secure setting, there is the possibility of bringing new life to an old urban street typology, the service alley, with their potential to (re)create a secondary vehicular and pedestrian network overlay. And in the case of e-commerce, in addition to addressing building access and growing package storage impacts, there is the potential to re-establish a community-building activity which could mitigate the societal anonymity that this new retail model reinforces.

15.7   Summary of Chapters 14 and 15 (and the role of pricing)

At the start of Chapter 14, we outlined how that much of what we have been examining could be considered as falling into the category of Rittel and Weber's "wicked problems," which they described in their "Dilemmas in a General Theory of Planning" [23]. When we look at the public realm, be it in a major metropolitan center or a modest rural town, we are confronted with how we should define it. If we see a street simply as a means for conveying vehicles and surface stormwater run-off, or being a right-of-way for shared utilities, then we have a relatively simple, or "tame," problem to be solved, requiring the application of mathematical rules solving for capacity and efficiency; if we see the street as being a place of social interaction, contributing to what we call the "quality of life," and which supports all modalities (with walking being one of those modes) and stakeholders (from shop owners opening onto the street to City transit agencies), and into all of this we add the mix of the "tame" problem, we do indeed have a very "wicked" problem.

That is not said in a negative manner, but simply to recognize that our first step is to understand the very nature of the problem and that in solving for it, we need to account for a wide range of variables that will change based upon geography, topography, climate, culture, priorities, budget and yes, conveyance. And in recognizing that we must also understand and accept that there probably is no "right" answer, but instead solutions that are appropriate for their specific location, needs, and demands. There will be no "one-size-fits-all" solutions but instead responses that strike a socially acceptable balance between competing demands in which we will need to propose an optimized solution for the specific place and reality being studied. This may sit uncomfortably with some people; however, it is essential that we recognize the limits of our ability to analyze and cleanly formulate all of the societal inputs, and accept that our outcomes may need to be more nuanced or fungible. Successful urban environments are often "messy" and somewhat ambiguous; however, this "messiness" or ambiguity that allows them to be flexible enough to adapt to the changes that society or technology brings.

In examining the precedents discussed earlier, we should recognize and accept that, as social beings, how we interact with a space is as critical to its "performance" as the efficiency of its throughput be that for traffic or stormwater. We started out in Chapter 14 by asking the question which of the two locations in Paris and Shanghai that the average resident would prefer to be in; most people would accept that the pre automobile neighborhood is that location. And further, as we define "efficiency" we need to recognize the bias that a "vehicles-per-hour" metric has in favor of the single-occupant vehicle; we need to acknowledge that this comes at the expense of the multioccupant transit vehicle (both are counted equally). When we changed the performance metric to "people-per-hour," we saw dramatic changes to the cross-section of Wilshire Boulevard, the quality of its sidewalks and active-mobility spaces as well as a significant increase in the potential throughput of people. Again, when people are asked which of the two illustrated Wilshire Boulevard scenarios they prefer, they unanimously select the revised scenario.

But we are not going to be able to realize these scenarios by accepting the status quo, or by responding to the rapidly evolving urban mobility environment through implementing reactive or ad hoc changes to our street designs and transportation policies. To date we have had the "disruptive" model whereby the modus operandi of the new mobility providers has typically been "implement first, ask permission later," which is very much a part of Silicon Valley's "fail-early; fail often" culture. While this may be appropriate for developing software whose impacts are confined to the virtual world, it is an entirely different proposition in the real world of streets, vehicles, and pedestrians. To be fair, the transformative changes that we are addressing would probably not be happening without this culture; however, there is a limit to the amount of unstructured experimentation that can and should take place in the public realm.

In anticipation of both automated vehicles and the evolving mobility environment, we have seen that there are likely to be very strong economic incentives for OEM's to radically change their business model. If, as appears likely, there is significant value to be captured by moving from the current commodity-based (car-for-sale) model to a service or subscription-based model, the mass-market vehicle producers will change. We have seen this radical change in the music and movie industries, and as consumers become more comfortable with surrendering "ownership" and shifting to a "pay-as-use" basis, there is no inherent reason to believe that the auto industry will be spared this shift. We also need to recognize that this may occur for no other reason than institutional investors, requiring a better and more secure return on their investment, will force OEM's too. As they react to the economic externalities of their investors, OEM's will have to acclimate their consumer base ahead of time to this new model (e.g. Audi's "Own the Experience; Not the Car" and Volvo's "You used to buy music ... the car you can subscribe to"). The net result of this "shared" or subscription-based model will likely be a significant reduction in the fleet size required to serve a given population, with the simultaneous knock-on effects of parking demand reduction and curbside demand increase.

As the new mobility industry matures, it needs to be integrated in a structured manner into the overall mobility ecology. And we need to have a clear understanding of what we want that ecology to look like. It is this understanding of what this new mobility ecology can potentially provide, and could literally look like, that situates the above-referenced scenarios. If this is what we would like to see, then we need to reverse engineer the required process of public policy, pricing and physical implementation so that we can identify a path to get from here to there. In reverse order, physical implementation can range from the sublimely simple examples that Janette Sadik-Khan gave in her book "Streetfight: Handbook for an Urban Revolution" [24] for the re-striping of Times Square and Harold Square; she started with leftover paint from the NY DPW and lawn furniture from Target. If implementation can start off simply, the next steps are anything but easy and will require a clear set of goals and policies to enable them along with a determined body politic to resist the inevitable push-back.

Pricing is a proven strategy throughout Europe and Asia, which is largely accepted by those communities where it has been implemented as bringing back a quality of life to the urban core and controlling the number of vehicles using the road at any given time. For instance, whereas the residents of Stockholm initially resisted the city center charge pilot program, when it came time to make it permanent and seeing the benefits it had brought, the residents voted to retain it. Something like this will probably be required if we are to achieve the scenarios illustrated earlier in this chapter. However, to date in North America road usage pricing experiments have not gained support from the general public, much less downtown congestion pricing. But with the increase in commute hour and dedicated toll roads, there is a willingness on the part of some consumers to accept road usage pricing if they can experience immediately tangible benefits. Some are willing to pay to get there faster even as this lays the groundwork for an inequitable two-tier mobility system; those who can afford the charges get there quickly; those that cannot, do not. This source of potential social inequity will need to be addressed by those setting and managing these charges. Congestion pricing is more stick than carrot and as a result is a tough political sell. So, what could the "carrot" look like? Tax rebates perhaps?

But that is in today's privately owned, internal combustion engine (ICE) powered automobile environment. With a decline in gas-tax revenue, both the Federal and State governments will need to change their method of acquiring revenue to pay for transportation infrastructure. As the fleet transitions from ICE to electric propulsion, the main source of revenue is likely to be road usage fees, that is, "pay for what you use." And further, as the OEM's are likely to transition to a subscription-based model, the road usage fees can be embedded in the subscription charges. This consumer payment model, combined with the embedded road user fees, could facilitate an add-on pricing structure that incentivizes route selection away from congestion, ridesharing, or multimodal trips. However, in using pricing to modify or incentivize trip choice and behavior, this needs to be structured in such a way that we do not create the two-tier inequitable mobility system noted above for "express lanes."

Where there is pricing there needs to be public policy with a clear vision laying out goals and outcomes for the general public to see and buy into; you pay this, you get that. This will require that Federal/National, State, County, and City Departments of Transportation (DoT's), and Metropolitan Planning Organizations (MPO's) start engaging all stakeholders before the introduction of the shared automated vehicle so as to understand the current state of transportation and land-use planning. So as not to be put in a reactive planning situation we need to start by asking where current trends could take us if no pre-emptive planning is adopted; we need to ask where this evolving array of mobility options has the potential to take us; and finally, we need to all agree upon what that "vision" might be like and what are the required pre-emptive steps that we all need to take to get us there.

Simple? No, "wicked"!

15.8 Conclusion to Chapters 14 and 15

The introduction of a shared automated vehicle fleet into our mobility ecology has the potential to reclaim the public realm of cities and reinvent suburbia. However, this is by no means a given and will only be realized if certain conditions are met. Key amongst these will be the following:

- Restructuring the ownership model: In order to reduce the fleet size required to serve a given population, the OEM's need to move from the current private ownership model to a subscription-based shared-fleet model. The trend toward this model is already seen with Audi's "Own the Experience, Not the Car" and Volvo's "You used to buy music .... the car you can subscribe to" advertising campaigns.
- Reclaiming the public realm: A reduced fleet size, combined with the inherent space utilization efficiencies of an AV, working in conjunction with the elimination of on-street parking will allow for a significant redesign of our streets. This benefits the community at large and does not simply allow the public realm to be monopolized by one group of users.
- Lower building costs increase the viability of lower cost buildings: The elimination of on-site (building) parking will reduce the cost of multifamily housing units, as well as increasing the financial viability of urban and suburban infill. This could lead to a densification of both our urban and suburban neighborhoods which will benefit transit utilization.
- Sprawl: The elimination of one of the points of friction for long distance commutes will need to be addressed by externalizing the true costs of driving a single-occupant vehicle. A holistic, or regional, cost of sprawl will need to be externalized, with an appropriate pricing structure implemented to disincentivize long commutes and single-occupancy vehicles, while rewarding shorter trips, for example, trips to transit and use of multioccupant vehicles.
- New development to reinforce existing transit infrastructure: Metropolitan Planning Organization (MPO) programs, such as MTC's Plan Bay Area, need

to become the norm whereby development is directed toward areas where high-occupancy transit infrastructure already exists or is planned.

- Equity and effectiveness: The guiding principle in all of this should be that the era of privileging one group of users over another has run its course. Going forward, the needs and benefits of all stakeholders must be accounted for. Whatever decisions we make concerning mobility in all its forms and applications, they must bring the maximum benefit to society at large. A successful metropolitan area or society cannot function without everyone having access to a safe, reliable, affordable, seamless, and effective mobility network. This network may involve several modalities, but each must play its role as supporting and not be in competition with the other modalities. The transit consumer is concerned about getting from one point to another and is agnostic to the underlying organizational structure; our mobility networks need to reflect this reality and understand that that the consumer would like to be able to treat mobility as a service (MaaS).

- Looking at current new developments such as Treasure Island and Mission Rock, as well as some of the findings of the street studies and scenario planning workshops held by both the University of Oregon and the University of San Francisco, we see that a start on reclaiming our public realm from the dominance of the private automobile can start immediately and that we do not need to wait for the automated vehicle to be the impetus. What this tells us is that the reclamation of our public realm is simply a rebalancing of priorities.

- For both our own wellbeing, as well as that of the cities and towns that most of us live in, we need to re-examine and re-state what we want our cities, towns, suburbs, or built environment to be. Fundamentally, this is a question about the livability and sustainability of where we live, work, and play, and is a question that needs to be asked and addressed regardless of automated vehicles. As noted at the start of Chapter 14, they are only a tool for living and should not become a crutch that we need to rely upon due to failed planning policies.

Author contribution

The author reviewed and approved the manuscript.

Declaration of conflicting interests

The author declared no potential conflict of interest with respect to publication of this chapter.

References

[1]   Economist (The). "AI: The gladiator's edge." *The Economist Magazine.* September 2018. p. 78. A review of Lee Kai-Fu. *AI superpowers: China, Silicon Valley and the new world order.* Haughton Mifflin Harcourt.

[2] Treasure Island Development Authority (TIDA). *Approved plans & documents, major phase 1 approved application–May 2015.* Available from https://sftreasureisland.org/majorphase1 [Accessed November 2018] and *Treasure Island/Yerba Buena Island design for development (D4D)– Approved 2011.* Available from https://sftreasureisland.org/ftp/devdocs/ D4D/12282011_FinalTI%20D4D%28Date06282011%29.pdf [Accessed November 2018]

[3] Port of San Francisco. *Mission rock.* 2018. Available from https://sfport. com/missionrock and http://www.missionrock.com [Accessed December 2018]

[4] Metropolitan Transportation Commission (MTC). *Vital signs.* 2017. Available from http://www.vitalsigns.mtc.ca.gov/transit-ridership [Accessed March 2019]

[5] Arup, Perkins+Will Mobility Lab. "A case study of autonomous vehicles in San Francisco." *New Mobility Street Design.* 2017. Available from https:// higherlogicdownload.s3.amazonaws.com/AUVSI/14c12c18-fde1-4c1d- 8548-035ad166c766/UploadedImages/2017/PDFs/Proceedings/Posters/ Tuesday_Pos ter%2023.pdf [Accessed March 2019]

[6] Castor-Warren E. *A new vision for Los Angeles streets.* Lyft, Nelson \Nygaard, Perkins+Will Mobility Lab. 2017. Available from https://med- ium.com/sharing-the-ride-with-lyft/a-new-vision-for-los-angeles-streets- 74613e2f0dba and https://www.curbed.com/2017/9/22/16349168/lyft- street-redesign-la-perkins-will [Accessed March 2019]

[7] Beiker, S., and Meyer, G. (eds.). "Autonomous vehicles and the built environment: Exploring the impacts on different urban contexts." *Lecture Notes in Mobility: Road Vehicle Automation 5.* Springer International Publishing AG. 2018

[8] Riggs, W. (moderator). University of San Francisco.

[9] University of Oregon. *Sustainable cities initiative: Urbanism next.* Available from https://sci.uoregon.edu/urbanism-next-0 [Accessed March 2019]

[10] San Francisco Planning and Urban Research (SPUR). Available from https:// www.spur.org/about/our-mission-and-history [Accessed March 2019]

[11] Arup and Perkins+Will Mobility Lab. "A study of 4th Street, San Francisco." *Poster for AUVSI Symposium 2017.* Available from https:// higherlogicdownload.s3.amazonaws.com/AUVSI/14c12c18-fde1-4c1d-8548- 035ad166c766/UploadedImages/2017/PDFs/Proceedings/Posters/Tuesday_ Poster%2023.pdf [Accessed January 2019]

[12] Available from https://sfplanning.org/project/central-soma-plan [Accessed January 2019]

[13] Arup and Perkins+Will Mobility Lab. "A study of 4th Street, San Francisco' *Poster for AUVSI Symposium 2017.* Available from https://high- erlogicdownload.s3.amazonaws.com/AUVSI/14c12c18-fde1-4c1d-8548-035a d166c766/UploadedImages/2017/PDFs/Proceedings/Posters/Tuesday_Poster% 2023.pdf [Accessed January 2019]

[14]   National Association of City Transportation Officials (NACTO). *Blueprint for autonomous urbanism*. NACTO. 2017. Available from https://nacto.org/publication/bau/blueprint-for-autonomous-urbanism/

[15]   Castor-Warren E. *A new vision for Los Angeles streets*. Lyft, Nelson \Nygaard Perkins+Will Mobility Lab. 2017. Available from https://medium.com/sharing-the-ride-with-lyft/a-new-vision-for-los-angeles-streets-74613e2f0dba and https://www.curbed.com/2017/9/22/16349168/lyft-street-redesign-la-perkins-will [Accessed March 2019]

[16]   Nygaard, N. *The future of parking*. Nelson Nygaard. 2017. Available from http://nelsonnygaard.com/our-views-on-autonomous-vehicles-and-the-future-of-parking/ [Accessed January 2019]

[17]   American Automobile Association (AAA). *Cost of operating a vehicle* (online). 2017. Available from https://newsroom.aaa.com/tag/cost-of-operating-a-vehicle [Accessed February 2019]

[18]   ApParkingSpot. *App based parking* (online). Stockholm. Available from http://apparkingspotnordic.com and https://apparkingspot.com/about?lang=en [Accessed February 2019]

[19]   Signature Properties, CityLift Parking. *Parking structure for the hive* (online]. Oakland, CA. Available from https://www.experiencehiveparking.com [Accessed February 2019]

[20]   Volley Automation Parking Technology. Fremont, CA. Available from https://volleyautomation.com [Accessed February 2019]

[21]   Statista. *Digital commercial sales growth platform* (online). Available from https://www.statista.com/statistics/319667/us-digital-commerce-sales-growth-platform/ [Accessed March 2019]

[22]   Austin, M., Leitner, M., Poirier, J., Snyder, M.C., and Tierney, G. "Adapting our neighborhoods, streets & homes to e-Commerce: A study workshop." *Urbanism Next 2019*. University of Oregon Sustainable Cities Initiative. May 2019. Available from https://www.urbanismnext.com/schedule-2019 [Accessed May 2019]

[23]   Rittel, H.W.J., and Webber, M.M. "Dilemmas in a general theory of planning." *Policy Sciences*. 1973; 4(2): 155–169.

[24]   Sadik-Khan, J., and Solomonow, S. *Streetfight: Handbook for an urban revolution*. Penguin Books. 2017

The views expressed are those of the authors.

Chapter 16

Economic factors

Ata M. Khan[1]

The shared automated vehicle (SAV) system is expected to meet the requirements of stakeholders for acceptance as a key mode of mobility. Although societal values go beyond economic factors, economic feasibility is a major consideration for investors who will require an acceptable rate-of-return. Stakeholders, their roles, and interest in economic factors provide the starting point for assessing the feasibility of the SAV system to serve customers. Analysis of business models leads to details on economic factors involved in implementing a new mobility system.

This chapter covers concepts, methods, and applications for providing answers to the economic feasibility question. Building blocks for business models are quantified, including estimates of costs and revenue for fleets consisting of vehicles of various sizes. Observations are drawn on the economic feasibility of the SAV system from after tax rate-of-return analysis of business models.

16.1 Introduction

For a new mobility service to become sustainable, economic and financial feasibility within a broader framework of stakeholder acceptance is necessary. In this chapter on economic factors, stakeholder interest in economic factors is described. Among the stakeholders, fleet owners, automotive industry, and city governments have a keen interest in economic factors. Indicators of feasibility are described and quantified using business model analyses.

For an SAV system to be recognized by stakeholders as an ongoing and stable business, it should consist of vehicles, stations for parking and charging, and among other functions, means to monitor and control automated electric vehicle fleets. Also, analysis of the SAV system should show economic feasibility of investment. In this chapter, a study of stakeholder interest and building blocks for business models provides the base for follow-up analysis of costs, revenue, and after-tax rate-of-return. For use in planning future SAV services, relevant vehicle sizes are employed in various versions of business models. The methodological

[1]Department of Civil and Environmental Engineering, Carleton University, Ottawa, Canada

framework and constituent methods serve the purpose of providing answers to the economic feasibility question. In this chapter, dollar estimates are in U.S. dollars.

16.2   Stakeholder interest: economic feasibility and social good

Stakeholder requirements shape the SAV system design, as well as its operation, maintenance, monitoring, and control functions. As expected, users of the system are keen on receiving acceptable quality of service at affordable price. Given the commercial nature of the initiative, the providers and operators of service aim for their acceptable after-tax rate-of-return. In accordance with common practice, they may acquire funds for investment at a favorable borrowing rate and use the interest on borrowed funds to reduce their taxable income. Likewise, capital cost or depreciation allowance offered by governments enables the renewal of assets.

The suppliers of technology can enhance their financial position by availing public sector incentives for the purchase of electrification technology. In many countries around the world, electric vehicle purchase incentives are available in order to lessen dependence on petroleum fuels and reduce emissions. Infrastructure and vehicle fleet owners and operators of service aim for favorable location and price for SAV battery charging and parking facilities. These are necessary for attracting customers and therefore enhancing the probability of economic feasibility.

The shared vehicle commercial service offered by electric automated vehicles will require durable and highly reliable equipment to function safely and without an outage. Specialized companies will evolve to service equipment and these will in turn offer training opportunities to new employees.

Governments are expected to provide public infrastructure required for shared mobility, namely, curb side and off-road loading and unloading sites. As custodians of social good, they have a role in sustainability of infrastructure and transportation services. Ensuring safety and avoiding adverse effects on the ridership of high capacity public transit are frequently cited subjects of interest to governments. On the positive impact side, there is the recognition that governments would like to use SAVs as a means to achieve public policy objectives.

16.3   Business models

Stocker and Shaheen [1] reviewed a number of business models for SAV service. Given the focus in this chapter on economic factors, a subset of stakeholders was selected for a study of their interest in offering economically feasible shared mobility services. In one business model, a private sector entity supplies vehicles, parking, and fast charging infrastructure. This entity can also serve as a system operator or can form an alliance with an experienced company to operate the system. A subscription-based customer pool can access vehicles on-demand. The private sector stakeholder can be a transportation network company (TNC) or an original equipment manufacturer (OEM). An example is Volkswagen's WeShare free-floating car sharing system [2]. It is likely that that fleet owners (i.e., TNC and OEM) will provide the framework within which the SAVs will operate.

Recent analyses of vehicle ownership data and future forecasts suggest that private ownership may not increase in the future at the same rate as in the past and instead the current phenomenon of TNC use of privately owned vehicles will continue and may even increase [3]. Even if fleet ownership will also increase, privately owned automated vehicles may also be used to serve the needs of a fleet owner on a part-time basis and extra seats in a vehicle may be shared during a trip.

A different business model can be formed as a public-private-partnership (PPP). That is, a public sector entity such as a public transit agency can partner with a private sector operator. The public sector partner will supply vehicles, the park and charge infrastructure, and the system control personnel and equipment. A private sector entity will serve as the operator of the system. Customers can join a subscription-based pool and/or services can be accessed in the open market. There is the option of sharing a ride with another traveler unrelated to the user, provided that sharing parties agree.

In this PPP model, a city government may apply a lower rate of return for capital recovery than in the private sector business model for the reason that it is intended as a social discount rate. However, economic feasibility is still an objective in order to meet the commercial objectives of the private sector partner.

16.4   Building blocks for business cases

16.4.1   Methodology for estimation of costs

The subject of cost estimation for shared mobility vehicles, infrastructure, and other components of the system has been of interest to a number of researchers. Most authors described the cost of the vehicle and in limited cases, infrastructure cost was discussed. However, the overall system level cost has not been reported. The cost estimation methodology for shared mobility used here takes into account all components of the system, including infrastructure (Figure 16.1).

For estimation of costs, detailed information on shared mobility type, system design, operation, maintenance, and other functions is required. Demand studies in a city of interest lead to the decision on the number and location of park and charge stations. Vehicle sizes are defined and the fleet size is estimated. The choice of vehicle sizes is guided by travel market characteristics.

Services provided by smaller vehicles (e.g. single seat solo, two seater mini) can be a better fit with shared on-demand point-to-point services offered to customers with party size not exceeding two persons as car share as well as use of a seat. On the other hand, a sedan with four seats offers the flexibility to serve party sizes of up to four persons. The load factors applied in the business cases can be somewhat higher than very low occupancy rates that are common for commuting trips made with conventional personal automobiles. Vehicles used in the business model described in this chapter have one to four seats. However, a van with eight seats included in the business model can be used as a shuttle or for on-demand deviations in real time.

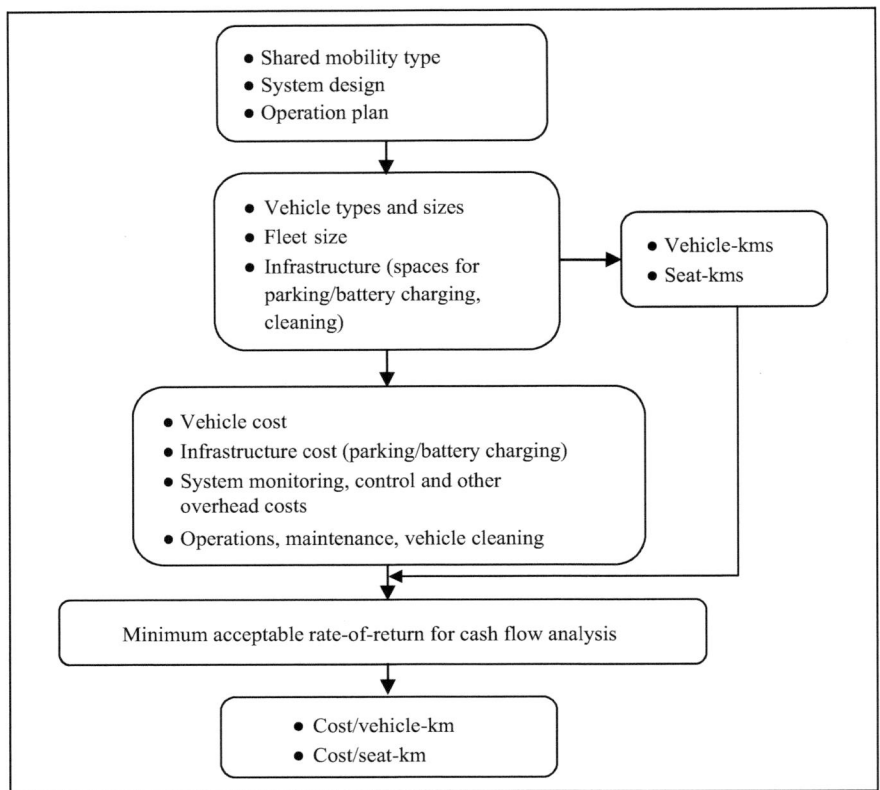

Figure 16.1   Estimation of costs

The inclusion of a number of vehicle sizes in a business model offers the opportunity to employ multiple vehicle types in different versions of the business model. Infrastructure requirements are defined in terms of stalls for parking/battery charging and cleaning. Also, the control system is defined in terms of personnel, equipment, algorithms, and communication system.

Next, cost estimates are developed for vehicles, infrastructure, operations, maintenance, control, and other functions. For applicable vehicles, costs are estimated for ownership, maintenance and operations. For infrastructure, cost estimates are required for building/leasing stalls and installing fast chargers. Infrastructure maintenance cost is accounted for and system control and other overhead costs are also included in the business plan. Estimates of vehicle-kms and seat-kms are necessary in order to find cost/vehicle-km and cost/seat-km.

In this study, cost estimates for components of the SAV system used are sourced from literature. When possible, multiple sources of cost information were examined and assumptions made by estimators were studied. The vehicle-kms and seat-kms are based on SAV system operating plan.

Two business models are described in this chapter. The first model is characterized by high productivity of vehicle use, and high cost-recovery factor using a minimum acceptable rate of return (MARR) of 10% (real). In the second model, a lower productivity of vehicle use is assumed, coupled with a lower MARR of 5% (real). The MARR used for capital recovery and other cash flow items is a business decision of the investor. The use of real rate of return implies constant dollar analysis.

A number of observations on costs are in order. Most researchers are of the opinion that the cost of electrification and automation in driving is likely to drop significantly. This observation applies to vehicles, infrastructure, and control technologies. The end result will be that the SAV services are likely to be cost-competitive with many existing urban transportation modes, including use of the private automobile.

16.4.2 *Methodology for estimating rate-of-return on investment*

For an investor, economic efficiency is a key objective since it is an indicator of the economic feasibility of services provided by the SAV system. The metrics of economic feasibility are:

- After-tax rate-of-return (ROR) (in real dollars) is higher than the minimum acceptable rate of return (MARR).
- Positive after-tax net equivalent annual worth (EAW) or net present worth (NPW), expressed in constant dollars.

If the after-tax ROR is greater than the MARR, the investment is feasible. Likewise, for the investment to be feasible, the after-tax net equivalent annual worth should be positive. Of course, it is understood that all constant dollar computations use the investor-specified MARR.

A high-level representation of the methodology for the estimation of capital cost recovery, yearly operations and maintenance costs, and revenue per year is presented in Figure 16.2. The variables for the revenue-cost study are the fleet size for each vehicle category, purchase cost of vehicles, operating and maintenance costs, service life (years and kms), station (parking/charging) cost (including charging infrastructure), maintenance and operating cost of facilities, MARR, ridership, and fare. The analysis is carried out using the rate of return, the equivalent annual worth, and present worth methods. Figure 16.3 shows the process for the estimation of the after-tax rate of return, which is considered as an indicator of economic feasibility. Also, the net equivalent annual worth and the net present worth are computed in constant dollars.

Cost efficiency of the SAV system is assessed with the following indicators:

- Cost per seat-km versus price charged per passenger-km.
- Vehicle kilometers versus passenger-kms.

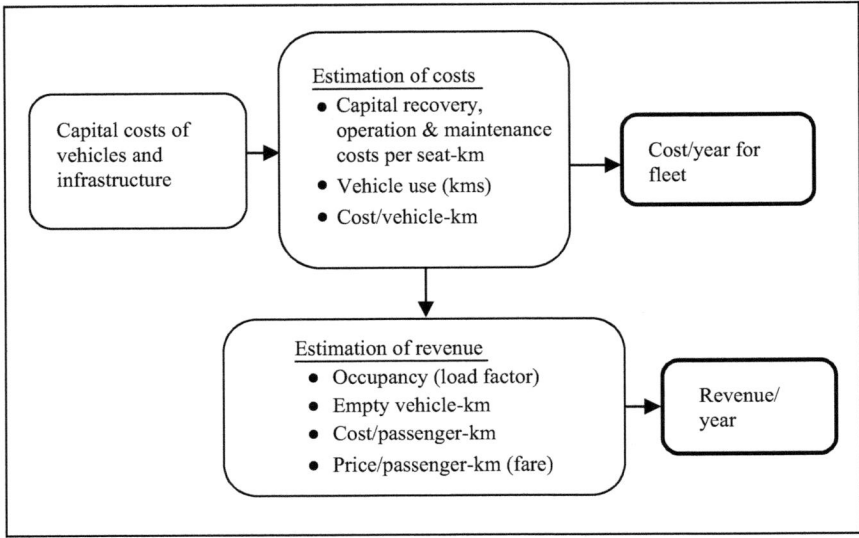

Figure 16.2   Estimation of costs and revenues

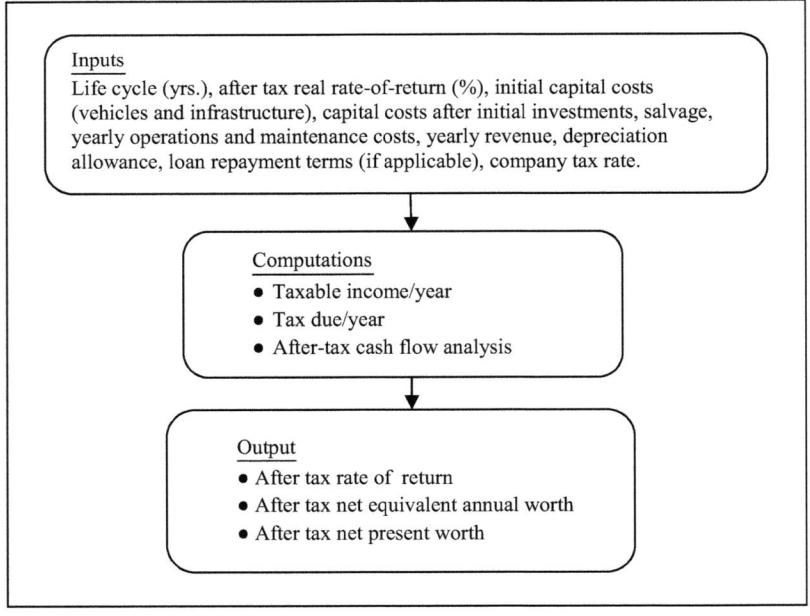

Figure 16.3   Estimation of after-tax rate-of-return

Although the cost of producing service is viewed carefully by the business owner to serve demand most efficiently, metrics such as price, trip details, and service quality are likely to affect the on-demand use of services.

The result of revenue-cost analysis establishes feasibility. If the computed after-tax rate-of-return is higher than the MARR, the SAV meets the economic feasibility criterion. Another approach is to examine the net equivalent annual worth and/or net present worth estimated by using the MARR (real). If the result is positive, the SAV system is economically feasible.

16.4.3 System design and operational factors

An SAV system with 100 electric automated vehicles and five park-charge stations (depots) strategically located in an urban area is used for business model analyses. The vehicles are at the SAE Level 4 or 5 automation and maturity of automation in driving technology is assumed [4,5]. Each park-charge station has 16 parking stalls and a cluster of four 150kW fast chargers are installed in four stalls [6]. The stalls with fast charges can also be used for parking, if required. The fast chargers are equipped with robot arms for charging an electric vehicle autonomously [7]. The composition of vehicle fleet in terms of vehicle sizes is defined in each version of the business model.

The vehicles include solo (one seat), mini (two seats), sedan (four seats), and van (eight seats). All vehicles are purpose-built for commercial service and have high capacity batteries which will enable them to provide service without frequent charging. The SAV system is monitored and controlled with the use of an advanced computer-communication system managed by humans. That is, qualified personnel oversee the monitoring and control system.

The key variables of a business model are costs, capital recovery of investment, revenue, availability of loan (if needed), capital cost or depreciation allowance, and the after-tax rate-of-return. The analyses are carried out in constant (real) dollars. The philosophy of the business model is to set the fare so that the resulting revenue will cover costs and ensure a specified profit level. The demand for service is generated by subscribing customers in the private sector models. The customer initially pays a membership fee, which is applied toward the fare if the service is used.

16.5 Cost structure

16.5.1 Vehicle ownership, operation, and maintenance

16.5.1.1 One-seat solo

Under special market conditions, a single-seater EV may be of interest in some cities around the world. Therefore one version of the business case presented in this chapter uses a solo vehicle. The following cost estimates are sourced from [8]. The purchase cost of this electric automated vehicle is estimated as US$10,920 (2016$). This estimate reflects a tax free purchase. Further, the purchase cost to the owner is set at 70% of the price due to fleet discount. The vehicle is replaced after 5 years of use and the estimated salvage is one-tenth the purchase cost.

The operating cost consisting of tires, fuel, registration, insurance, and overhead is estimated to be \$0.185/km. The contribution to overhead (used for system level monitoring and operation, and maintenance of parking stall and the battery charging infrastructure) is \$0.14/km. The maintenance cost of this automated EV (including cleaning) is estimated as \$0.039/km. The total O&M cost/veh-km (overhead included) is \$0.224/km. In the high productivity business case, 60,000 kms/year is applied in computations. Total O&M/year cost for this one-seat automated EV is \$13,440/vehicle (2016 \$).

16.5.1.2   Two-seat mini

Cost estimates for this vehicle are based on [8]. The ownership cost of the mini is estimated to be US \$20,580 (2016 \$). This estimate takes into account zero tax for EV and 70% fleet discount is applicable to this vehicle. The operating cost consisting of tires, fuel, registration, insurance, and overhead is estimated to be \$0.190/km. The contribution to overhead is \$0.14/km. The maintenance cost of this automated EV (including cleaning) is estimated as \$0.063/km. The total O&M cost/veh-km (overhead included) is \$0.253. Based on 60,000 kms/year, total O&M cost/year for the mini is US\$15,180/veh (2016 \$s). The vehicle is replaced after 5 years and the salvage is one-tenth the purchase cost.

16.5.1.3   Four seat EV sedan

The purchase cost of the sedan is sourced from [8,9]. Other costs are obtained from [8]. The estimated ownership cost of the four seat sedan is \$30,240 (2016\$). This amount takes into account zero tax for EV purchase and 70% fleet discount is applicable to this vehicle. The operating cost consisting of tires, fuel, registration, insurance, and overhead is estimated to be \$0.198/km. The contribution to overhead is \$0.14/km. The maintenance cost of this automated EV (including cleaning) is estimated as \$0.088/km. The total O&M cost/veh-km (overhead included) is \$0.286/km. Based on 60,000 kms/year, total O&M cost/year for the sedan is \$17,160/veh (2016 \$). The vehicle is replaced after 5 years and the salvage is one-tenth the purchase price.

16.5.1.4   Eight seat EV van

The van costs are based on [8]. The estimated ownership cost of the eight seat van is \$55,400. This estimate takes into account zero tax for EV purchase and 70% fleet discount applicable to this vehicle. The operating cost consisting of tires, fuel, registration, insurance, and overhead is estimated to be \$0.22/km. The contribution to overhead is \$0.14/km. The maintenance cost of this automated EV (including cleaning) is estimated as \$0.16/km. The total O&M cost/veh-km (overhead included) is \$0.38/km. Based on 60,000 kms/year, total O&M cost/year for the van is \$22,800/veh (2016 \$). The vehicle is replaced after 5 years and the salvage is one-tenth the purchase price.

16.5.2   *Infrastructure for charging and parking*

As previously noted, according to the system plan, five stations are located in the central area of the city. In total, 100 stalls are provided that include 20 stalls equipped with 150 kW fast chargers with robotic arms. Equal distribution of stalls among stations is assumed. These stalls are intended to serve the 100 vehicle fleet that at times will require fast charging and/or space for parking between service calls.

For economic analysis, a 10-year life of infrastructure is used for capital recovery and other cash-flow analyses. For the high productivity business model, 10% real rate of return is applied (for constant dollar calculations). The salvage value is assumed as zero. For a low productivity business model, 5% real rate of return is applied (for constant dollar analysis) and salvage is assumed as zero.

Capital cost of infrastructure consists of two items. These are parking stalls and the battery chargers. The parking stalls are located/installed on leased property. The cost of installation and maintenance is set equal to yearly parking charges in central area, which is set at $5,600/stall (2016 US constant dollars) [10]. The cost for 100 stalls that are to be used for 10 years amounts to $5.6 million.

The cost of a 150 kW fast charger equipped with a robotic arm installed in a stall is estimated to be $100,000 for charger, $18,000/charger installation, and $18,000 for robotic system per charger. The total in constant dollars is $136,000/charger. The cost of a cluster of four chargers/station is $0.544M and therefore cost for 5 clusters is $2.72 million. These charger costs are based on [6] and industry sources.

The total capital cost for 100 stalls and 20 fast chargers amounts to $8.32 million. No salvage value is assumed for the infrastructure due to the reason that land is leased and the chargers are assumed to have no market value after 10 years of use.

16.5.3   *Total cost of vehicle and infrastructure*

Useful insights can be obtained from an examination of cost/seat-km for each vehicle category and each cost item. Also, total cost/seat-km for each vehicle is very meaningful for planning SAV system for different travel markets that are served with various vehicle sizes.

In Sections 16.4 and previous parts of this Section 16.5, the methodology for estimating $/vehicle-km was described. In the high productivity version of the business model, a 5-year life span is assumed for vehicles and 10 years for infra-structure. In the low productivity version of the business model, vehicles are used for 10 years and retired at zero salvage value.

In the high productivity version of business model, the usage of vehicles is set at 60,000 kms/year for each vehicle and the 30,000 km/year is applied in the low productivity version. The operations and maintenance costs for vehicles are accounted for as a distinct cost items. Likewise, infrastructure maintenance cost is covered through contracts included in infrastructure cost. The overhead is intended to cover system level monitoring and control as well as unusual infrastructure maintenance. The overhead generated by the 100 vehicle fleet amounts to $0.84 million/year (2016$).

Table 16.1, and Figures 16.4–16.6 provide insights on the cost structure of the SAV system. Notable observations are presented as follows:

1. Cost per seat-km is an ingredient that contributes to the economic efficiency of service. This cost indicator drops as seats per vehicle increase. For example, the four seater sedan is three-time more efficient than a single-seat solo car.

Table 16.1   Cost structure of vehicles and infrastructure (2016 US$) (cost recovery @10% real)

Cost item	Solo (one seat) $/seat-km %		Mini (two seats) $/seat-km %		Sedan (four seats) $/seat-km %		Van (eight seats) $/seat-km %	
Ownership cost	0.043 8.7		0.041 14.6		0.030 19.0		0.028 26.9	
Maintenance (includes cleaning)	0.039 7.9		0.032 11.4		0.022 13.9		0.020 19.3	
Operation (includes registration, insurance, overhead)	0.185 37.5		0.095 33.8		0.050 31.6		0.028 26.9	
Infrastructure (parking and charging)	0.226 45.8		0.113 40.2		0.056 35.5		0.028 26.9	
Total	$0.493 100.0		$0.281 100.0		$0.158 100.0		$0.104 100.0	

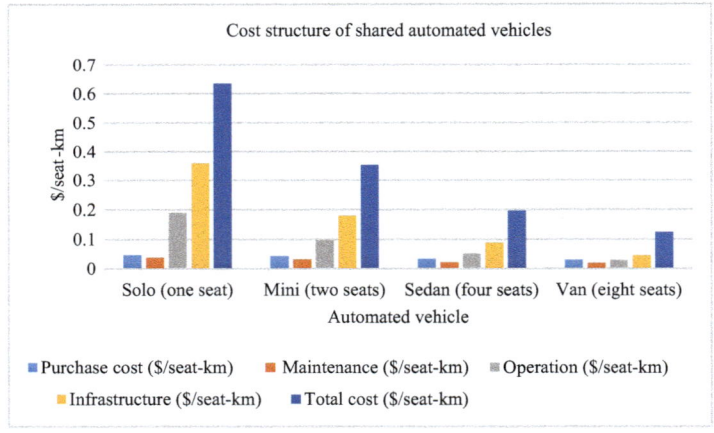

Figure 16.4 $/seat-km for shared automated vehicles

2. Since all vehicles use common stalls, smaller vehicles have much higher infrastructure use cost.
3. As expected, all cost items in terms of dollar per seat-km drop with increasing number of seats per vehicle.
4. In percentage terms, the share of infrastructure cost dominates other cost items. The degree of dominance of infrastructure cost drops with increasing seating capacity of a vehicle.

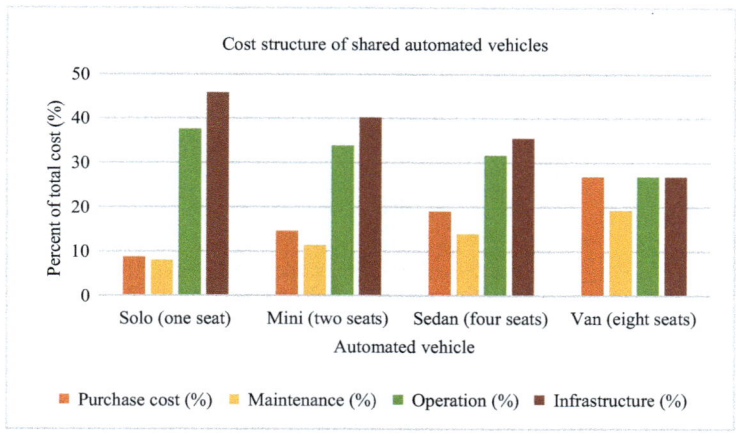

Figure 16.5   Cost structure of shared automated vehicles

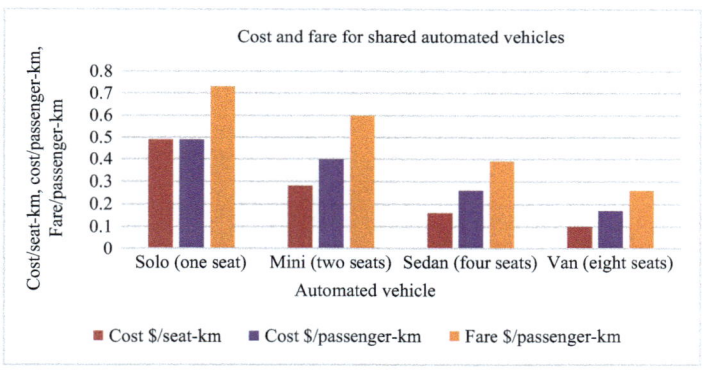

Figure 16.6   Cost and fare for shared automated vehicles: High productivity business case

5. The ownership cost per seat-km is the lowest among other items for the one-seater solo. For larger vehicles, namely, sedan and van, the maintenance cost/seat-km is the lowest cost item.

16.6   Cost, fare, and revenue

System level equivalent annual cost (EAC) and the present worth (PW) costs are found using a real dollar interest rate (in this study, 10% was used for the high productivity business model version and 5% was used for the low productivity version) for capital recovery and other cash flow items. From the EAC, cost/vehicle-km and the cost/seat-km are obtained.

The variables quantified so far can be applied to estimate the equivalent annual cost in $/year. As noted above, this estimate takes into account the capital recovery cost as well as yearly costs in providing the SAV service.

The cost/passenger-km is computed by applying the occupancy of seats in a vehicle. From demand–supply interaction study, occupancy estimates are obtainable. For SAV service, a 60% occupancy level for sedan and van is a reasonable estimate. For mini with two seats, a 70% occupancy is obtainable, and for the solo with one seat, 100% occupancy applies. The use of occupancy estimate enables the computation of cost per passenger-km (i.e., cost/passenger-km = (cost/seat-km) divided by occupancy level in fractional terms).

Next, the price/passenger-km (i.e., a proxy for fare) that the system owner may wish to charge can be found by applying the following factors: cost/passenger-km, profit margin, fare payment transaction fees, and the sale tax. The following formula was suggested by [8]:

Price/passenger-km or fare = (cost/passenger-km)/[(1–profit margin)
(1–payment transaction fee)] × (1+ sale tax)

In this study, the profit margin used is 20% for the high productivity business model and 10% for the low productivity model. The payment transaction fee is 5% and the sale tax is 13%.

The price/passenger-km estimate in association with passenger-kms/year is used to find revenue per year. The equation used is noted below:

Estimation of yearly revenue in constant dollars = [(yearly kms/vehicle)
× (seats/veh) × (vehicles in the fleet) × (occupancy in fractional terms)
× (fare or price/pass.-km)] × (percent nonempty productive kms)

The occupied runs are assumed as 80% and the occupancy estimates are noted above. In addition to occupancy level (i.e., occupied seats/available seats) and percent occupied runs, another indicator of the productivity of the SAV service is the percentage of revenue seats:

Percent of revenue seats = (passenger-kms/seat-kms) × 100

The revenue/year is found with the following formula:

Revenue/year = (price/passenger-km) × (passenger-kms/year)

Figure 16.6 enables observations on cost/seat-km, cost/passenger-km (at a specified occupancy level) and price or fare/passenger-km. The computations are based on the values of variables that are noted are noted above:

1. As previously noted, the cost per seat-km drops with increasing vehicle size.
2. Except for one-seater solo, the difference between cost/passenger-km and cost/seat-km is a reflection of occupancy of seats (occupancy level).

3. The cost per passenger-km drops with vehicle size.
4. The price or fare/passenger reflects the combined effect of the following variables: cost/passenger-km, percent occupied vehicle-kms, transaction cost, tax, and profit margin.
5. The price/passenger-km or fare drops with an increase in vehicle size, provided that the occupancy levels for SAV service are maintained at reasonably high levels (i.e., 60% sedan and van, 70% for mini, and 100% for single-seater solo).

16.7 Comparison with economic factors reported in the literature

At the outset, the reader is advised that literature sources on the economic factors of SAV service are very scarce. When available, these usually do not cover infrastructure cost items. However, an attempt is made here to summarize economic factors sourced from literature.

The cost for an automated medium sedan was estimated by [11] for use in urban and suburban areas. The assumption was that the use of this vehicle will replace existing transportation services in a small to medium town. According to their analysis, such a system could offer improved mobility experience at lower cost than a number of existing services. The cost of the automated (driverless) purpose-built vehicles was found to be $0.15 per trip-mile (2016 US$). This estimate is much lower than found in this study.

The scope of analysis was widened by [12] by considering the external costs of mobility (e.g., crash or congestion cost) provided by a nonautomated private transport system and applied these to estimate AVs' potential benefits. As expected, they found such benefits to be substantial. Although this observation is useful, a comparison cannot be made with the present study for the reason that external costs are out of the scope of this study.

In another study, Fagnant and Kockelman [13] reported possible prices for users of a centrally organized, shared automated vehicle system. Based on an assumed investment cost of US$70,000 and operating costs of US$0.50 per mile for the automated vehicle only, their analysis showed that even at a fare of US$1.00 per trip-mile for an automated taxi, the owner could make a profit. Although this is a higher price level than assumed by [11], it is comparable to present transport options.

A real-world perspective on shared mobility costs was adopted by [10,14]. For example, the author provided the cost of parking in central locations and the cost of cleaning shared vehicles. His cost estimates indicate that shared autonomous vehicles will cost more than car-sharing (US$0.60–US$1.00 per mile), but the cost for the automated system is likely to be lower than driver-operated taxis (US$2.00–US$3.00 per mile).

The price (i.e., fare) of shared automated vehicles estimated by [15] amounted to US$0.44 per trip-mile. This estimated price was based on operating cost plus a 30% profit margin. For a shared autonomous vehicle system operating as pooled taxis, the price per trip-mile was estimated to be only US$0.16. Detailed estimates of cost items were used to find the total cost. An attempt was made to compare and validate calculations against the then-current private cars.

Other estimates were reported by [16] and [17]. For fully autonomous vehicle intended for ride-sharing, [16] found the cost to range from US$0.20 (or even lower) to US$0.30 per passenger-mile. This range of cost compares well with that of [17], who estimated €0.15 per passenger-km for a ride-sharing scheme in an urban area in Germany. It should be noted that [16] treated private and commercially offered vehicles alike, and [17] based their cost analysis on the ride-sharing service only. The cost of US$0.30 per passenger-mile (2016$) was also estimated by [18]. An estimate on the rather low side was provided by [19] who found that fully autonomous vehicles in a ride-sharing scheme in the Netherland can be operated at €0.09 per passenger-km. In the estimation of costs noted above, it is not clear if overhead costs of shared services were taken into account.

It was observed by [8] that the above-noted approaches to estimate costs for automated vehicles did not cover the diversity of applicable operational models and also not all cost components were included in cost estimates. To improve cost estimates, they addressed gaps by conducting a comprehensive, bottom-up calculation of the respective cost structures of fully autonomous vehicles (Level 5 as defined by the SAE International) [4] used for dynamic ride-sharing, taxi, and shared vehicle fleets. In their methodology [8], the most applicable cost components are treated, but their estimates do not appear to include all infrastructure cost items.

16.8   Analysis of business models

Two cases of a private sector-based business model are analyzed that are applicable to a fleet operator as well as the MaaS arm of an OEM. The first version is a high productivity model that features high utilization of vehicles and a high MARR. Four versions of the model, each with two scenarios are analyzed. Table 16.2 shows common inputs for this model. It is based on a fleet of 100 electric Level 4 or 5 automated vehicles, consisting of solo, mini, sedan, and van. The fleet composition is varied from one version of the model to another. The capital, operations, and maintenance costs (including cleaning) for each vehicle type are in accordance with estimates provided in Sections 16.5 and 16.6.

The low productivity business case differs from the high productivity case in terms of vehicle utilization (i.e., kms/year), profit margin, and rate of return for capital recovery and other cash flow items. Otherwise, as can be seen in Table 16.3, the two cases are identical. In this case, the vehicles are used for 30,000 kms/year. Following their use for 10 years, the salvage is zero. The MARR is 5% (real) for constant dollar cash flow analysis. The overhead of $0.42 million/year is identical to the high productivity case.

The business cases are compared in terms of cost ($/seat-km, $/passenger-km) and price ($/passenger-km) (Table 16.4). The price can serve as a proxy for fare and therefore it is used for the estimation of revenue. The cost/seat-km and the

Table 16.2   System cost structure for high productivity case (2016 US$)

Input variable	Description/value
Fleet size	100 electric Level 4 or 5 automated vehicles
Vehicle sizes	Solo (1 seat), mini (2 seats), sedan (4 seats), van (8 seats)
Stations for fast charging and parking stalls	Five stations (located in a central area of the city). Each station has 20 stalls that include four stalls equipped with 150 kW fast chargers).
Number of years for feasibility study	10 years
Capital cost of vehicles (fleet prices) (2016 US$)	Solo (one seat) $10,920; mini (two seats) $20,580; sedan (four seats) $30,240; van (eight seats) $55,540
Usage and salvage	60,000 kms/year; vehicles are replaced after 5 years; salvage is 10% of purchase price
Infrastructure cost (2016 US $) (stalls for fast charging and parking)	$56,000/stall built on leased property; $136,000/fast charger including robotic arm; salvage after 10 years is zero
Capital recovery (real rate)	10% (real) for constant dollar analysis
Profit margin	20%

Note: For operation and maintenance costs, please see Section 16.5.

Table 16.3   System cost structure for low productivity case (2016 US$)

Input variable	Description/value
Fleet size	100 electric Level 4 or 5 automated vehicles
Vehicle sizes	Mini (two seats), sedan (four seats), and van (eight seats)
Stations for fast charging and parking stalls	Five stations (located in the central area of the city). Each station has 20 stalls that include four stalls equipped with 150 kW fast chargers
Capital cost of vehicles (fleet prices) (2016 US$)	Mini (two seats) $20,580; sedan (four seats) $30,240; van (eight seats) $55,540
Usage and salvage	30,000 kms/year; zero salvage after 10 years
Infrastructure cost (2016 US $) (stalls for fast charging and parking)	$56,000/stall built on leased property; $136,000/fast charger including robotic arm; salvage after 10 years is zero
Capital recovery (real rate)	5% (real) for constant dollar analysis
Profit margin	10%

Note: (1) For operation and maintenance costs, see Section 16.5. (2) Ten years used for feasibility study.

cost/passenger-km are higher for the low productivity model than for the high productivity model due to relatively lower utilization of vehicles. The price ($/passenger-km) is higher for the low productivity case as compared to the high productivity case, which reflects the higher cost/passenger-km for this business case.

Table 16.4   Comparison of business cases: cost and price

Business case and vehicle	Cost: $/seat-km	Cost: $/passenger-km	Price: $/passenger-km
High productivity case			
Solo one seat	0.493	0.493	0.733
Mini two seats	0.281	0.401	0.595
Sedan four seats	0.158	0.263	0.391
Van eight seats	0.104	0.172	0.256
Low productivity case			
Mini two seats	0.353	0.504	0.666
Sedan four seats	0.196	0.327	0.432
Van eight seats	0.124	0.207	0.274

Note: In the low productivity case, the solo with one seat is not used.

16.8.1   After-tax rate-of-return analysis of high productivity business case

Four versions of this business case are defined in terms of vehicle sizes used to serve the market. For the analysis of after-tax performance of the shared vehicle system, two scenarios are used for each version of the business case. In one scenario, no loan is assumed and in another scenario, the effect of a loan is illustrated in the form of improved after-tax rate-of-return. The interest paid on loan reduces taxable income. The allowable depreciation and corporate tax are based on Canada Revenue Agency guidelines, which are in general similar to those of U.S. Internal Revenue Service guidelines. Other inputs to the after-tax cash flow analysis have already been explained in this chapter.

A number of observations are drawn from the after-tax (real) rate-of-return results shown in Table 16.5.

1. The percent revenue seats are relatively higher for smaller vehicles as compared to larger vehicle due to the application of realistic assumption for load factors and empty vehicle-kms.
2. In spite of pessimistic estimate of percent revenue seats, the ROR (real) exceeds the MARR.
3. If a loan is obtained at a favorable rate, the after-tax ROR (real) improves.

16.8.2   After-tax rate-of-return analysis for low productivity business case

Three versions of this business case are analyzed in terms of vehicle sizes employed to serve the market. Here, one seater solo vehicle is not used. For after-tax performance of the shared vehicle system, two scenarios are defined for each version of the business case. No loan is assumed in one scenario and in another scenario, the improvement in after-tax rate-of-return is illustrated due to reduction

Table 16.5 High productivity business model after-tax rate-of-return

Version	Scenario	Fleet	Price/pass.-km (2016 US$)	Load factor (%)	% revenue seats	After tax ROR% real
1	No loan	100 Solo	0.733	100	80	16.2
1	Loan@7%	100 Solo	0.733	100	80	23.4
2	No loan	80 Mini	0.595	70	56	16.7
		20 Sedan	0.391	60	48	
2	Loan@7%	80 Mini	0.595	70	56	23.0
		20 Sedan	0.391	60	48	
3	No loan	100 Sedan	0.391	60	48	17.0
3	Loan@7%	100 Sedan	0.391	60	48	22.8
4	No loan	40 Mini	0.595	70	56	17.0
		50 Sedan	0.391	60	48	
		10 Van	0.256	60	48	
4	Loan@7%	40 Mini	0.595	70	56	22.9
		50 Sedan	0.391	60	48	
		10 Van	0.256	60	48	

Table 16.6 Low productivity business cases: after-tax rate-of-return

Version	Scenario	Fleet	Price/pass.-km (2016 US$)	Load factor (%)	% revenue seats	After tax ROR (% real)
1	No loan	80 Mini	0.666	70	56	6.2
		20 Sedan	0.432	60	48	
1	Loan@5%	80 Mini	0.666	70	56	7.5
		20 Sedans	0.432	60	48	
2	No loan	100 Sedan	0.432	60	48	6.1
2	Loan@5%	100 Sedan	0.432	60	48	7.2
3	No loan	40 Mini	0.666	70	56	6.2
		50 Sedan	0.432	60	48	
		10 Van	0.274	60	48	
3	Loan@5%	40 Mini	0.666	70	56	7.3
		50 Sedan	0.432	60	48	
		10 Van	0.274	60	48	

in taxable income made possible by payment of interest on loan. The allowable depreciation and corporate income tax used in the analysis are based on Canada Revenue Agency guidelines, which are in general similar to those of the U.S. Internal Revenue Service. Other inputs to the analysis of after-tax cash flow have already been explained in this chapter.

A number of observations are drawn from the after-tax (real) rate-of-return results shown in Table 16.6.

1. For the two seater mini, the percent revenue seats is relatively higher as compared to the larger sedan and the van due to the application of realistic load factors and empty vehicle-kms.
2. In spite of pessimistic estimate of percent revenue seats, the ROR (real) exceeds the minimum acceptable rate of return.
3. If a loan is obtained even at the rate equal to the MARR, the after tax ROR (real) improves.

16.9   Conclusions

For the shared vehicle system based on electric automated vehicles to succeed in meeting stakeholder objectives, economic viability is essential. This is the reason why literature sources have emphasized the need to enhance the knowledge of economic factors. Investor confidence can be gained with detailed analyses such as the ones reported in this chapter that treat all cost components of the system, including vehicle, infrastructure, operation, maintenance, monitoring, and control.

Fleet owner and the *Maas* arm of the OEM are likely to be the first stakeholders to implement business models based on the use of a variety of vehicles, ranging from one seater solo to van with eight seats. Another observation on the business model is the importance of having a network of stations, each containing sufficient number of stalls for parking. Some of these stalls at each station can be equipped with fast chargers.

The cost structure of the shared electric automated vehicle system reported in this chapter contributes notable observations.

- Cost per seat-km is an important variable for the study of economic efficiency of service. Analyses show that this cost indicator drops as seats per vehicle increase. For example, the four seater sedan exhibits one-third the cost/seat-km of the single seater solo car.
- At reasonably high occupancy level, the cost per passenger-km as well as price per passenger-km drop with increasing vehicle size.
- Since all vehicles use common stalls, smaller vehicles show much higher infrastructure use cost than larger vehicles.
- All cost items expressed as dollar per seat-km decrease with increasing number of seats per vehicle.
- The proportion of infrastructure cost out of total cost dominates other cost items. However, with the increasing seating capacity of a vehicle, the degree of dominance of infrastructure cost drops.
- For the one seater solo, the ownership cost per seat-km is the lowest among other items. For larger vehicles, namely sedan and van, the lowest cost item is maintenance cost/seat-km.
- The cost/passenger-km and price/passenger-km reported in this chapter appear reasonable as compared with costs and fares of existing modes.

- Literature-sourced costs, in general, do not cover all cost items. Therefore, direct comparison with findings of this study is not appropriate. One cost item that is usually absent from estimates reported in the literature is the infrastructure cost. However, when most cost items are covered by other researchers, the comparisons improve.

  Analysis of business model versions leads to the following conclusions.

- The percent revenue seats are relatively higher for smaller vehicles as compared to larger vehicle due to the application of realistic assumptions for load factors and empty vehicle-kms.
- In spite of pessimistic estimate of percent revenue seats, the after-tax analysis results show that the ROR (real) exceeds the minimum acceptable rate of return.
- As expected, if a loan is obtained at a favorable rate, the after tax ROR (real) improves due to the reason that interest on loan reduces the taxable income.

As the analyses reported in this chapter show, depending upon the market and SAV system attributes, business model productivity can range from high to low, but still these show acceptable economic performance. The results reported in this chapter suggest that even the low productivity business case is feasible in terms of after-tax rate-of-return.

Predicting economic factors for an SAV system is challenging due to technological and market uncertainties. Although costs sourced from literature are reasonable, nevertheless these are not based on real world experience. The market acceptance of SAV services is also uncertain. However, the absence of the usual car ownership trend in forecasts, consumer acceptance of on-demand ride hailing services, growing fleet owner and OEM company interest in the SAV system, and favorable public policy toward electric automated vehicles collectively elevate the credibility of a future consumer market for the services of the SAV system. This chapter on economic factors in intended as a contribution to knowledge for developing and implementing business models.

Acknowledgments

The author wishes to thank Seth Gatien for reviewing a draft of the chapter.

Author contribution

The author reviewed and approved the manuscript.

Declaration of conflicting interests

The author declared no potential conflict of interest with respect to publication of this chapter.

References

[1]   Stocker A., and Susan Shaheen. *Shared automated vehicles: Review of business models*. Discussion Paper No. 2017-09. The International TransportForum (OECD).

[2]   Volkswagen. "WeShare launched in Berlin as full-electric service." *VW Newsroom* (https://www.volkswagen-newsroom.com/en/press-release), June 27, 2019. Also, see "VW's new car sharing service aims to be a little different from the rest." Posted by Trevor Mogg on June 27, 2019 and "Volkswagen launches WeShare all-electric car sharing service," by Darrel Etherington, June 27, 2019.

[3]   City of Toronto. *Automated vehicles tactical plan*. IE87 – Attachment 1. 2019.

[4]   SAE International. *Taxonomy and definitions for terms related to on-road motor vehicle automated driving systems*. Warrendale, PA; 2014; j3016 edn., January 2014 http://standards.sae.org/j3016_201401/

[5]   U.S. Department of Transportation. *Preparing for the future of transportation automated vehicles 3.0*. 2018. Washington, D.C.

[6]   Ribberink, H., Wilkens, L. Abdullah, R., McGrath, M., and Wojdan, M. "Impact of clusters of DC fast charging stations on the electricity distribution grid in Ottawa, Canada." *EVS30 Symposium*, Stuttgart, Germany, October 9–11, 2017.

[7]   Hawkins, A. "VW wants to test robotic EV chargers for self-driving cars: The future is electric and autonomous." *The Verge* (a newsletter about computers) August 01, 2019.

[8]   Bösch, P.M., Becker, F., Becker. H., and Axhausen. K.W. "Cost-based analysis of autonomous mobility services." *Transport Policy*. 2018; 64: 76–91.

[9]   U.S. Dept. of Energy, Office of Energy Efficiency and Renewable Energy. *Costs associated with non-residential electric vehicle supply equipment*. Also see *Enabling fast charging: A technology gap assessment*. Washington, D.C., October 2017.

[10] Litman, T. 2015). *Automated vehicle implementation predictions*. Victoria Transport Policy Institute, 2015. Accessed March 16, 2017. www.vtpi.org/avip.pdf.

[11] Burns, L.D., Jordan, W., and Scarborough, B. *Transforming personal mobility. Technical Report*. The Earth Institute, Columbia University, 2013.

[12] Fagnant, D., and Kockelman, K. "preparing a nation for autonomous vehicles: Opportunities, barriers and policy recommendations for capitalizing on self-driven vehicles." Paper presented at the *93rd Annual Meeting of the Transportation Research Board*. Washington, DC, January 2014.

[13] Fagnant, D., and Kockelman, K. "Dynamic ride-sharing and optimal fleet sizing for a system of shared autonomous vehicles." Paper presented at the *94th Annual Meeting of the Transportation Research Board*. Washington, DC, January 2015.

[14]   Litman, T. *Autonomous vehicle implementation predictions implications for transport planning*. Victoria Transport Policy Institute, 18 March 2019.

[15]   Johnson, B. *Disruptive mobility, research report*. Barclays; 2015.

[16]   Stephens, T., Gonder, J., Chen, Y., Lin, Z., Liu, C., and Gohlke, D. *Estimated bounds and important factors for fuel use and consumer costs of connected and automated vehicles*. Technical Report National Renewable Energy Laboratory, U.S. Department of Energy, Golden, CO (2016)

[17]   Friedrich, M., and Hartl, M. *MEGAFON - Modellergebnisse geteilter autonomer Fahrzeugflotten des oeffentlichen Nahverkehrs*. Research report. Universität Stuttgart, Institut für Strassen- und Verkehrswesen, Stuttgart, December 2016 (originally cited by Reference 8).

[18]   Johnson, C., and Walker, J. *Peak car ownership*. Research Report. Rocky Mountain Institute. 2016.

[19]   Hazan, J., Lang, N., Ulrich, P., Chua, J., Doubara, X., and Steffens, T. *Will autonomous vehicles derail trains?* Boston Consulting Group (BCG) Perspectives. September 2016.

The views expressed are those of the authors.

Chapter 17

The impacts of shared and automated mobility

Susan A. Shaheen[1] and Adam P. Cohen[2]

Shared mobility is an innovative transportation strategy enabling users to gain short-term access to transportation modes on an "as-needed" basis. The ecosystem of shared services continues to grow and includes an array of services such as carsharing; microtransit; app-based for-hire ride services (e.g., sometimes referred to as transportation network companies (TNCs), ridesourcing, or ridehailing); moped-style scooter sharing; shuttles; taxis; urban air mobility; and public transportation. Shared mobility also encompasses shared micromobility, the shared use of a bicycle, scooter, or other low-speed mode. Shared micromobility includes various service models and transportation modes that meet the diverse needs of travelers, such as station-based and dockless bikesharing and standing electric scooter sharing. Shared mobility also includes last-mile delivery services such as app-based deliveries (commonly referred to as courier network services), robotic delivery, drones, and other last-mile delivery innovations.

A number of social, environmental, and behavioral impacts have been attributed to shared mobility, and an increasing body of empirical evidence supports many of these relationships—although more research is needed. The various effects can be grouped into four categories: (1) travel behavior, (2) environmental, (3) land use, and (4) social. Due to its potential impacts, shared mobility can be a strategy to mitigate congestion, improve air quality, aid in parking management, and facilitate multimodal integration. In recent years, climate action planning has further raised awareness about shared mobility as a transportation strategy among public agencies.

Documenting the comparative impacts of shared modes can be challenging due to differences in service models, data collection, and study methodologies, which can produce inconsistent results based on limited samples and aggregate-level analyses that may not accurately reflect regional differences. For these reasons, it can be challenging to provide a comprehensive and unbiased picture. While automated traveler activity data can offer a rich understanding, these data typically do not capture changes in auto ownership, travel behavior across all modes, and

[1]Department of Civil and Environmental Engineering and Transportation Sustainability Research Center, University of California, Berkeley, CA, USA
[2]Transportation Sustainability Research Center, University of California, Berkeley, CA, USA

respondent perceptions over time. Beyond operator surveys, many large transportation surveys have begun to assess shared mobility including the American Community Survey and the California Household Travel Survey; however, these instruments also collect self-reported data. While travel behavior surveys have validity issues, such as respondents exaggerating travel behaviors, underreporting the extent or frequency of travel or reporting inaccurately, and sample bias, they can still offer a source of behavioral understanding [1]. Despite these limitations, insights into shared mobility can help public agencies understand the impacts of shared modes on public infrastructure, the environment, and the community.

This chapter reviews findings from shared mobility studies including ridesharing (carpooling and vanpooling), carsharing, shared micromobility (bikesharing and scooter sharing), and TNCs. We conclude with a discussion of shared automated vehicles (SAVs) and their potential impacts.

17.1   Ridesharing impacts

Ridesharing (also known as carpooling, vanpooling, and pooling) allows travelers to share a ride to a common destination. Ridesharing includes carpooling and vanpooling. While a number of social, environmental, and behavioral impacts have been attributed to ridesharing, more research is needed as this mode can be challenging for researchers to observe and evaluate. Ridesharing can provide numerous benefits to society and individual users.

17.1.1   Societal impacts of ridesharing

Empirical and anecdotal evidence indicates that ridesharing provides numerous societal benefits, such as:

• Reduced vehicle miles/kilometers traveled and congestion mitigation,
• Reducing energy consumption and emissions, and
• Reduced parking and roadway infrastructure demand.

One impact commonly associated with ridesharing is a reduction in vehicle miles traveled (VMT) or vehicle kilometers traveled (VKT). VMT/VKT is a travel demand metric that measures the sum of the number of miles traveled by each vehicle. A study by the Federal Highway Administration during the 1970s energy crisis found a 23% reduction in VMT [2]. Employee-based trip reduction (EBTR) and transportation demand management (TDM) programs are recognized as best practices to support VMT reduction goals. However, many of these programs lack performance monitoring and assessment and only a handful of empirical studies have examined the VMT impacts of these policies. One study found that employees participating in the program had 4.2% to 4.8% lower VMT than employees at the same worksite who did not participate [3]. Studies assessing the implementation of Washington State's Commute Trip Reduction Law have found similar effects [3]. Another study [4] found an average VMT reduction of 6% for employees at worksites subject to the law. Boarnet *et al.* [5] estimate that these programs can reduce VMT for workplace commutes by 4% to 6% (or approximately 1% regionally).

Only two studies estimated VMT reduction for the entire region or metropolitan area [6]. Hillsman *et al.* [3] used survey data from employers to estimate the number of commute trips eliminated by Washington State's commute trip reduction (CTR) program. The study estimated declines in total VMT of 1.33% on all roadways and a reduction in freeway VMT of 1.07% for the four central counties in metropolitan Seattle. It is important to note that this is a smaller impact than other studies because the authors examined all travel during the morning peak, including commute trips to nonparticipating sites and noncommute trips. A separate study by the CTR Task Force using different years in the same data set analyzed by Hillsman *et al.* [3] estimated a 1.6% reduction in total VMT [7]. Despite these reductions, ridesharing can also lead to induced demand (e.g., encouraging more people to drive) due to reduced travel times and costs.

In addition to the potential for reduced VMT/VKT, studies have found that ridesharing can save fuel and reduce greenhouse gas (GHG) emissions [8]. Noland *et al.* [9] concluded that enacting policies to increase pooling is the most effective strategy to reduce energy consumption besides driving prohibitions. A study by Jacobson and King [10] estimated a potential fuel savings of 0.80 to 0.82 billion gallons of gasoline per year in the United States (U.S.), if one additional passenger was added to every 100 vehicles. The same study also estimates a potential annual fuel savings of 7.54 to 7.74 billion gallons per year in the U.S., if one additional passenger was added to every 10 vehicles. Another study found that carpooling could save 33 million gallons of gasoline daily if each average commuting vehicle carried one additional passenger [11].

By reducing fuel consumption, carpooling can reduce GHG emissions. Using a simulation model, Hillsman *et al.* [3] forecast that individual carpoolers reduce personal commute GHG emissions by approximately 4% to 5% after joining an employer trip reduction program. The authors estimate savings of 7.2 million tons of GHG emissions annually in the U.S., if one additional passenger were added to every 100 vehicles [3]. Further, the Jacobson and King study approximated a savings of 68 million tons of GHG emissions annually in the U.S., if one passenger were added to every ten vehicles [10]. In another study, the SMART 2020 report estimated that employing information and communication technology (ICT), such as app-based carpooling, to optimize roadway performance could abate 70 to 190 million metric tons of carbon dioxide emissions [12].

17.1.2   Individual impacts of ridesharing

Studies suggest that ridesharing is a flexible strategy that can be employed by many users. Although research is limited, ridesharing participants frequently benefit from pooling in a number of ways. Individually, ridesharing users can benefit from:

- Cost savings (through shared travel costs),
- Travel time savings from high occupancy vehicle (HOV) lane access and preferential parking at destinations, and
- Reduced commute stress (through shared driving responsibilities).

A number of studies have found that casual carpooling participants (i.e., informal carpooling among strangers) are more likely to be single or married without children. Teal [13] found that carpool participants were more likely to have lower incomes and be the second worker in a household. Another study of casual carpooling users between the ages of 25 and 34 in Houston found that they were more likely to make commute trips (96%) versus noncommute trips (80%) [14]. This study also found that HOV lane users tended to belong to larger households with over 60% of carpools comprising family members.

For low-income households, the inability to access a private vehicle can be a limiting factor to economic mobility. In the nation's largest metropolitan areas, 7.5 million predominantly lower-income households do not have access to an automobile [15]. Only 40% of these public transit dependent households can access metro-wide jobs with a commute of 90 min or less [15]. Long commutes and limited job access via public transit in most metropolitan regions can leave many jobs out of reach for carless households. Some studies have shown that carpooling can provide job access to households with lower incomes and households with more workers than vehicles [15]. More recent data from the National Household Travel Survey and the American Community Survey show that ridesharing users tend to have lower incomes, and Hispanics and African Americans carpool more than other racial and ethnic groups. These surveys and other studies indicate that ridesharing may serve an important role in enhancing mobility in low-income, immigrant, and nonwhite communities where travelers are more likely to be unable to afford personal automobiles and obtain driver's licenses [16].

Teal [13] found that some ridesharing users can have longer commute distances and therefore have higher commute costs. Ridesharing can be an important cost-saving travel strategy for commuters. For example, in the San Francisco Bay Area, commuters often use casual carpooling to get from the East Bay to downtown San Francisco during the morning commute. Carpooling, which uses the HOV lanes of the San Francisco-Oakland Bay Bridge, allows travelers to take advantage of a toll discount and shorter waits at the toll plaza. According to a 1998 survey, approximately 9,000 commuters (6,000 riders and 3,000 drivers) used casual carpooling each morning [17].

Commuters who participate in ridesharing frequently have access to preferential parking and HOV lanes, which also contribute to pooling's convenience and time savings. Several studies of casual carpooling have documented travel time savings, cost savings, and convenience as key motivators to share a ride [14,18–20]. One study of casual carpooling in the San Francisco Bay Area found that convenience, time savings, and monetary savings were key motivators to carpool [21]. Another study of casual carpooling in Washington D.C. and Northern Virginia found that driver departure flexibility was a primary reason for driving instead of riding as a carpool participant. This study also found that the top reason for choosing to be a rider was the desire to save on gasoline cost, followed by a preference to do other things during the drive [22]. This study found that 60% of casual carpooling participants in Washington, D.C. and

Northern Virginia only participated as passengers, 12% only as drivers, and 28% as both passengers and drivers.

The inability to have access to a private vehicle during the workday is often cited as a common drawback associated with traditional carpooling and ride-matching programs. However, a variety of on-demand shared modes are creating a network effect to help overcome this challenge. Today an increasing number of shared mobility options, such as carsharing, shared micromobility, and others, are providing carpooling users innovative options for getting around during the work-day. App-based carpooling is also providing increased flexibility by allowing car-poolers to have different morning and evening carpool matches. This allows travelers to share a ride who may not have been able to previously due to variable or irregular work schedules. Section 17.2 explores the impacts of carsharing.

17.2 Carsharing impacts

Carsharing provides members access to a fleet of autos for short-term use throughout the day, reducing the need for one or more personal vehicles. A number of academic and industry studies of shared mobility have documented carsharing impacts, predominantly employing self-reported survey data. These studies collectively show the following carsharing outcomes:

- Sold vehicles or delayed or foregone vehicle purchases;
- Increased use of some alternative transportation modes (e.g., walking, biking);
- Reduced VMT/VKT;
- Increased access and mobility for formerly carless households;
- Reduced fuel consumption and GHG emissions; and
- Greater environmental awareness.

While the environmental, behavioral, and economic impacts of carsharing services have been well studied, the magnitude of impact varies. Variations in measured impacts can be due to a variety of factors such as: (1) region; (2) density; (3) built environment; (4) public transit accessibility; and (5) carsharing service (e.g., roundtrip and one-way) and business model. The vast majority of carsharing impact studies have examined carsharing's impact based on business-to-consumer (B2C) and peer-to-peer (P2P) business models.

- **Business-to-consumer (B2C) carsharing:** In a B2C model, a carsharing provider offers individual consumers access to a business-owned fleet of vehicles through memberships, subscriptions, user fees, or a combination of pricing models. Examples of B2C carsharing providers include Zipcar and Enterprise CarShare (roundtrip) and GIG Car Share (free-floating one-way). In October 2016, there were approximately 10.3 million roundtrip and 4.7 million one-way carsharing members worldwide [23].
- **Peer-to-peer (P2P) carsharing:** In a P2P model (sometimes referred to as personal vehicle sharing), carsharing providers broker transactions among

vehicle owners and guests by providing the organizational resources needed to make the exchange possible. Members access vehicles through a direct key transfer from the host (or owner) to the guest (or driver) or through operator-installed, in-vehicle technology that enables unattended access. Pricing and access terms for P2P carsharing services vary, as they are typically determined by vehicle hosts listing their vehicles. The P2P carsharing operator generally takes a portion of the P2P transaction amount in return for facilitating the exchange and providing third-party insurance. Examples of P2P carsharing providers in the U.S. include Turo (formerly RelayRides) and Getaround. For example, Turo takes a 25% commission from the host along with 10% from the guest, and Getaround takes 40% from the host for its services. As of January 2017, 2.9 million members shared 131,336 vehicles as part of a P2P carsharing program in North America [23].

17.2.1   Impacts of business-to-consumer (B2C) carsharing

One documented impact of carsharing is a reduction in vehicle ownership. Studies and surveys in the U.S. indicate that 11% to 26% of roundtrip carsharing participants sold a personal vehicle and 12% to 68% postponed or entirely avoided a car purchase [24–26]. In another study, 30% of City CarShare members in the San Francisco Bay Area shed one or more personal vehicles, and two-thirds chose to postpone the purchase of another vehicle after using the service for two years [27]. Several Canadian studies and member surveys suggest that between 15% and 29% of roundtrip carsharing participants sold a vehicle after joining carsharing programs, while 25% to 61% delayed or had forgone a vehicle purchase [25,28,29]. An aggregate-level study of 6,281 people in the U.S. and Canada documented 25% of members selling a vehicle due to carsharing and another 25% postponing a vehicle purchase due to roundtrip carsharing [30]. U.S. and Canadian aggregate data also reveal that each roundtrip carsharing vehicle removes between six to 23 cars on average from roads [24,25,31]. Martin and Shaheen [30] concluded that one carsharing vehicle replaces nine to 13 vehicles among carsharing members (on average across this aggregate-level study). According to European studies, a roundtrip carsharing vehicle reduces the need for four to ten privately owned vehicles on average [32].

Studies of one-way carsharing have also documented a reduction in vehicle ownership. A study of free-floating one-way carsharing members across five cities in the U.S. and Canada found that 2% to 5% of participants sold a vehicle after joining carsharing, and 8% to 10% on average delayed or had foregone a vehicle purchase [33]. This study also found that each free-floating one-way carsharing vehicle removed seven to 11 vehicles on average from the road in the cities studied. A study of station-based one-way carsharing participants in France found that each Autolib vehicle removed three private vehicles from the road (on average), resulting in a 23% reduction in private vehicle ownership after joining the carsharing program [34].

Roundtrip and one-way carsharing also have an impact on modal shift. Studies have examined the impact of these carsharing service models on public transit and nonmotorized travel [30]. Generally, while these studies have found a slight overall decline in public transit use, carsharing members tend to increase their use of alternative modes, such as walking. Location-specific variations—including urban density, public transit service and availability, sociodemographics, and cultural norms—contribute to these modal shifts, and they are likely to result in varying impacts depending on the specific context in which carsharing is deployed.

In France, the French National Survey comparing roundtrip and station-based carsharing showed differing modal shift impacts [34]. The study found that both forms of carsharing reduced private vehicle use, with roundtrip carsharing having a greater reduction impact. The study also found that roundtrip carsharing increased public transit use slightly, whereas station-based one-way carsharing reduced it. While the study found that both forms of carsharing reduced private bicycle use, roundtrip carsharing increased bikesharing ridership.

A reduction in vehicle ownership may also contribute to lower VMT/VKT, reduced parking demand, and increased use of other transport modes (such as cycling and walking) in lieu of vehicle travel. Carsharing is thought to lead to lower VMT/VKT by emphasizing variable driving costs, such as per hour and/or mileage charges [35]. Reductions range from as little as 7.6% to as much as 80% of a member's total VMT/VKT on average in both Canada and the U.S. for roundtrip carsharing. Estimates differ notably between members who gave up vehicles after joining carsharing programs and those that gained vehicle access through carsharing [24,31,36]. European studies of roundtrip carsharing also indicate a large reduction in VKT ranging from 28% to 45% on average [37]. Martin and Shaheen [30] also documented roundtrip carsharing reductions in VMT/VKT from 27% to 43% in the U.S. and Canada. One-way studies have also documented reductions in VMT/VKT. A study of one-way station-based carsharing in France recorded an 11% reduction in VKT [34]. Martin and Shaheen [33] found VMT reductions ranging from 6% (in Calgary, Alberta) to 16% (in Vancouver, British Columbia and Washington, D.C.) for free-floating one-way carsharing in the U.S. and Canada. This percentage reduction considers an estimate of the total driving by households on average, as derived from annual VMT/VKT responses and broader driving reductions computed for the population.

Reduced vehicle ownership rates and VMT/VKT can also lead to lower GHG levels, as trips are shifted to other modes. In Europe, carsharing is estimated to reduce the average user's carbon dioxide emissions by 40% to 50% [32]. In an aggregate study across North American cities, Martin and Shaheen [30] estimated an average GHG emission reduction of 34% to 41% per household or an average reduction of 0.58 to 0.84 metric tons per household for roundtrip carsharing. Studies of free-floating one-way carsharing estimate that each car2go vehicle reduced GHG emissions by 4% (Calgary) to 18% (Washington, D.C.) on average [33]. In addition, many carsharing organizations include low-emission vehicles— such as electric, plug-in hybrid, and gasoline-electric hybrid cars—in their fleets; use of these vehicle types can result in additional GHG emission decreases. Carsharing members also report a higher degree of environmental awareness after joining a carsharing program [24].

Finally, empirical evidence demonstrates that carsharing has a range of beneficial social impacts. Households can gain or maintain access to vehicles without bearing the full costs of car ownership. Depending on the location and the organization operating the carsharing program, the maximum user mileage where carsharing is more cost effective (in comparison to owning or leasing a personal vehicle) is between 6,200 and 10,000 miles [35]. Low-income households and college students can also benefit from participation in carsharing programs [38,39]. Numerous studies of roundtrip carsharing in North America have found that carsharing households saved an average of $154 US to $435 US per month in contrast to private vehicle use [40].

17.2.2   Impacts of peer-to-peer (P2P) carsharing

A few studies have examined P2P carsharing impacts. Shaheen *et al.* [41] found that P2P carsharing encourages some households to reduce, delay, or even avoid a vehicle purchase. P2P carsharing also enables hosts to reduce their ownership costs, monetize otherwise idle assets, or both. The authors surveyed 1,151 guests and hosts from three U.S. P2P carsharing companies and documented four key findings:

- **Vehicle ownership:** Forty-six percent of P2P carsharing members were from carless households that joined P2P carsharing to gain additional mobility. Another 20% enrolled to earn money sharing their vehicle, while 14% of respondents indicated that they held off on a vehicle purchase due to their carsharing membership. A small percentage (3%) noted that they had sold a vehicle because of their membership.
- **Ease of use:** Forty-eight percent of respondents felt that P2P carsharing was easier than expected to use compared to 15% who said that vehicle sharing was more challenging to employ than anticipated.
- **Travel mode changes:** Most respondents reported no major change in their public transit use as a result of P2P carsharing, with 9% increasing bus ridership and 10% decreasing it. Similarly, 7% of respondents reported increasing rail use, while 8% reported a decrease. Taxi use showed a net decline among all respondents. Survey respondents that use TNCs were split, as 9% reported an increase and another 9% noted a decrease. In contrast, carpooling showed a net increase (6%) among the sample, suggesting that P2P carsharing users were likely traveling with multiple occupants.
- **Super sharers:** In addition, P2P carsharing was used in conjunction with other shared mobility services. Respondents reported that 14% were members of at least one other P2P carsharing service, 43% were members of at least one other carsharing organization, and 78% had used at least one other shared mobility service. Many P2P carsharing members were also frequent Lyft and Uber users, broadly suggesting that they used an array of shared mobility modes to meet their mobility needs [41].

Additionally, the study unveiled motivations and barriers for using P2P carsharing. For vehicle owners, key opportunities and motivations included:

(1) earning revenue on existing, often underused vehicles and (2) contributing to the "sharing economy" by providing mobility access to others. Common barriers for vehicle owners included: (1) concerns about their inability to use their personal vehicle when it is accessed by a guest, (2) potential vehicle damage, and (3) complex insurance requirements that vary by jurisdiction.

For vehicle guests, key opportunities and motivations to participate in P2P carsharing were: (1) accessing a wide array of vehicles, including luxury and zero-emission models and (2) avoiding the costs and hassles associated with private vehicle ownership, such as parking, maintenance, and insurance. Common barriers for vehicle owners include: (1) first- and last-mile connections to access P2P carsharing vehicles, (2) key pick-up and drop-off, and (3) lack of reliable response from a car host following a sharing request. Access to/from vehicles and other challenges suggest expanding P2P carsharing outside of urban areas could be more challenging.

Other studies have focused on the impacts of P2P carsharing on low-income household vehicle access. A study of P2P carsharing use in Portland, Oregon found that 37% of families in poverty live in a census block group that contains at least one P2P vehicle, but only 13% live in a census block that has a roundtrip carsharing vehicle. In parts of East Portland, which is a lower-income area of Portland, P2P vehicles are the only type of carsharing vehicles available [42]. Finally, the authors predict that P2P carsharing will have more pronounced impacts on below-median income consumers than above-median income users. Section 17.3 focuses on more active forms of transportation including shared bikes, scooters, and other low-speed modes.

17.3   Shared micromobility impacts

Shared micromobility—the shared use of a bicycle, scooter, or other low-speed mode—is an innovative transportation strategy that enables users to have short-term access to a low-speed mode on an as-needed basis. Shared micromobility includes various service models and transportation modes that meet the diverse needs of travelers, such as station-based bikesharing (a bicycle picked-up from and returned to any station or kiosk) and dockless bikesharing and scooter sharing (a bicycle or scooter picked up and returned to any location) [43].

Although before-and-after studies documenting shared micromobility impacts are limited, a few North American programs have conducted user surveys to record program outcomes. These studies suggest that a number of social, environmental, and behavioral impacts are attributable to shared micromobility, and an emerging body of empirical evidence supports many of these relationships—although more research is needed as studies on dockless modes (bikesharing and scooter sharing) are limited. In general, impact studies of shared micromobility tend to demonstrate the following outcomes:

- Shared micromobility may help to bridge gaps in the transportation network and encourage multimodal journeys;

- Increased use of some alternative and nonmotorized modes of transportation (e.g., walking and biking);
- Potential for reduced fuel consumption and GHG emissions;
- Opportunities to burn calories [44]; and
- Greater environmental awareness.

This section reviews key study findings of station-based bikesharing and dockless standing electric scooter sharing.

17.3.1   Station-based bikesharing

At present, studies on dockless bikesharing are more limited. A few North American programs have conducted before-and-after studies documenting the impacts of station-based bikesharing. Many of these studies have been completed by bikesharing operators and represent findings from a single city or region. Only a limited number of studies have researched the impacts of station-based bikesharing across multiple cities. A much larger body of literature has studied optimization issues associated with equipment balancing and lifecycle analysis to assess the environmental impacts of bikesharing associated with the product's life from raw material extraction through materials processing, manufacturing, distribution, use, repair and maintenance, and disposal or recycling of bikesharing equipment.

Documented user impacts of bikesharing include: increased mobility, reduced GHG emissions, decreased automobile use, economic development, and health benefits. For example, Boston's Bluebikes estimates 267,000 users completed more than 1.7 million trips, traveled 2.1 million miles, and offset three million pounds of GHG emissions in 2018 [45]. Similarly, in Fort Worth, Texas, the local station-based bikesharing program estimates that approximately 15,000 unique riders completed 59,000 trips covering 266,000 miles, which offset 251,000 pounds of GHG emissions in 2017 [46].

A number of studies have shown that station-based bikesharing can reduce driving and taxi use while increasing cycling in many cities. One study found that half of all bikesharing members report reducing their personal automobile use [47,48]. Shaheen *et al.* [47] conducted an online survey of annual bikesharing members and 30-day subscribers (n=1238) in four metropolitan regions (Montreal, Quebec; Toronto, Ontario; Washington, D.C.; and Minneapolis, Saint Paul, Minnesota) between November 2011 and January 2012. In Minneapolis-Saint Paul, more people shifted toward rail (15%) than away from it (3%) in response to bikesharing. For walking, more respondents shifted toward walking (38%) than away from it (23%) due to bikesharing. However, the study found a slight decline in bus ridership: 15% of respondents increased their use of buses compared to 17% that decreased it. In Washington, D.C., more people shifted away from rail (47%) than to it (7%), and more respondents shifted away from walking (31%) than to it (17%) as a result of bikesharing. Similar to the Twin Cities, the study also found a decline in bus ridership, with just 5% of respondents increasing bus ridership compared to 39% that decreased it [47,49].

A geospatial analysis of this study data involved mapping modal shifts and found that shifts away from public transportation were most prominent in urban environments within high-density urban cores. Shifts toward public transportation in response to bikesharing tended to be more prevalent in lower-density regions on the urban periphery, suggesting that station-based bikesharing may serve as a first- and last-mile connector in smaller metropolitan regions with lower densities and less robust public transit networks. The findings also suggest that in larger metropolitan regions with higher densities and more robust public transit networks, station-based bikesharing may offer faster, cheaper, and more direct connections compared to short distance transit trips. Additionally, public bikesharing may be more complementary to public transportation in small and medium metropolitan regions and more substitutive in larger metropolitan areas, perhaps providing relief to crowded transit lines during peak periods [49]. Another study of bikesharing in New York City found a notable decrease in bus ridership coincident with station-based bikesharing. This study estimates that every thousand bikesharing docks along a bus route are associated with a 1.69% to 2.42% reduction in daily unlinked bus trips on routes in Manhattan and Brooklyn (with and without controlling for bicycle infrastructure, respectively) [50].

Studies also indicate that bikesharing can have measurable effects on economic activity, public health, helmet use, and safety. A study of 1,197 users by Schoner [51] found that Nice Ride Minnesota users spent an average of $1.25 per week on new economic activity that would likely have not occurred without the bikesharing system; this resulted in approximately $29,000 of new economic activity per season in the Twin Cities. The findings suggest that bikesharing stations: (1) increase accessibility to station areas, (2) users may alter destinations or make additional trips, and (3) users spend more money in the immediate vicinity around bikesharing kiosks [51].

In addition to the economic impacts of bikesharing, a number of programs have also documented public health impacts. Several station-based bikesharing programs have attempted to quantify aggregate calories used while cycling. Boston's Bluebikes estimates that its users expended nearly 159 million calories riding on its bicycles in 2018 [45]. Similarly, Citi Bike in New York estimates that its users burned 4.5 billion calories between 2013 and 2018 [52]. Capital Bikeshare in Washington, D.C. reported that its users expended 186 million calories over a one-year period in 2013 [44]. A study by Alberts *et al.* [53] found that 31.5% of Capital Bikeshare users reported reduced stress, and about 30% indicated they lost weight due to bikesharing use. However, a key limitation of these bikesharing health impact assessment studies is that they do not examine potential negative health impacts associated with ridership, such as the costs associated with increased exposure and risks related to injuries and collisions [53].

With respect to safety, helmet usage tends to be lower among shared micro-mobility users, including station-based bikesharing. A study by Buck *et al.* [54] found that only 6% of short-term Capital Bikeshare users (in the Washington, D.C., area) wore helmets, while 37% of annual users wore helmets. Similarly, Shaheen *et al.* [47] found that a high number of respondents in four North American cities never wear helmets (62% in Montreal; 50% in Minneapolis-Saint Paul; 45% in Toronto, and 43% in Washington, D.C.).

In spite of relatively low helmet usage, annual crash rates are relatively low among North American station-based bikesharing operators. Although differences in data collection make it difficult to compare bikesharing crash rates among operators, Shaheen *et al.* [47] document an average collision rate of 4.33 crashes per year among operators with more than 1,000 bicycles, with rates decreasing among operators with smaller fleets. A study of bikesharing safety in Minneapolis-Saint Paul, the San Francisco Bay Area, and Washington, D.C., using data on bicycle and bikesharing activity and bicycle collisions, found that the number of bicycle collisions was generally rising in bikesharing regions, but this increase was very likely due to bicycle activity growth in all regions [55]. For example, between 2006 and 2013, the estimated number of people commuting to work by bicycle in Washington, D.C. increased 162%, while bicycle collisions increased 121%. In San Francisco, the estimated number of bicycle commuters increased 98%, and collisions increased 40% over this same period. Only in Minneapolis-Saint Paul were collisions relatively flat (a 1% increase), while bicycle commuters increased an estimated 65% [55]. However, identifying comparative safety outcomes is challenging for researchers to determine. If trips were diverted from automobiles, buses, or rail, then the risk to individual bikesharing users as well as overall transportation safety could be expected to increase—based on statistics of the per-trip fatality rates of bicycle ridership in comparison to car, bus, or train travel.

17.3.2   *Standing electric scooter sharing*

Research is beginning to emerge on the impacts of standing electric scooter sharing. In Portland, the Bureau of Transportation (PBOT) initiated an e-scooter sharing pilot that ran for 120 days between 23 July and 20 November 2018 with three companies: Bird Rides Inc., Lime, and Skip Transport Inc. Each company started with 100 scooters, which expanded up to 683 scooters per company through the pilot duration [56]. During the pilot, PBOT collected quantitative and qualitative data including: activity data; accident tracking; user surveys; observational studies; focus groups; and online engagement tools, such as webforms, emails, and polls.

Over the course of the four-month pilot in Portland, 700,369 scooter trips were made (averaging approximately 5,885 per day) covering a total of 801,888 miles (averaging 1.15 miles per a trip). The study found that 34% of local users would have used a motor vehicle had scooter sharing not been available. Nineteen percent said they would have driven a personal vehicle, and 15% stated they would have used a for-hire service, such as a taxi, Uber, or Lyft in the absence of scooter sharing. Among visitors, 48% said they would have used a motor vehicle (driving or a for-hire service) without the availability of scooter sharing. Additionally, the study found that 6% of local users sold a vehicle, and 16% considered selling a vehicle due to standing electric scooter sharing [56].

While the study found that a number of respondents replaced motor vehicle travel with scooter sharing, the study also reported that scooter sharing replaced some lower emission active transportation trips. Forty-two percent of respondents said they would have either walked (37%) or ridden a bicycle (5%), if scooter sharing had not been available. Additionally, scooter sharing added some vehicular trips to retrieve and redistribute scooters throughout the day; however, the overall impact of this behavior was not quantified as part of this study [56]. Using a Monte Carlo analysis, Hollingsworth *et al.* [57] estimate an average value of lifecycle global warming impacts of 202 grams (g) CO_2-eq/passenger-mile, driven by materials and manufacturing (50%), followed by daily collection of scooters for charging (43%). The study further estimates the potential to reduce lifecycle emissions through a variety of scooter collection and charging approaches such as (1) using fuel-efficient vehicles for collection (yielding 177 g CO_2-eq/passenger-mile), (2) limiting scooter collection to those with a low battery state of charge (164 g CO_2-eq/passenger-mile), and (3) reducing the driving distance per scooter for e-scooter collection and distribution (147 g CO_2-eq/passenger-mile). However, the study concludes that when scooter sharing use replaces average personal automobile travel, there is almost universally a net reduction in environmental impacts [57].

In addition to the travel behavior and environmental impacts of scooter sharing, a few studies have attempted to quantify scooter impacts on safety and curb space management. During the Portland pilot, the study attempted to measure the safety impacts of scooter sharing by reviewing reported scooter incidents with the Multnomah County Health Department. However, precise numbers are difficult to quantify because emergency room visits could include a variety of other scooter accidents, such as mopeds and nonmotorized standing scooters. During the pilot period, the study identified 176 scooter-related emergency room visits compared to 16 during the same period a year earlier (prior to the pilot). On average, emergency room visits increased from less than one a week prior to the pilot to approximately ten a week during the pilot period. The study also suggests that scooter-related emergency room visits accounted for approximately 5% of total traffic crash injury visits during the pilot. There were no scooter sharing fatalities reported during the pilot. Although the number of scooter emergency room visits was lower than the number of bicycle-related visits ($n=429$), the study lacked comparable data on how many trips were taken and distance traveled while bicycling and therefore could not compare injury rates across modes [56].

Of the entire sample of scooter-related emergency visits, 83% did not involve another mode compared to 13.6% involving a motor vehicle and 2.8% including a pedestrian. Only one collision (0.6%) was reported involving two scooters. Additionally, intoxication was reported in 16% of the collisions. PBOT staff observations suggest that approximately 90% of riders do not wear helmets [56].

In addition to the Portland pilot, a retrospective study of scooter sharing safety in Los Angeles collected data on emergency department visits at two University of California Los Angeles medical facilities. This study queried electronic medical records for clinician notes for the terms "scooter," "bird," or "lime." Note the latter two terms are shared micromobility operator names. The retrieved records were then reviewed by researchers to confirm relevance and inclusion in the data set. Records that included other scooter accidents (e.g., mopeds and nonmotorized scooters) may have been included, and records that did not include the above search terms may have been omitted from the study [58]. The study found that over a one-year period between September 2017 and August 2018, 249 patients sought medical treatment at two medical center emergency rooms. The mean age of patients was 33.7 years, and 58% of patients were male. Ninety-two percent were injured as riders compared to 8% as nonriders. Approximately 11% of patients were under 18 years of age, and less than 5% reported wearing a helmet. Additionally, 5% were intoxicated at the time of their medical treatment. Of the emergency room visits, only 6% (15 reports) were admitted patients, and only two cases (less than 1%) were admitted to the intensive care unit. The study suggests that scooter-related injuries are common with varying severity, low rates of adherence to rider age requirements, and low rates of helmet use.

A few studies have also attempted to document the impacts of standing electric scooter sharing on curb space management. For example, during the Portland pilot, PBOT received over 1,600 complaints of illegal sidewalk riding, representing approximately 27% of public comments. These complaints generally indicated that sidewalk use made pedestrians and persons with disabilities feel unsafe or uncomfortable. Additionally, survey respondents also indicated a strong preference for protected bicycle and/or scooter infrastructure. Another study by the Mineta Transportation Institute attempted to better understand how users park scooters using observational data in San Jose, California. In Summer 2018, researchers observed and photographed 530 parked scooters and categorized key attributes about where and how they were parked and whether they likely impeded pedestrian flow [59]. The researchers defined "well parked" scooters as meeting three key criteria: (1) standing upright; (2) placed on the periphery of pedestrian paths or in areas that are already obstructed, such as by street furniture; and (3) not blocking pedestrian access.

Based on the study's observations, 97% of scooters were parked upright. Seventy-two percent of scooters were parked within a foot of some other vertical object, such as a wall or street furniture, avoiding parking scooters in the middle of open spaces. Less than 2% of scooters were parked in automobile parking spaces. Only 3% of scooters were parked on unpaved surfaces, such as vegetation or dirt. The researchers note that although these scooters did not block pedestrian flow, these parking practices could raise concerns about aesthetics and the impact on landscaping. The study concluded that fewer than 2% of scooters blocked access for people with disabilities, and 90% were parked out of the way of pedestrian traffic, either on the edge of sidewalks or in already obstructed street furniture zones [59]. The authors acknowledge that more research is needed to better understand the impacts of scooters on different types of neighborhoods, such as communities with narrower sidewalks and higher pedestrian flow, aesthetics, and maintenance activities. Section 17.4 is focused on TNC impacts.

17.4   Transportation network company (TNC) impacts

TNCs provide prearranged and on-demand transportation services for compensation, connecting drivers of personal vehicles with passengers. Smartphone applications are used for booking, ratings (for drivers and passengers), and electronic payment. TNCs can provide both pooled services, sometimes referred to as ridesplitting, and nonpooled services. In a pooled service, a TNC ride serves separate parties with similar routes. In a nonpooled service, a TNC ride serves only one party. TNCs differ from taxicab services in that taxis are permitted to pick-up street hails where TNCs are not in most jurisdictions [60]. As TNCs have gained popularity, policymakers, advocates, and researchers have sought to understand how these services are changing travel behavior and affecting society and the environment.

A number of studies assessing the impact of TNC services on modal shift have found that passengers are either substituting a trip they formerly made with another transportation mode (public transit, driving, walking, biking, etc.) or making a new trip they otherwise would not have made without the availability of TNC services (i.e., induced demand). There are conflicting conclusions regarding the extent to which TNCs compete with public transit. While some studies conclude that TNCs are largely not substituting for public transit trips [61–63], several others suggest that a significant portion of travelers do substitute TNCs for public transit, biking, and walking [64–69]. Past surveys show that the degree to which TNCs substitute for other travel modes varies by city and the built environment. Denser cities such as New York City, Boston, San Francisco, and Washington, D.C. exhibited some of the highest proportions of passengers who would have used public transit for their last TNC trip had TNCs been unavailable.

It is important to note that aggregated cross-city studies may obscure city-specific differences in TNC impacts. Also, studies may frame the question aimed at analyzing modal shift differently. Some ask in a general manner what transportation mode travelers might have taken instead of a TNC, while others may ask what mode travelers would have used for their last TNC trip. Depending on how this question is presented, responses may be less representative. The results of existing modal shift studies are shown in Table 17.1, along with the survey question asked in each study.

Broadly, TNC impacts can be summarized in terms of the impacts on public transportation ridership, automobile ownership, and VMT/VKT and GHGs. Each of these impacts is summarized below. Social equity impacts are discussed in Chapter 6: Shared mobility services: prioritizing social good.

17.4.1   Impacts on public transportation

Studies on the impacts of TNCs on public transit ridership vary, which can be attributed to local differences in transit service, urban density, and the built environment. A few studies have investigated the effect that TNCs have on aggregate public transit ridership in U.S. metropolitan areas. One of these studies examined the impact of Uber's entry on public transit ridership between 2004 and 2015 across the 196 U.S. Metropolitan Statistical Areas (MSAs) where Uber was operating. This study found that Uber is a complement for the average public transit agency, increasing ridership by 5% after two years [70]. A similar study, by Feigon and Murphy [62], examined TNCs and public transit ridership trends in Chicago, Illinois; Washington, D.C.; Los Angeles, California; Nashville, Tennessee; Seattle,

Table 17.1 TNC mode substitution impacts[^,†,‡]

Study Authors/ Location/ Survey year	Rayle et al.* San Francisco, CA 2014 [68]	Henao* Denver and Boulder, CO 2016 [67]	Gehrke et al.* Boston, MA 2017 [65]	Clewlow and Mishra† 7 U.S. Cities†† Two Phases, 2014–16 [63]	Feigon and Murphy‡ 7 U.S. Cities††2016 [62]	Hampshire et al.** Austin, TX 2016 [61]	Alemi et al. ‡‡ California 2015 [69]	NYCDOT ‡‡ New York City, NY 2017 [66]
Drive (%)	7	33	18	39	34	45	66	12
Public transit (%)	30	22	42	15	15	3	22	50
Taxi (%)	36	10	23	1	8	2	49	43
Bike or walk (%)	9	12	12	23	18	2	20	15
Would not have made trip (%)	8	12	5	22	1	–	8	3
Carsharing/car rental (%)	–	4	–	–	24	4	–	–
Other/other TNC (%)	10	7	–	–	–	42/2	6/0	–

[^]Survey question: "If Lyft and Uber were not available, how would you have made your most recent trip instead?"

*Survey question: "How would you have made your last trip, if TNC services were not available?"

†Survey question: "If TNC services were unavailable, which transportation alternatives would you use for the trips that you make using TNC services?"

‡Survey crosstab and question, for respondents that use TNCs more often than any other shared mode: "How would you make your most frequent (TNC) trip if ridesourcing was not available?"

**Survey question: "How do you currently make trips like the last one you took with Uber or Lyft, now that these companies no longer operate in Austin?"

††The impacts in these studies were aggregated across Austin, Boston, Chicago, Los Angeles, San Francisco, Seattle, and Washington, D.C.

‡‡These studies allowed multiple responses to the question: "How would you have made your most recent TNC trip (if at all), if these services had not been available?" Therefore, the percentages add up to more than 100%, making it challenging to directly compare it to the other studies.

Washington; and San Francisco, California from 2010 to 2016. The authors concluded there was no relationship between the peak-hour TNC trip share and changes in public transit ridership in these cities. However, a three-city study by Martin *et al.* [64] found a greater portion of respondents would have used public transit (bus and rail) than would have driven or rode in a personal vehicle, if TNCs were not available in San Francisco and Washington D.C. (Los Angeles being the exception). Another study using data from 2002 to 2018 found that the entry and presence of TNCs cumulatively decreased heavy-rail ridership by 1.29% per year and bus ridership by 1.70% per year [71]. Gehrke *et al.* [65] found that passengers with lower incomes and those who possess a weekly or monthly transit pass were more likely to have substituted TNC services for public transit. In addition, relatively low TNC service cost, low TNC trip times, poor weather, and the unavailability of public transit were also predictive of public transit substitution. It is important to account for aggregate trends and individual modal choices in assessing TNC impacts on public transit. Additional research is needed to better understand evolving TNC impacts on public transportation ridership.

17.4.2   Impacts on automobile ownership

Research on TNC impacts on personal vehicle ownership—including the decision to either sell or forgo purchasing a personal vehicle—is growing but requires further study. One study [63] found that 9% of respondents sold one or more household vehicles due to TNCs. A New York City of Department of Transportation study [66] found that approximately 13% of respondents reported owning fewer cars due to the availability of TNCs in Denver and Boulder, Colorado. Another study of rail transit users found that 5% of respondents in Atlanta, 12% in the San Francisco Bay Area, and 21% in Washington, D.C. either postponed a purchase, decided not to purchase, or sold a personal vehicle due to TNCs [62]. Hampshire *et al.* [61] asked respondents about the effect of the mid-2016 Lyft and Uber service suspension in Austin, Texas on their personal vehicle acquisitions. It is important to note that this study is unique because the suspension offered an opportunity to measure vehicle suppression using revealed preference survey data. It found that 9% of respondents acquired a personal vehicle due to the Austin suspension, and another 9% considered purchasing one but ultimately did not. Although Lyft and Uber were not operating in Austin from mid-2016 to mid-2017, other smaller TNC services continued to operate in their place. An even larger portion of respondents may have acquired a personal vehicle, if all TNC services had exited the region.

17.4.3   Impacts on VMT/VKT and GHG emissions

A few studies have assessed the impacts of TNCs on VMT and tripmaking decisions. The most comprehensive studies have employed trip-level TNC activity data in San Francisco [72,73] and New York City [74–76] to analyze mileage, trip metrics, and impacts. Schaller [74] conducted an analysis with publicly available taxi and for-hire vehicle trip and mileage data in New York City. This study found that, after accounting for mileage declines in yellow cabs and personal vehicles, TNCs and other on-demand ride services (including Uber, Lyft, Via, Gett, and Juno) contributed 600 million additional miles of vehicle travel to the city's roads between 2013 and 2016. These additional miles equate to an estimated 3.5% increase in citywide VMT and a 7% increase in VMT in Manhattan, western Queens, and western Brooklyn in 2016. Another study, conducted by Schaller [76], found that usage rates among taxis and TNC vehicles declined in New York City between 2013 and 2017, while the number of unoccupied taxi and TNC vehicles increased by 81% over this time period. This study also found that total taxi and TNC weekday mileage in the central business district (CBD) increased by 36% from 2013 to 2017.

Martin *et al.* [64] studied the impacts of TNCs on VMT and GHG emissions in Los Angeles, San Francisco, and Washington, D.C. In San Francisco and Los Angeles, the study found the VMT produced by Lyft and Uber was larger than the VMT reductions that occurred due to passenger behavior and vehicle ownership changes. In Washington, D.C., the balance of impacts resulted in a net VMT and GHG reduction. The authors hypothesize that these differences may reflect land use and built environment factors that led to lower VMT traveled by TNCs. The study also estimated the potential substitution effect of TNCs in place of personal vehicle ownership. This research suggests that between 2016 and 2017, TNCs may have contributed to a reduction in individual vehicle registrations of 1.7% to 4.2%. It is important to note that the timing of this study coincided with a reduction in vehicle registration growth, which was not associated with any other major economic shocks that would reduce vehicle demand (e.g., economic recession, higher fuel prices, etc.). Thus, the results suggest that this decline in vehicle demand could be in part associated with the rise in TNC popularity [64].

The San Francisco Country Transportation (SFCTA) has also conducted two studies of TNC impacts on the City of San Francisco [72,73]. This study collected TNC trip data from one month in late-2016 and found that TNC trips made up 15% of average weekday vehicle trips within San Francisco and 9% of average weekday person trips within the city [72]. In terms of mileage, this study found that TNCs represented 20% of average weekday intra-San Francisco VMT (trips that originate and end within city limits only) and 6.5% of total VMT (including regional trips starting or ending within city limits) on an average weekday [73]. This study found that around 20% of all TNC miles were deadheading (or zero-occupancy travel) miles. The studies by Schaller [74] and [76] are based on TNC activity data, and the SFCTA study is based on data scraped from a TNC application programming interface (or API) and did not include user surveys. The findings of these three studies are summarized in Table 17.2. Note that these studies do not assess the impact of pooled TNC services.

17.5   Potential impacts of shared automated vehicles (SAVs)

The convergence of shared mobility and automation is predicted to have a trans-formative effect on mobility and goods delivery. Shared automated vehicles

Table 17.2   TNC VMT and trip metrics in the U.S.

City *Study Author* Data, Time period	Key trip metrics	Key mileage metrics
San Francisco, CA *SFCTA* 1 month, late-2016	TNC trips comprise… **15% of vehicle trips** (intra-SF, avg. weekday) **9% of person trips** (intra-SF, avg. weekday) [72]	TNC mileage comprises… **20% of intra-SF VMT** (avg. weekday) **6.5% of total VMT** (avg. weekday) **10% of total VMT** (avg. Saturday) [73]
New York City, NY *Schaller Consulting* Full year, 2016 [76]	TNC trips comprise… **80 million vehicle-trips** (in 2016) **133 million person-trips** (in 2016)	TNC mileage comprises… **7% of total VMT** (in 2016 TNC mileage equates to an *estimated increase of*… **3.5% citywide VMT** (in 2016) **7% VMT in Manhattan, western Queens, and western Brooklyn** (in 2016)
New York City, NY *Schaller Consulting* June 2013 and June 2017 [74]	TNC/taxi trips increased by… **15% between June 2017 and June 2013** (Manhattan CBD, avg. weekday)	TNC/taxi mileage increased by… **36% between June 2017 and June 2013** (Manhattan, avg. weekday)

(SAVs) powered by clean energy can further reduce air pollutants and GHG emissions, mitigating many of the transportation-related impacts associated with vehicle travel [77]. SAE International, a global mobility standards organization, has established five levels of vehicle automation. Level 1 describes vehicles that automate only one primary control function (e.g., self-parking or adaptive cruise control). Level 2 describes a vehicle with automated systems that provides full control of specific functions, such as accelerating, braking, and steering. With Level 2, the driver must still monitor driving and be prepared to immediately resume control at any time. Level 3 allows the driver to engage in non-driving tasks for a limited time. With Level 3, the vehicle handles situations requiring an immediate response; however, the driver must still be prepared to intervene within a limited amount of time when prompted. With Level 4, a human operator does not need to control the vehicle as long as it is operating under the specific conditions the vehicle was intended to function. Level 5 describes vehicles capable of driving in all environments without human control.

As Level 4 and 5 automated vehicles become more mainstream, it is possible that a variety of shared mobility services (e.g., carsharing, car rental, TNCs, and taxis) may converge to create a new model comprised of SAVs. A number of studies have attempted to model the potential impacts of electric SAVs (sometimes referred to as shared automated electric vehicles (SAEVs)); however, these studies do not always account for nuances in business models, vehicle ownership, the built environment, public transit accessibility, urban density, challenges associated with pooling, and other factors. Thus, predicting generalizable potential impacts of SAEVs is challenging. The table below provides a summary of the current modeling literature on the expected environmental impacts of large-scale SAEV deployments. In general, these studies have found that SAEV fleets could: (1) reduce vehicle emissions compared to gasoline-powered fleets; (2) replace privately owned vehicles; and (3) require less charging infrastructure than previously anticipated, particularly when electric vehicles (EVs) are shared (Table 17.3).

In addition to reducing vehicle ownership, SAEVs have the potential to lower costs and offer flexible public transportation systems. Pakusch *et al.* [83] predict SAVs would draw users from public transit as opposed to conventional vehicles. Shaheen and Cohen [84] emphasize the importance of understanding the potential relationships between AVs and public transportation. Shaheen *et al.* [85] identified five potential emerging use cases for SAEVs including:

- **Closed campus:** SAEVs could provide short-distance, point-to-point travel in closed campus environments that can be easily mapped by software. These locations include theme parks, resorts, malls, business parks, college campuses, airport terminals, construction sites, downtown centers, real estate developments, gated communities, industrial centers, and others.
- **First-mile and last-mile connectivity:** Traditionally, public transit has been limited by fixed routes and fixed schedules. Due to these limitations, travelers may find it difficult to complete the first- or last-mile of their journey using public transit. SAEVs may be able to help bridge first- and last-mile gaps in the public transportation network.
- **Low-density service:** SAEVs have the potential to provide lower cost and more frequent or responsive public transit strategies in rural, exurban, and low-density suburban areas where low ridership and high labor costs often contribute to inefficient or cost-prohibitive fixed route service.
- **Off-peak/late night service:** Similarly, SAEVs may be able to complement public transit by providing service during off-peak times, especially late at night when service is difficult and costly to provide.
- **Paratransit:** Paratransit services could be provided by SAEVs to meet the needs of persons with disabilities; nevertheless, human assistance may still be required.

These applications have the potential to bridge gaps in the transportation network, such as first- and last-mile connections to public transportation, late-night transportation, and service for low-density communities. While early exploratory studies provide insights into how SAVs could impact communities, more research is needed to better understand the complex factors that could impact user adoption, such as land use, the built environment, and public transit accessibility. In the future, SAEVs could have a transformative effect in

Table 17.3   Impacts of shared, automated, and electric vehicles (SAEVs)

Publication	Methods	Expected impacts
Greenblatt and Saxena [78]	Estimated per-mile GHG emissions of a fleet of SAEVs	• SAEV fleet emissions expected to be 87% to 94% lower than a fleet of 2014 conventional gasoline vehicles • SAEV fleet emissions expected to be 63% to 82% lower than a fleet of projected 2030 HEVs
Chen *et al.* [79]	Modeled management of a fleet of SAEVs under various charging infrastructure and vehicle range scenarios	• Fleet size is dependent upon battery recharging time and vehicle range • An 80-mile range SAEV could replace 3.7 privately owned vehicles • A 200-mile range SAEV could replace 5.5 privately owned vehicles • Faster charging equipment could increase vehicle shedding
Biondi *et al.* [80]	Optimized parking location and capacity for a carsharing service with an EV fleet	• Most charging stations would require less than four spots • Only a few large charging stations (up to 15 spots) are needed to be placed in areas of high turnover • Possible impact on peak electricity demand can be mitigated by using fast-charging technologies
Harb *et al.* [81]	Explored AV-related travel behavior shifts for underserved populations (older adults/retirees and children)	• The study estimates an 83% overall VMT increase, 21% of which was zero-occupancy travel VMT (or deadheading)
Harper *et al.* [82]	Explored AV-related travel behavior shifts for underserved populations (non-drivers, older adults, and people with disabilities)	The study estimates an increase of total annual light-duty VMT by 14%. Sixty-five percent of this VMT increase would be from current adult non-drivers, while 16% is from older drivers without a medical condition, and 19% is from adult drivers with travel-restrictive medical conditions.

enhancing social equity and accessibility by bridging gaps in the transportation network through the extension of geographic coverage and service availability. Public policy (e.g., pricing, incentives) and other interventions will likely be needed to encourage higher rates of SAV pooling [86,87], particularly among strangers.

17.6   Conclusion

Better understanding the impacts of shared mobility can help policymakers and public agencies leverage the positive transportation and environmental impacts of shared and automated mobility services, as well as tame the unintended or negative impacts [1]. To better prepare for shared and automated mobility, the public sector should:

- Understand the sociodemographic trends associated with shared mobility services, along with their impacts on cities and regions;
- Encourage multimodality and the bridging of first- and-last-mile connections;
- Reduce fossil fuel consumption and support climate action and air quality goals;
- Provide education and training on shared modes for all users, including subsidies (when appropriate);
- Avoid, reduce, and/or mitigate the negative health, social, and economic effects of the transportation network on underserved populations and communities, such as minority and low-income communities, rural and suburban populations, older adults, persons with disabilities, and zero-vehicle households; and
- Ensure affordable, full, and equitable transportation access and mobility to all users (discussed in Chapter 6).

Requiring data sharing to conduct impact assessments is a critical first step toward better understanding the impacts of shared and automated mobility on communities. Additionally, administering pilot studies, testing innovative policy approaches, and conducting scenario analyses can provide new avenues for evaluating and understanding the travel, social, equity, labor, and economic impacts of shared and automated mobility services.

Author contributions

The authors confirm contribution to this chapter. All authors reviewed and approved the manuscript.

Declaration of conflicting interests

The author(s) declared no potential conflict of interest with respect to publication of this chapter.

References

[1]   Cohen, A., and Shaheen, S. *Planning for shared mobility*. Chicago: American Planning Association; 2016.

[2] Pratsch, L. *Commuter ridesharing.* Englewood Cliffs, NJ: Pretince Hall, 1979.

[3] Hillsman, E., Reeves, P., and Herzog, B. "Estimation of effects of Washington state's trip-reduction program on traffic volumes and delays: Central Puget sound region." *Transportation Research Record.* 2001; 1765: 16–19.

[4] Lagerberg, B. "Washington state's commute trip reduction program: Phase 1: Assessment and implications for program design." *Transportation Research Board.* 1997;1598: 36–42,

[5] Boarnet, M.G., Hsu, H-P., and Handy, S. *Policy brief on the impacts of employer-based trip reduction on a review of the empirical literature.* Sacramento: California Air Resources Board, 2014. (Online). Available from https://ww2.arb.ca.gov/sites/default/files/2020-06/Impacts_of_Employer-Based_Trip_Reduction_Programs_and_Vanpools_on_Passenger_Vehicle_Use_and_Greenhouse_Gas_Emissions_Policy_Brief.pdf.

[6] Shaheen, S., Cohen, A., and Bayen, A. *The benefits of carpooling.* Berkeley, CA: Transportation Sustainability Research Center; 2018.

[7] Interagency Commute Trip Reduction Board. *Joint comprehensive commute trip reduction plan.* State of Washington, 2011.

[8] Minett, P., and Pearce, J. "Estimating the energy consumption impact of casual carpooling." *Energies.* 2011; 4(12): 126–139.

[9] Noland, R., Cowart, W., and Fulton, L. "Travel demand policies for saving oil during a supply emergency." *Energy Policy.* 2006; 34(17): 2994–3005.

[10] Jacobson, S., and King, D. "Fuel saving and ridesharing in the US: Motivations, limitations, and opportunities." *Transportation Research Part D: Transport and the Environment.* 2009;14(1).

[11] PACommutes. "Eco-impact." 2016. (Online). Available from www.pacommutes.com/ridesharing/car-pooling/eco-impact/.

[12] The Global e-Sustainability Initiative. "SMART 2020: Enabling the low carbon economy in the information agency." 2008. (Online). Available from http://bnrg.eecs.berkeley.edu/~randy/Courses/CS294.F09/Smart2020Report_lo_res.pdf.

[13] Teal, R. "Carpooling: Who, how, and why." *Transportation Research Part A.* 1987; 21(3): 203–214.

[14] Burris, M., and Winn, J. "Slugging in Houston: Casual carpool passenger characteristics." *Journal of Public Transportation.* 2006;9(5). DOI: 10.5038/2375-0901.9.5.2

[15] Tomer, A. "Transit access and zero-vehicle households." 2016. (Online). Available from https://www.brookings.edu/wp-content/uploads/2016/06/0818_transportation_tomer.pdf.

[16] Liu, C.Y., and Painter, G., "Travel behavior among Latino immigrants: The role of ethnic neighborhoods and ethnic employment." *Journal of Planning Education.* 2012; 31(1): 62–80.

[17] Metropolitan Transportation Commission, 1999. (Online). Available from www.nctr.usf.edu/wp-content/uploads/2011/04 /Casual-Carpool-Report-1998.pdf.

[18] Maltzman, F., and Beroldo, S. "Casual carpooling: An update." RIDES for Bay Area Commuters, San Francisco, CA. 1987

[19] Reno, T., Gellert, W., and Verzosa, A. "Evaluation of Springfield instant carpooling." *Transportation Research Record.* 1989; 1212: 53–62.

[20] Beroldo, S., "Casual carpooling in the San Francisco bay area." *Transportation Quarterly.* 1990; 44(1): 133–150.

[21] Shaheen, S., Chan, N., and Gaynor, T. "Casual carpooling in the San Francisco bay area: Understanding characteristics, behaviors, and motivations." *Transport Policy.* 2016; 51: 165–173.

[22] Oliphant, M. *The native slugs of Northern Virginia: A profile of slugging in the Washington DC Region, Blacksburg, Virginia.* Virginia Tech, 2008.

[23] Shaheen, S., Cohen, A., and Jaffee, M. *Innovative mobility carsharing outlook.* Berkeley, CA: Transportation Sustainability Research Center, University of California, Berkeley, 2018.

[24] Lane, C. "PhillyCarShare: First-year social and mobility impacts of carsharing in Philadelphia, Pennsylvania." *Transportation Research Board.* 1927; 2005, 158–166.

[25] Martin, E., Shaheen, S., and Lidicker, J. "Impacts of carsharing on household vehicle holdings." *Transportation Research Board.* 2010; 2143: 150–18.

[26] Arlington County Commuter Services. "Arlington pilot carshare program." 2005. (Online). Available from http://mobilitylab.org/wp-content/uploads/2012/02/ArlingtonCarshareProgram.pdf.

[27] Cervero, R., and Tsai, Y. "City CarShare in San Francisco: Second-year travel demand and car ownership impacts." *Transportation Research Record.* 2004; 1887: 117–127.

[28] Communauto. "Potentiel de L'Auto-Partage Dans Le Cadre d'Une Politique de Gestion de La Demande en Transport." *Potentiel de L'Auto-Partage Dans Le Cadre d'Une Politique de Gestion d Forum de L'AQTR, Gaz à Effet de Serre: Transport et Développement, Kyoto: Une Opportunité d'Affaires,* Montreal, Canada, 2000.

[29] Jensen, N. *The co-operative auto network social and environmental report.* Vancouver, BC:Cooperative Auto Network; 2001.

[30] Martin, E., and Shaheen, S. "Greenhouse gas emission impacts of carsharing in North America." *IEEE Transactions on Intelligent Transportation Systems.* 2011; 12(4): 1074–1086.

[31] Zipcar. "Zipcar customer survey shows car-sharing leads to car shedding." 2005. (Online). Available from www.autorentalnews.com/75124/zipcar-releases-survey-on-car-sharing-impact.

[32] Ryden, C., and Morin, E. *Mobility services for urban sustainability: Environmental assessment.* Stockholm, Sweden, Report WP 6. 2005.

[33] Martin, E., and Shaheen, S. *Impacts of car2go on vehicle ownership, modal shift, vehicle miles traveled, and greenhouse gas emissions: An analysis of five North American cities.,* Berkeley, CA: Transportation Sustainability Research Center; 2016.

[34]   6t. "One-way carsharing: Which alternative to private cars?." 2014. (Online). Available from https://www.6-t.co/wp-content/uploads/2014/05/ AD_ExecutiveSummary_140708-copie-2.pdf.

[35]   Shaheen, S., Cohen, A., and Roberts, D. "Carsharing in North America: Market growth, current developments, and future potential." *Transportation Research Record*, 1986; 2006, 116–124.

[36]   Cooper, G., Howe, D., and Mye, P. *The missing link: An evaluation of CarSharing Portland Inc., Portland, Oregon.*, Portland, OR: State of Oregon Department of Environmental Quality; 2000.

[37]   Shaheen, S., and Cohen, A., "Worldwide carsharing growth: An international comparison." *Transportation Research Record*. 2007; 1992: 81–89.

[38]   Kyeongsu, K. "Can carsharing meet the mobility needs for the low-income neighborhoods? Lessons from carsharing usage patterns in New York City." *Transportation Research Part A: Policy and Practice*. 2015; 77: 249–260.

[39]   Stocker, A., Lazarus, J., Becker, S., and Shaheen, S. *North American college/university market carsharing impacts: Results from Zipcar's college travel study 2015*. Berkeley, CA: Transportation Sustainability Research Center: 2016.

[40]   Shaheen, S., Mallery, M., and Kingsley, K. "Personal vehicle sharing services in North America." *Research in Transportation Business and Management*. 2012; 3: 71–81.

[41]   Shaheen, S., Martin, E., and Bansal, A. *Peer-to-peer (P2P) carsharing: Understanding early markets, social dynamics, and behavioral impacts*. Berkeley, CA: Transportation Sustainability Research Center: 2018.

[42]   Dill, J., Howland, S., and McNeil, N. "Peer-to-peer carsharing: A preliminary analysis of vehicle owners in Portland, Oregon, and the potential to meet policy objectives." *Transportation Research Board Annual Meeting*, Washington, D.C., 2014.

[43]   Shaheen, S., and Cohen, A. *Shared micromobility policy toolkit: Docked and dockless bike and scooter sharing*. Washington, D.C.: The Smart Cities Lab and the International Council on Clean Transportation; 2019.

[44]   Freed, B. "Capital bikeshare users burned 186 million calories last year, and other fun facts." 17 October 2015. (Online). Available from https://www. washingtonian.com/2014/10/17/capital-bikeshare-users-burned186-million-calories-last-year-and-other-fun-facts/.

[45]   Bluebikes. "Media kit." 04 January 2019. (Online). Available from https:// www.bluebikes.com/about/media-kit.

[46]   Camareno, K., and Brennan, M. *Fort Worth bikesharing 2017 Annual Report*. Fort Worth: Forth Worth Bikesharing; 2017.

[47]   Shaheen, S., Martin, E., Cohen, A., and Finson, R. *Public bikesharing in North America: Early operator and user understanding*. San Jose, CA: Mineta Transportation Institute; 2012.

[48]   Shaheen, S., Martin, E., Chan, N., Cohen, A., and Pogodzinski, M. *Public bikesharing in North America during a period of rapid expansion:*

Understanding business models, industry trends and user impacts. San Jose, CA: Mineta Transportation Institute; 2014.

[49] Shaheen, S., and Martin, E. "Unraveling the modal impacts of bikesharing." *Access*. 2015; 8–15.

[50] Campbell, K., and Brakewood, C. "Sharing riders: How bikesharing impacts bus ridership in New York City." *Transportation Research Part 1*. 2017; 264–282.

[51] Schoner, J. "Sharing to grow: Economic activity associated with nice ride bike." *23rd Annual Center for Transportation Studies Research Conference*, Minneapolis, Minnesota, 2012.

[52] Motivate. "New milestone: Citi bike reaches 60 million rides since launch." 06 June 2018. (Online). Available from https://www.motivateco.com/new-milestone-citi-bike-reaches-60-million-rides-since-launch/.

[53] Alberts, B., Palumbo, J., and Pierce, E. *Vehicle 4 charge: Health implications of the capital bikeshare program*. Portland, OR: The George Washington University; 2012.

[54] Buck, D., Buehler, R. Happ, P., Rawls, B., Chung, P., and Borecki, N. "Are bikeshare users different from regular cyclists? A First look at short term users, annual members, and area cyclists in the Washington DC region." *Transportation Research Record: Journal of the Transportation Research Board*. 2013; 2387: 112–119.

[55] Martin, E., Cohen, A., Botha, J., and Shaheen, S. *Bikesharing and bicycle safety*. San Jose, CA; Mineta Transportation Institute, 2016.

[56] Portland Bureau of Transportation. *2018 E-scooter findings report*. Portland, OR: Portland Bureau of Transportation; 2018.

[57] Hollingsworth, J., Copeland, B., and Johnson, J. "Are E-scooters polluters? The environmental impacts of shared dockless electric scooters." *Environmental Research Letters*. 2019; 14(8).

[58] Trivedi, T., Liu, C., Antonio, A.L., *et al.* "Injuries associated with standing electric scooter use." *Journal of the American Medical Association*. 2019; 1–9.

[59] Fang, K., Weinstein Agrawal, A., Steele, J., Hunter, J.J., and Hooper, A.M. *Where do riders park dockless, shared electric scooters?* San Jose, CA: Mineta Transportation Institute; 2018.

[60] Shaheen, S., and Cohen, A. "Shared ride services in North America: Definitions, impacts, and the future of pooling." *Transport Reviews*. 2018; 39(4):427–442

[61] Hampshire, R., Simek, C., Fabusuyi, T., Di, X., and Chen, X. "Measuring the impact of an unanticipated disruption of Uber/Lyft in Austin, TX." 2017. (Online). Available from https://papers.ssrn.com/sol3/papers.cfm?abstract_id=2977969&download=yes.

[62] Feigon, S., and Murphy, C. "Shared mobility and the transformation of public transit." 2018. (Online). Available from https://s3-us-west-2.amazonaws.com/cdn.sudbury.ma.us/wp-content/uploads/sites/391/2018/12/Shared-Mobility-and-the-Transformation-of-Public-Transit-2.pdf?version=adeae0a529ec275303f2ed4d8109c3d0.

[63] Clewlow, R., and Mishra, G. *Disruptive transportation: The adaption, utilization, and impacts of ride-hailing in the United States.*, Davis, CA: Institute of Transportation Studies; 2017.

[64] Martin, E., Shaheen, S., and Stocker, A. *impacts of transportation network companies on vehicle miles traveled, greenhouse gas emissions, and travel behavior: Analysis from the Washington, D.C., Los Angeles, and San Francisco markets.*, Berkeley, CA: Transportation Sustainability Research Center; 2021 (Forthcoming).

[65] Gehrke, S., Felix, A., and Reardon, T. *Fare choices: A survey of ride-hailing passengers in Metro Boston*. Boston, MA: Metro Area Planning Council; 2018.

[66] New York City Department of Transportation. *Citywide mobility survey*. New York City, NY, 2017.

[67] Henao, A. *Impacts of ridesourcing – Lyft and Uber–On transportation including VMT, mode replacement, parking, and travel behavior*. Denver, CO: University of Colorado at Denver; 2017.

[68] Rayle, L., Dai, D., Chan, N., Cervero, R., and Shaheen, S. "Just a better taxi? A survey-based comparison of taxis, transit, and ridesourcing services in San Francisco." *Transport Policy*. 2016; 45: 168–178.

[69] Alemi, F., Circella, G., Handy, S., and Mokhtarian, P. "what influences travelers to use Uber? Exploring the factors affecting the adoption of on-demand ride services in California.' *Travel Behaviour and Society*. 2018; 13:88–104

[70] Hall, J., Palsson, C., and Price, J. "Is Uber a substitute or complement for public transit?" *Journal of Urban Economics*. 2018; 108: 36–50.

[71] Graehler, M., Mucci, R.A., and Erhardt, G., "understanding the recent transit rideship decline in major US cities: Service cuts or emerging modes?." *98th Annual Meeting of the Transportation Research Board*, Washington, D.C., 2018.

[72] San Francisco County Transportation Authority. *TNCs & congestion*. San Francisco, CA: SFCTA; 2018.

[73] San Francisco County Transportation Authority. *TNCs today: A profile of San Francisco transportation network company activity.*, San Francisco, CA: SFCTA; 2017.

[74] Schaller, B. *Empty seats, full streets: Fixing Manhattan's traffic problems*. New York City, NY: Schaller Consulting; 2017.

[75] Schaller, B. *The new automobility: Lyft, Uber and the future of American cities*. New York City, NY: Schaller Consulting; 2018.

[76] Schaller, B. *Unsustainable? The growth of App-based ride services and traffic, travel and the future of New York City*. New York City, NY: Schaller Consulting; 2017.

[77] Shaheen, S., and Bouzaghrane, M.A. "Mobility and the energy impacts of shared automated vehicles: A review of recent literature." *Current Sustainable/Renewable Energy Reports*. 2019; 6: 1–8.

[78] Greenblatt, J., and Shaheen, S. "Autonomous taxis could greatly reduce greenhouse-gas emissions of US light-duty vehicles." *Nature Climate Change*. 2015; 9(860).

[79]   Chen, D., Kockelman, K., and Hanna, J. "Operations of a shared, autonomous, electric vehicle fleet: Implications of vehicle and charging infrastructure decisions." *Transportation Research Part A: Policy and Practice.* 2016: 94: 243–254.

[80]   Biondi, E., Boldrini, C., and Bruno, R. "The impact of regulated electric vehicle fleets on the power grid: The car sharing case." *2016 IEEE 2nd International Forum on Research and Technologies for Society and Industry (RSTI) Leveraging a Better Tomorrow*, 2016.

[81]   Harb, M., Xiao, Y., Circella, G., Mokhtarian, P., and Walker, J. "Projecting travelers into a world of self-driving vehicles: Estimating travel behavior implications via a naturalistic experiment." *Transportation.* 2018; 45: 1671–1685. https://doi.org/10.1007/s11116-018-9937-9.

[82]   Harper, C., Hendrikson, C., Mangones, S., and Samaras, C. "Estimating potential increases in travel with autonomous vehicles for non-driving, elderly and people with travel-restrictive medical conditions." *Transportation Research Part C: Emerging Technologies.* 2016: 72: 1–9.

[83]   Pakusch, C., Stevens, G., and Bossauer, P. "Shared autonomous vehicles: Potentials for a sustainable mobility and risks of unintended effects." *ICT Express Series.* 2018. 245–258.

[84]   Shaheen, S., and Cohen, A. "Is it time for a public transit renaissance?." *Journal of Public Transportation.* 2018; 67–81.

[85]   Shaheen, S., Cohen, A., Yelchuru, B., and Sarkhili, S. *mobility on demand operational concept report.* Washington D.C.: U.S. Department of Transportation; 2017.

[86]   Shaheen, S., Lazarus, J., Caicedo, J., and Bayen, A. *To pool or not to pool? Understanding the time and price tradeoffs of on-demand ride users – Opportunities, challenges, and social equity considerations for policies to promote shared-ride services.* Berkeley, CA: UfC Office of the President: University of California Institute of Transportation Studies; 2021.

[87]   Lazarus, J., Bauer, G., Greenblatt, J., and Shaheen, S. *Bridging the income and digital divide with shared automated electric vehicles.* Berkeley, CA: Transportation Sustainability Research Center; 2021.

The views expressed are those of the authors.

Chapter 18

Future directions: maximizing the social and environmental benefits of shared and automated mobility services

Ata M. Khan[1] and Susan A. Shaheen[2]

Converging innovations—including the sharing economy, digitization, autonomy, and electrification—are key building blocks of future mobility systems. These innovations are continuously impacted by socio-demographic and mobility trends, along with other disruptive forces (e.g., COVID-19 pandemic and climate change). In this chapter, we discuss eight focus areas (central to this book) for optimizing the public good and guiding this convergence: (1) advanced, safe, secure, and efficient shared mobility; (2) policy and regulations; (3) social equity and justice; (4) environmental and financial sustainability; (5) transportation system and land use planning; (6) system design, operations, and management; (7) implementation, urban development, and other impacts; and (8) measures to adapt to long-term pandemic impacts. We recognize that automated vehicle (AV) adoption will be gradual, following a phased evolution over the next decade or longer. We also acknowledge that COVID-19's impacts on innovative mobility services will require adaptive measures over the short- and longer-term (e.g., contactless systems and hygiene). Innovations and unexpected disruptions will continue into the future. As such, we will need to revisit research, policies, and approaches to optimize the social and environmental benefits of autonomy, shared mobility, electrification, digitization, and other emerging technologies and services.

18.1 A starting point for moving to future directions

Shared mobility services include: (1) carsharing; (2) microtransit; (3) for-hire ride services, such as transportation network companies (TNCs) (also known as ride-sourcing and ridehailing); (4) shared micromobility (bikesharing and scooter sharing); (5) shuttles, taxis, and pedicabs; (6) public transit; and (7) shared urban air mobility. Shared mobility also includes last-mile delivery services, such as

[1]Department of Civil and Environmental Engineering, Carleton University, Ottawa, Canada
[2]Department of Civil and Environmental Engineering, University of California, Berkeley, CA, USA

app-based deliveries (commonly referred to as courier network services), robotic delivery, and drones. Taken together, these services meet user needs that can include previously unmet demand.

Shared mobility's evolution and adoption reflect the influence of several causal factors, noted in Figure 18.1. Advances in smartphone and communication technologies associated with apps have played a major role in facilitating on-demand services. The availability of battery-powered shared micromobility vehicles (i.e., e-bikes and e-scooters) is another technological advance creating a market for local, short-distance trips previously made by other modes or previously unmade trips (also known as latent demand). Changing socio-demographic factors (e.g., licensing rates, the increasing average age of marriage, urbanization trends) and technological advances are also shaping innovative mobility trends.

Technology and socio-demographic trends are driving many converging mobility innovations, shown in Figure 18.1. Today's shared mobility ecosystem is evidence of these dynamic innovations. Although not fully developed, evolving services have the potential to fill service gaps and replace inefficient public transit routes within a multimodal transportation system. A key enabling technology of shared mobility services is digitization. This includes digital information and fare payment to support seamless routing, booking, and payment. The goal of integrated services is to facilitate "complete trips" (i.e., all aspects of a trip or tour, including logistics and supporting infrastructure from start to finish) across modalities and travel choices, including e-commerce. Growth in the commodification of

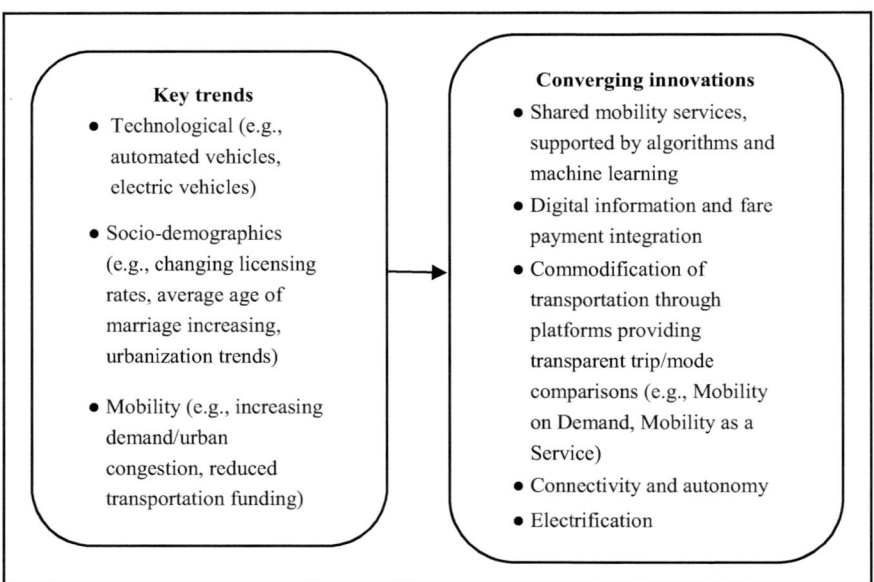

Figure 18.1   Trends and converging innovations impacting future mobility.
Source: Based on Chapter 2 of this book

transportation services through mobility on demand (MOD) and Mobility as a Service (MaaS) platforms: (1) supports multimodal mobility; (2) offers opportunities to influence travel behavior (e.g., nudging and incentives); (3) compete with private vehicle ownership and use; and (4) advance transportation sustainability (i.e., social equity, financial, and environmental).

The last two mobility innovations noted in Figure 18.1, connectivity/autonomy and electrification, are expected to advance shared mobility in terms of performance and sustainability—social equity, finance, and the environment. While the electrification of bikes, scooters, and electric vehicles (EVs) in North America, Europe, and Asia started some time ago, global mainstreaming of electrification and charging networks coupled with shared mobility services will usher in a new era of mobility services [1,2].

Although many medium- and large-size urban areas acknowledge shared mobility in their official and master plans, proactive planning for future transportation needs is at a preliminary stage [3]. In contrast to current-day shared mobility services, the future could be more disruptive due to connected and automated vehicle (CAV) technologies, particularly depending upon their deployment (e.g., gradual and incremental vs. more disruptive). In planning for the future of shared mobility services across the globe, a large number of policy and technical actions are needed to advance social and environmental goals. Many action areas are crosscutting and complex, with uncertain impacts. More research is needed to fill critical gaps in understanding.

18.2   Future directions

Eight focus areas are identified for guiding future developments, ranging from technology advances and service innovations to transportation sustainability to safeguarding public health (Figure 18.2). Since CAVs will gradually gain acceptance in real-world applications, current shared mobility services are key to informing the transition to fully automated services. Recent experiences with COVID-19 are likely to require a longer-term response and measures to adapt to the impacts of the pandemic (e.g., hygiene and sanitation).

Existing shared mobility services can serve as a baseline for understanding how best to achieve public policy goals and meet private-sector investor requirements. In the future, MOD and MaaS will be integrated with other emerging technologies, including electrification and CAVs. Vehicle right-sizing and new form factors, including active transportation modes, will provide opportunities for improving efficiency.

The eight key focus areas discussed in this chapter and throughout this book are complex, requiring a multidisciplinary perspective for successful deployment and best practices (see Figure 18.2). They include: (1) advanced, safe, secure, and efficient shared mobility; (2) policy and regulations; (3) social equity and justice; (4) environmental and financial sustainability; (5) transportation system and land use planning; (6) system design, operations, and management; (7) implementation,

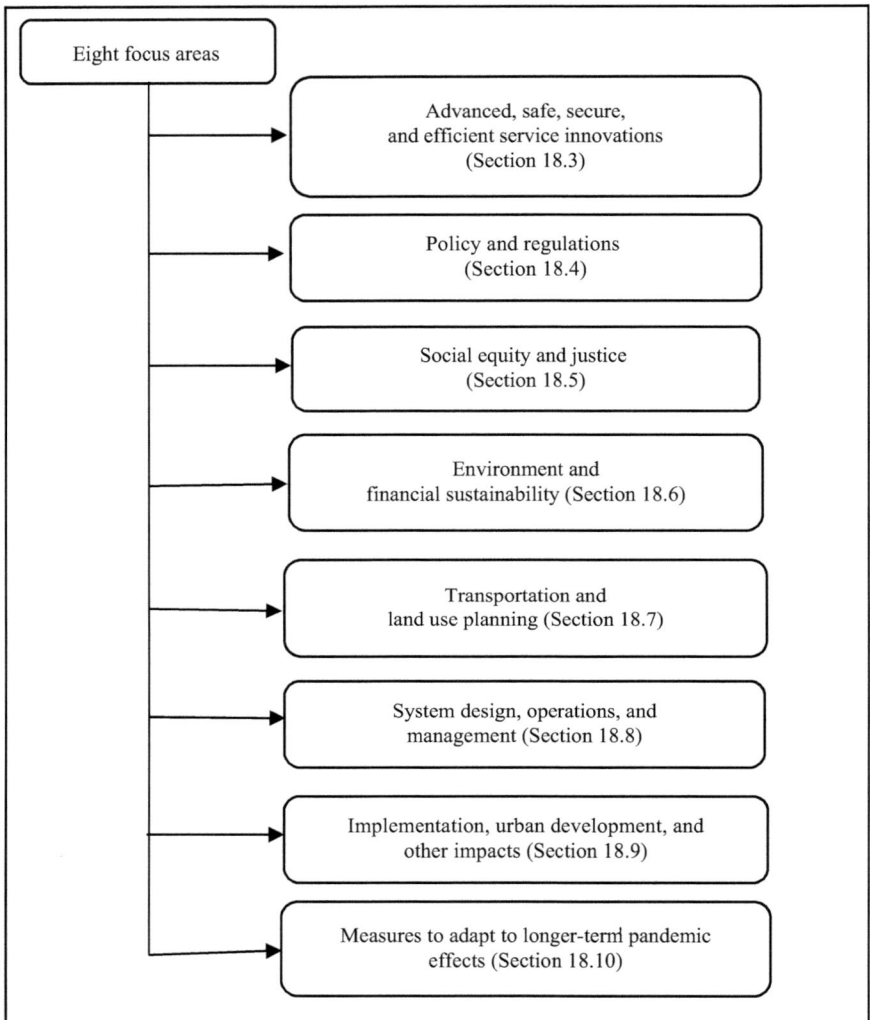

Figure 18.2   Future directions focus areas

urban development, and other impacts; and (8) measures to adapt to long-term pandemic impacts.

While upgrading services are desirable actions, there are cross-cutting impacts that require ongoing attention, such as labor, job retraining and transition, and social equity. Public-sector transportation agencies are eager to understand how to avoid the unintended, negative consequences of shared mobility and CAVs (e.g., empty passenger miles/kilometers, congestion, labor impacts, reduced active transportation, etc.). To address complex impacts and negative externalities, research is needed to support the policy and planning process. The public sector recognizes that shared automated vehicles (SAVs) could help to improve safety and efficiency, increase social equity and public transit access, fill mobility gaps, and provide increased transportation choices if these services are managed to optimize social and environmental goals.

It is important to note that encouraging people to share rides with strangers may continue to be a challenge in the future, as it is today. This could impact the viability and potential emission benefits of concurrent SAV services in the future. Prior to COVID-19, there was a strong preference for single (or sequential) versus pooled (or concurrent) TNC trips. Due to concerns about virus transmission, pooled trips were not offered during the first 16+ months of the pandemic. What will stick following the pandemic with respect to user preferences and traveler behavior? These are important questions for scholars, planners, and policymakers to consider across the globe. This will impact public transit, work from home/telework, shared mobility, SAV trends, and more.

In the following sections (18.3–18.10), each focus area is discussed in the context of strategic actions of key stakeholders. Some action areas have the potential to contribute to more than one societal objective (e.g., environmental sustainability and reduced operating costs). In the future, we assume that all AVs are electric and connected, employing the latest communication technologies.

18.3   Advanced, safe, secure, and efficient service innovations

Shared mobility services have become more common in cities across the globe. Municipalities throughout Europe, the U.K., Asia, and North America are employing a range of policy and planning initiatives to maximize the benefits of technological advances and emerging services to the traveling public. Over time, shared mobility systems will continue to evolve to better serve the user and operational needs, including the integration of technological advancements in autonomy, artificial intelligence (AI), digitization, and electric propulsion. Future directions include: (1) upgrading shared mobility services to address public safety, security, and efficiency and (2) public transit improvements. Actions in support of these developments include:

- Safe and secure automation;
- Electrification;
- SAV services, including cost reductions due to autonomy;
- Availability of on-demand mobility services in response to traveler needs (e.g., paratransit);
- Information technology or app-based platform services (e.g., MOD and MaaS) to support multimodal tripmaking;
- Flexible microtransit services (right-sized vehicles reflecting demand) supported by autonomy, including first- and last-mile connections, filling service gaps, and public transit replacement; and

- Micromobility—both shared and privately owned (including bikes, scooters, and new form factors), linked to public transit and supporting an individual's point-to-point tripmaking.

Acceptance of SAVs will depend upon the prerequisites of safety, security, convenience of access and use, and cost attributes. Safety can be viewed as crash safety, cybersecurity safety, and personal safety (e.g., if pooling is allowed). If SAVs are perceived to be safe, secure, right-sized for trip making, accessible without much delay, and competitive with private vehicle ownership costs, they could be widely accepted. Cost information, as reported in Chapter 16, is favorable for SAVs in contrast to private vehicle ownership and use. A number of other characteristics noted in this chapter are receiving consumer and policy attention, such as social equity and privacy.

SAVs are expected to be capable of fully automated driving. According to SAE International, at Level 5 of automation, there is no need for a human driver or safety attendant to be in the vehicle to safely complete travel from an origin to a destination even in a complex urban traffic network [4]. Although progress has been made in developing fully automated vehicles, AV safety will require ongoing research into the future [5,6]. The security of automated driving (e.g., cybersecurity) is also essential and is expected to continue to receive research attention [7,8].

Studies focused on smart communities and smart infrastructure require favorable safety, security, and efficiency capabilities as building blocks for future transportation. Applications of 5G communication services and automated connected vehicle technology are expected to strengthen safety and privacy attributes (i.e., protection of personal identifiable information). Nevertheless, there are many unanswered questions that require further study. This includes monitoring and remote safety intervention to protect travelers in the case of crashes, cybersecurity issue, and personal safety (e.g., pooling with a stranger). This also includes privacy protections.

When fully developed and thoroughly tested, CAVs are expected to be safer in comparison to vehicles controlled by human drivers. Naturally, there is a need to monitor and, if warranted, to initiate remote control of an AV. This topic is of special importance in the context of shared mobility services, particularly pooling with strangers, if allowed, as noted earlier. The role of human monitoring and control needs further investigation, designs have to be developed for this function, and the cost of these activities should be included in business model development [9]. As a part of system design, it is important to identify technological and methodological factors to protect against crashes, cyber threats, and personal attacks on individuals, alongside protecting personal identifiable information.

To improve operations, SAVs can be efficiently routed in congested networks to fulfill on-demand requests or to reach a charging station. Dynamic routing algorithms can be used for this purpose, and additional methodologies (e.g., Bayesian decision analysis) can be developed to address uncertainty in estimating travel times. For additional information, see Chapter 12.

Public policy can influence technology initiatives needed to support shared mobility services and information technology platforms (e.g., geofencing and pricing). As shared mobility services grow and gain acceptance, the automotive and digital sectors will need guidance on the specialized needs of partially and highly automated SAVs to inform vehicle and digital platform design. For example, shared electric mobility fleets could be expected to be in service around the clock with different design characteristics (e.g., right sizing based on trip purpose and time of day) in contrast to AVs intended for private mobility alone.

Over the last decade, apps and algorithms have spawned transportation service innovations, such as pooled rides through transportation network companies (TNCs, also known as ridesourcing and ridehailing) and taxi apps (also known as ridesplitting), along with shared micromobility services (docked and dockless bikes and e-scooters). In recent years, shared mobility operators have partnered in the delivery of paratransit, first- and last-mile services, filling service gaps, and offering public transit replacement through MOD and MaaS pilot projects.

A number of cities across the world have gained experience in offering flexible transit services via pilot projects operated by public transit operators and private companies [10]. For example, the City of Calgary, in association with RideCo, is demonstrating the merits of a first- and last-mile connection with high-capacity public transit hubs in the city [11,12]. Another MOD service is operating in the City of Innisfil, Ontario. This subsidized service, operated by Uber, is an example of how to serve mobility needs in lower-density areas where conventional public transit cannot be sustained [13].

In the United States (U.S.), microtransit companies are also providing dynamic and fixed route services in a range of environments (both higher and lower density). Some services are outsourced and subsidized, and in other cases, the public transit agencies provide the service with backend support from private-sector companies [14]. Given financial and operational challenges, some early TNC and microtransit providers already have left the market [10]. Microtransit has the potential to make public transportation even more accessible with CAVs, as well as routing and matching algorithms. Automation is likely to reduce service and travel costs and serve a wider range of land-use and built environments across the world (e.g., suburban and more rural locations). This could result in more equitable access for all travelers. In the future, it is possible that a number of shared mobility service models (e.g., carsharing, car rental, TNCs, taxis, and microtransit) could converge into SAV systems.

While app technologies and multimobility platforms, such as MOD and MaaS, have disrupted the urban travel market, many factors will continue to shape future directions and outcomes, including technological advancements, public policies (e.g., pricing and finance), and traveler response. In addition, a number of challenges will continue to impact the future marketplace including cybersecurity; data privacy; traveler safety/security; social equity and labor impacts; land-use considerations (rural, urban, and suburban); and the unintended consequences of machine learning/AI. More research is needed to better understand how to shape and maximize the benefits of shared and automated mobility through public policy and transportation planning.

18.4   Policy and regulations

At present, a number of policies and regulations related to permits, curbside management, insurance, and data sharing are generating controversies. These topics will continue to require future attention by stakeholders. Key policy and regulatory actions rely upon:

- Information to support permits, curbside management, and insurance decisions; and
- Data sharing and management between the public and private sectors, which involves privacy concerns and cybersecurity.

18.4.1   Permits and insurance

The process of granting permits to shared mobility operators involves many checks to ensure safety and convenience for the traveling public. At present, TNC and taxi drivers have to meet conditions defined by regulatory agencies (e.g., licensing, fingerprinting, and driver benefits, as applicable). Also, insurance requirements have to be met. When SAVs are ready to serve the traveling public, permitting and associated insurance requirements will reflect many new considerations. Many nations are considering policies to guide the introduction of AVs and services.

In the near- if not longer-term future, auto insurance policies will assume vehicle collisions are mainly caused by human error. Shared mobility companies and their drivers will continue to secure the required insurance prior to serving a market [15]. For insurance purposes, the use of CAVs implies that the automated systems will be responsible for collisions. According to the Insurance Bureau of Canada (IBC) [16], current auto insurance policies will require a redefinition regarding liability and benefit rules. Both the vehicle manufacturer and the automated technology provider will become a part of the equation.

The IBC has proposed an insurance framework that enables the use of automation, while co-existing with current policies [15]. The main elements of the framework include: (1) a single insurance policy devised to cover both driver error and the automated technology; (2) a data sharing arrangement negotiated with vehicle manufacturers, vehicle owners, and auto insurers for determining the cause of a collision; and (3) new federal vehicle standards developed that include technology and cybersecurity standards.

The logic behind the IBC insurance framework can be explained as follows [15]. First, automation will reduce collisions due to the absence of human error and distraction. Further, this capability can only be achieved with well-designed expensive systems, but these will incur high repair costs. A 2015 KPMG[1] study [17] estimated that with AVs accident expenses could rise significantly due to increased crash severity and AV pricing—on average from US $14,000 to US $35,000 by 2040. New sources of risk also play a role; these include cyberattacks,

---

[1]A global risk and opportunities assessment firm.

along with software and hardware malfunctions [18]. Another factor that will influence auto insurance cost is access to driving data provided by AV data loggers. A 2018 study by Deloitte [19] suggests that the availability of logged data will be of great value to auto insurers in assessing crashes and pricing insurance premiums.

While insuring shared mobility vehicles currently entails considerable complexity, further complexity will be added with autonomy. Since future SAV deployment issues are not fully understood, further studies will be necessary.

18.4.2   Data sharing and privacy concerns

Due to technological innovations, there is a growing abundance of valuable travel behavior data for transportation planning and operations. In the case of shared micromobility (bikes and scooters) services, data are being shared between private operators and public agencies through a software tool, called the Mobility Data Specification (MDS). MDS is a standard for exchanging data between mobility operators and public agencies via application programming interfaces (APIs). In the case of TNCs, however, these data are not readily available. Although customers of shared mobility services allow companies to collect information on their service use, including trip origins and destinations, these data often cannot be easily accessed by public sector planners.

Public agencies could use these data for informing decisions to help regulate public infrastructure use or planning for future transportation investments. Since these data are collected while shared mobility vehicles are operating on public rights-of-way, a condition for regulatory approval—across modalities—by governmental agencies could be data sharing and safeguarding traveler privacy.

Reasons for the reluctance of some private companies to readily share information with the public sector are related to a number of issues. Many companies regard these data as proprietary and essential to their business success. Sharing data on how their services are used could theoretically affect their competitive edge (e.g., revealing the markets they serve by time and location). Another reason is the requirement to protect user privacy (or personal identifiable information). In some instances, cities have been given access to TNC data, but in many cases, conditions were placed on sharing these data with other public agencies, such as public transit operators.

As noted earlier, data gathered by shared mobility services could be a key resource for the policy and planning activities of governmental agencies. If data are shared, city agencies and public transit operators would benefit from the disaggregate data regarding origins and destinations of on-demand trips. Given the detailed nature of these data, city planners could better understand traveler movements. These data can offer important insights into improving operations; enhancing infrastructure (e.g., mobility hubs, docking stations, corrals, etc.); designating curbside pickup/drop off locations; and first- and last-mile connections to public transit.

A number of emerging practices and strategies for data access and sharing have been reported in the literature [10]. These include negotiating access to shared mobility usage data, engaging third parties to coordinate data sharing, and incorporating data sharing provisions into partnerships and regulatory agreements.

The MDS approach has been deployed by the Los Angeles Department of Transportation [20,21]. This system allows for two-way communications between mobility companies and local governments. However, legal complaints have been filed by a shared mobility operator and the American Civil Liberties Union due to privacy concerns [20]. While MDS is not universally popular with the private sector, shared micromobility companies are highly supportive and have agreed to use it [21]. Future studies are needed to better understand these issues, document MDS best practices, and suggest improvements.

18.5   Social equity and justice

Social equity and justice in transportation are complex subjects and require the attention by policy makers, planners, and others in a position to define the nature and extent of these multifaceted issues and initiate measures to address it at all levels of governance, along with the private sector. It is critical to tackle the hardships faced by persons with disabilities and underserved populations, particularly those without access to a private vehicle or living in areas with inadequate public transit. Addressing gender biases in service provision is also important.

Historically, paratransit services have helped fill mobility gaps. In today's app-based environment, however, there is notable potential to improve paratransit similar to the services offered by on-demand microtransit providers. Paratransit systems are already adopting apps that are making services more efficient in serving persons with disabilities [22]. In the era of automation, special purpose vehicles should be designed (and right sized) so that these do not present technological barriers to serving mobility-challenged individuals.

Fair access to shared mobility services in various parts of the city requires ongoing attention. Providing mobility justice in various parts of a city is a policy goal that calls for disaggregate level planning. Travel markets with high demand usually get priority for infrastructure provision and public transit services. At present, pilot projects are underway in U.S. cities to better understand how shared modes and services can address inequities in serving transportation needs across a city and diverse use cases (e.g., first-last mile, gap filling, public transit replacement, and paratransit) [23,24].

According to some studies, transportation service models, including the choices of human drivers, along with pricing and matching algorithms, may further social inequities [25–27]. It is important to note that in the absence of safeguards, emerging mobility services and enabling technologies may contribute to inequity. Improving algorithmic fairness in AI is important with shared mobility services and in the future with SAV services.

Social equity concerns in transportation can also be impacted by outside factors, such as a lack of affordable housing pushing lower-income households into less central and/or dense locations and spatial mismatch (i.e., the discrepancy between where low-income households reside and suitable job opportunities are located). Understanding these relationships and impacts will also play a critical role in improving transportation equity and justice.

18.6   Environmental and financial sustainability

In addition to social equity and justice, as described in Section 18.5, environmental and financial sustainability are necessary requirements for the future of shared mobility, along with mobility justice. Strategies include: electrification, automation, vehicle pooling, pricing (incentives and disincentives), vehicle right sizing, flexible financing mechanisms in public transportation provision (e.g., subsidies), and an emphasis on new business model.

18.6.1   Reducing emissions and energy use

Transportation electrification is a policy strategy being deployed across the globe to reduce emissions and energy use. For shared mobility services, this strategy is being encouraged at all levels of governance. For example, many cities in Canada and the U.S. have started to electrify transportation systems. Toronto and other major cities have adopted electrification as a policy, and steps are underway to work on details such as building standards and charging infrastructure [3]. E-scooters and e-bikes powered by batteries are being deployed across the world. Naturally, these devices require supportive charging infrastructure at vehicle staging locations.

Vehicle electrification in the shared mobility marketplace is evolving in North America, and the policy environment is becoming more favorable, such as the Clean Miles Standard (SB 1014) in California [28]. California's SB 1014 directs state agencies to enact stricter emission targets for TNCs to promote zero emission vehicles (or ZEVs) by 2021 [29]. Expanding charging infrastructure for TNCs and taxis is one way in which public agencies can support the electrification of for-hire services. Designing and locating fast chargers for use by shared vehicles is complex [30]. Another complexity is balancing the number of chargers with demand for fast charging over time.

The economics and impacts of fast charging on the power grid is a subject that also needs further investigation. It is important to note due to the high costs of fast charging equipment and installation, these units are not likely to be cost effective at private dwellings or apartment complexes. Thus, there is a need to locate fast charging stations at accessible, high-demand locations in cities [31].

Electrification of shared mobility vehicles requires information on public fast charging station requirements, locations, specifications, grid effects, and economics. Areas for future study include: public fast charging requirements for TNCs, electrification of shared micromobility vehicles and their charging requirements, specifications for fast charging clusters at a station, location of fast charger stations, and cost-effective operations [32]. Electric shared mobility vehicles will require fast charging stations within the geographic area of the travel market, but there are many unexplored factors that may hinder planning and investment decisions. Research studies are needed to address these knowledge gaps.

Along with electrification and automation, shared mobility services are expected to be a part of mobility as a service (MaaS) and mobility on demand (MOD) platforms that could ultimately support a mobility wallet, including a unified payment platform for various public transit systems and other public and private elements of the mobility network. MOD, as defined by the U.S. Department of Transportation (US DOT), is a new concept based on the principle that transportation is a commodity where modes have economic values that are distinguishable in terms of cost, journey time, wait time, number of connections, convenience, and other attributes [33].

MOD enables consumers to access mobility, goods, and services on demand by dispatching or using shared mobility, delivery services, and public transportation strategies through an integrated and connected multimodal network. The most advanced forms of MOD passenger services incorporate trip planning and booking, real-time information, and fare payment into a single user interface. While there are some similarities, MOD differs from the emerging European concept of MaaS. MOD focuses on the commodification of passenger mobility and goods delivery and transportation systems management, whereas MaaS primarily focuses on passenger mobility aggregation and subscription services. Brokering travel with suppliers, repackaging, and reselling it as a bundled package is a distinguishing characteristic of MaaS [34].

Across the world, concerns have been expressed that shared mobility services that use internal combustion engine vehicles could result in higher pollution and greenhouse gas emissions due to increased vehicle miles/kilometers traveled (VMT/VKT) in areas with rising congestion. This can happen if shared mobility services are not guided by public policy and result in greater VMT/VKT, lower pooling rates, and/or declines in public transit ridership. Therefore, a lack of policies and planning initiatives could cause adverse emission and energy use impacts.

Policies and plans, such as California's SB 1014 (Clean Miles Standard) that are favorable to vehicle electrification, shared micromobility modes (active transportation), and pooling have the potential to integrate shared modalities into the transportation system and avoid negative impacts. Optimistic views are often expressed due to the potential of connected, automated, and electrified shared mobility fleets integrated into multimodal transportation systems that result in higher public transit usage and active transportation (micromobility and walking). To realize this future, public policies are needed to guide deployment and to optimize the social and environmental benefits, including taming empty VMT/VKT in SAVs to the extent possible [10,35–37].

While it is challenging to predict the precise impacts of electric SAVs, the current modeling literature, cited in Chapter 17, indicates that SAV fleets have the potential to reduce vehicle emissions and energy use compared to gasoline-powered fleets; lead to reduced private vehicle reliance; and if shared, require less charging infrastructure.

18.6.2 Financial feasibility

Economic factors and emerging business models, discussed in Chapter 16, indicate that electric SAVs have the potential to become more cost effective than current modes. However, further studies are needed to evaluate different system designs and to run sensitivity tests to produce a better understanding of the financial feasibility of private sector investments.

Regional and local governments can employ their authority to regulate the use of public rights-of-way including parking and the curb, which are commonly used by on-demand vehicles. Moving forward, it is important for municipalities to develop approaches for prioritizing and regulating public infrastructure use for traffic and emission/energy management (e.g., curb, parking, and road pricing). Such pricing policies could also lead to revenue generation, which could be redirected into public transit and active transportation modes [3,38].

Use of the public infrastructure has generated active dialogues regarding the use of regulatory mechanisms for achieving transportation planning and management goals. Since public agencies require funding to build and maintain transportation infrastructure, curb fees, for instance, could be used to recover capital investment in facilities and operational costs associated with transportation system management, including enforcement. This starts with managing private vehicle use, along with shared modes and public transit. While some policy analysts and planners view shared mobility and SAVs as a source of potential new revenue streams, others argue that pricing approaches must be applied across modes, prioritizing access to active transportation and higher occupancy modes (e.g., public transit and pooling). Revenues from pricing could also be used to provide subsidies for low-income individuals. Further research is needed to inform policymakers and planners in developing strategies and best practices for managing public infrastructure use across modalities.

18.7 Transportation system and land use planning

Many public sector policy makers and planners are keen to maximize the positive impacts of shared mobility services and avoid adverse effects. Stakeholders recognize that the integration of SAVs into a multimodal transportation system could improve accessibility across a range of land uses and mobility justice. As noted earlier, public policy is key to guiding shared mobility and SAV integration into comprehensive planning efforts to maximize the social and environmental benefits. In the absence of this coordination, it will be challenging for cities to plan for shared mobility and achieve desired outcomes. Key actions include: (1) strengthening public transit through first/last mile connections, right-sizing service, and filling gaps in the public transit network; (2) integrating shared mobility into multimodal transportation through platforms such as MaaS and MOD; (3) integrating shared mobility into modeling and forecasting [10,39]; (4) encouraging favorable land use impacts; and (5) managing changes in parking and opportunities for infill development and affordable housing.

According to a U.S. DOT study, failure to understand and integrate shared mobility within the legacy systems of roads, public transit, and other travel modes could limit their potential benefits and lead to greater future challenges. Well planned, careful adaptation of emerging mobility modes and services is needed to achieve public policy goals (e.g., safety, social equity, environmental benefits, and congestion management). In-depth knowledge of opportunities and challenges is important to advancing planning efforts [10,40–42]. Further, it is critical that cities proactively plan for the policy, physical, and virtual adaptations needed to ensure that testing and deployment of electric, connected, and SAV services focus on maximizing their social and environmental benefits [3,43].

Research studies carried out by the International Transport Forum [43] and other organizations have generated insights that can be further refined by transportation agencies and used as a starting point for planning SAV and shared micromobility services as part of the metropolitan/regional transportation system. Further research is needed to fine-tune knowledge on land use/built environment challenges and opportunities for improving planning practices for integrated future mobility services to increase access, social equity, and environmental benefits.

Shared mobility services have resulted in a suite of transportation options. In combination with existing modes, there is potential for more seamless multimodal transportation services that can serve all travelers. The combined complementary effect of public transit and shared mobility services could ultimately challenge the advantages of private vehicle ownership. However, guiding policies and proactive planning are the prerequisites for such integration to materialize [36,41].

While some scenario-based studies have modeled the replacement of public transit with private sector MOD services, policy makers and planners argue that more traditional public transit services (e.g., rail and buses) must be protected to provide access for transit-dependent users. The US Department of Transportation's MOD Sandbox demonstration program has helped public transit agencies initiate pilot projects with private shared mobility companies, and this program has informed the public transit industry's understanding of MOD services [14].

Literature is available on emerging practices and strategies for incorporating shared mobility into the regional multimodal transportation planning process [10,44–46]. Several studies have attempted to model demand in response to CAVs [3]. Chapters 10 and 11 of this book present demand models that incorporate SAVs. A scan of best practices suggests the following recommendations:

- Long-range strategic planning for transportation and land use should incorporate shared modes.
- Future plans should reflect a comprehensive approach that includes infrastructure for electric, shared, and automated vehicles.
- The transition from nonautomated to automated driving should be carefully considered.

Shared mobility should be integrated into modeling and forecasting. To support this activity, shared mobility services should be incorporated into travel surveys, and longitudinal data collection should become a routine part of the planning and modeling process. While models are being studied, "off-model" approaches can be used to estimate shared mobility impacts. Off-model adjustments are used to estimate emission reductions when regional travel demand and land use models do not include emerging modes.

18.8 System design, operations, and management

System design, operations, and management focus on future infrastructure for electric SAVs. This includes: (1) electric vehicle (EV) fleet accommodations; (2) adapting (or "future proofing") infrastructure for connected automated driving; and (3) multimodal curbside management to incorporate shared modes, micromobility, and SAVs into street design and management. While this subject has a wide scope and appears to be on the priority list of many cities, more research is needed. Key topics include intelligent traffic control, public fast chargers, curb management, building standards, etc.

EV fleet accommodations require infrastructure planning and design. According to emerging practices in Asia, Europe, and the U.S., large EV fleets are already being introduced in cities or will be in the future by automakers and on-demand mobility providers for use in the shared mobility marketplace [30,47,48]. Since many fleet owners are keen on fast charging opportunities to support service, it is likely that they will participate in the development of a fast charging station network [3]. This is already happening in China [30]. Airport authorities also have responded to the need for TNC vehicle charging at staging areas close to passenger terminals. More research on fast charging needs and digital networks for managing supply and demand interaction should be investigated in future studies.

Future proofing of infrastructure is an area for prospective action. Research studies in the UK, Europe, and the U.S., for instance, have highlighted the need to investigate transportation infrastructure changes to ensure the safe and efficient operation of shared mobility services [49–52]. Such changes can enhance convenience in using on-demand transit [53]. In a recent report, the City of Toronto has identified the need to take steps to future proof infrastructure for use by CAVs [3]. Of special importance to planners is the use of the curb by shared mobility vehicles, as well as freight delivery services. In addition to physical design factors, curb management is a subject that requires further investigation [3,15].

Cities need to develop proactive readiness plans and undertake activities to ensure readiness for electric SAVs. Among other readiness activities, developing plans for CAV pick-up and drop-off zones is needed for safety and efficient operations. Research is needed on the location and design of these facilities to ensure passenger safety and convenience. Also, accessibility for persons with disabilities and underserved users should be emphasized. Another use of these facilities is to provide safe harbors or staging areas for CAVs in the event of a malfunction or an unusual driving situation that necessitates vehicles to park and wait for assistance.

Preparing to manage congestion attributed to on-demand modes, including SAVs, is another important planning task [54]. Recent studies have observed that TNCs are adding to traffic congestion in major metropolitan areas [3,55]. This observation is in line with the scenario that expanded shared mobility results in more single use of TNCs. According to current practice, single-use TNCs are more dominant than TNC pooling or ridesplitting [56,57]. If policy innovations do not change this trend (e.g., California's SB 1014 – Clean Miles Standard), TNCs could increase VMT/VKT and congestion could worsen.

However, if guided by proactive public policies to restrict deadheading miles to the extent possible, shared mobility services could result in more pooling, public transit, and active transportation. Under favorable conditions, shared mobility could potentially contribute to congestion reduction.

Other infrastructure adaptation actions are needed, along with research. These include facilities for dockless or free-floating vehicles, including shared e-bikes and e-scooters. Studies are needed to better understand how sidewalks and other public rights-of-ways can be shared by all users [58,59]. Given the disruptive nature of future shared mobility modes, practitioners would benefit from the development of guidelines to help identify appropriate infrastructure treatments on a corridor or smaller area, for instance.

18.9   Implementation, urban development, and other impacts

To support future policy and planning decisions for shared mobility services, studies on implementation, including impacts are needed. Due to limited real-world experience, research on these subjects, including land use impacts, is not available. This book makes contributions to understanding the implementation and fostering favorable land use development impacts due to shared mobility services. Chapters 14, 15, and 17 provide reference material on this theme.

To maximize policy goals, cities of all sizes should jointly incorporate shared mobility and SAV services into land use and transportation systems planning within a comprehensive planning framework [46]. While there are diverse views on shared mobility impacts on urban land use, there are even more diverging perspectives on the integration of connected and automated shared mobility vehicles on a range of land uses [9].

In the future, SAVs could reduce the need for people to own and maintain a personal vehicle, as this asset is parked and unused for the majority of the day. Should this happen, the built environment in metropolitan areas will likely be affected due to the large amounts of land dedicated to parking use. Thus, cities could reduce parking requirements and allocate valuable urban land for affordable housing or open greenspace. This positive effect of SAVs could counter increases in VMT/VKT, if privately owned AVs are used for commuting. AV use for low occupancy commuting has been addressed by many scholars due to the increased travel (or induced) demand effects resulting from lower trip costs and convenience, alongside shifting demand for more suburban and rural land uses [9]. This travel pattern, if not countered by antisprawl urban development policies, will likely lead to increased congestion and emissions.

Another subject of interest is the availability and location of AV staging/ parking facilities after dropping off passengers at work sites. This and many other subjects require research to fill knowledge gaps on how shared mobility might influence future land use development.

The above observations suggest that at present there is much uncertainty about shared mobility impacts on future land use [10,46]. To fill these gaps, better understanding of the following factors is needed. [9].

• Study of the marginal cost of commuting with electric SAVs as compared to current costs in constant dollars.
• Study of the cumulative effects of future shared mobility vehicles on repurposed parking land/space in the central areas of major cities (e.g., land price, developments caused by land price, and land re-zoning).
• Impacts of future shared mobility technologies and services on household decision-making to locate or relocate to a lower density area.
• Staging facilities (e.g., parking and charging locations) of electric SAVs when not in service.

18.10   Measures to adapt to longer-term pandemic effects

The COVID-19 pandemic has resulted in the wide adoption of work from home/ telework and social distancing practices. These shorter-term effects have adversely affected public transportation use, and offices have been lightly used or not at all. In contrast, e-commerce and courier services have been widely employed for food delivery and other items.

While the short-term impacts on many sectors of the economy are now well known, the longer-term effects of the pandemic are still uncertain [60–62]. Given the impacts of the pandemic, it is critical to add preparedness for potential pandemics to the list of future directions. The following is a list of measures to be investigated to adapt to the longer-term effects of a pandemic.

• Vehicles intended for on-demand service, including shared micromobility, can be designed to automatically spray disinfectant at the end of a service cycle and before the start of another.
• Ultraviolet technologies to disinfect vehicles can be deployed (e.g., in-vehicle and handlebars).
• During a pandemic, on-demand vehicle services can be offered, but pooled services can be canceled or limited in party size to ensure physical distancing.
• Reservation capabilities are integrated into public transit and shared mobility services to ensure physical distancing and reduce crowding.
• Passive temperature checks and contact tracing can be employed at turnstiles and in-app, alongside digital vaccine passports (when applicable).
• Contactless payment and vehicle entry can be deployed to reduce the spread of the virus.
• New vehicle designs may be introduced to reduce contact, including partitions.

18.11   Conclusions

Today, the rapidly changing and complex transportation ecosystem needs re-envisioning so that it better serves all travelers across a wide range of land use and built environments. To plan for an integrated multimodal system, an inclusive approach is needed that involves all stakeholders, most notably the communities they serve, to ensure that a balance is achieved between public policy goals and the interests of the public and private sectors. Additionally, to ensure longer-term sustainability, high-capacity public transit services (e.g., rail and express bus services) are needed to provide mobility to all travelers, including underserved populations, in dense urban areas.

This chapter identifies future directions for improving shared mobility services and integrating them into a seamless, multimodal transportation system while avoiding the adverse impacts on public transit, emissions, social and racial equity, and land use. The suggested actions cover service innovations, as well as policies and regulations that support social, environmental, and fiscal sustainability goals. Suggestions include changes in transportation and land use planning, as well as infrastructure modifications that accommodate shared mobility services, active transportation, and SAVs. Implementation context, impacts, and longer-term changes to urban development are considered. Suggested action areas can assist policymakers and planners responsible for integrating shared mobility into multimodal transportation planning and decision making. Actions to enhance fiscal sustainability and economic feasibility also are addressed. The revenue potential of public infrastructure access is noted as a measure of fiscal sustainability and equitable access. Furthermore, resilient actions for adapting to the longer-term impacts of the pandemic to protect public health and to mitigate adverse economic effects are noted.

While significant uncertainties and knowledge gaps are identified in various parts of the book, research can increase our understanding of how to best approach shared mobility, active transportation, and SAVs in planning for a multimodal transportation system in a range of land use and built environments. Since we are entering a new era of inclusive planning for emerging transportation technologies and services, all stakeholders, including developers and insurance providers, must improve their knowledge of how best to provide a safe, efficient, equitable, and sustainable transportation system for the traveling public. As noted at the beginning of this chapter, the future landscape of shared mobility and SAVs will continue to evolve, with unexpected disruptions, opportunities, and challenges. Ongoing research, policy development, and innovative approaches will be needed to guide new developments and to optimize the social and environmental benefits of these emerging technologies and services.

Author contributions

The authors confirm contribution to this chapter. All authors reviewed and approved the manuscript.

Declaration of conflicting interests

The authors declared no potential conflict of interest with respect to publication of this chapter.

References

[1]   Volkswagen. *WeShare launched in Berlin as full-electric service.* Wolfsburg/Berlin. 27 June 2019. Available from https://www.volkswagen-newsroom.com

[2]   General Motors. *GM technology paves the way for an all-electric future. Technology.* 14 June 2020. Available from https://www.gm.com

[3]   City of Toronto. *Automated vehicle tactical plan and readiness 2022.* IE8.7 Report for Action. 2019.

[4]   U.S. Department of Transportation (US DOT). *Preparing for the future of transportation: Automated vehicles 3.0.* Washington, D.C. October 2018. Available from https://www.transportation.gov/av

[5]   Kalra, N. *Challenges and approaches to realizing autonomous vehicle safety and mobility benefits: Testimony.* CT-475 Testimony presented to the House Appropriations Committee, Subcommittee on Transportation, Housing and Urban Development, and Related Agencies on May 18, 2017. RAND Corporation, Santa Monica, CA, USA.

[6]   Khan A. "Bayesian artificial intelligence-based driver for fully automated vehicle with cognitive capabilities" in Stanton N. (ed.) *Advances in Human Factors of Transportation, AHFE 2019, Advances in Intelligent Systems and Computing.* Vol. 964. Springer, Cham. https://doi.org/10.1007/978-3-030-20503-4_6

[7]   European Union Agency for Cybersecurity (ENISA). *ENISA good practices for security of smart cars.* European Union Agency for Cybersecurity (ENISA); November 2019. DOI 10.2824/17802

[8]   Deloitte Canada. *Cybersecurity for connected and autonomous vehicles, considerations and opportunities for growth.* Commissioned by Ontario Centres of Excellence (OCE), AVIN (Ontario) and the Automotive Parts Manufacturers (APMA); 2019.

[9]   MIT Energy Initiative. *Insights into future mobility.* Cambridge, MA: MIT; 2019. Available from http://energy.mit.edu/insightsintofuturemobility

[10] Federal Highway Administration (FHWA). *Integrating shared mobility into multimodal transportation planning: Improving regional performance to meet public goals.* FHWA-HEP-18-033, U.S. DOT, Washington, DC; 2018.

[11] Fast, W. "Calgary transit case study: Rideco solves first-last mile gap." *Medium.* 06 November 2019.

[12] Calgary Transit. "Calgary transit on-demand." *News*, Calgary-transit-demand. Calgary, Canada; 2019.

[13] Conference Board of Canada. *My ride, your ride, our ride. Public transit and shared mobility.* Ottawa, Canada; 2019.

[14] Lucken, E., Frick, K.T., and Shaheen, S. "Three Ps in a MOD: Role for mobility on demand (MOD) public-private partnerships in public transit provision." *Research in Transportation Business & Management.* Vol. 32; September 2020. https://doi.org/10.1016/j.rtbm.2020.100433

[15] Autonomous Vehicle Innovation Network (AVIN) (Ontario). *Autonomous vehicles reshaping the future, cross-sector opportunities and consideration.* Toronto; October 2019.

[16] Insurance Bureau of Canada (IBC). *Auto insurance for automated vehicles: Preparing for the future of mobility.* Ottawa, Canada; 2018. Available from https://tinyurl.com/ycrwfjsm

[17] KPMG. *Marketplace of change: Automobile insurance in the era of autonomous vehicles.* Ottawa; 2015. Available from https://tinyurl.com/yxtgh9ak

[18] European Parliamentary Research Service. *A common EU approach to liability rules and insurance for connected and autonomous vehicles.* Brussels, Luxembourg; 2018. Available from http: tinyurl.com/y56ovaj6

[19] Deloitte. *Connected and autonomous vehicles in Ontario: Implications for the insurance industry.* Ottawa; 2018. Available from https://tinyurl.com/y5f8ahnn

[20] Hawkins, A.J. "The ACLU is suing Los Angeles over its controversial scooter tracking system mobility data specification (MDS) used by LADOT to track all of the city's electric scooters." *Theverge*; 2020. Available from https://www.theverge.com/2020/6/8/21284490/aclu-ladot-mds-lawsuit-scooter-tracking-uber

[21] Descant, S. "Los Angeles DOT in federal lawsuit over scooter data." *Government Technology, News.* The Future of Community Design; 22 June 2020.

[22] Ellis, B., Rodman, W., Detamore, I., and Scalise, P. "TRB's Para Transit Committee: A forum for mobility innovation since 1974." *Transportation Research Board (TRB) Centennial Paper.* Standing Committee on Paratransit (AP060). Washington, D.C.; 2020.

[23] Brown, C.T. "12 strategies for centering and prioritizing equity in transportation." *ITE Journal.* 2020; 42–45.

[24] Stowell, H.G. "Making micromobility equitable for all." *ITE Journal.* 2020: 46–49.

[25] Ge, Y., Knittel, C. R., MacKenzie, D., and Zoepf, S. *Racial and gender discrimination in transportation network companies*; 2016. Available from https://www.nber.org/papers/w22776.pdf

[26] Pandy, A., and Caliskan, A. *Iterative effect-size bias in ridehailing: Measuring social bias in dynamic pricing of 100 million rides.* Licence CC BY-NC-SA 4.0; 2020.

[27] Lee, T. N. "Detecting racial bias in algorithms and machine learning." *Journal of Information, Communication and Ethics in Society.* 2018; 16(3): 252–260. https://doi.org/10.1108/JICES-06-2018-0056

[28] Congressional Research Services. *Electric vehicles: A primer on technology and selected policy issues*. CRS Report R46231; Washington, DC; 14 February 2020.

[29] Shaheen, S., Cohen, A., Randolph, M., Farrar, E., Davis, R., and Nichols, A. *Shared mobility policy playbook*. Retrieved from https://escholarship.org/uc/item/9678b4xs; 2019

[30] Gordon, B. Zheng, C., Greenblatt, J.B., Shaheen, S., and Kammen, D. "On-demand automotive fleet electrification can catalyze global transportation decarbonization and smart urban mobility." *Environmental Science & Technology*. DOI: 10.1021/acs.est.0c01609; 2020 Available from https://pubs.acs.org/doi/10.1021/acs.est.0c01609?ref=pdf

[31] Ribberink, H., Wilkens, L., Abdullah, R., Matthew McGrath, M., and Wojdan, M. "Impact of clusters of DC fast charging stations on the electricity distribution grid in Ottawa, Canada." *EVS30 Symposium Stuttgart, Germany, 9–11* October 2017.

[32] Shaheen, S., Cohen, A., Yelchuru, B., and Sarkhili, S. *Mobility on demand operational concept report*. U.S. Department of Transportation. Report # FHWA-JPO-18-611; 2017. Available from https://rosap.ntl.bts.gov/view/dot/34258

[33] Shaheen, S., and Cohen, A. "Mobility on demand in the united states: From operational concepts and definitions to early pilot projects and future automation." *Analytics for the Sharing Economy: Mathematics, Engineering and Business Perspectives*. DOI: 10.1007/978-3-030-35032-1_14; 2020. pp. 227-254. Available from https://escholarship.org/uc/item/14f893rv

[34] Shaheen, S., and Cohen, A. "*Mobility on demand (MOD) and nobility as a service (MaaS): How are they similar and different?*" Available from https://medium.com/move-forward-blog/mobility-on-demand-mod-and-mobility-as-a-service-maas-how-are-they-similar-and-different-a853c853b0b8. 07 March 2019.

[35] Ryus, P., Perk, V., Bunker, J., and Danaher, A. "Transit capacity and quality of service. Standing Committee on Transit Capacity and quality of service (AP015)." Transportation Research Board (TRB) *Centennial Papers,* Washington, D.C.; 2020.

[36] Institute of Transportation Engineers (ITE). "Mobility as a service as part of smart community." *Quick Bites*. ITE; 2019.

[37] Deloitte Insights. *Forces of change: The future of mobility*. Ottawa; 2018.

[38] Grush, B. "Ignoring current needs? Understanding deployment." *ReNews Canada: The Infrastructure Magazine*; May/June 2019.

[39] Shaheen, S., Cohen, A., and Farrar, E. Shared mobility policy and modeling workshop. *UC Berkeley: Transportation Sustainability Research Center*. Retrieved from https://escholarship.org/uc/item/1b51g9p3; 2019.

[40] U.S. Department of Transportation (DOT). *Shared mobility, current practices and guiding principles: Federal Highway Administration*. Report FHWA-HOP-16-022. Authors: Shaheen, S., Cohen, A., and Zohdy, I; 2016.

[41] U.S. Department of Transportation (DOT). *Our new mobility future*. A U.S. DOT Volpe Center Thought Leadership Series, Volpe Center; January 2020.

[42] Transportation Research Board (TRB). *Forum on preparing for automated vehicles and shared mobility, mini workshop on the roles of government and the private sector*. July 15, 2019. Orlando, Florida.

[43] International Transport Forum (ITF). *Shared mobility, Innovation for liveable cities*. Corporate Partnership Board Report. OECD, Paris. 2016.

[44] National Cooperative Highway Research Program (NCHRP). *Updating regional transportation planning and modelling tools to address impacts of connected and automated vehicles: Volumes 1 & 2*. Washington, DC; 2018.

[45] Shared Mobility Center. *Shared mobility and the transformation of public transit, Research Analysis*. *TCRP J-11/Task 21*. Submitted to the National Academies TRB Transit Cooperative Research Program; 2019.

[46] National Academies of Sciences, Engineering, and Medicine. *Foreseeing the impact of transformational technologies on land use and transportation*. Washington, DC: The National Academies Press; 2019. Available from https://doi.org/10.17226/25580.

[47] Mistele, B. "Automated, connected, and shared (ACES) vehicles and the future of driving." 09 July 2019. In U.S. Department of Transportation Our new mobility future, A U.S. DOT Volpe Center Thought Leadership Series, Volpe Center; January 2020.

[48] International Council on Clean Transportation (ICCT). *Electric vehicle capitals: Showing the path to a mainstream market, Briefing*. Washington. DC, November 2019.

[49] Catapult Transport Systems. *Future proofing infrastructure for connected and automated vehicles: Exploring intelligent mobility*. Technical Report; February 2017. www.ts.catapult.org.uk

[50] Hourdoc, J. *How locals need to prepare for future of V2V/V2I connected vehicles*. Final Report 2019-35. MN/RC 2019-35; August 2019.

[51] National Institute for Transportation and Communities. *Emerging technologies and cities, assessing the impacts of new mobility on cities*. Authors: Lewis, R. and Steckler, R. University of Oregon, Portland, OR.; January 2020.

[52] Hallmark, S. *Preparing local agencies for the future of connected and autonomous vehicles*. Local Road Research Board, Final Report 2019-18, Minnesota Department of Transportation. 2019.

[53] Pike, H. "On-demand transit is changing how some Calgary-area commuters get around: App-based service brings transit to your door, or close enough." *CBC News*, Posted February 11, 2020.

[54] Ungemah, D. *"Congestion pricing: Driving efficiency through innovation, research, and deployment."* Standing Committee on Congestion Pricing (ABE25). *Centennial papers*. Transportation Research Board (TRB). Washington, DC.; 2020.

[55] Shaheen, S., Chan, N., Bansal, A., and Cohen, A. "Chapter 13: Sharing strategies: Carsharing, shared micromobility (bikesharing and scooter sharing), and innovative mobility modes." *Transportation, land use, and environmental planning*. 2020, pp. 237–262. http://dx.doi.org/10.1016/B978-0-12-815167-9.00013-X. Available from https://escholarship.org/uc/item/0z9711dw

[56] Shaheen, S. "Shared mobility: The potential of ridehailing and pooling." *Three revolutions: Steering automated, shared, and electric vehicles to a better future.* Washington, DC: Island Press, pp. 55–76; 2018.

[57] Shaheen, S., and Cohen, A. "Shared ride services in North America: Definitions, impacts, and the future of pooling." *Transport Reviews.* Available from https://doi.org/10.1080/01441647.2018.1497728; 2018, pp. 1–17.

[58] ITE Journal. "First evaluation of DC dockless bike and scooter pilot program released." *Industry News*; February 2019.

[59] Shaheen, S., and Cohen, A., "Shared micromobility policy toolkit: Docked and dockless bike and scooter sharing." *Transportation Sustainability Research Center*, 31 pages; 2019. https://doi.org/10.7922/G2TH8JW7

[60] AVIN Ontario. *The auto sector and the COVID-19 pandemic, recovery support and opportunities.* AVIN Specialized Report. Toronto; June 2020.

[61] Descant, S. "The pandemic will mean big, lasting changes for urban mobility." *Government Technology, Future Structure, Transportation*; 08 May 2020. Available from https://www.govtech.com

[62] McKinsey & Company. *The impact of COVID-19 on future of mobility solutions: A global company.* Available from https://www.mckinsey.com; May 2020.

The views expressed are those of the authors.

Glossary of technical terms and acronyms

Susan A. Shaheen[1]

Agent-based traffic simulation: A computational model employed to simulate actions and interactions of agents or individuals. The tool is employed to assess the individual agents' impacts on the overall traffic system.

Algorithm: A set of rules employed in calculations or problem solving, especially in the context of computers.

American Disability Act (ADA) is a United States (U.S.) civil rights law that prohibits discrimination based on disability. It affords similar protections against discrimination to Americans with disabilities as the U.S. Civil Rights Act of 1964 that made discrimination based on race, religion, sex, national origin, and other characteristics illegal. Additionally, ADA requires public entities (e.g., public transit providers) and businesses providing public accommodations (e.g., transportation services) to provide reasonable accommodations (e.g., wheelchair access) or equivalent transportation services (e.g., paratransit) to people with disabilities.

Application programming interface (API): An interface used by developers to create software applications, which defines data formats and processes to coordinate interactions among multiple software agents.

Architecture: A framework for characterizing elements of a system.

Artificial intelligence (AI): Human intelligence simulated through computer programs that mirror human learning and problem-solving abilities.

Automated vehicle (AV) (see levels of automation): Vehicles that employ technologies ranging from driver-assist capabilities to full automation to help move passengers or freight. There are five levels of vehicle automation, as defined by the National Highway Traffic Safety Administration and SAE International. These levels include: (1) level 0 (no driving automation), (2) level 1 (driver assistance), (3) level 2 (partial driving automation), (4) level 3 (conditional driving automation), (5) level 4 (high driving automation), and (6) level 5 (full driving automation).

Bayesian method: The interpretation of a probability distribution as a representation of knowledge. It is a method of statistical inference in which Bayes' theorem is used to update probabilities.

[1]Department of Civil and Environmental Engineering, and Transportation sustainability Research Center, University of California, Berkeley, CA, USA

Bikesharing: Provides users with on-demand access to bicycles at a variety of pick-up and drop-off locations for one-way (point-to-point) or roundtrip travel. Bikesharing users access bicycles using one of three bikesharing models: (1) station-based bikesharing (users access bicycles via unattended stations), (2) dockless (users may access (unlock) a bicycle and park it at any location within a predefined geographic region), and (3) hybrid bikesharing systems (users may check out and return bicycles either through a station or nonstation location). Bikesharing fleets are commonly deployed in a network within a metropolitan region, city, neighborhood, employment center, and/or university campus.

Binary logit model: A modal split model that results in the probability of using one out of two competing modes for tripmaking. It is a special case of the multinomial logit model that calculates the probability of using a specific mode out of a number of competing modes for tripmaking.

Business to consumer (B2C) is a vehicle service network provided to consumers by a business.

Car rental: A service that provides the traveler with access to a shared fleet of commercially owned motor vehicles for short time periods (typically one day to a few weeks) usually for a fee. Gasoline (and/or electric vehicle charging) and insurance may be provided for an additional fee. Car rental companies typically own and maintain the vehicle fleet.

Carpooling (see ridesharing): The formal or informal sharing of rides between drivers and travelers with similar origin-destination pairings using vehicles for two to six passengers.

Carsharing: Individuals gain the benefits of private vehicle use without the costs and responsibilities of ownership. Individuals typically access vehicles by joining an organization that maintains a fleet of cars and light trucks deployed in lots located within neighborhoods and at public transit stations, employment centers, and colleges and universities. Typically, the carsharing operator provides gasoline, parking, and maintenance. Generally, participants pay a fee each time they use a vehicle.

Cell transmission model: A model that predicts macroscopic traffic behavior on a corridor. It evaluates the flow and density at a finite number of intermediate points at different time steps. The corridor is divided into homogeneous sections (or cells), which are numbered starting downstream. The length of the cell is chosen to be equal to the distance traveled by free-flowing traffic in one time step. The traffic behavior is evaluated every time step. Initial and boundary conditions are set to iteratively evaluate each cell.

City Center: A development framework with the highest concentration of jobs comprised of central business districts and surrounding neighborhoods.

Community planning: This is also known as a Subarea Plan. It focuses on a smaller area, such as a part of a district and can include a number of neighborhoods. The purpose of community plans is to focus on more specific issues and allows public agencies to identify the locations, service availability, and coverage gaps of shared mobility services within particular neighborhoods.

Comprehensive planning: This is also known as a General or Master Plan, which establishes a set of long-term goals and policies that communities use to guide development decisions. These plans typically establish a high-level policy vision and often focus on key elements including transportation, land use, housing, conservation and climate, open space, noise, and public safety. Within the context of transportation or innovative mobility, sections of the comprehensive plans can catalog the number of shared mobility options within a community.

Concept level design: An early phase of the design process that outlines the general forms and purpose of a concept. This also includes the design of interactions, processes, and strategies.

Conditional use permits: The permitting of land uses under specifically approved conditions. Conditional use permits may be appropriate where zoning may otherwise prohibit shared mobility from legal operations. For example, a zoning ordinance may prohibit commercial activities in a residential zone, but a conditional-use permit could provide specific exceptions for shared mobility services, such as carsharing, to operate in residential neighborhoods.

Connected and automated vehicle: A vehicle that is both connected to other vehicles/infrastructure and automated.

Connected vehicle: A vehicle that employs a number of wireless technologies to communicate with other vehicles, infrastructure, and the Internet/"cloud" (computing resources provided via the Internet). Connected vehicles include driverless and nonautomated vehicle technologies that can improve safety, fuel efficiency, and mobility by providing real-time connectivity across the transportation system.

Courier network services: Provides for-hire delivery services for monetary compensation via an online application or platform (such as a website or smartphone app) to connect couriers using their personal vehicles, bicycles, or scooters with freight (e.g., packages, food).

Cyber security: Protocols and methods to protect against criminal or unauthorized use of sensitive electronic data.

Data architecture: Describes how data are collected, stored, distributed, converted, and employed. This includes rules that govern how databases and filing systems operate, along with processes for connecting users with data.

Deadhead (Zombie) miles/kilometers: Empty miles/kilometers that do not account for passengers or goods. They are essentially the excess miles/kilometers associated with passenger and goods delivery.

Dedicated short-range communications: One- or two-way short- to medium-range wireless communication channels that are used in automotive transportation that correspond to dedicated protocols and standards.

Density bonus: This generally means that more housing units can be built on a given site. Policies that allow increased density (or density bonuses) can include: (1) greater floor-to-site area ratios, (2) more dwelling units permitted per acre, and (3) greater height allowances. Policies that provide density bonuses for developers leverage a similar principle of aiming to make development more lucrative for developers and real estate investors. Rather than reducing per unit or overall project costs, these policies increase the overall cash flow of development projects. Allowing density bonuses for the inclusion of shared mobility can be an attractive strategy for cities wanting to increase overall urban density, residential density, or both.

Dial-a-ride-problem (DARP): Vehicle routes are modified in real time in response to trip requests arriving in time.

Direct current fast charging (DC fast charging): DC fast chargers convert alternating current (AC) power to DC within the charging station, which allows DC power delivery directly to the battery of an EV. This allows the battery to charge faster.

Discretionary review: A process that permits local officials and planning commissions (or other designated bodies) to review specific development proposals and attach specific conditions or deny approval.

Drones: A short-range unmanned aerial vehicle (or UAV) that can transport small packages, food, or other goods.

Edge city: An urbanization pattern presenting some features of city center employment mixed with suburban form. Edge cities tend to have large concentrations of office and retail space often paired with multifamily residences.

Electric vehicles (EVs): Vehicles/automobiles propelled by one or more electric motors, which employ energy stored in batteries that are recharged.

Equivalent annual cost (EAC): This reflects the total annual costs of owning, operating, and maintaining an asset over its lifetime. The time value of money is taken into account in calculating this annuity.

Exurban: A low-density residential development within the commute shed of a larger and denser urbanized area.

Fast chargers: These DC fast chargers use high power levels to charge electric vehicle batteries. In the future 100–350 kilowatts (kW) chargers are likely to be implemented. Also, depending upon charging needs, clusters of these fast chargers may be needed at specific sites.

Future proofing: This entails anticipating the future and developing methods to minimize the impacts of future events.

Geofence: Application of a virtual perimeter to a geographic region in the real world.

Geographic information system (GIS): A framework for collecting, handling, and visually analyzing spatial data through the use of maps and 3D imagery.

Global positioning system (GPS): A navigational system that uses satellite signals to assign the location of a radio receiver on the earth's surface or above it.

Hadoop distributed file system (HDFS): A distributed file system that is designed to run on commodity hardware. It is designed to be fault-tolerant and deployed with low-cost hardware, while providing high throughput access to application data with large data sets.

Incentive zoning: Incentive zoning is typically incorporated into local municipal codes. It allows a developer to build a larger, higher-density project than permitted under existing zoning.

Infrastructure to infrastructure communication: The wireless exchange of information between roadway-based infrastructure devices.

Intelligent transportation systems (ITS): Represent a group of technologies that can improve transportation system management and public transit, as well as individual decisions surrounding many aspects of travel. ITS technologies include state-of-the-art wireless, electronic, and automated technologies with the goal to improve surface transportation safety, efficiency, and convenience.

Key performance indicator (also known as performance metric): Tracking and measurement of targeted goals and performance, which are typically linked to success.

Levels of automation (see automated vehicle): SAE has defined five levels of automation from Level 0 to Level 5. *Level 0:* Vehicles are not automated and drivers perform all of the tasks. *Level 1:* Vehicles automate only one primary control function (e.g., self-parking or adaptive cruise control). *Level 2:* Vehicles with automated systems have full control of specific vehicle functions, such as accelerating, braking, and steering, but drivers must still monitor driving and be prepared to immediately resume control at any time. *Level 3:* Vehicles that allow drivers to engage in nondriving tasks for a limited time. Vehicles will handle situations requiring an immediate response; however, drivers must still be prepared to intervene within a limited amount of time when prompted to do so. *Level 4:* A human operator does not need to control the vehicle as long as it is operating in the specific conditions in which it was intended to function. *Level 5:* Vehicles are capable of driving in all environments without human control.

LIDAR: This stands for light detection and ranging (or LIDAR). This is a remote sensing method used in vehicle automation. It employs light in the form of a pulsed laser to measure ranges or variable distances.

Machine learning: A method of data analysis that automates analytic model development. It is based on the idea that systems can learn from data, patterns, etc. and make decisions without human input. It is a subset of artificial intelligence.

MATLAB®: This is an abbreviation for "matrix laboratory." It is a proprietary multiparadigm programming language and a numerical computing environment.

MATSIM: An open-source agent-based traffic simulation platform, which can be used for shared vehicle systems design.

Metropolitan Planning Organization (MPO): An organization designated to oversee the transportation planning process for a metropolitan region. They are required to represent areas with populations over 50,000, as determined by the U.S. Census.

Micromobility: This includes active transportation modes, including scooters and bikes. These devices can be privately owned or shared (known as shared micromobility).

Microscopic traffic simulation: A modeling tool that can accurately represent traffic conditions on road networks via a computer. They include two key components: (1) traffic lights and detectors, changeable message signs, etc. and (2) detailed modeling of traffic behavior that reflects dynamics in individual vehicles, drivers, and vehicle types. States of movement of all vehicles are simulated on a second or even at finer temporal resolution (e.g., one-tenth of a second).

Microtransit: Privately or publicly operated, technology-enabled transit services that typically use multipassenger/pooled shuttles or vans to provide on-demand or fixed-schedule services with either fixed or dynamic routing.

Mileage-based pricing: These include policies where fees are paid on a per-distance basis with the typical aim of reducing congestion and vehicle miles/kilometers traveled (VMT/VKT) and to provide additional infrastructure funding. Since transportation services, including shared automated vehicles (SAVs), comprise some percentage of deadheading miles/kilometers, mileage-based strategies could incentivize operators to reduce these empty miles/kilometers. Reducing deadheading miles/kilometers is critical to mitigating the potential negative impacts of AVs and SAVs on congestion and VMT/VKT. One study demonstrated that AVs could more than double vehicle travel to, from, and within dense, urban cores since cruising could become less costly than parking.

Minimum acceptable rate of return (MARR): The lowest rate of return a project must earn to offset the investment costs. It is the effective annual rate of return on investment that must satisfy the investor's threshold of acceptability.

Mixed integer programming (MIP) problem: This is a problem in which key decision variables are constrained to whole numbers, for example, -1, 0, 1, 2, at the optimal solution. The scope of useful optimization problems that can be defined and solved is greatly expanded by employing integer variables.

Mixed use development: A zoning type that combines residential, commercial, entertainment, and institutional uses into one space, which provides pedestrian connections. This type of development can reduce vehicle-kilometer of travel.

Mobility as a Service (MaaS): MaaS restructures the mobility distribution chain by integrating the products and services of mobility providers and supplying them to users as a single service. Typically, a digital platform creates and manages trips that users can pay for via a single account. A distinguishing feature of MaaS is giving users the option to purchase MaaS products, such as monthly subscription plans that best fit a user's or household's needs. These subscriptions can include a certain amount of each transportation service (e.g., public transportation, bike-sharing, carsharing, taxis, etc.) and are similar to other service bundles, such as mobile phone plans, where the user pays one price for the combination of multiple service elements (e.g., talk, text, data, roaming, long distance, etc.). Brokering travel with suppliers, repackaging, and reselling it as a bundled package is a distinguishing characteristic of MaaS.

Mobility data specification (MDS): A data and API standard that allows a city to gather, analyze, and compare real-time and historical data from shared mobility service providers. An API allows the creation of applications that access data from another service or application. The specification also serves as a measurement tool that helps enable enforcement of local regulations. In addition, MDS allows service providers and public agencies to communicate with each other about their services because it consists of two APIs: (1) a service provider API and (2) a public agency API. MDS includes data such as: (1) mobility trips (and routes); (2) location and status of equipment (e.g., available, in-use, and out-of-service); and (3) service provider coverage areas. As of Fall 2019, cities using MDS include Los Angeles, CA; Santa Monica, CA; Austin, TX; and Ulm, Germany.

Mobility on demand (MOD): MOD is an innovative concept based on the principle that transportation is a commodity where modes have economic values that are distinguishable in terms of cost, journey time, wait time, number of connections, convenience, and other attributes. MOD enables consumers to access mobility, goods, and services on demand by dispatching or using shared mobility, delivery services, and public transportation strategies through an integrated and connected multimodal network.

Mode choice: Process for determining means of travel in tripmaking.

Monte Carlo simulation: The simulation of possible outcomes through the application of a range of values or probability distribution for key factors that reflect uncertainty/risk. It repeatedly calculates results for a set of random values, reflecting a number of uncertainties and ranges. This could involve thousands of calculations. The final simulation exercise reflects the distributions of possible outcomes for the key values modeled.

Multistation system: Shared vehicles are used among multiple stations to go from one activity center to another.

Near-field communication: Short-range wireless technology that provides quick connectivity among devices with a single touch.

Occupancy-based pricing: User fees can be charged based on the number of passengers in a vehicle, which could in turn encourage more pooling and increase vehicle occupancies, possibly reducing congestion and VMT/VKT. The discounted per-passenger fee for pooled transportation network company (TNC) rides in New York City is one example of this. Similar policies could be enacted for future SAV services.

On-board diagnostics II (OBDII): Refers to a vehicle's self-diagnostic and reporting capability. OBD systems provide technicians with information about the status of vehicle sub-systems (e.g., fuel).

Open street map (OSM): A collaborative and crowd sourced initiative resulting in a free and editable worldwide map. Users collect data using manual surveys, GPS devices, aerial images, and other free resources. It has over two million registered users.

Original Equipment Manufacturer (OEM): Automaker.

Overlay zoning: Providing parking reductions and/or density bonuses for the inclusion of shared mobility into specific areas known as "overlay zones." Overlay zones provide an additional layer of zoning standards applied over part of a zoning district or multiple zoning districts.

Parking reduction: A policy that can be employed in urban areas with high housing or parking construction costs. It can help reduce housing costs (by reducing per-unit construction costs) and can encourage neighborhood revitalization by making it easier for projects to be financially viable through higher capitalization rates on real estate projects.

Parking substitution: Parking substitution can be applied to both new and existing developments. Substituting private vehicle parking for shared modes can contribute to an overall network effect for MOD by enhancing access to a network of alternative modes across an urban area.

Pedicab: A for-hire ride service in which a cyclist transports traveler(s) on a tricycle with a passenger compartment. This service may be concurrently or sequentially shared.

Peer to peer (P2P): A peer-to-peer vehicle network; it can be automated or non-automated. See personal vehicle sharing.

Performance metric (also known as key performance indicator): Tracking and measurement of targeted goals and performance, which are typically linked to success.

Personal vehicle sharing: A service that provides the traveler on-demand, short-term access to a fleet of personally owned motor vehicles for a fee for use. Vehicle hosts own and maintain the vehicle fleet. Vehicle hosts and drivers broker transactions using an online-enabled application or platform (i.e., smartphone app) provided by a personal vehicle sharing company. The company may provide resources and services to make the exchange possible (e.g., an online platform to facilitate the transaction, customer support, etc.). Personal vehicle sharing companies do not own or maintain a fleet of vehicles. Personal vehicle sharing is also referred to as peer-to-peer carsharing.

Planned unit development (PUD): These are developments that are typically built by one or more developers comprised of a mixture of land uses and densities.

Pooled shuttle: See microtransit definition.

Pooling: Concurrent vehicle sharing in a commercial vehicle or incidental trip (e.g., carpooling or vanpooling).

Present worth: (1) The equivalent value at present, taking into account time value of money. (2) The monetary equivalent at present sum equivalent to a future sum or sums using time value of money. (3) The discounted value of future sums using the specified discount/interest rate.

Rate of return (ROR): (1) The interest rate earned by an investment. (2) The interest rate at which the present worth of costs and revenues/benefits become equal.

Relational database: A database designed to recognize relationships among stored informational items.

Ridehailing (also known as TNC and ridesourcing): Provides prearranged and on-demand transportation services for compensation, which connect drivers of personal vehicles with passengers. Smartphone mobile applications facilitate booking, ratings (for both drivers and passengers), and electronic payment. Ridehailing also includes "ridesplitting" in which customers can choose to split a ride and fare in a ridehailing vehicle when available.

Ridesharing (carpooling and vanpooling): Facilitates formal or informal shared rides between drivers and passengers with similar origin-destination pairings.

Ridesplitting: Customers split a ride and fare in a TNC, ridehailing, or ridesourcing vehicle.

Ridesourcing (also known as TNC and ridehailing): Provides prearranged and on-demand transportation services for compensation, which connect drivers of personal vehicles with passengers. Smartphone mobile applications facilitate booking, ratings (for both drivers and passengers), and electronic payment. Ridesourcing also includes "ridesplitting" in which customers can choose to split a ride and fare in a ridesourcing vehicle when available.

Rights-of-way: Rights-of-way is a term used to describe the legal passage of people (and their means of transportation) along public and sometimes private property (the latter typically through licenses and easements). Rights-of-way often encompass most ground transportation facilities including streets, bicycle lanes, and sidewalks. A number of local governments and public agencies have developed a combination of formal and informal policies to dedicate public rights-of-way for shared mobility, such as curb space and parking.

Robotic delivery: Offer short-range unmanned ground-based delivery of packages, food, or other goods using a small conveyance robot.

Rural: The lowest density development pattern characterized by low-density light industrial, agricultural, and other resource-based employment.

Scooter sharing: Users gain the benefits of a private scooter without the costs and responsibilities of ownership. Individuals typically access scooters by joining an organization that maintains a fleet at various locations. The scooter operator usually provides gasoline, parking, and maintenance. Generally, participants pay a fee each time they use a scooter. Scooters can be accessed via unattended stations or accessed (unlocked) and returned (parked) to any location within a predefined geographic region. Scooter sharing includes two types of services:

1. Standing electric scooter sharing that employs shared scooters with a standing design with a handlebar, deck, and wheels that are propelled by an electric motor. The most common scooters today are made of aluminum, titanium, and steel.
2. Moped-style scooter sharing uses shared scooters with a seated-design, either electric or gas-powered, that generally have a less stringent licensing requirement than motorcycles designed to travel on public roads.

Shared micromobility: This includes active transportation modes, including scooter sharing and bikesharing. These devices are typically shared through a third-party operator. See scooter sharing and bikesharing definitions for more detail.

Shared mobility: The shared use of a vehicle, bicycle, or other mode on a short-term basis.

Sharrow: A road marking that appears in the form of two inverted V-shapes above a bicycle. This image indicates the part of a road that should be used by cyclists when the road is shared with automobiles and other vehicles.

Spatiotemporal pricing: Pricing when entering or circulating in specified areas or during certain times of day. Areas can include central business districts, particular lanes, or designated curb spaces. These strategies could keep AVs from entering busy downtowns or other areas during peak times and could help mitigate congestion. However, these pricing strategies could have a variety of unintended social equity implications, such as impacting workers with particular schedules and displaced workers who are forced to travel across pricing zones due to broader equity considerations, such as the lack of affordable housing in close proximity to one's workplace.

Specific planning: This is also known as functional plans, which are the most detailed plans. They are used to implement particular planning provisions. Specific plans generally include special development standards that apply to limited geographical areas. In some cases, specific plans can be used in lieu of zoning ordinances. Specific plans are a good way for public agencies to illustrate how shared mobility can be deployed at specific sites and support urban design that connects people and places through a cohesive mobility vision.

State of charge: Indicates the charging level of an electric battery relative to its overall capacity.

STEPS framework: The STEPS social equity framework is defined as spatial – temporal – economic – physiological – social.

Stochastic demand: Demand is uncertain or unknown, reflecting a random probability distributions.

Stochastic user equilibrium (SUE): These models allow the representation of perceptual and preferential differences, which exist when drivers compare alternative routes along a transportation network. Typical applications of SUE are based on the assumption that all available routes have a positive probability of being chosen regardless of their attractiveness.

Stochastic variable: A variable whose value depends upon the outcome of a random event.

Suburb: A built environment characterized by high levels of low-density residential uses with fewer jobs than residences.

Taxis: Provide prearranged and on-demand vehicle services for compensation through a negotiated price, zone pricing, or a taximeter. Trips can be made by advance reservations (booked through a phone, website, or smartphone application); street hail (by raising a hand or standing at a taxi stand or specified loading zone); or e-Hail (dispatching a taxi driver using a smartphone application).

Telematics: A field that encompasses vehicle technologies, wireless and electronic communications, and computer science. Telematics data are collected onboard through a vehicle tracking device, which can receive and transmit the data wirelessly. Data typically include location, speed, idling time, acceleration, braking, fuel consumption, etc.

Transportation on-demand (TOD): An on-demand mobility service.

Transport demand management (also known as transportation demand management): Strategies aimed at providing traveler choice, often with the goals of reducing single occupant vehicle use, particularly during commute times. This has often included carpooling and vanpooling modes, among other options, such as park-and-ride, telework, work from home, etc.

Transportation network companies (TNCs, also known as ridehailing and ridesourcing): Provides prearranged and on-demand transportation services for compensation, which connect drivers of personal vehicles with passengers. Smartphone mobile applications facilitate booking, ratings (for both drivers and passengers), and electronic payment. TNCs also include "ridesplitting" in which customers can choose to split a ride and fare in a TNC vehicle when available.

Trip detour factor (TDF): Represents the ratio between the customer's trip distance under the ridesharing mode and under the exclusive riding mode.

Trip-based pricing: These entail fees that are applied to each trip taken using a particular transportation service, which could be applied to SAV services (especially single-occupant trips) to encourage use of more sustainable modes, like active transportation or public transit.

Unbanked refers to individuals that do not have access to an account at a financial institution.

Underbanked means that a household has a checking or savings account but does not have sufficient funding to access mainstream financial services. This means they often rely upon alternative financial services, such as money orders, check-cashing services, and payday loans, rather than traditional lines of credit (e.g., credit cards) and loans to manage their finances and fund transactions.

Urban air mobility (UAM): A safe and efficient system for air passenger and cargo transportation within an urban area, inclusive of small package delivery and other urban Unmanned Aerial Systems (UAS) services, which supports a mix of onboard/ground-piloted and increasingly autonomous operations.

Vanpooling: The formal or informal sharing of rides between drivers and travelers with similar origin-destination pairings using vehicles of seven to 15 passengers who share the cost of a van and operating expenses, and they may share driving responsibility.

Variances: Special permission granted to a land owner on a case-by-case basis.

Vehicle to everything (or X) **(V2X):** V2X is a technology that allows vehicles to communicate with elements throughout the traffic system through short-range wireless signals. This includes V2V and V2 infrastructure (or V2I) communications. V2X is primarily used to address safety issues, but it can also be used to pay automated tolls.

Vehicle to infrastructure (V2I): V2I communication enables vehicles to wirelessly exchange information with road infrastructure. It is typically bidirectional in nature.

Vehicle to vehicle (V2V): V2V communication enables vehicles to wirelessly exchange information about their speed, location, and direction. Messages exchanged within a range of 300 m or more can be used to assess a potential crash threat.

Zoning: This entails institutionalizing parking reductions and/or density bonuses for the inclusion of shared mobility into local land-use laws and ordinances universally across an entire land use or zone type.

Acronyms

Americans with Disability Act (ADA)
Application programming interface (API)
Artificial intelligence (AI)

Automated mobility on demand (AMoD)
Automated vehicle (AV)
Battery electric vehicle (BEV)
Business to consumer (B2C)
Connected and automated vehicle (CAV)
Commute trip reduction (CTR)
Direct current fast charging (DCFC)
Electric vehicle (EV)
Employee-based trip reduction (EBTR)
Equivalent annual cost (EAC)
Geographic information system (GIS)
Global positioning system (GPS)
Greenhouse gas (GHG)
Hadoop distributed file system (HDFS)
High occupancy vehicle (HOV)
Hybrid-electric vehicle (HEV)
Internal combustion engine (ICE)
Key performance indicator (KPI)
Multiagent transport simulation (MATSIM)
MATLAB (a programming and numeric computing platform; note: it is not an acronym)
Metropolitan Planning Organization (MPO)
Minimum acceptable rate of return (MARR)
Mixed integer programming (MIP)
Mobility data specification (MDS)
Mobility on demand (MOD)
Mobility as a Service (MaaS)
Near-field-communication (NFC)
On-board diagnostics II (OBDII)
Open street map (OSM)
Operation and maintenance (O&M)
Optimal ride matching with demand-side cooperation (ODC)
Operational design domain (ODD)
Original equipment manufacturer (OEM)
Peer to peer (P2P)
Planned unit development (PUD)
Present worth (PW)
Public–private partnership (P3)
Rate of return (ROR)
Relational database management system (RDBMS)
Road usage charging (RUC)
Shared automated mobility (SAM)
Shared automated vehicle (SAV)
Shared electric and automated mobility (SEAM)
State of battery charge (SOC)

STEPS (spatial – temporal – economic – physiological – social) equity framework
Transport demand management (TDM)
Transportation Network Company (TNC)
Transportation on demand (TOD)
Urban air mobility (UAM)
Value of travel time (VOTT)
Vehicle miles traveled (VMT)
Vehicle kilometers traveled (VKT)
Vehicle to vehicle (V2V)
Vehicle to everything (V2X)

Author contribution

The author reviewed and approved the manuscript.

Declaration of conflicting interests

The author declared no potential conflict of interest with respect to publication of this chapter.

Literature for further readings

Ata M. Khan[1]

In Chapter 1 of the book, eight focus areas were described that served as the framework for the subjects covered in the book. The contents of the final chapter (Chapter 18) are organized according to these focus areas. The literature for further readings, presented in this chapter, is also classified by these eight focus areas. These literature items are not used as references in Chapters 1 to 18. The readings included in each section are organized alphabetically using last name of the first author or the name of the agency.

Most literature items included in this part of the book cover a number of topics (e.g., technology, safety, economic factors, and environmental effects). Given such a broad coverage, the decision to include a particular literature item in one section as opposed to another is based on contribution of the item to knowledge in the field. The reader is advised that the author(s) of the works included in this chapter on *literature for further readings* are not included in the Author Index.

A.1 Advanced, safe, secure, and efficient shared mobility

Arbib, J., and Seba, T. *Rethinking Transportation 2020–2030*. RethinkX Sector Distribution Report; 2017. https://www.rethinkx.com/.

The authors of this report suggest that due to economic trends, connected and automated vehicles (CAVs) are likely to disrupt the petroleum fuel-based transportation and oil markets. According to their projections, it will become costly to own and operate conventional privately owned vehicles and by 2030. A very high percentage of passenger miles will be served by on-demand electric CAV fleets in a transport-as-a-service mobility model. In addition to cost savings, safety benefits will be achieved.

Autonomous Vehicle Innovation Network (AVIN) (Ontario). *Autonomous vehicles reshaping the future, cross-sector opportunities and consideration*. Toronto; October 2019.

Cross-sectoral opportunities are defined from the perspective that automated vehicles (AVs) can serve mobility-linked sectors. A number of sectors, including

---

[1]Department of Civil and Environmental Engineering, Carleton University, Ottawa, Canada

the following, are discussed: autonomous delivery, health care, urban planning, and auto insurance. The report elaborates on the need to be prepared to avail benefits generated by transformations that will be made possible by technological advances. When the AV technologies become available for use in real-world applications, well-defined prior plans should be in place.

Biondi, F., Alvarez, I., and Jeong, K-A. "Human–vehicle cooperation in automated driving: A multidisciplinary review and appraisal." *International Journal of Human-Computer Interaction*. 2019. https://doi.org/10.1080/10447318.2018.1561792.

Other than in fully automated driving, the capability for human-vehicle cooperation is essential in developing designs. This review article describes how the human driver and the vehicle can interact to overcome issues in driving, taking into account human factors and machine capabilities. The objective of the study is to describe advances in research that can assist designers in modeling so as to enhance human-machine cooperation for safety and efficiency in automated driving.

Blumenthal, M.S., Fraade-Blanar, L., Best, R., and Irwin, J.L. *Safe enough: Approaches to assessing acceptable safety for automated vehicles*. Santa Monica, CA: RAND Corporation; 2020. https://www.rand.org/pubs/research_reports/RRA569-1.html.

Automated vehicle (AV) safety is an essential attribute for a shared mobility service. Due to lack of enough experience in real-world applications, it is a challenge to establish if automated vehicles are acceptably safe. RAND Corporation researchers acknowledge that such an evaluation is not straightforward and ongoing technology changes add further challenges. Based on the analysis of measurements, processes, and thresholds, they found the different kinds of evidence associated with each and when different approaches jointly applied. Also, the evaluation becomes challenging due to a lack of information on the extent to which stakeholder groups agree with the use of these approaches. This report on approaches to assess safety of AVs is intended for the use of the general public, the industry, and government.

Chajka-Cadin, L., Petrella, M., and Plotnick, S. *Intelligent Transportation Systems Deployment: Findings from the 2019 Connected Vehicle and Automated Vehicle Survey*. U.S. Department of Transportation Volpe National Transportation Systems Center, FHWA-JPO-20-807; June 2020.

The 2019 ITS JPO Connected Vehicle (CV) and Automated Vehicle (AV) Survey results are presented in summary form. The respondents were agencies responsible for freeway, arterial, and transit systems belonging to 78 large metropolitan areas and 30 medium-size cities. The survey was intended to obtain information on topics including deployment levels for CVs and AVs, the nature of CV applications, AV tests underway, communication technologies used to support CVs, CV and AV readiness, technical challenges faced, and resources required in deploying CVs and AVs, and resources needed to support CVs and AVs. The survey methodology and key findings are covered in the report. Additional tables, open-ended responses to questions, and the survey tool are described in the Appendices.

City of Boston. *Boston Transportation Department. Go Boston 2030: Vision and Action Plan.* Boston, MA; 2017.

In this report, the City of Boston is reporting measures to improve future mobility choices and enhance user travel experience. A clear intent is noted on electric and new technology vehicles for moving persons and goods. The plan includes infrastructure for use by future vehicles and encompasses advances in traffic management that will be necessary to accommodate and benefit from the use of new nonpolluting vehicles. There is emphasis on demonstration projects to study the role of advanced technologies and strategies for serving the future person and goods movements.

Coppola, P., and Esztergar-Kiss, D. *Autonomous vehicles and future mobility (1st edition).* Elsevier; 2019.

This book covers new future mobility forms, and their impacts, that will be enabled by autonomous vehicles. These include workplace mobility needs, demand-responsive travel service, and mobility as a service. Related topics covered include the change in travel behavior and transport policy. The contents are intended for use by researchers, persons engaged in professional practice, and policy makers.

Hancocka, P.A., Nourbakhsh, I., and Stewart, J. "On the future of transportation in an era of automated and autonomous vehicles." *PNAS.* 2019; 116(16): 7684–7691. Colloquium paper, www.pnas.org/cgi/doi/10.1073/pnas.1805770115.

This colloquium paper addresses a number of current questions about the deployment and impacts of automated vehicles (AVs). The questions include impact of AVS on evolving transportation systems, safety and social effects, the effect of various levels of automation, shared use of infrastructure by CAVs and human-driven vehicles, and dissemination of information to the public on evolving capabilities of CAVs. The authors attempt to address issues and suggest solutions.

Hannon, E., McKeracher, C., Orlandi, I., and Rajkumar, S. *An integrated perspective on the future of mobility.* McKinsey & Company, A Global Company; October 2016.

This report describes how the converging social, economic, and technological trends will shape future mobility and related impacts. The on-demand door-to-door mobility will be served within the framework of a multimodal transportation system with the capability to offer a choice of modes, automated shared vehicles that will serve high-quality public transit. Electric-connected automated vehicles and smart software platforms will offer mobility as a service (MaaS). It is likely that a higher level of travel need will be served without an increase in the number of vehicles.

Hussain, R., and Zeadally, S. "Autonomous cars: Research results, issues, and future challenges." *IEEE Communications Surveys and Tutorials.* 2019; 21(2): 1275–1313.

The authors review the state of development of the autonomous vehicle technology and discuss constraints to autonomous vehicle's development and large-scale deployment. They describe benefits to consumers and other sectors obtainable from advanced technology vehicles. Challenges to making these new technology vehicles cost-effective, safe, and efficient are described and also helpful ideas are advanced to overcome hurdles.

Iberraken, D., Adouane, L., and Denis, D. "Reliable risk management for autonomous vehicles based on sequential Bayesian decision networks and dynamic intervehicular assessment." in *Proceedings of IEEE Intelligent Vehicles Symposium (IV)*, Paris, France; 2019, pp. 2344–2351.

This study on safety of autonomous vehicles (AVs) describes a probabilistic approach to risk assessment and management of AV in highway driving. The Bayesian Decision Network and Kalman Filter are used for modeling safety factors. Simulation results of the proposed control architecture are described in terms of the ability to handle stochastic decision-making scenarios.

International Transport Forum (ITF). "Transition to shared mobility: How large cities can deliver inclusive transport services." OECD, Paris; May 31, 2017.

Research reported in this publication investigated the approach that cities can follow to adopt shared mobility services. The city of Lisbon, Portugal, was used as a case study to assess issues associated with the offer of shared mobility services at the city level. Impacts on existing high-capacity public transit were studied as well as accessibility for users to work locations, schools, and health facilities.

International Transport Forum (ITF). *Blockchain and Beyond: Encoding 21st Century Transport*. Corporate Partnership Board Report. OECD, Paris; 2018.

The effect of digital technology on the transport industry and services is well known. However, new arrivals in this field, namely blockchain and other distributed ledger technologies (DLTs) require studies on applications in transportation. This report advances ideas on how blockchain/DLTs can be used in the urban travel market for the coordination of seamless urban mobility services and the delivery of Mobility as a Service (MaaS). The report describes how transportation systems could benefit from these technologies in the form of decentralized deployments for running peer-to-peer networks and in agent-based transactions.

INRIX. Micromobility Potential in the US, UK, and Germany; September 2019. Author: Trevor Reed, Transportation Analyst.

Shared micromobility technology has enabled the offer of new efficient, accessible, and convenient mobility options. The resulting micromobility-as-a-service, based on shared bikes, e-bikes, and e-scooters, is well received by consumers and commercial entities. This study report provides information on two major factors that will contribute to the success of micromobility. The first factor is the high proportion of trips in major urban markets that can be served by micromobility due to suitable distances. The second factor is that in some neighborhoods, a disproportionately high percentage of short car trips are made that could benefit from micromobility deployments.

Khan, A.M. "Cognitive connected vehicle information system design requirement for safety: Role of Bayesian artificial intelligence." *Systemics, Cybernetics and Informatics*. 2013; 11(2): 54–59.

ITS research is a high-priority area for the research and development of connected vehicles. This development is focused on making connected vehicles operate in a safe, efficient, and eco-conscious manner through ITS advancements. The cognitive features and required architecture of information system requirements are presented, with focus on specifying information processing capabilities and the role of Bayesian artificial intelligence for data fusion. The role of the information system and Bayesian artificial intelligence are highlighted, for the use in the design of a new generation of cognitive connected vehicles.

Khan, A.M. "Cognitive vehicle design guided by human factors and supported by Bayesian artificial intelligence." Part of the Advances in Intelligent Systems and Computing book series (AISC, Volume 597). Springer International; 2017.

The development of cognitive vehicle features has been expected to be a stepping-stone to fully autonomous driving and researchers and automotive industry experts favor its development. This requires integrating intelligent technology and human factors. The current advances in driver-vehicle interface are presented in this study in addition to highlights of research in design. Bayesian Artificial Intelligence and its application to the designs are presented for overcoming issues in in-vehicle telematics systems.

Khan, A. "Autonomous vehicles: reliability of their perception of the world around them and the role of human driver." *Advances in Human Aspects of Transportation, AHFE 2018*. pp 57–66. Springer International; 2018.

To technology promotors, fully autonomous vehicles do not need a driver even in complex situations. However, technological, legal, social, and economic constraints may keep this from being available for many years. A "cognitive vehicle" is one that has a seamless nondistracting interface with the driver and keeps them involved for liability and cost. The driver can take control if needed or the vehicle can handle situations if the driver is distracted or disabled. This acts as a step toward full automation.

McKinsey Center for Future Mobility. *The race for cybersecurity: Protecting the connected car in the era of new regulation*. McKinsey & Company; 2019.

Cyber security has become an important subject due to the car industry's digital transformation. This report highlights what the original equipment manufacturers (OEMs) can do to counter the actions of hackers and therefore protect vehicles and users. Without cyber security, numerous facets of automation in driving and vehicle electrification are vulnerable to hacker actions.

Metrolinx. *Shared mobility in the Greater Toronto and Hamilton Area: A backgrounder, on industry trends and a summary of stakeholder discussions. Planning and policy, and sustainability*. Toronto; 2017. Authors: Chan, K., Sadway, M., Tzventarny, J., and Gulecoglu, E.

This report summarizes background information on a number of shared mobility-related subjects, including legislated facets of the Regional Transportation Plan, workshop discussions, and key themes of the feedback. Among other themes, transportation market segments, roles of the public and private sectors are articulated.

National Academies of Sciences, Engineering, and Medicine. *Between public and private mobility: Examining the rise of technology-enabled transportation services*. Washington, DC: The National Academies Press; 2016. https://doi.org/10.17226/21875.

The report notes that the convergence of information and communication technologies, in association with the availability of smartphone applications supplemented with the global position system-provided location data, have enabled new transportation services at affordable prices (e.g., carsharing, bikesharing, and microtransit). The report also covers the emergence of transportation network companies (TNCs) such as Uber and Lyft that play a role in offering some popular on-demand services.

National Academies of Sciences, Engineering, and Medicine. *The role of transit, shared modes, and public policy in the new mobility landscape.* Washington, DC: The National Academies Press; 2021. https://doi.org/10.17226/26053.

The report focuses on the potential benefits of travel when shared mobility modes and public transportation are used as a combined system. Opportunities available to private sector mobility companies in the prepandemic era are reviewed in terms of coordinating services with public transportation agencies and also the effects of the pandemic are described. The contents of the report are based on the latest literature and recent blog posts that reflect changing developments. The report can serve as a resource for public agencies in managing future developments in transportation.

National Academies of Sciences, Engineering, and Medicine. *Low-speed automated vehicles (LSAVs) in public transportation.* Washington, DC: The National Academies Press. Transit Cooperative Research Program (TCRP) Research Report 220; 2021. https://doi.org/10.17226/26056.

This report is intended to be a source of information on current applications of low-speed automated vehicles (LSAVs) in a number of urban areas. On the basis of research and demonstration of the technology, information is compiled that serves as a Practitioner Guide for use in planning and implementing LSAV-based public transportation services such as shuttles. Potential users of the information are public transit agencies, communities, and private sector mobility companies. Information provided can serve as a guide for incorporating automated vehicles such as LSAVs into public transportation services.

Ohnemus, M., and Perl, A. "Shared autonomous vehicles: Catalyst of new mobility for last mile?" *Built Environment.* 2016; 42, 589–602.

The authors argue that shared autonomous vehicles (SAVs) can provide door-to-door travel service while avoiding costs and congestion caused by the use of single-occupant vehicles. They suggest that SAVs have the potential to serve the first and last mile of trips in low-density areas, and connect with public transport systems. Service provided to low-density land use could maintain accessibility to auto-dependent locations.

Rohr, C., Ecola, L., Zmud, J., Dunkerley, F., Black, J., and Baker, E. *Travel in Britain in 2035: Future scenarios and their implications for technology innovation.* The RAND Corporation, Santa Monica, California, and Cambridge, Innovate UK; 2016.

This research study describes findings on the effect of new technologies on travel and ways to make future travel more efficient. A number of scenarios were analyzed to find potential effects on travel for work and business, shopping, healthcare, long distance journeys, and freight movement. Information was obtained from expert opinions and workshops. Analyses carried out used statistical methods and results were used to define innovation investments and policies. The findings reported may be used for advanced studies on future travel and influence of technology.

Sperling, D., van der Meer, E., and Pike, S. "Vehicle automation: Our best shot at a transportation do-over?" in D. Sperling, *Three revolutions: steering automated, shared, and electric vehicles to a better future* (Chapter 4). Washington, D.C.: Island Press; 2018. https://islandpress.org/books/three-revolutions.

The authors predict that the combination of shared services, automation, and electrification technology will have a strong influence on enhancing passenger service, reducing cost and favorable environmental impact. They elevate these influences to the level of three revolutions in transportation.

Zmud, J., Ecola, L., Phleps, P., and Feige, I. *The future of mobility scenarios for the United States in 2030.* Institute for Mobility Research, RAND Corporation; 2013.

The RAND and the Institute for Mobility Research (ifmo) collaborated in studying long-term scenarios for passenger travel that include travel by car, transit, domestic air, and intercity rail. These scenarios take into account demographics, economics, energy, transportation funding and supply, and technology. The authors recognize the influence of policy makers and other decision makers on the assumptions made in the scenarios regarding paths leading to alternative futures.

A.2 Policies and regulations

American Planning Association. *Knowledgebase collection: Autonomous vehicles; 2018.* Retrieved from https://www.planning.org/knowledgebase/autonomousvehicles/

Resources on autonomous and connected vehicles are provided, including videos, briefing papers, functional plans, staff reports, guides, and background repositories. These resources provide background on autonomous vehicles (AVs), policy recommendations for communities, and examples of impacts to equity and access, the transportation network, land use, and the built environment.

Anderson, J.M., Kalra, N., Stanley, K.D., Sorensen, P., Samaras, C., and Oluwatala, O.A. *Autonomous vehicle technology: A guide for policy makers.* Rand Corporation; 2016. Accessed April 09, 2017 at www.rand.org/content/dam/rand/pubs/research_reports/RR400/RR443-2/RAND_RR443-2.pdf.

This report was written as a guide for state and federal policy makers on the issues of autonomous vehicle technology. From a survey, the researchers found the benefits likely outweigh the disadvantages. However, these benefits will not all be seen directly by the technology purchasers. Areas in policy issues, communications, regulation and standards, and liability issues are explored. The report concludes with guidance for policy makers on the allowance of the technology and possible encouragement when it is superior to an average human driver.

Beer, R., Brakewood, C., Rahman, S., and Viscardi, J. "Qualitative analysis of ride-hailing regulations in major American cities." *Transportation Research Record: Journal of the Transportation Research Board.* 2017; 84–91. doi:10.3141/2650-10.

A qualitative analysis of ridesharing regulations for TNCs with a focus on driver and company-related regulations. Driver-related included background checks, driver's licenses, vehicle registration, and external vehicle display. Company-related included data sharing, proving a list of drivers, and numerical limits on fleet size. The findings from this show that driver-related regulations vary by city, of which fingerprint background checks lessen the likelihood of TNC operations and the city of Atlanta is the only city to have numerical limits on TNCs.

Canada Senate. *Driving change – Technology and the future of the automated vehicle. 2018.* Retrieved from https://sencanada.ca/content/sen/committee/421/TRCM/Reports/COM_RPT_TRCM_AutomatedVehicles_e.pdf.

This report covers the Standing Senate Committee on Transport study on the regulatory and technical issues related to the deployment of automated (i.e., driverless) and connected vehicles. The findings are based on opinions of Canadian and US experts. The report acknowledges substantial benefits including safety. However, the report notes a number of concerns in terms of job losses, privacy, cybersecurity, urban sprawl, and infrastructure. Due to the pressing need for government action, the Committee made 16 recommendations to the federal government in support of a coordinated national strategy on automated and connected vehicles.

Collier, R., Dubal, V., and Carter, C. *Disrupting regulation, regulating disruption: The politics of Uber in the United States.* San Francisco, CA: University of California, Hastings College of the Law; 2018.

Uber's growth in the ride-hailing market has disrupted the highly regulated transportation sector where before taxis benefited from anticompetitive barriers to entry and price control. Safety and labor provisions also protected the public. This study raises questions of the current and future regulation for Uber and other TNC services and its relation with taxi regulations.

Congressional Research Service. *Electric vehicles: A primer on technology and selected policy issues*; February 14, 2020. R46231. Washington, D.C. https://crsreports.congress.gov.

This report covers the expansion of the electric vehicle (EV) passenger or light-duty vehicle market by providing a primer of the expansion. The discussion addresses factors of EV adoption, categories of EVs and related technology, and the current federal policies.

Economic Commission for Europe Inland Transport Committee World Forum for Harmonization of Vehicle Regulations 180th session Geneva. *Framework document on automated/autonomous vehicles*; March 10–12, 2020.

This document provides guidance by identifying key principles for the safety and security of autonomous vehicles level 3 and higher. The technical provisions, guidance resolutions, and evaluation criteria for automated vehicles should be performance-based, technology-neutral, and state of the art while not restricting future innovation. Autonomous vehicles need to operate in a manner that ensures road user safety and complies with road traffic regulations.

European Commission. *Public support measures for connected and automated driving.* Tender No. GROW-SME-15-C-N102 within the Framework contract No. ENTR/300/PP/2013/FC May – 2017. Final Report.

This report analyses different country's strategies, funding programs, standards and regulations, and value chains for connected and automated driving (C&AD). Additionally, related technologies and the support measures are reviewed and analyzed. The countries involved in the study are the USA, Japan, South Korea, China, France, Germany, Italy, Spain, Sweden, and the United Kingdom. A comparative study shows Europe's current position in the field of C&AD.

Fagnant, D.J., and Kockelman, K. "Preparing a nation for autonomous vehicles: opportunities, barriers and policy recommendations." *Transportation Research Part A.* 2015; 77: 167–181.

The new autonomous vehicle (AV) technology has the potential to impact vehicle safety, congestion, and travel behavior and produce savings of up to US $2000 per year per AV. However, barriers exist in the form of initial costs and licensing and testing standards. The study suggests that leaving the efforts to the state level could create inconsistencies. Liability and privacy are also a concern. The study recommends a national approach to establishing a licensing framework for AVs and creating a standard for liability, security, and data privacy.

Goodin, G., and Moran, M. *Transportation network companies testimony to the Texas House Committee on Transportation.* Austin, TX; 2016.

This testimony presents background information on policy and legislation surrounding driving requirements, TNCs, and taxis. The standards, data collection, operational features, and insurance requirements for drivers are commonly regulated at the state level, along with TNCs, while taxis are handled at the local level. This difference between the TNC and taxi regulation is compared, as they operate under different requirements for insurance, permits, and fleet sizes.

Government Office for Science (UK). "A time of unprecedented change in the transport system: The future of mobility." *Foresight.* January 2019.

This report is focused on presenting the challenges and opportunities for mobility in the UK. For success, industry, science, policymakers, and citizens will need to collaborate. The UK has leading transportation expertise and knowledge that it can use with commercial opportunities to develop its transportation system for the future.

InterVISTAS. *Transportation Network Companies and Car-Sharing at Airports.* ACINA/AAAE Airport Board & Commissioners Conference. Indianapolis, IN; 2016.

A conference presentation on TNCs and carsharing, a service that is growing in popularity at airports, discussing the timeline for TNC regulations, reasons for passengers choosing TNCs, typical airport commercial ground transportation requirements, and challenges in TNC regulation. It defines that the reasons passengers choose TNCs include reliability, cost, convenience, and accountability but these have the challenges of collecting fees, usage of curbside and staging areas, signage and wayfinding, auditing of self-reported trips, and competing services.

Khan, A.M., Bacchus, A., and Erwin, S. "Policy challenges of increasing automation in driving." *IATSS Research*. 2012; 35(2): 79–89.

Information and communication technologies (ICT) and automotive technologies already exist and are expected to grow to meet consumer demand, dropping costs of components, and improved reliability. Current features are mainly for information and warning, but future developments are expected to show cognitive vehicle features, increasing the level of automation. Autonomous driving, though, is a long-term scenario. Knowledge on forecasts regarding automation, policy challenges, and cost-effectiveness for policy analysis are presented in the study. Costs are measured on a per vehicle based on meeting policy and improving vehicle factors. The study features a probabilistic and utility-theoretic capability to account for uncertainties.

Kim, S., and Puentes, R. *Taxing new mobility services: What's right? what's next?* Washington, D.C.: Eno Center for Transportation; 2018.

This report examines how cities and states are currently imposing taxes and fees on TNCs. The information on the taxes and fees, when they were implemented, and disposition of funds for seven cities and ten states is provided in a table. From this, policy questions on the effects TNC taxes and fees have on congestion, infrastructure and public transit funding, traditional taxi services, and funding for regulatory costs and community needs are addressed while suggesting that the focus should be on the reduction of single occupancy vehicles over targeting fees on TNCs.

Kortum, K., Lindley, J., and Norman, M. "Transportation research board forum on preparing for automated vehicles and shared mobility." *Mini-Workshop on the Importance and Role of Connectivity*. Transportation Research Circular E-C247, Transportation Research Board, Washington, D.C.; February 14, 2019.

This report provides a summary of key takeaways, panelist remarks, summaries of breakout group discussions, and a set of proposed research questions for the workshop. From the specific research questions from the workshop, two notable takeaways were (1) connected automation is critical to enable the transportation system to function most safely and effectively, and (2) connectivity may require a different legislative and regulatory framework than exists today, particularly with respect to public and private sector relationship and to funding and risk assignment liability.

KPMG. *The chaotic middle, the autonomous vehicle and disruption in automobile insurance*. White Paper, KPMG International; June 2017.

This is a follow-up study to KPMG's previous study on insurance titled "Marketplace of change: Automobile insurance in the era of autonomous vehicles." According to a message from the insurance task force, the core business models followed by traditional automobile insurance carriers may not be relevant in the future and that the automobile manufacturers may assume the role to cover driving risk. Therefore, the insurance industry may undergo transformation.

Lee, J., Chang, H., and Park, Y.I. "Influencing factors on social acceptance of autonomous vehicles and policy implications." in *Proceedings of the Portland International Conference on Management of Engineering and Technology (PICMET)*. 2018.

This report is a study in South Korea on the factors of autonomous vehicles that affect consumers and if the people recognize the differences in levels of autonomous driving technology. The study involved 459 people over 20 years of age and was conducted in response to South Korea's perceived bias toward suppliers and lack of preparation for the new technology. Multiple regression analysis results showed general acceptance was affected by usefulness, reliability, and legality while partial autonomous vehicle acceptance was concerned with safety, anxiety, ease of driving, driver convenience, driving education, and driver's carelessness. Full autonomous vehicle acceptance was concerned with safety, user convenience, and extra expenses. The report suggests different strategies for the acceptance of different levels of automation.

Metz, D. "Developing policy for urban autonomous vehicles: Impact on congestion." *Urban Science*. 2018; 2(33); doi:10.3390/urbansci2020033 www.mdpi.

The allowance of shared use driverless vehicles, according to the author, could provide a way to reduce the cost of taxis and make more public transportation vehicles economic. If these services also reduce congestion, improve safety, and operate on electric drive to reduce emissions, policymakers will be challenged with allowing the application of automation in driving to gain these benefits.

Moran, M., and Lasley, P. "Legislating transportation network companies." *Transportation Research Record: Journal of the Transportation Research Board*. 2017; 2650: 163–171. http://dx.doi.org/10.3141/2650-19163-171; 2017.

This report presents a review of the TNC legislation in the United States and developed a database of state-level TNC legislation to guide policy makers using the 34 states and Washington, D.C. that had passed TNC legislation, as of May 2016. It found key questions about the regulation of TNCs to ensure public safety while maintaining competition and integration with existing taxi policies.

National Academies of Sciences, Engineering, and Medicine. *Impacts of laws and regulations on CV and AV technology introduction in transit operations*. Washington, DC: The National Academies Press; 2017. https://doi.org/10.17226/24922.

The potential barriers to new technology due to the old operating policies, agency regulations, and governmental laws on the transit environment are explored in this report. The project gives a roadmap for use by industry groups, legislatures, and federal government to adjust for automated transit.

National Academies of Sciences, Engineering, and Medicine. *Partnerships between transit agencies and transportation network companies*. Washington, DC: The National Academies Press; 2019. Authors: Curtis, T., M. Merritt, C. Chen, D. Perlmutter, D. Berez, and B. Ellis. https://doi.org/10.17226/25425.

This report is a combination of knowledge from active and inactive partnerships between transit agencies and TNCs. The information presented is focused on enhancing the understanding of project development and structure. The information presented was gathered from dozens of transit agency surveys which were in turn supported by follow-up interviews, past literature, and interviews with TNC policy and staff, industry experts, and FTA representatives. It provides a Partnership Playbook for a more deliberate approach to working with TNCs.

National Academies of Sciences, Engineering, and Medicine. *Data sharing guidance for public transit agencies now and in the future.* Washington, DC: The National Academies Press; 2020. https://doi.org/10.17226/25696.

This study covers practical guidance, in a how-to guide format, of data sharing for transit agencies and others, how to evaluate benefits, costs, and risks is included. This report identified two types of models for sharing public transit agency data: Public and Private Data Sharing. The legal context for data sharing is addressed and the report also addresses the other major challengers, how transit data sharing is expected to evolve in the future, and provides topics for future research.

National Association of City Transportation Officials (NACTO). *Guidelines for the regulation and management of shared active transportation version 1.* New York; 2018. www.nacto.org.

This document acts to support Shared Active Transportation Company regulation by public entities and covers two broad categories of policy where cities should be in alignment and where policy should be decided at a local level. The guidance includes a state of the practice overview for key issues as well as to inform cities with information as they decide how to manage and regulate Shared Active Transportation Companies. A key highlight found was that entities want guidelines and frameworks that establish key roles and responsibilities while keeping consistent and flexible regulatory approaches for technology development, policy adoption, and new responsibilities. Those providing the structure are hesitant to adopt practices in case it hinders innovation, and most states indicate a preference to keep an open format that still protects data privacy.

New York Times, Business Section. *Where self-driving cars go to learn.* November 2017. Retrieved from https://www.nytimes.com/2017/11/11/technology/arizona-tech-industry-favorite-self-drivinghub.html.

Arizona is provided as an example state that is using less regulation to attract the self-driving car industry. Its plan is to attract companies to capture the economic growth. This has come with setbacks, though, from a high-profile fatal self-driving vehicle crash and concern from public safety advocates.

Pew Research Center, Internet & Technology. *Shared, collaborative and on demand: The new digital economy.* May 2016. Washington, DC, USA.

A survey on opinions, attitudes, and behaviors toward TNCs to provide the market for ridesharing. Findings show TNC services are preferred by young adults, urban residents, college graduates, those not owning or unlikely to drive a car and more likely to use transit. TNC services are seen as software companies over transportation companies, view the drivers as independent contractors, and feel strongly that TNCs should not be regulated the same as taxis.

RAND Corporation. *Why waiting for perfect autonomous vehicles may cost lives.* November 2017. Retrieved from https://www.rand.org/blog/articles/2017/11/why-waiting-for-perfect-autonomous-vehicles-may-cost-lives.html.

While the timing for introducing autonomous vehicles (AVs) is in question to make sure the technology is safe, this research shows that allowing nonperfect AVs on the roads could save hundreds of thousands of lives due to long-term traffic safety benefits.

Reeder, V., Schmidt, S., and Kortum, K. "Forum on preparing for automated vehicles and shared mobility." *Mini-Workshop on the Roles of Government and the Private Sector.* Transportation Research Board; 15 July 2019, Orlando, Florida. Transportation Research Circular E-C258.

While the new transportation technologies have the potential to improve safety and traffic conditions, there is also a concern in meeting the goals or causing unintended consequences. This document summarizes a workshop on the roles of government and the public sector for meeting these goals and gave the key take-aways of (1) current roles will pave the way, but partnerships and relationships will need to change; (2) the key is to acknowledge what we do not know; (3) everyone wants guidelines and frameworks; (4) consistency and standardization are critical, but it is unclear who should establish that; and (5) vehicle manufacturers will continue to push agencies to improve existing traffic control devices.

Taeihagh, A., and Lim, H.S.M. "Governing autonomous vehicles: Emerging responses for safety, liability, privacy, cybersecurity, and industry risks." *Transport Reviews.* 2019; 39(1): 103–128, DOI: 10.1080/01441647.2018.1494640. https://doi.org/10.1080/01441647.20.

This article examines autonomous vehicles (AVs) and the different categories of technological risks associated with them, the strategies that can be adapted to address these risks, and explore responses by governments. The findings show governments have mostly chosen nonstringent and nonbinding measures that promote AV developments while using councils or working groups to explore AV implications. Countries have been introducing legislation covering privacy and data security or have acknowledged this issue. Less attention has been given to environmental and employment risks.

Transport Canada. *Report Presented to the Council of Ministers of Transportation and Highway Safety by Canadian Council of Motor Transport Administrators, Engineering and Research Support Committee, Transportation Association of Canada.* January 21, 2019. TP No. 15408E Cat. No. T42-13/2019E-PDF ISBN 978-0-660-29330-1.

This report is focused on providing information for Canadian transportation officials by giving an overview of short, medium, and long-term policy implications related to the introduction of autonomous and connected vehicles (AV/CVs) on public roads. The report recommended developing a Policy Framework for Canada, as part of preparing Canada for the safe deployment of automated technology, which provides foundational principles to help government, industry, and academia.

Transport Canada. *Canada's Safety Framework for Automated and Connected Vehicles*; February 2019. TP 15403E. TC 1006019E PRINT Cat. No. T86-53/2018E ISBN 978-0-660-28714-0 PDF Cat. No. T86-53/2018E-PDF ISBN 978-0-660-28713.

The safety framework for automated and connected vehicles (AV/CV) described in this document builds upon Canada's current legislative and regulatory regimes and standards and provides policy guiding principles for a national approach to enhanced safety and security. It provides a policy direction for the safe deployment of these vehicles on public roads. Among other subjects covered are nonregulatory tools to support safe testing and deployment, and guidance to government and industry stakeholders in the various facets of research, development, demonstration, and deployment of these advanced technology vehicles.

Wyman, K. *The novelty of TNC regulation.* New York, NY: New York University School of Law; 2018.

This report discusses the highlights of TNC regulations of the 48 state legislatures that have passed legislation that legalizes and regulates TNCs, as of 2017. TNCs need to comply often with both city and state regulations. Safety is the focus of regulation, with fares and vehicle fleets not regulated by jurisdictions.

Zmud, J., Goodin, G., Moran, M., Kalra, N., and Thorn, E. *Advancing Automated and Connected Vehicles: Policy and Planning Strategies for State and Local Transportation Agencies.* NCHRP Research Report 845, National Cooperative Highway Research Program, Transportation Research Board, National Academies of Sciences, Engineering, and Medicine; 2017.

This is a detailed policy report describing strategies to advance applications of automated (AV) and connected vehicles (CV). The findings are intended as aids for decision-making for AVs and CVs services, while considering the high level of uncertainty.

A.3   Social equity and justice

ARUP. Mobility-as-a service, the value proposition for the public and our urban systems. Toronto; 2018.

This study based on the Greater Toronto and Hamilton Area discusses the requirements and impacts of the Mobility as a Service (MaaS). The study looked at MaaS in the context of the private sector providing the needed mobility choices. The study discusses the subject of inequities, mobility needs for all segments of society, the evolution of MaaS, and how MaaS has changed how we navigate our cities.

Bayless, S.H., and Davidson, S. *Driverless Cars and Accessibility, Designing the Future of Transportation for People with disabilities.* The Intelligent Transportation Society of America (ITS America); 2019.

A report on the challenges faced by people with disabilities and what requirements may be necessary for a fully autonomous vehicle to be considered accessible. Gaps currently exist in accessibility and our current stage of developing new autonomous vehicle designs provides an opportunity to reexamine the needs of people with disabilities, especially for demand-responsive passenger service.

Bloom, C., Tan, J., Ramjohn, J., and Bauer, L. "Self-driving cars and data collection: Privacy perceptions of networked autonomous vehicles." *Proceedings of the Thirteenth Symposium on Usable Privacy and Security (SOUPS 2017)*; Santa Clara, CA, USA; July 12–14, 2017.

In the era of connected and automated vehicles, protection of privacy is gaining importance. The motivation of this research study was to find out public expectations and concerns about privacy so that findings can be used to develop measures for enhancing the acceptance of connected automated vehicles and to advance technological and policy mechanisms for protecting privacy. Studies were carried out in cities with and without Uber autonomous vehicles. Experiments were carried out for obtaining information from residents regarding their perception of privacy violations. Survey results and implications in terms of guidelines for the protection of privacy are presented.

Canadian Urban Transit Association (CUTA). *Integrated Mobility Implementation Toolbox.* Prepared By Dillon Consulting, Ottawa; 2017.

This document provides concrete and actionable tools to guide systems towards integrated mobility. It covers three major areas of mobility management: (1) design for movement in conjunction with urban planning, (2) demand management, and (3) mobility management through partnerships with other mobility actors. Social equity is covered.

Childress, S. Nichols, B., Charlton, B., and Coe, S. "Using an activity-based model to explore the potential impacts of automated vehicles." *Transportation Research Record: Journal of the Transportation Research Board, Washington, D.C.* 2015; 99–106. https://doi.org/10.3141/2493-11.

Automated vehicles (AVs) may enter the market at various levels in 10 years or sooner while some planning agencies are planning for their arrival for 2040 or later; therefore, this research was done to help decision makers understand the effects of AVs. The research used the activity-based travel model for Seattle, Washington to test travel behavior impacts from AV technology development. The model was modified to account for AVs in the roadway capacity, user values of time, and parking costs but larger structural model changes were not considered. Results suggest increased capacity and quality may lead to large increases in vehicle miles traveled but per mile usage charges could counteract this change.

Cohn, J., Ezike, R., Martin, J., Donkor, K., Ridgway, M., and Balding, M. "Examining the equity impacts of autonomous vehicles: A travel demand model approach." *Transportation Research Record: Journal of the Transportation Research Board.* 2019; 2673(5): 23–35. https://doi.org/10.1177/0361198119836971.

This study considers the effect of autonomous vehicle (AV) technology on transportation inequities in Washington, D.C. The factors examined were job accessibility, trip duration, trip distance, mode share, and vehicle miles traveled. The model compared disadvantaged and nondisadvantaged communities under scenarios of low or high occupancy shared vehicles and different public transit responses. It found a high occupancy scenario with enhanced transit mitigated existing gaps or reduced the effect the changes had on expanding the gap.

Conference Board of Canada (The). *My ride, your ride, our ride: Public transit and shared mobility.* Ottawa. Authors: Olateju, B., Markovich, J., and Francis, R. 2019.

The combined effect of shared, connected, autonomous, and electric vehicles create uncertainty for long-term transit investments and operations in Canada as transit ridership varies by location and the impact these have is in debate. This report reviews literature and stakeholder interviews to provide insights and strategies for transit authorities when considering transit-shared mobility relationships. A section on equity in shared mobility platforms is included.

Guo, Y., Chen, Z., Stuart, A., Li, X., and Zhang, Y. "A systematic overview of transportation equity in terms of accessibility, traffic emissions, and safety outcomes: From conventional to emerging technologies." *Transportation Research Interdisciplinary Perspectives.* 2020; 4: 100091. Journal homepage: https://www.journals.elsevier.com/transportation-research-interdisciplinary-perspectives.

This study reviews the methods found in existing literature for assessing the equity of accessibility, traffic emissions, and safety. It also identifies existing challenges of analyzing equity for the new technologies and unifies the existing methodologies into a three-step framework that includes population, cost/benefit measurement, and equity assessment. Literature from emerging transportation technologies is included. Major findings are summarized with promising directions for the development of better equity assessment methodologies. The study contributes a framework for assessing the equity of transportation systems and can be an overview resource. Research gaps are also identified and provide directions for equity research.

International Transport Forum (ITF). *Accessibility and Transport Appraisal, Summary and Conclusions.* ITF Roundtable 182; 2020. OECD, Paris. www.itf-oecd.org.

A roundtable on the differences between the analysis of transport issues based on mobility perspectives and accessibility. The use of accessibility metrics was found to be related to equity in its ability to improve the understanding of distributional issues. It works better than traditional cost-benefit analysis to get the level of accessibility of groups and the combination of accessibility and cost-benefit analysis will improve the information available to decision makers.

Kodransky, M., and Lewenstein, G. *Connecting low-income people to opportunity with shared mobility.* Institute for Transportation & Development Policy (ITDP); December 2014.

A report on a survey of existing shared mobility strategies for expanding services to low-income individuals. The findings are intended to inform operators, government agencies, funders, non-profit organizations, and other with using shared mobility strategies to improve conditions for low-income individuals by providing highlights of the potentials of the systems.

National Academies of Sciences, Engineering, and Medicine. *Critical Issues in Transportation.* Washington, DC: The National Academies Press; 2019. https://doi.org/10.17226/25314.

A report that presents challenging questions about potential critical issues of transportation trends for the next 10 to 20 years and gives high-level questions that can be addressed during the next 5 to 10 years through research, policy analysis, and debate. Questions include interest in the implications of developing trends that may not fully manifest until after the 20-year time period and equity is included as a critical issue.

Shaheen, S., Totte, H., and Stocker, A. *Future of Mobility White Paper.* California Department of Transportation; 2018. Retrieved from http://www.dot.ca.gov/hq/tpp/offices/osp/future-of-mobility.pdf.

A report assessing automated vehicles, zero emission vehicles, carsharing, ridesourcing, equity considerations, shared mobility public-private partnerships, and data sharing. Factors for the analysis included research coverage, state of development, and degree of variance in prediction. It found a growing popularity in shared trips for UberPOOL and Lyft Line services and found information and communications technology had an important role in enabling shared and automated vehicles.

Transportation Research Board (TRB), National Academies of Sciences, Engineering, and Medicine & Association for Unmanned Vehicle Systems International. Transportation Research Circular E-C232. *Automated Vehicles Symposium 2017. Summary of a Symposium;* July 11–13, 2017, Hilton San Francisco Union Square, San Francisco, CA. Proceedings prepared by Texas A&M.

A collaboration and information sharing of potential public policy, safety and security, ethical, equity concerns, and more. Technical innovations and applications were also discussed.

U.S. Department of Transportation. *Beyond Traffic 2045*. Final Report; January 2017. Washington, D.C.

This report contains a collection of information expanding a draft report using comments submitted by engineers, researchers, transportation planners, and other transportation members. It contains a chapter dedicated to considering equity issues in transportation policies prioritizing needs for accommodating diverse users and underserved communities.

A.4 Environmental and financial sustainability (economic factors and business models)

Baptista, P., Meloa, S., and Rolima, C. "Energy, environmental and mobility impacts of car-sharing systems: Empirical results from Lisbon, Portugal." *Procedia – Social and Behavioral Sciences*. 2014; 111: 28–37.

The state of road transport is at high levels for their environmental impacts and many approaches have been tried with minimal effect on mobility patterns. Promising results can be found in car-sharing. This research shows how car-sharing gives improvements from reducing the demand on both the use of roads and on consumption of physical and economic resources through the change in vehicle use and ownership. A case study in Lisbon, Portugal showed the impacts of the use of car sharing and technology changes can result in reductions in energy consumption and CO_2 emissions, if a shift to hybrid or electric vehicles is promoted. The report also estimates the impacts of reducing vehicle ownership and an analysis to find the economic break-even point.

Boston Consulting Group. The Reimagined Car: Shared, Autonomous, and Electric. Boston, USA; December 2017.

The researchers propose, in this report, that shared autonomous vehicles (SAVs) will have an effect of lowering transportation costs as much as 50% in large cities. They state that the high costs of private vehicle ownership incurred from insurance, fuel, maintenance, and parking and the impacts to traffic congestion, air pollution, and automobile fatalities can be reduced from SAVs. Stakeholders, market opportunity, underlying technologies, and implementation are discussed.

City of Toronto. *Electric Vehicle Strategy Supporting the City in achieving its TransformTO transportation goals.* Submitted to: City of Toronto Environment & Energy Division. Prepared by Dunsky Energy Consulting (Montreal). *Attachment 1 – City of Toronto Electric Vehicle Strategy*; 09 December 2019.

An identification of actions for the city of Toronto to help achieve its 2050 goal of all transportation powered by zero carbon energy sources, a key goal in the city's climate action strategy. The report covers the electrification of passenger light duty vehicles and initiatives to increase the use of sustainable transportation modes such as walking, cycling, or public transit. The use of electric vehicles aims to reduce greenhouse gas emissions, air quality improvements, noise pollution reduction, a reduction in the urban heat island effect, and strengthening of the local economy through reduced operating costs. A pathway was developed to help understand if the City is on track for its 2050 goal.

Greenblatt, J.B., and Shaheen, S. "Automated vehicles, on-demand mobility, and environmental impacts." *Current Sustainable/Renewable Energy Reports* 2015; 2: 74–81. https://doi.org/10.1007/s40518-015-0038-5.

A review of the history, current developments, projected future trends, and environmental impacts of automated vehicles (AVs) and on-demand mobility. Automobile manufacturers plan to release AVs between 2017 and 2020 with potential effects on energy use and greenhouse gas emissions ranging from around an 80% decrease to a threefold increase, the report argues the net decrease is likely. Meanwhile, on-demand mobility exists in many cities and is growing in popularity with advances in mobile technology and can provide transportation, land use, environmental, and social benefits while also tending to decrease both vehicle ownership and annual vehicle miles traveled. A combination of both AVs and on-demand mobility could increase the benefits seen by each, of both lower energy use and greenhouse gas emissions.

International Transport Forum (ITF). *Good to go? Assessing the environmental performance of new mobility.* OECD, Paris; September 17, 2020.

A report examining the climate impact of personal and shared electric kick-scooters, bicycles, e-bikes, electric mopeds, and car-based ride-sharing services. A rapid growing mobility form and one that is sold as a "green" solution. This study analyses these services at a life-cycle scale, as compared to privately owned cars and public transport, for their real impact on energy demand and greenhouse gas emissions and identifies solutions so the new mobility modes can be used while reducing energy use and limit climate change.

Kukreja, B., Cheng, A. *Life Cycle analysis of electric vehicles, quantifying the impact.* City of Vancouver and the University of British Columbia, Canada; 2018.

A lifecycle analysis of an ICEV and EV vehicle, the Ford Focus and Mitsubishi i-MiEV, respectively, of their carbon emissions and energy consumption while considering raw material production, vehicle manufacture, transportation, operation, and decommissioning at a vehicle life of 150,000 km. The analysis found the EV to have lower carbon emissions and energy use, 203.0 g CO2-eq/km and 2.0 MJ/km, over the ICEV, 392.4 g CO_2-eq/km and 4.2 MJ/km. From a sensitivity analysis, a longer vehicle life showed more favor for the electric vehicle.

Laurischkat, K., Viertelhausen, A., and Jandt, D. "Business models for electric mobility." *Procedia CIRP*. 2016; 47: 483–488.

A review of the business models for the electric mobility market and a presentation of a framework for analyzing e-mobility business models by defining central business model patterns, customer segments, and essential key values of electric mobility and aims to systematically identify e-mobility business model potentials. Acts to overcome existing barriers to entry. Included conducting interviews with mobility, energy, and infrastructure providers to apply the theoretical framework and answer the study's questions.

Litman, T. *Autonomous vehicle implementation predictions: implications for transport planning*. Victoria, Canada: Victoria Transport Policy Institute; 2018.

A discussion on the benefits, costs, impacts, and timeline for AV development and adoption on the subjects of self-driving and microtaxi services in urban areas, traffic and parking congestion reduction, increased mobility, increased safety, and reduced air pollution. States an uncertainty to the extent of the benefits, costs, impacts, and timelines due to double-counting, complexities of regulation, and the extent of benefits if not all are AVs.

Migliore, M., D'Orso, D., and Caminiti, D. "The environmental benefits of carsharing: The case study of Palermo." *Transportation Research Procedia*. 2020; 48: 2127–2139.

This study suggests finding solutions that reduce the number of cars in cities and supporting sustainable modes of transport to address the traffic emissions issue in cities. It focuses on the environmental benefits related to carsharing and uses the COPERT methodology for estimating road-transport-related pollutant emissions. It used the city of Palermo as a case study of the emissions benefits of the use of carsharing, since a carsharing service already exists within the city.

National Academies of Sciences, Engineering, and Medicine. *Assessment of technologies for improving light-duty vehicle fuel economy 2025–2035*. Washington, D.C: The National Academies Press; 2021. https://doi.org/10.17226/26092.

This newly released technology assessment report credits zero emission light-duty vehicles for future improvements in energy efficiency and reduction in petroleum fuels and emissions (including greenhouse gases) in the 2025–2035 period. In this U.S. congressionally mandated report, recommendations are advanced that the U.S. Department of Transportation, the U.S. Department of Energy, and the U.S. Environmental Protection Agency should play a role in facilitating the development and deployment of zero emission vehicles. Another recommendation is made that the National Highway Traffic Safety Administration (NHTSA) should take into account the predicted sales of zero emission vehicles by 2035 in setting vehicle efficiency standards that should be compatible with these predictions. The sales predictions are based on the likelihood that EV costs will reach parity with conventional internal combustion engine vehicles during the 2025–2035 period.

Nordic Council of Ministers. *Mobility as a service and greener transportation systems in a Nordic context.* Authors: Laine, A., Lampikoski, T., Rautiainen, T., Bröckl, M., Bang, C., Stokkendal Poulsen, N., and Kofoed-Wiuff, A. TemaNord. 2018:558. http://dx.doi.org/10.6027/TN2018-558.

This study estimated the potential of shared mobility services through a combination of reviewing available research and modeling and calculations. The more available subjects of research found was that of car sharing. This research found that Nordic households replacing a private car with car sharing reduce their annual VKT and greenhouse gas emissions. The study assessed the potential impact of mobility as a service in the Nordic countries and identified the barriers of creating a more transport-efficient society through these services. The study presents potential solutions to these barriers, for action by governments, cities, and private companies. Policy recommendations are also presented that include how to reduce dependence on car ownership, reduce vehicle kilometers driven, and stimulated demand for smart mobility services and greener mobility systems.

Organization for Economic Co-operation and Development (OECD). *Integrating environmental and economic policies exploring the impact of shared mobility services on CO2.* ENV/EPOC/WPIEEP (2020)6/FINAL Report, 26. Environment Directorate Environment Policy Committee Working Party. Paris; June 2020.

This report covers the use of shared mobility within cities as an alternative to both personal vehicle ownership and public transport. The effect of this service compared to that of personal vehicle ownership and public transport is unknown, though, and should be explored before being adopted. This study explores the impact that widespread uptake of shared mobility services could have on the carbon footprint of urban transport. The study uses simulations of the share of each transport mode and aggregate emissions from 247 cities between 2015 and 2050. Simulations found, on average, to eliminate 6.3% of passenger transport emissions from shared mobility but vary widely and depend on the current modal split. Situations where personal vehicles or public transit alone heavily dominate the modal split do not see benefits. The report suggests policies that can be used to encourage the uptake of shared mobility services.

Rocky Mountain Institute. *Racing to accelerate electric vehicle adoption, decarbonizing transportation with ridehailing.* Basalt, CO, USA; 2021. www.rmi. org

Electrification of Transportation Network Company (TNC) vehicles has the potential to significantly reduce emissions. The vehicles used for ridehailing service are well suited for electrification because of their high mileage per year and favorable economics. Since a single fulltime ridehailing vehicle can average roughly 40,000 miles per year, it can potentially reduce emissions equal to emission reduction by three privately owned electric vehicles. However, accelerating electrification of TNC vehicles will require stakeholders (i.e., TNCs, original equipment manufacturers (OEMs), utility agencies and their regulators, cities, charging equipment and infrastructure providers, and drivers) to mount joint efforts to achieve the goal of electrification of TNC vehicles.

Rodier, C., Broaddus, A., Jaller, M., Song, J. Joschka Bischoff, J., and Zhang, Y. *Cost-Benefit Analysis of Novel Transit Access Modes: A Case Study in the San Francisco Bay Area. 2020.* Mineta Transportation Institute College of Business San José State University San José, CA. Report No. 20–46; November 2020. DOI: 10.31979/mti.2020.1816.

A report on investigating the use of automated vehicles (AVs) for the use of connecting transit riders on the first-mile, last-mile portion of their trip through ridehailing, ridesharing, and microtransit. It models travel and revenue impacts of a fleet of AVs in the San Francisco Bay Area for different ranges of fares and three different service models.

Santos, G. "Sustainability and shared mobility models." *Sustainability.* 2018; 10: 3194. doi:10.3390/su10093194 www.mdpi.com/journal/sustainability.

This report compiled a literature review and found the following for emerging models for shared mobility: (1) peer to peer provision where individuals can rent their cars when not in use; (2) short term rental of vehicles by a provider; (3) companies sign up car owners as drivers; and (4) on-demand private vehicles shared by passengers with common direction of travel. It identifies the fourth model as promising for addressing congestion and emissions concerns while also noting it as the least attractive to individuals. The report also presents potential incentives to encourage shared mobility and outlines research needs.

Santos, G., and Davies, H. "Incentives for quick penetration of electric vehicles in five European countries: Perceptions from experts and stakeholders." *Transportation Research Part A: Policy and Practice.* 2020; 137: 326–342.

A report presenting the results and assessment of an interview of 143 experts and stakeholders from Germany, Austria, Spain, the Netherlands, and the UK on incentives for the uptake of electric vehicles. The incentives, in largest impact to least, were found to be: charging infrastructure, purchase subsidies, demonstrations of electric vehicles, and tax incentives. Climate change, air quality policies, consumer information schemes, and differential taxation applied to various fuels and energy vectors were also found to have a positive influence.

U.S. Department of Energy. *Costs associated with non-residential electric vehicle supply equipment: Factors to consider in the implementation of electric vehicle charging stations.* Prepared by New West Technologies, LLC for the U.S. Department of Energy Vehicle Technologies; 2015.

A report on the need for nonresidential plug-in charging stations for electric vehicles covering the costs associated with purchasing, installing, and ownership in workplace, public, and fleet settings. The information presented is supported from studies around the world and input from owners, manufacturers, installers, and utilities of electric vehicle supply equipment.

U.S. Environmental Protection Agency (EPA). Shared mobility, Green vehicle guide. Viewed on December 02, 2020. www.epa.gov › green vehicles

The development of smartphone has led to new shared mobility options in transportation. These include carsharing, ridesourcing, microtransit, and micromobility. The smartphone or computer gives access to these services and the ability to conduct reservations, access, real-time location tracking, and payment.

Wadud, Z. "Fully automated vehicles: A cost of ownership analysis to inform early adoption." *Transportation Research Part A*. 2017; 101: 163–176.

A study identifying the vehicle sectors that will likely be the earliest adopters of full automation. It uses a total cost of ownership analysis to compare vehicle automation for private vehicles and commercial vehicles in the taxi and freight sectors in the UK. The study finds commercial operators benefit more from automation and households with the highest income benefit due to higher driving distances and perceived value of time.

A.5   Transportation system and land use planning

Alemi, F., Circella, G., Mokhtarian, P., and Handy, S. "What drives the use of ridehailing in California? Ordered probit models of the usage frequency of Uber and Lyft." *Transportation Research Part C*. 2019; 102: 233–248. doi:https://doi.org/10.1016/j.trc.2018.12.016.

A report on a statistical model built from survey data of millennials and generation X showing sociodemographic variables are predictors of adoption but not the change in frequency. Users of TNCs are those more willing to pay to reduce travel time, use smartphone apps to manage their travels, and fly when traveling far for leisure. Private vehicle owners and those unsure of the security of the service use TNCs less.

Bansal, P., Kockelman, K.M., and Singh, A. "Assessing public opinions of and interest in new vehicle technologies: An Austin perspective." *Transportation Research Part C*. 2016; 67: 1–14.

An Internet-based survey, recent as of 2016, surveying people in Austin about their opinions on smart-car technologies and strategies, including shared autonomous vehicles (SAVs). The results for SAVs include estimates of adoption rates under different pricing scenarios, choice dependence on friends' and neighbors' adoption rates, and home-location decisions when SAVs are a common mode of transport.

Bischak, C.A. *The impact of transportation network companies on urban transportation systems*. Austin, TX: The University of Texas at Austin; 2019.

An analysis of national household travel survey data on TNCs and their effect on transportation. The analysis found TNCs are supplementing urban transportation services instead of changing them and that users only used the service a few times a month or less. Main reason for using TNCs over public transportation or taxis was convenience.

Brakewood, C., Ghahramani, N., Peters, J., Kwak, E., and Sion, J. "Real-time riders: A first look at user interaction data from the back end of a transit and shared mobility smartphone app." *Transportation Research Record: Journal of the Transportation Research Board*. 2017; 2658: 56–63. http://dx.doi.org/10.3141/2658-07.

This study looks at data collected from smartphone applications as they are able to capture many travel modes at a relatively cheaper cost than traditional methods. The smartphone data are also able to capture bikesharing, carsharing, and ride-hailing which are not easily captured by traditional methods. Data used for the study came from the application called Transit and was compared to traditional data sources. Results show the new data source has potential advantages over the traditional sources.

Canalys. *The road to autonomous vehicles.* Palo Alto, CA; 2018.

A discussion of autonomous vehicles and their ability to improve mobility. Included are topics of addressing problems with traditional driving, the progression of vehicle autonomy toward full automation, the role of legislation and regulations, and business strategies for companies.

Centre for Aviation. *Airports and Uber 2016: Transportation network companies now more welcome at airports.* CAPA, Sydney NSW 2001, Australia; 2016.

A report summarizing the findings from surveying airports on their relations, attitudes, and operations of TNC services shows that millennials continue as the leading group using TNCs with frequent business passengers increasing their use due to accessibility. Included is a discussion on the future direction, emerging negative impacts to car parking revenues, and other impacts.

City of Los Angeles Office of the Mayor and Department of Transportation. *Urban mobility in a digital age: A transportation technology strategy for Los Angeles*; August 2016.

According to this official document, the Los Angeles Department of Transportation (LADOT) acknowledges the importance of planning for the impacts and benefits of shared mobility, automated vehicles, and other transportation technologies. For this purpose, a platform for transportation innovation was developed so that a number of policies, short and long-term actions, as well as pilot and demonstration projects can be defined, studied, and implemented. The new direction for LADOT is intended to analyze and manage new mobility marketplace. The defined new transportation technology strategy will assist in understanding changing travel behavior and defining new regulatory frameworks.

Coleman, M. "Portland International Airport's TNC Experience and Plans for the Future." *Transportation Research Board Annual Meeting 2018.* Washington, DC: National Academies of Sciences, Engineering, and Medicine; 2018.

A discussion on the impacts to ground transportation operations from the growth of TNC operations at Portland International Airport from 1.7% in 2015 to 12.4% in 2017. It found pick-up and drop-off still grew but more slowly, rental car traffic remained steady, while taxi, fixed-schedule shuttles, and light rail declined.

Conway, M. W., Salon, D., and King, D. A. "Trends in taxi use and the advent of ridehailing, 1995–2017: Evidence from the US National Household Travel Survey." *Urban Science.* 2018; 2(3): 79. doi:10.3390/urbansci2030079.

An investigation of data from the National Household Travel Survey on TNC services in the United States showed growth is greater in dense urban areas and among young users and wealthier households. The growth in denser areas calls for the need for cities to plan for this growth.

Cyganski, R. "Automated vehicles and automated driving from a demand modeling perspective." Chapter 12 of Maurer, M., J. Gerdes, J.C., Lenz, B.,

Winner, H. (Editors). *Autonomous driving, technical, legal and social aspects.* Springer-Verlag GmbH Berlin Heidelberg: Springer Open, Springer Nature; 2015.

A chapter on incorporating automated vehicles with choices of transport mode in modeling of demand in passenger transport. The chapter covers factors on individual processes when weighing different transport modes and an introduction to transport demand modeling's manner of operation. The chapter then discusses how automated vehicles will change behaviors in mode choice. Factors from the geographic context and potential users are used to discuss the resulting competitive situation between transport modes due to automated vehicles.

Dia, H., and Javanshour, F. "Autonomous shared mobility-on-demand: Melbourne pilot simulation study." *Transportation Research Procedia.* 2017; 22: 285–296.

A study presenting the results of a simulation-based study on the impacts of shared autonomous vehicles. The research outlines a framework for the development and evaluation of solutions from automated self-driving and on-demand shared mobility services. The study focuses on model development for travel demand in a connected and autonomous age. Results from a pilot study in Melbourne, Australia are used to demonstrate the feasibility of the approach. It compared a base case to two autonomous mobility-on-demand scenarios, showing a decrease in the needed number of vehicles and parking spaces but with an increase in vehicle kilometers traveled.

Economist (The). "Special report. Autonomous vehicles: Reinventing wheels." London, UK; March 2018.

This report presents information on how autonomous vehicles (AVs) could affect traffic safety and also its effects on urban areas and its subsequent effects on saving the need to use space for right-of-way. It gives examples of how this space could be repurposed in an AV environment and public policies that could be used to encourage, or enforce, this change.

Fagnant, D.J., and Kockelman, K. "The travel and environmental implications of shared autonomous vehicles, using agent-based model scenario." *Transportation Research C.* 2014; 40: 1–13.

A report on many case studies of an agent-based model of a city using shared autonomous vehicles (SAVs). The model scenarios varied by trip generation rates, trip distribution patterns, network congestion levels, service area size, vehicle relocation strategies, and fleet size but were made to reflect realistic patterns. Each scenario was run over 100 days and the SAVs adjusted their locations every 5-min interval to try to best serve the traffic demand. The results show that an SAV can replace around eleven conventional vehicles but adds up to 10% more travel distance. This additional travel distance effect on emissions, though, was found to not outweigh the benefits once fleet-efficiency changes and embodied versus in-use emissions were assessed.

Fagnant, D.J., and Kockelman, K.M. "Dynamic ride-sharing and fleet sizing for a system of shared autonomous vehicles in Austin, Texas." *Transportation.* 2018; 45: 143–158. https://doi.org/10.1007/s11116-016-9729-z.

The use of shared autonomous vehicles and a dynamic ride-sharing (DRS) system are explored in this report. DRS allows the service to pool trips of similar origins, destinations, and departure times together. Simulation results suggest an overall reduced average service time and cost for users. The base case, however, suggests this may come with an increase in vehicle-miles traveled, but this could be reduced if travelers became more flexible in trip timing and routing. The simulations also suggest the service is profitable while remaining cheaper per mile than the study area's current taxicab fare.

Gnann, T., Funke, S., Jakobsson, N., Plötz, P., Sprei, F., and Bennehag, A. "Fast charging infrastructure for electric vehicles: Today's situation and future needs." *Transportation Research Part D: Transport and Environment.* 2018; 62: 314–329.

This study analyses real-world fast charging data from Sweden and Norway. It created a model to determine the ratio of battery electric vehicles and fast charging points and found that they could be similar to that of conventional fuel. It was also found that charging power and the battery size of vehicles determine these results. The use of fast charging can become economically viable quickly.

Graehler, M., Mucci, A., and Erhardt, G.D. "Understanding the recent transit ridership decline in major US cities: Service cuts or emerging modes?." *2019 Annual Meeting of the Transportation Research Board.* Washington, DC: National Academy of Sciences; 2019.

A review of the potential reasons behind the lack of growth and decline of public ridership in North American cities, looking at the ridership levels of public rail versus bus as well. Changes in service levels, gas price, and auto ownership were found to be insufficient to explain the decline. Introducing bikeshare showed increases in rail use but decreases in bus ridership. Cities with TNC services also show a steady decline in both rail and bus ridership.

Haboucha, C.J., Ishaq, R., and Shiftan, Y. "User preferences regarding autonomous vehicles." *Transportation Research Part C.* 2017; 78: 37–49.

A study on survey results asking people in Israel and North America if they would continue to use a regular vehicle, buy a new privately owned autonomous vehicle, or switch to using shared-autonomous vehicle (SAV) on-demand services. The study found the relevant variables for attitudes toward the choices and found overall hesitation towards autonomous vehicles. SAV service was also not chosen by everyone even if it was free. The study also found that individuals in Israel are more likely to shift to autonomous vehicles than those in North America.

International Transport Forum (ITF). Understanding consumer vehicle choice: A new car fleet model for France. OECD, Paris; 2019.

A report on a model of consumer choice in vehicle type in France, separated by conventional, plug-in hybrid, battery-electric, and fuel cell cars. The model parameters are based on findings from a preference survey. Private and company fleets are both considered as well as nonmonetary factors. Information on the uncertainty in electric vehicle uptake is covered, as their growth is uncertain, and forecasts differ widely.

Krueger, R., Rashidi, T.H., and Rose, J.M. "Preferences for shared autonomous vehicles." *Transportation Research Part C: Emerging Technologies.* 2016; 69: 343–355.

An article giving the results of a survey analyzed using a mixed logit model. The research was on finding characteristics of likely users of shared autonomous vehicles (SAVs) and dynamic ridesharing (DRS). The results give the different factors that are critical to SAV and DRS services and found that SAVs with and without DRS are seen as separate mobility options. Young individuals with multimodal travel patterns were found to be more likely to adopt SAVs.

Laidlaw, K., Sweet, M., and Olsen, T. *Forecasting the outlook for automated vehicles in the Greater Toronto and Hamilton area using a 2016 Consumer Survey.* Prepared for Metrolinx and the City of Toronto. School of Urban and Regional Planning, Faculty of Community Services. Ryerson University, Toronto; March 09, 2018.

A report on a survey done in the Greater Toronto and Hamilton Area on the consumer demand to help understand the conditions under which consumers will adopt automated vehicles. The report also explores future travel behavior changes, mode share, and vehicle market shares.

Litman, T. *Autonomous vehicle implementation predictions, implications for transport planning.* Victoria Transport Policy Institute; October 27, 2019.

A report on exploring the future timeline of economic, traffic, and environmental benefits of autonomous vehicles. The report uses a comparison to previous uptakes in vehicle technologies to make estimates on the level of impacts autonomous vehicles will have including when each aspect will be affected. It predicts increased mobility for affluent nondrivers may come soon but other benefits to congestion, safety, or energy consumption may come later when autonomous vehicles become more affordable.

Martin, E., Shaheen, S., Zohdy, I., *et al. Understanding travel behavior: Research scan.* Federal Highway Administration, US Department of Transportation. FHWA-PL 17-025; 2016.

A report giving the current state of knowledge in transportation as measured today supported by a literature review in travel behavior research. The report further explores socio-demographics of Americans, emerging technology, and emerging methodologies, as well as new forms of data, in behavior information. Gaps in understanding are identified that could be addressed with the emerging data and technological resources.

McKinsey & Company. *Private autonomous vehicles: The other side of the Robo-taxi story.* Automotive and Assembly Practice; Global Company; November 2020.

A detailed mobility-market model for more than ten modes of transport, including, the autonomous vehicle (AV) market. The model predicts the use and factors of private and shared transport through 2030. There is a focus on the private market and how original equipment manufacturers (OEMs) can use this market for growth. Scenarios were created to help OEMs, suppliers, and investors make decisions on opportunities for growth.

Michel, C., and Mai, A. *Lessons from autonomous shuttles & multi-modal shifts.* Keolis Group, Paris, France; March 2018.

A report discussing autonomous vehicle (AV) shuttle services for Lyon, Paris, London, and Las Vegas and the needed actions to be taken to support these services. The report states a need for close collaboration with city safety services and cities for operation needs. Ridesharing and its effect on public transportation and emissions are also covered.

National Academies of Sciences, Engineering, and Medicine. *Updating Regional Transportation Planning and Modeling Tools to Address Impacts of Connected and Automated Vehicles, Volume 1: Executive Summary. Volume 2: Guidance.* The National Academies Press, Washington, DC. NCHRP Research Report 896; 2018. https://doi.org/10.17226/25319.

A presentation of information and guidelines to support transportation planners on updating modeling and forecasting tools for connected and automated vehicles (CAVs). The report covers factors affecting both personal and goods movements that it expects will be affected by CAVs.

National Academies of Sciences, Engineering, and Medicine 2019. *Partnerships between transit agencies and transportation network companies (TNCs)*. Washington, DC: The National Academies Press; 2019. https://doi.org/10.17226/25576.

TNCs are a new private mobility service that transit agencies can partner with to assist in delivering fixed-route and paratransit services. This report is for decision makers and provides guidance for making partnerships with TNCs. Where, when, and how the partnerships should be made are covered.

Qiu, H., Li, R., and Jinhua Zhao, J. "Dynamic pricing in shared mobility on demand." *IEEE Transactions on Intelligent Transportation Systems*; February 10, 2018. pp 1–9.

This study gives guidance on pricing strategies for the provider of an urban mobility-on-demand service that owns its own fleet. The problem is defined by combining parametric rollout policy and stochastic optimization. Discrete-choice-based price optimization was applied and evaluated on the traffic network in Langfang, China showing high profits. This model, though, showed higher congestion levels and lower service capacity and suggested the need for policies that balance private profit making ability with meeting service needs.

Rayle, L., Dai, D., Chan, N., Cervero, R., and Shaheen, S. "Just a better taxi? A survey-based comparison of taxis, transit, and ridesourcing services in San Francisco." *Transport Policy*. 2016; 45: 168–178. doi:http://dx.doi.org/10.1016/j.tranpol.2015.10.004.

TNCs are not only replacing taxi trips, but their impact on other modes is not clear. This study finds that over half of TNC trips are replacing modes other than taxis and the service is expanding in dense city areas. The study provides suggested future research needed to better understand the impacts.

Transportation Research Board (TRB). *TRB Special Report 319: Between Public and Private Mobility: Examining the Rise of Technology-Enabled Transportation Services.* Washington, DC: The National Academies Press; 2016. doi:10.17226/21875.

This report presents areas that need attention due to TNCs. The report suggests regulations, employee classifications, integration of services, and safety requirements all need to be reassessed for TNCs. The impact of TNCs is not fully known as data from the TNCs is limited and the unknown balance of added distance for pick-up and drop-off detours compared to the saved distance from higher vehicle occupancy.

Transportation Research Board (TRB). *Advancing Automated and Connected Vehicles: Policy and Planning Strategies for State and Local Transportation Agencies.* NCHRP Research Report 845. Washington, DC: National Academies of Sciences, Engineering, and Medicine; 2017. doi:10.17226/24872.

A discussion of policy and planning strategies covering safety risks, encouraging SAV use, addressing liability/insurance issues, reducing congestion, and improving air quality for influencing decisions on automated and connected vehicles. Different examples are provided to promote the new technology while giving caution that new policies would still need to adapt to emerging problems.

Transportation Research Board (TRB). *U.S. Department of Transportation's Mobility on Demand Initiative: Moving the Economy with Innovation and Understanding.* Washington, DC: National Academies of Sciences, Engineering, and Medicine; 2018.

A discussion on mobility-on-demand and the modes it affects. The goals of mobility-on-demand are to increase service while reducing costs. The use of alternative shared modes is facilitated by combining the trip planning, booking, and payment together into a single application. The study covers the Los Angeles Metro's interest in partnering with TNCs and providing equal access without the need for smartphones.

Transportation Research Board (TRB). *Preparing for Automated Vehicles and Shared Mobility: State-of-the-Research Topical Paper #8 Implications for Planning and Modeling.* NCHRP Project 20-113F. Washington, DC; September 21, 2020. www.trb.org.

Research published as of July 10, 2020 is reviewed. The focus is on (1) gaps in knowledge and defining problem statements regarding planning and modeling research, taking into account the unknown; (2) critical information needs; (3) performance-based planning and related subjects; (4) building on research-in-progress; and (5) feedback mechanisms for demonstration projects for evaluation purposes.

Transportation Research Board (TRB). *Preparing for Automated Vehicles and Shared Mobility: State-of-the-Research. Topical Paper #9 Impacts and Opportunities Around Land Use and Automated Vehicles and Shared Mobility. Automated Vehicles and Shared Mobility Forum.* NCHRP Project 20-113F Washington, DC; September 21, 2020. www.trb.org.

This study discusses the role that land use plays in defining the role of automated vehicles (AVs) as a part of future transportation exhibiting smart mobility innovations. Research in automated vehicles (AVs), land use impacts, and other related topics are reviewed. A number of common themes are discussed. These include increasing acceptance of AVs, shared mobility modes and services, land use impacts in the urban area, rural area effects, curb/sidewalk management in urban areas, management of the public right-of-way, pricing of right of way, investment requirement, affordable housing, equity and accessibility issues, the effect of 5G communication system, and regulatory factors.

U.S. Department of Transportation, Federal Highway Administration. *Analysis of travel choices and scenarios for sharing rides.* FHWA-HOP-21-011. March 2021.

This study contributes information on factors that influence traveler decisions regarding private mobility (i.e., driving) versus selecting a shared ride. Also, the study generated knowledge on trade-offs involved among attributes of available travel options including trip price. The study enhances understanding of how mode-shifting incentives and disincentives could be applied as instruments to influence traveler decision in favor of shared mobility and active mode trips in order to reduce vehicle miles of travel and associated congestion. The data used in the study was sourced from a survey carried out by a large transportation network company (TNC) of its users. Also, information was contributed by two developers of application tools on carpooling incentives and scenarios of varying cost and time differentials for enhancing the likelihood of sharing rides.

Wang, M., and Mu, L. "Spatial disparities of Uber accessibility: An exploratory analysis in Atlanta, USA." *Computers, Environment and Urban Systems*. 2017; 67; 169–175.

A study into the questions of social inequality around TNC services, specifically Uber. Factors were investigated using a quantitative neighborhood-level analysis of Uber accessibility. Wealth, race, and unemployment were not significant while higher road network density, population density, and shorter commute time increased accessibility. Different Uber services correlated with public transit differently.

Yap, M.D., Correia, G., and Arem, B. "Preferences of travellers for using automated vehicles as last mile public transport of multimodal train trips." *Transportation Research Part A: Policy and Practice*. 2016; 94: 1–16.

A study comparing traveler preference for automated vehicles (AVs) compared to existing modes to understand the AV market position. The study looked at different attributes, with a focus on AVs as an egress mode of train trips. The study used a discrete choice model and showed first-class train travelers were the most likely to use AVs as their egress mode. Results also show that in-vehicle time was perceived more negatively compared to when driving and that trust and sustainability affect AV attractiveness. This suggests attention is needed toward the psychological factors when considering implementing AVs.

Zahraei, S.M., Kurniawan, J.H., and Cheah, L. "A foresight study on urban mobility: Singapore in 2040." *Foresight*. 2020; 22(1): 37–52. doi 10.1108/FS-05-2019-0044.

A study of future mobility, focused on Singapore, for use in making future policy. The study used key drivers of change and produced two scenarios each describing key features in terms of dominant transport modes of passengers and freight. This was used to discuss the implication for individuals, society, industry, and government. This research is to be used by transport planners to understand how current decisions will be affecting the long-term mobility system and to prepare stakeholders for uncertain futures.

A.6   System design, operations, and management

AAA Foundation for Traffic Safety (AAAFTS). *Users' understanding of automated vehicles and perception to improve traffic safety: Results from a national survey*. Research Brief; American Automobile Association; December 2019.

Survey results from people's understanding and concerns for autonomous vehicles (AVs). The results showed that people viewed higher levels of automation capable of better crash prevention than lower levels. However, the higher levels of automation were also seen with distrust due to unfamiliarity and perceived unreliability. The study suggests a need to increase the safety and reliability of the technology and an increase in public awareness to increase people's trust.

Ait-Ouahmed, A., Josselin, D., and Zhou, F. "Relocation optimization of electric cars in one-way car-sharing systems: modeling, exact solving and heuristics algorithms." *International Journal of Geographical Information Science.* 2018; 32(2): 367–398. doi: 10.1080/13658816.2017.1372762. https://doi.org/10.1080/13658816.2017.137.

The use of shared electric cars was studied in this report using model simulations. The focus of the simulations was to solve the issue of vehicle stock imbalance across vehicle stations, at which people can go to pick up a vehicle. The model was built so it could handle large problems, so a greedy algorithm and a tabu search algorithm were proposed. This was applied to the Auto Bleue network in Nice, France. Results found showed near-optimal solutions within good computing time.

Beckera, H., Becker, F., Abe, R., *et al.* "Impact of vehicle automation and electric propulsion on production costs for mobility services worldwide." *Transportation Research Part A: Policy and Practice.* 2020; 138: 105–126.

Research on the production costs of transportation modes for both today and in an automated-electric future. The study was conducted in 17 cities around the world. Results show high-income countries will benefit the most from automation due to the labor cost of current taxi and bus operations. The study suggests in the future transportation costs may not depend on an area's income level, which favors higher-income areas.

Cambridge Systematics, Inc. *Effects on intelligent transportation systems planning and deployment in a connected vehicle environment.* FHWA-HOP-18-014. U.S. Department of Transportation, Federal Highway Administration Office of Operations. Washington, DC; July 2018.

A study assessing the impacts of connected vehicles (CVs) on intelligent transportation systems (ITS) planning and implementation. The report covers the consideration of CVs in ITS planning, and deployment, while further covering ITS functions. Functions that may be used by ITS practitioners in the future of automated CVs are also covered. The report states a need for expanded data management, security, operational policies, and practices in the future due to CV technology.

Cepolina, E.M., and Farina, A. A new shared vehicle system for urban areas. *Transportation Research Part C: Emerging Technologies.* 2012; 21(1): 230–243.

A study on a proposed transport system based on a fleet of eco-sustainable personal intelligent city accessible vehicles for a pedestrian area. The study proposes a new methodology to optimize the fleet. The proposed methodology accounts for both the transportation system and user waiting time costs. Microsimulation is used to assess the use of each vehicle and the waiting time of each user second-by-second.

Ciari, F., Weis, C., and Balac, M. "Evaluating the influence of carsharing stations' location on potential membership: A Swiss case study." *EURO J Transp Logist*; 2016; 5: 345–369.

A study of round-trip carsharing services using data from Switzerland. This study looks at the link between the spatial distribution of the stations and potential memberships. The research uses two parts. First, a binary logistic model of round-trip-based carsharing memberships is estimated. Second, the model is run on a population based on full census data and validated against actual carsharing membership data. The results show the ability to reproduce the spatial distribution of carsharing fairly well.

Davol, A. "A new model for airport ground transportation: Transportation network companies at San Francisco International Airport." *Journal of Airport Management*; 2016: 147–153.

A review of the use of TNCs at the San Francisco International Airport, as it was one of the first to permit TNCs. The review showed an operational issue is administrative fines and enforcement. Early operations faced trade dress and airport placard fines while more recent are unauthorized parking or staging. A real-time tracking system is used to support auditing and enforcement. The review recommends engaging in the development of regulations, learning about TNC operations and technology, and learning from the passenger's perspective.

Electric Autonomy. *Toronto City Council votes to adopt electric vehicle strategy, joining other major Canadian cities*. 30 January 2020. https://electricautonomy.ca

A look at the plan the city of Toronto has approved to reach 100% zero emissions from personal vehicles by 2050. These measures include public charging networks, financial incentives, and building code updates.

Eno Center for Transportation. *Emerging Technology Trends in Transportation*. USA; February 2016.

Developed to inform a Federal Highway Administration workshop, the study provides an overview of select developing transportation technologies and includes a discussion of the policy implications. Autonomous vehicles were a focus of the study discussing driving capabilities of the different levels of automation, from none to full self-driving. Identified emerging issues are the impact to travel demand, vehicle ownership, urban spaces, highway design, existing infrastructure, insurance, liability, and cyber security.

Erhardt, G., Roy, S., Cooper, D., Sana, B., Chen, M., and Castiglione, J. "Do transportation network companies decrease or increase congestion?" *Science Advances*, 2019; 5(5). doi:10.1126/sciadv.aau2670.

Research examining if TNCs actually do reduce congestion in major cities as they claim. The existing research provides conflicting results and there is a lack of data. This research combined data scraped from the application programming interfaces of two TNCs and observed travel data. The results show TNCs increased the vehicle delay in the city compared to a scenario without TNCs. This provides insight on how to make informed decision on integrating TNCs.

Fagnant, D.J., Kockelman, K., and Bansal, P. "Operations of shared autonomous vehicle fleet for the Austin, Texas market." *Transportation Research Record:*

Journal of the Transportation Research Board. April 28, 2019. https://doi.org/10.3141/2536-12.

This report investigated the implications of shared autonomous vehicles (SAVs) at a low market penetration rate. To see the effects, a simulation of a fleet of SAVs in the core region of Austin, Texas was used and further used a sample of trips from the region's planning model. The results showed the SAV could be capable of replacing about nine conventional vehicles within the service area but produce more vehicle miles traveled due to unoccupied miles driving to the next traveler or the next anticipated service location.

Hermawan, K., and Regan, A. "On-demand, app-based ride services: A Study of emerging ground transportation modes serving Los Angeles International Airport (LAX)." *Journal of the Transportation Research Forum (JTRF).* 2017; 111–128.

A study on the impact of travel time and cost for travelers choosing Uber and Lyft at the Los Angeles International Airport. The study states that fares are a key finding and that demand for TNCs would drop if fares raised to match current taxi costs.

Hermawan, K., and Regan, A. "Impacts on vehicle occupancy and airport curb congestion of transportation network companies at airports." *Transportation Research Board Annual Meeting 2018.* Washington, DC: National Academies of Sciences, Engineering, and Medicine; 2018.

This study looks at the impacts of TNCs on current shared modes of transport at airports. TNC trips are primarily low-occupancy and findings show TNCs are replacing shared modes more than supporting them. Low occupancy may lead to TNCs causing higher congestion, San Francisco International Airport saw net shared trips decrease by 215,000 from TNCs and projected that number to raise to 840,000 in 2020.

International Energy Agency (IEA). *Shared, automated... and electric?* IEA, Paris; March 28, 2019. https://www.iea.org/commentaries/shared-automated-and-electric.

A study looking at the implications for vehicle electrification and the broader electricity system. The study reviews the assumption that all automated driving and shared mobility will adopt electric vehicles. Electric vehicles have higher purchasing costs but lower operating and maintenance costs, favoring a cheaper overall alternative for shared and autonomous vehicle use, but their use is not guaranteed. The opportunities and challenges for fleets of today and the future are discussed with attention brought to how electric vehicle policies may need to be looked at for the combination of electric, shared, and autonomous vehicles.

International Transport Forum (ITF). *Blockchain and beyond: Encoding 21st century transport.* Corporate Partnership Board Report. OECD, Paris; 2018.

A review of distributed ledger technologies (DLTs) and their potential to support broader coordination of seamless urban mobility services and Mobility as a Service. DLT would mean, with the use of a common language and syntax, that agents could interact directly without a central authority. This report frames principal policy considerations for application of DLT. The report finds urban mobility today is separate and independently regulated while DLT would lead to mobility more aligned to other "as-a-service" models.

International Transport Forum (ITF). *The shared-use city: Managing the curb.* OECD, Paris; May 24, 2018.

This report discusses street curbside redesign for use after a large-scale introduction of ride-sharing services and other innovative mobility options, pricing implications are also discussed. The potential to change from using curb space for street parking to use for pick-up and drop-off zones are looked at and quantitative modeling results are presented with a discussion on their relative efficiency, contribution to wider policy objectives, and implications on city revenues.

International Transport Forum (ITF). *Governing Transport in the Algorithmic Age*. Corporate Partnership Board Report. OECD, Paris; 2019.

A report exploring the impacts of automated decision-making systems to transport activity and seeks to improve algorithmic literacy of policy makers. The opportunities and challenges from these changes are presented and findings can be used by transportation authorities. The report outlines how a new framework of public governance may take shape based on discussions from a workshop in December 2018.

Johnson, C., and Walker, J. *Peak car ownership: The market opportunity of electric automated mobility services*. Rocky Mountain Institute; 2016.

A report of the market impact of electric automated mobility services as they become plausible and part of a large market share currently attributed to privately owned vehicles. The findings show automated mobility services will grow substantially, oil companies are projected to lose revenue, automobile manufacturers to be split, and electrical utilities to gain revenue.

Kaplan, S., Gordon, B., El Zarwi, F., and Zilberman, D. "The future of autonomous vehicles: Lessons from the literature on technology adoption. *Applied Economic Perspectives and Policy*. First published on December 04, 2019. https://doi.org/10.1093/aepp/ppz005.

A review of current literature and past technology adoptions to predict potential issues with the introduction of autonomous vehicles (AVs). It found private ownership of AVs may continue past a transition period and as technology progresses the price to own one may decline. It also found potential for more vehicles on the road, higher miles traveled, and an expansion of the driving population to include those now limited by mobility. This results in congestion depending on increased vehicle miles traveled compared to the increased efficiency of AVs, which could lead to increased greenhouse gas emissions. The study states that AVs will need policies for metropolitan areas and for TNCs for a sustainable future for transportation.

Kerr, C., and McKenna, D. "The impact of TNCs at airports: Operational consequences and future considerations." *Airport Magazine*; January 2018.

A report on the challenges and opportunities to airport landside facilities due to the growth of TNCs. TNCs could increase curbside congestion and be detrimental to revenue but they could also address curbside congestion in ways other modes cannot. TNC data could be used to find traveler access patterns and preferences.

Kester, J., Zarazua de Rubens, G., Sovacool, B.K., and Noel, L. "Public perceptions of electric vehicles and vehicle-to-grid (V2G): Insights from a Nordic focus group study." *Transportation Research Part D.* 2019; 74: 277–293.

A study on the public opinion of electric vehicles (EVs) and vehicle-to-grid (V2G) mobility. The study looked at public perceptions of both EVs and V2G across five Nordic countries. The data used are original to the study and found eight themes of relevance for future research and policy. These themes cover commonly covered topics for EVs but also include themes such as social status, sound, and acceleration. The effect of V2Gs on EV desirability was also asked in the study.

KPMG. The rise of electric, shared and autonomous fleets. Reimagine the future of how people and goods will move. UK; February 11, 2019. kpmg.com/uk/mobility2030.

A report on the impacts of the new technologies being introduced to transportation. The report looks at electric vehicles, alternative fuels, connected and autonomous vehicles, and Mobility-as-a-Service and how they could address current issues in transportation. Both economic and environmental opportunities are presented. How organizations can be prepared to take advantage of these opportunities are also outlined.

Loeb, B., and Kockelman, K.M. "Fleet performance and cost evaluation of a shared autonomous electric vehicle fleet: A case study for Austin, Texas." *Presented at the Autonomous Vehicles Symposium 2017 (San Francisco) and published in Transportation Research Part A.* 2019; 121: 374–385.

The range and recharging time of electric vehicles (EVs) will have an effect on the mileage, vehicle productivity, response times, and costs of shared autonomous electric vehicles (SAEVs). This study presents estimates of cost for a SAEV fleet for the costs associated with operation and administration. Low, high, and mid-cost scenarios are estimated and used in a simulation of SAEVs across the Austin, Texas 6-county region. The study found a gasoline hybrid-electric fleet performed better than EV fleets and either the cost of gasoline needs to increase or the cost of EVs needs to decrease for EVs to be the most profitable. Long-range fast-charging EVs, though, were found to be the most profitable and provide the best service.

McKinsey & Company. *How charging in buildings can power up the electric vehicle industry.* Authors: Hoover, Z., Nägele, F., Polymeneas, E., and Sahdev, S. Global management consulting company, Washington, D.C.; January 2021.

In most advanced electric vehicle (EV) markets, such as the European Union, UK, USA, and China, millions of electric vehicles will require charging infrastructure. Urban planners, real estate developers, designers, electrical equipment suppliers, grid operators, and public agencies will face the challenge to make EV chargers more accessible and affordable. The article suggests that EV charging infrastructure must be integrated into the design of building and building complexes. Also, the effects of expanding EV charging on infrastructure planning are described. The considerations and recommendations advanced are intended for informed decision-making by stakeholders.

Mitchell, W.J., Borroni-Bird, C.E., and Burns, L.D. *Reinventing the automobile: Personal urban mobility for the 21st century.* Cambridge, MA: MIT Press; 2010.

A book on the future direction of automobile development and design. The book covers four ideas to make future vehicles that are green, smart, connected, and fun. They state a reinvention of the automobile is required to make urban mobility more sustainable while bringing the automobile industry out of a crisis.

Nair, R., and Miller-Hooks, E. "Equilibrium network design of shared-vehicle systems." *European Journal of Operational Research*. 2014; 235: 47–61.

A model and presentation of results to determine the optimal configuration of a vehicle sharing program (VSP). A VSP involves multiple modes and allows users to check out a vehicle at the start of their trip and return it to a station near their destination but users will only likely use VSP resources if their travel utilities improve. The model is a bi-level, mixed-integer program and provides exact solutions from the model properties.

Nicholas, M., and Hall, D. *Lessons learned on early electric vehicle fast-charging deployments*. White Paper, International Council on Clean Transportation, Washington, DC, USA; July 2018.

Findings about fast-charging deployments for electric vehicles are presented in this report. The report reviews developments in the technology, the distribution across major electric vehicle markets, summarizes research on the impacts to the electric grid, and mitigation for associated issues. Planning issues, cost, and business cases of fast-charging stations are also reviewed. Policy implications are presented based on the conclusions.

Papa, E., and Ferreira, A. "Sustainable accessibility and the implementation of automated vehicles: Identifying critical decisions." *Urban Science*. 2018; 2(1). doi:10.3390/urbansci2010005.

A scenario-based approach identifies critical accessibility and societal changes that are emerging with fully AVs. The disruptive potential of AVs comes with risks; one of the leading risks is the management of an implementation that may reinforce car dependency. This could lead to negative health, environmental, and societal consequences. Governmental intervention is required to protect public interest because of the tendency for profit and corporate goals in the competitive environment that AVs will operate in. The role of local government is critical to prepare the legal, transportation, and urban systems for AVs integration.

Pavone, M. "Autonomous mobility-on-demand systems for future urban mobility." Chapter 19 of Markus, M., *et al*. (Editors). *Autonomous driving technical, legal and social aspects*. Springer Open, Springer Nature: Springer-Verlag GmbH, Berlin, Heidelberg; 2020.

A chapter discussing the operational and economic aspects of autonomous mobility-on-demand (AMoD) covering two case studies, one replacing New York City taxi services and the other for Singapore personal mobility. Three dimensions of AMoD systems are addressed: modeling, control, and economics. The results of the case studies show AMoD can replace the taxi services in Manhattan with only roughly 70% of the fleet size and the personal mobility needs in Singapore with one-third of the vehicles.

Sharp, R., Delmonte, E., and Jenkins, R. *Smart electric vehicle charging: What do drivers and businesses find acceptable?* TRL Limited. The Future of Transportation, Report PPR903. TRL Crowthorne House, United Kingdom; 2019.

The attitudes and perceptions of energy use, electric vehicle (EV) charging, and ways to make these options more acceptable and compatible with their needs from people driving EVs, conventional vehicles, and business representatives are covered. The report discusses these topics with five questions on peoples' understanding to become more flexible in energy use, how acceptable they find various smart charging options, the barriers to uptake of these options, information needs, and required provisions for increased acceptance.

Smith, T.J. "Protecting your bottom line from the 'Uber Effect'." *Airport Improvement;* April 2018.

The expansion of transportation network companies (TNCs) is expected to increase with the introduction of driverless vehicles. This change may negatively affect parking revenue at airports, as TNCs are suggested as the cause of the drop in both parking and rental car revenues. The report provides methods on how airport facility and curbside management could change to preserve revenue sources by adapting to the increased TNC use.

Transportation Research Board (TRB). *Transportation Network Companies: Challenges and Opportunities for Airport Operators.* ACRP Synthesis 84. Washington, DC: The National Academies Press; 2017. doi:10.17226/24867.

The presence of TNCs at airports to pick-up and drop-off passengers is controlled by permit. Most airports use a dedicated area for TNC operations and provide traffic officers or staff to enforce TNC rules and regulations. Airports make their revenue from TNCs through various fees ranging from annual permit fees to per trip-based fees. The impact on airport revenue remains unknown since the loss to other modes due to TNCs remains unknown. Impacts on airport operations are known and discussed in this study with examples.

A.7   Implementation, urban development, and other impacts

Alessandrini, A., Campagna, A., Delle Site, P., Filippi, F., and Persia, L. "Automated vehicles and the rethinking of mobility and cities." *Transportation Research Procedia.* 2015; 5: 145–160.

A report on the future of transport systems using different cyber mobility scenarios. The current status of automated vehicles and main trends are covered for a view of a preliminary future using mainly automated vehicles. The study identifies positive impacts from this future, including car sharing, reduced parking, increased accessibility, improved safety, and improved efficiency.

Autonomous Vehicle Innovation Network (AVIN). *Smart mobility and the future of cities, opportunities and readiness tactics.* AVIN Specialized Reports. Toronto, Canada; December 2020.

Smart mobility technologies present opportunities for transport systems. This report discusses the major opportunities that will impact the future of cities and areas that need to change to take advantage of these opportunities. These opportunities include improved efficiency, safety, optimizing freight operations, and reducing environmental impacts. Benefits can be seen in the short and long term.

Bloomberg Philanthropies & the Aspin Institute. *Taming the autonomous vehicle: A primer for cities.* Long Island City, NY: Bloomberg Philanthropies and the Aspen Institute Center for Urban Innovation; March 2017.

This report covered the different benefits, challenges, and impacts to cities from autonomous vehicle (AV) technology. Cities are preparing for AVs on their local streets as development shifts from autonomous high-speed highways to urban driving. The urban AV looks to replace the lightweight urban vehicles and will increase mobility for many groups of people. The report provides the timeline for AV adoption, critical steps in development are included.

Henao, A., Sperling, J., Garikapati, V., Hou, Y., and Young, S. "Airport analyses informing new mobility shifts: Opportunities to adapt energy-efficient mobility services and infrastructure: Preprint." *Intelligent Transportation Society of America, 2018 Annual Meeting.* Golden, CO & National Renewable Energy Laboratory; 2018.

This study describes case studies for four major cities in the USDOT Smart City Challenge. The report analyses the changes to parking revenues, parking growth rates, and rental car revenues during TNC operations. A difference in the parking rate to passenger rate growth at airports shows increased air travel will depend on curb demand. Areas of future study are suggested.

Institute of Transportation Studies, UC Davis. *The adoption of shared mobility in California and its relationship with other components of travel behavior.* A Research Report from the National Center for Sustainable Transportation; March 2018.

The use of shared mobility services is growing among millennials and in dense central parts of cities but the reasons for this growth and its impacts remain unclear. This report looks to provide insights on how shared mobility services are changing demand and supply. The research uses online survey responses about the shared mobility services in California, focusing on factors affecting ridehailing services. The survey included more than 2,000 respondents, including millennials and Generation X.

International Transport Forum (ITF). *Urban mobility system upgrade: How shared self-driving cars could change city traffic?* Corporate Partnership Board Report. OECD, Paris; 2015.

This study looks at two different self-driving vehicle concepts in an environment of large-scale uptake of shared and self-driving vehicle fleets. These two concepts are that of a shared autonomous vehicle that can service multiple trips at once and one that services only one passenger at a time. The concepts were examined under the conditions of servicing the same number of trips needed as of current origin, destination, and timing while replacing all car and bus trips. Impacts on fleet size, volume, and parking demand were examined for both the 24-h average and peak hours.

International Transport Forum (ITF). *Shared mobility simulations for Dublin.* October 10, 2018. OECD, Paris.

This report describes the results of a study on the effect of new shared mobility services on mobility in the Greater Dublin Area (Ireland). Results based on simulations of shared transport scenarios cover the effect on congestion, CO_2 emissions, and public space use. Additional results are reported on service quality, travel cost, access to opportunities, and use of public transportation. The study results can potentially be used by decision makers to enhance opportunities and overcome challenges that relate to the deployment of new forms of shared transport. The ITF has published a series of studies on shared mobility in different urban areas.

International Transport Forum (ITF). *Shared mobility simulations for Lyon*; April 07, 2020. OECD, Paris.

Using Lyon, France as a case study, simulation results of five scenarios provide observations on the effect of shared mobility services that replace privately-owned cars. The simulation results suggest a reduction in congestion, CO_2 emissions reductions, and vacant public spaces. Other travel-related effects studied include cost and service quality, access to opportunities, and interaction of shared mobility modes with public transit.

Kellerman, A. *Automated and autonomous spatial mobilities*. March 30, 2018. https://doi.org/10.4337/9781786438492 www.elgaronline.com

A book on the development of automated spatial mobilities and their social and urban implications. The book distinguishes between automation and autonomy, the distinction between spatial mobility and automated spatial mobility is also covered. The effect on automation processes for transportation and communications media from these differences are discussed. The book uses these discussions to address the emergence of autonomously mobile cities. The content is intended for those in the fields of transportation, urban planning, geography, and sociology.

Maurer, M., Gerdes, J.C., Lenz, B., and Winner, H. (Eds.). *Autonomous driving technical, legal and social aspects*. Springer; 2016.

The contents of this book look at discussing the many open questions of fully automated vehicles. Key issues are defined by experts from Germany and the United States and discuss decisions of automated vehicles that have ethical consequences. They identify the expectations and concerns for acceptance of autonomous driving. Further, they determine safety and traffic improvements will only be realized if the concepts are considered at the base of their design.

Meyer, J., Becker, H., Bosch, P.M., Axhausen, K.W. "Autonomous vehicles: The next jump in accessibilities?' *Research in Transportation Economics*. 2017; 62: 80–91.

This report uses the Swiss national transport model to simulate the impact of autonomous vehicles on accessibility. The introduction of autonomous vehicles is expected to lower travel prices and increase road capacity; results from the model show it could also cause a leap in accessibility. The findings also show spatial distribution of the accessibility favors urban sprawl and could make public transport unnecessary outside of dense urban areas.

National Academies of Sciences, Engineering, and Medicine. *Transportation network companies (TNCs): Impacts to airport revenues and operations reference guide*. Washington, DC: The National Academies Press; 2020. https://doi.org/10.17226/25759.

This reference guide identifies strategies and tools for the use of airport land-side access and includes impacts to other airport revenues and operations, from the new transportation modes of TNCs and autonomous vehicles. Strategies in the guide include long-term sustainable revenue models, managing congestion, customer service impacts, forecasts for demand, and technology and socioeconomic factors. The guide also addresses operational and business considerations, while considering making progress for sustainability initiatives. Supporting documentation includes a statistical database.

Schaller Consulting. *The new automobility: Lyft, Uber and the future of American cities.* Brooklyn, NY: Schaller Consulting; 2018.

TNC use, as estimated in this study, is expected to surpass local bus ridership by 2018. The report provides detailed profiles of TNC ridership, users, and usage showing dense areas have high TNC ridership, while suburban and rural areas continue using traditional methods. The TNCs were found to increase mileage, even shared ride services, but can be managed through trip fees, bus lanes, congestion pricing, and traffic signal timing. TNCs are competing with public transportation, walking, and biking. The study emphasizes that public policy intervention will be needed.

A.8   Measures to adapt to longer-term pandemic impacts

Arthur D. Little. Global mobility post. Turning the crisis into an opportunity to accelerate towards more sustainable, resilient and human-centric urban mobility systems. Brussels, Belgium: International Management Consulting Company; 2020.

This report covers the impacts of COVID-19 on future mobility patterns, a summary of actions that are being taken, and the major changes that could be used to move toward a sustainable future. The report sees the current situation as a way to make major changes to the transportation and mobility systems. Many stakeholders in transportation would have a role and would need to both change and reinvent their service offerings. According to this report, now is the best opportunity to make these large changes.

Boston Consulting Group. *How COVID-19 will shape urban mobility?* Authors: Bert, J., Schellong, D., Hagenmaier, M., Hornstein, D., Wegscheider, A. K., and Palme, T. Boston, USA; June 16, 2020.

COVID-19 has not changed the needed improvements to the transportation system. Cities are still in need of public transit, biking infrastructure, and new shared mobility modes. The value from these improvements to user economics and environmental health has not changed, but business models may need to be modified. The study states that the restoration of economic health to cities depends on these developments.

Council of Canadian Academies, 2021. *Choosing Canada's automotive future*, Ottawa (ON). The Expert Panel on Connected and Autonomous Vehicles and Shared Mobility, Council of Canadian Academies, Ottawa.

This report, prepared under the guidance of a Panel of interdisciplinary experts, describes a 10-year outlook on a number of key areas of interest to the Canadian automotive industry, government, and the public. While the study was in progress, the panel recognized that in addition to covering other developments, there should be a recognition of the future influence of the COVID-19 pandemic and therefore it should be taken into account in future policy and planning endeavors. At the outset of the study, the panel adopted the assumption that *"connected and autonomous vehicles would most likely be shared and electric"* and chose to maintain this assumption in the main body of the report. However, in a manner similar to other studies, the panel also recognized that the pandemic warrants a special consideration by way of an addendum to the report. Among a number of ideas outlined in the addendum, a suggestion is noted that investments in infrastructure for EVs (e.g., charging stations) could enhance post-COVID economic recovery.

Deloitte, Insights. *The futures of mobility after COVID-19 scenarios for transportation in a post coronavirus world*. Ottawa, Canada; May 22, 2020.

An article on the possible future direction of mobility after the pandemic, considering the severity of the pandemic and the degree of cooperation between governments. The study used high-level scenarios as their starting point and was developed further to consider possible futures over the next 3 to 5 years. Scenarios showed wide variance; however, commonalities were still found in needing better hygiene in vehicles and the growing importance of last-mile delivery. These scenarios are presented in the hope to be referenced as the pandemic continues so stakeholders can make informed decisions in the continued development of mobility.

Future Bridge. Impact of COVID-19 on shared mobility. Industry pulse. Future outlook – Challenges for automotive & shared mobility players. Global Company. North American Office: Morristown, NJ, USA. November 2020.

See https://www.futurebridge.com/industry/perspectives-mobility/impact-of-covid-19-on-shared-mobility/. November 2020.

In general, the impact of COVID-19 on automotive and the mobility industry has been negative. The resulting economic shifts and current measures have long-term impacts on work and trade that are unclear. This study presents several questions that need to be addressed for stakeholders to answer so they can adjust their strategies in response to the pandemic. These questions include needed vehicle features in shared vehicles, best technologies to enable these features, processes to ensure public health, regulations, and policies around acquiring consumer health data in a shared mobility scenario, the impact to goods delivery market, and latest specific technologies.

McKinsey & Company. McKinsey Center for Future Mobility. *Why shared mobility is poised to make a comeback after the crisis*. Global Company; July 2020. www.mckinsey.com

A report on results from a global auto consumer survey to show how shared mobility services can recover from the effects of COVID-19. These results show which features could be implemented or introduced sooner than planned. These features include providing vehicles with minimum distances between passengers or protective shields, interior surfaces that offer antibacterial properties, and other technologies currently used in other fields. These technologies are aimed at making shared mobility safe even during a global pandemic.

McKinsey & Company. McKinsey Center for Future Mobility. *The future of micromobility: Ridership and revenue after a crisis. The COVID-19 crisis is causing serious disruptions to the multibillion dollar micromobility industry.* Global Company; July 2020.

This report states that a full recovery from the pandemic is possible if companies prepare for what is to come. Three time horizons are examined in this study and the state of micromobility for all three are presented. Micromobility is currently declining, but long-term consumer action and policy action could be used for the industry to grow.

McKinsey & Company. McKinsey Center for Future Mobility. *Five COVID-19 aftershocks reshaping mobility's future.* Global Company; September 2020.

The long-term effects of COVID-19 may differ by location; this article looks at how mobility responses could also differ by location and highlights defining trends. The article states that development has been focused around autonomous driving, connected cars, electrified vehicles, and shared mobility but due to the pandemic will need to consider factors beyond these.

McKinsey & Company. McKinsey Center for Future Mobility. *Compendium 2020–21. From no mobility to future mobility: Where COVID-19 has accelerated change?* Global Company; December 2020.

Chapters within this compendium cover the effects of COVID-19 on the future of shared mobility. All mobility stakeholders will need to adjust their strategies for the near future due to the pandemic. Investment into the new mobility technologies, including autonomous technology, connectivity, and EVs, has also continued despite the pandemic and their presence will become stronger. The chapters cover how the new mobility solutions will affect the local movements and account for regional variations. During all of this, regulators will play a major role in helping mobility recover while ensuring sustainability.

Megahed, N.A., and Ghoneim, E.M. "Antivirus-built environment: Lessons learned from Covid-19 pandemic." *Sustainable Cities and Society.* 2020; 61.

The study describes how the antivirus-built environment could look like for stopping the virus from spreading. Architecture and urban design approaches are hypothesized for increasing the protection of the built environment. The authors describe lessons drawn from the current pandemic in searching for answers regarding adding more security layers to overcome future virus like-attacks.

University of South Florida (USF), National Center for Transit Research (NCTR). *Impact of COVID-19 on travel behavior and shared mobility systems.* Final Report. Prepared by USF Center for Urban Transportation Research. Authors: Menon, N., Keita, Y., and Bertini, R.L.; October 2020.

The COVID-19 pandemic has affected both commercial and private mobility. Shared mobility has seen a decrease as users have chosen to use private vehicles to maintain social distancing to protect themselves. This research aims to provide an understanding of the impacts of COVID-19 to travel behavior and shared mobility systems. The research used a web-based review to find the publicly available strategies that have been used and changes during the pandemic that may continue afterward. Key findings include: public transit and ride-hailing have decreased, bikesharing have increased, robot taxis have been identified as important for future pandemics to maintain mobility options while maintaining social distancing, stricter health standards in shared mobility, the need for additional space for micromobility, change in residential and work location decisions, a change to vehicle occupancies, changed vehicle miles traveled, and telecommuting.

A.9   Acknowledgments

The author wishes to thank Seth Gatien for reviewing a draft of the chapter.

A.10   Author contribution

The author reviewed and approved the manuscript.

A.11   Declaration of conflicting interests

The author declared no potential conflict of interest with respect to the publication of this chapter.

Author Index

Subject Index